Mass

The **gram** is the basic unit.

1 **kilo**gram	= 1000 grams	1 kg	= 1000g
1 **hecto**gram	= 100 grams	1 hg	= 100g
1 **deka**gram	= 10 grams	1 dag	= 10g
1 **deci**gram	= $\frac{1}{10}$ of a gram	1 dg	= .1g
1 **centi**gram	= $\frac{1}{100}$ of a gram	1 cg	= .01g
1 **milli**gram	= $\frac{1}{1000}$ of a gram	1 mg	= .001g

Some Approximate Comparisons Between English and Metric Systems
(the symbol $\approx$ means "is approximately equal to.")

Length

1 inch $\approx$ 2.5 centimeters   1 centimeter $\approx$ .4 of an inch

1 foot $\approx$ 30.5 centimeters   1 centimeter $\approx$ .03 of a foot

1 yard $\approx$ .9 of a meter   1 meter $\approx$ 1.1 yards

1 mile $\approx$ 1.6 kilometers   1 kilometer $\approx$ .6 of a mile

Volume

1 pint $\approx$ .5 of a liter   1 liter $\approx$ 2.1 pints

1 quart $\approx$ .9 of a liter   1 liter $\approx$ 1.1 quarts

1 gallon $\approx$ 3.8 liters   1 liter $\approx$ .3 of a gallon

Mass

1 ounce $\approx$ 28.35 grams   1 gram $\approx$ .04 of an ounce

1 pound $\approx$ 453.6 grams   1 gram $\approx$ .002 of a pound

1 pound $\approx$ .45 of a kilogram   1 kilogram $\approx$ 2.2 pounds

Elementary Algebra

Elementary Algebra

FIFTH EDITION

Jerome E. Kaufmann

PWS Publishing Company

I(T)P **An International Thomson Publishing Company**

Boston • Albany • Bonn • Cincinnati • Detroit • London • Madrid • Melbourne • Mexico City
New York • Paris • San Francisco • Singapore • Tokyo • Toronto • Washington

PWS PUBLISHING COMPANY

20 Park Plaza, Boston, Massachusetts 02116-4324

International Thomson Publishing
The trademark ITP is used under license.

For more information, contact:

PWS Publishing Co.
20 Park Plaza
Boston, MA 02116

International Thomson Publishing Europe
Berkshire House 168-173
High Holborn
London WC1V 7AA
England

Thomas Nelson Australia
102 Dodds Street
South Melbourne, 3205
Victoria, Australia

Nelson Canada
1120 Birchmount Road
Scarborough, Ontario
Canada M1K 5G4

International Thomson Editores
Campos Eliseos 385, Piso 7
Col. Polanco
11560 México D.F., Mexico

International Thomson Publishing GmbH
Königswinterer Strasse 418
53227 Bonn, Germany

International Thomson Publishing Asia
221 Henderson Road
#05-10 Henderson Building
Singapore 0315

International Thomson Publishing Japan
Hirakawacho Kyowa Building, 31
2-2-1 Hirakawacho
Chiyoda-ku, Tokyo 102
Japan

Library of Congress Cataloging-in-Publication Data

Kaufmann, Jerome E.
    Elementary algebra / Jerome E. Kaufmann. -- 5th ed.
        p.   cm.
    Rev. ed. of: Elementary algebra for college students. 4th ed. 1991.
    Includes index.
    ISBN 0-534-94860-X
    1. Algebra.   I. Kaufmann, Jerome E.   Elementary algebra for college students.   II. Title.
OA152.2.K38   1995
512.9--dc20   95-21468
                                                                                   CIP

Sponsoring Editor: David Dietz
Developmental Editor: Mary Beckwith
Production Coordinator: Robine Andrau
Marketing Manager: Marianne C. P. Rutter
Manufacturing Coordinator: Marcia A. Locke
Production: Susan Graham
Interior/Cover Designer: Julia Gecha

Interior Illustrator: Network Graphics
Typesetter: American Composition & Graphics, Inc.
Cover Photo: ©David W. Hamilton/
                    The Image Bank
Cover Printer: New England Book Components
Text Printer: Quebecor/Hawkins

Printed and bound in the United States of America
95 96 97 98 99—10 9 8 7 6 5 4 3 2 1

CONTENTS

4 Formulas and Problem Solving 136

5 Exponents and Polynomials 182

9 Square Roots and Radicals   374

10 Quadratic Equations   408

11 Additional Topics   448

PREFACE

When preparing *Elementary Algebra*, Fifth Edition, I attempted to preserve the features that made the previous editions successful while at the same time incorporating a number of improvements suggested by reviewers. Some of this new material addresses the emerging reforms in the mathematics curriculum.

This text is for students who have never had an elementary algebra course, as well as for those who need a review before taking additional mathematics courses. The basic concepts of elementary algebra are presented in a simple, straightforward manner. Concepts are developed through examples, continuously reinforced through additional examples, and then applied in problem solving situations.

Algebraic ideas are developed in a logical sequence, but in an easy-to-read manner, without excessive technical vocabulary and formalism. Whenever possible, the algebraic concepts are allowed to develop from their arithmetic counterparts. The following are two specific examples of this development.

1. Manipulation with simple algebraic fractions begins early (Sections 2.1 and 2.2) when we review operations with rational numbers.

2. Multiplying monomials—without any of the formal vocabulary—is introduced in Section 2.4 when we work with exponents.

There is a common thread running throughout the book, namely, *learn a skill*, next *use the skill to help solve equations*, and then *use equations to solve word problems*. This thread also influenced the following additional decisions:

- Approximately 550 word problems are scattered throughout the text. Every effort was made to start with easy ones, thereby building confidence in solving problems. Numerous problem-solving suggestions are offered with special discussions in several sections. *The key is to work with various problem solving techniques and not to become overly concerned with whether all of the traditional types of problems are studied.*

- Newly acquired skills are used as soon as possible to solve equations and word problems. Therefore, the concept of solving equations is introduced early and developed throughout the text. The concepts of factoring, solving equations, and solving word problems are tied together in Chapter 6.

Approximately 700 worked-out examples demonstrate a large variety of situations while leaving some things for students to think about in the problem sets. Examples are used to guide students in organizing their work and to help them decide when a shortcut may be tried. The progression from showing all the steps to demonstrating a suggested shortcut format is gradual.

As recommended by the American Mathematical Association of Two-Year Colleges, some geometric concepts are integrated in a problem solving setting that shows the connections between algebra, geometry, and the real world. Approximately 25 examples and 180 problems are designed to review basic geometric ideas. The following sections contain the bulk of the geometry material:

Section 2.5: Linear measurement concepts

Section 3.3: Complementary and supplementary angles; the sum of angles of a triangle equals 180°

Section 4.3: Area and volume formulas

Section 6.3: Pythagorean theorem

Section 10.1: More on the Pythagorean theorem, including work with isosceles right triangles and 30°–60° right triangles

New in This Edition

- Problems called **Thoughts into Words** are now included in every problem set except the review exercises. These problems are designed to give students an opportunity to express in written form their thoughts about various mathematical ideas. See, for example, Problem Sets 3.2, 3.4, 5.2, 5.4, and 6.4.

- Miscellaneous problem sections, now called **Further Investigations**, have been enhanced by adding more problems that lend themselves to small group work. These problems remain as "extras" but add flexibility for the instructor. See, for example, Problem Sets 1.2, 4.2, 5.3, and 9.2.

- The **chapter introductions** have been rewritten in an effort to provide more motivation for students to study algebra. Each introduction begins with at least one application that leads into the material of the chapter.

- A **Chapter Test** has been included at the end of each chapter. Along with the Chapter Review Problem Sets, these practice tests should provide the students with ample opportunity to prepare for the "real" tests.

- **Applications** have been added in several sections, including the following:
    1. Some applications have been added to sections of Chapters 1 and 2 that review arithmetic skills.
    2. In Chapter 5, examples and problems were added that connect geometry and the study of polynomials.
    3. In Section 8.3, several applications of slope were added as examples and problems.
    4. In Sections 9.1 and 9.5, applications involving radicals and radical equations were added.

5. In Section 10.1, some additional applications of the Pythagorean theorem were added.

- **Chapters 3 and 4** have been reorganized as suggested by some users of the previous edition. The work with ratios, proportions, and percents has been moved from the end of Chapter 3 to the beginning of Chapter 4. Inequalities and problem solving have been moved from the end of Chapter 4 to the end of Chapter 3. Old Section 4.6 (Equations and Inequalities Involving Absolute Value) has been moved to Chapter 11.

- At the request of users of the previous edition, a few more **motivating paragraphs** have been added. See, for example, the paragraphs at the ends of Sections 3.4, 4.4, and 7.6.

Other Special Features

- **Cumulative Review Problem Sets** appear at the ends of Chapters 2, 4, 6, 7, and 10. The one at the end of Chapter 10 could serve as a review of all of the basic topics of elementary algebra.

- All **answers** for Chapter Review Problem Sets, Chapter Tests, and Cumulative Review Problem Sets appear in the back of the text.

- **Problem Sets** have been constructed on an even-odd basis; that is, all variations of skill-development exercises are contained in both the even- and odd-numbered problems. Furthermore, problem sets are a focal point of every revision. We constantly add, subtract, and reword problems as suggested by users of the previous editions.

- **Chapter 11** was originally designed to be an "extra" chapter for those students who probably are going to take more mathematics. It remains that type of chapter with the addition of old Section 4.6 (Equations and Inequalities Involving Absolute Value) and old Section 8.8 (Graphing Linear Inequalities). Each section is a continuation of some topic presented in an earlier chapter.

- The text allows **flexibility in the continuity pattern**. For example, if an earlier introduction to graphing and solving systems of linear equations is desired, then Chapter 8 could be covered immediately after Chapter 4. This would provide the opportunity to use graphing and systems of linear equations to solve problems.

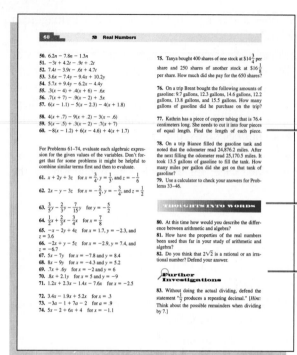

Well-crafted word problems reflect true-to-life situations.

"Thoughts into Words" problems encourage students to express their mathematical understanding verbally.

"Further Investigations" problems, which require skills learned in the section, are especially appropriate for group work.

Many worked examples show careful, step-by-step problem solving.

The "Check" feature in worked examples and problems reminds students to complete this important problem-solving step.

Annotations make clear each step of the problem.

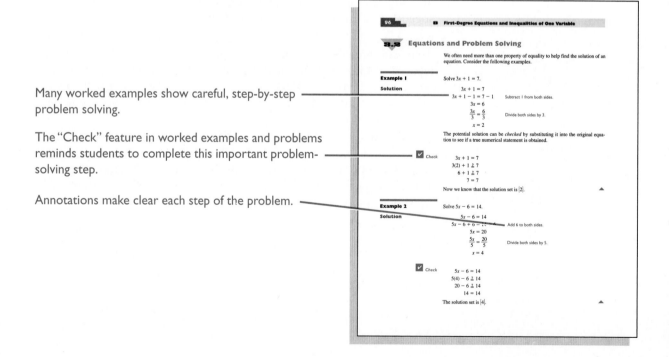

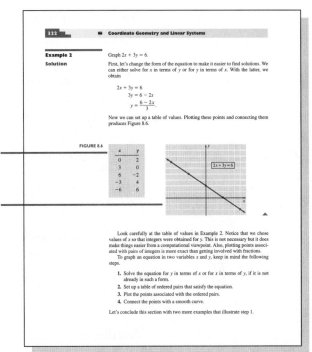

In sections on graphing, tables of values show numerical approaches to problems.

Large, clear graphs depict slopes accurately and restate the equation being graphed.

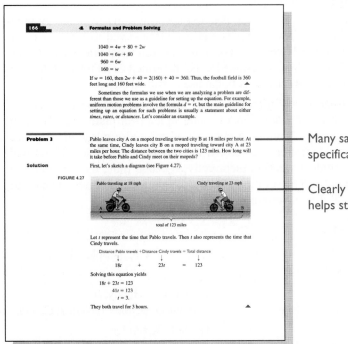

Many sample word problems are fully solved in sections specifically emphasizing problem solving.

Clearly rendered representational art lends interest and helps students visualize the problem.

Ancillaries for Instructors

The following supplements are available to adopters of this text:

- **Annotated Instructor's Edition** includes answers to all problems in the text—most printed adjacent to the problem.

- **Instructor's Solutions Manual** contains solutions for even-numbered problems and answers for all odd-numbered problems.

- **Test Bank with Chapter Tests** contains all questions and answers from the computerized test bank and three sample tests (two multiple choice, one open ended) for each chapter. These tests may be duplicated for student testing by instructors using the text.

- **Computerized testing software** is available for the IBM and compatibles and for the Macintosh. The computerized testing programs contain multiple-choice and open-ended questions that allow users to edit, rearrange, and add to the question bank.

- **Videotape Series** is text-specific, following the organization and style of the textbook. Video lectures include basic instruction and worked examples.

Ancillaries for Students:

- **Student's Solutions Manual** contains complete worked-out solutions for all odd-numbered exercises.

- **Worksheets and Study Guide** is a text-specific study resource in work-text format. It includes examples and exercises keyed to sections in the text so that students have the opportunity for additional practice and study assistance. The manual is designed to be integrated as an interactive component to lectures or for instructional use outside the classroom.

- **MathQuest Tutorial Software** is an interactive, text-specific intuitive tutorial that runs on both Windows and Macintosh platforms. The program provides fill-in, multiple-choice, and true/false questions. If a student answers a question incorrectly, the program will first respond with hints; if the student answers incorrectly a second time, the program will supply a step-by-step solution. Record-keeping capabilities enable students to monitor their progress.

Acknowledgments

I would like to take this opportunity to thank the following people who served as reviewers for this text:

Helen Banes
Kirkwood Community College

Corinna M. Goehring
Jackson State Community College

Laurence C. Huddy, Jr.
Horry-Georgetown Technical College

Gayle L. Krzemien
Pikes Peak Community College

Alisa Carter Lewis
Tyler Junior College

Katherine R. Struve
Columbus State Community College

Mary Lee Seitz
Erie Community College

Molly Sumner
Pikes Peak Community College

Karen Sharp
Mott Community College

Richard D. Townsend
North Carolina Central University

I am very grateful to the staff of PWS, especially David Dietz and Mary Beckwith, for their continuous cooperation and assistance throughout this project. I would also like to express my sincere gratitude to Robine Andrau and to Susan Graham. They continue to make my life as an author so much easier by carrying out the details of production in a dedicated and caring way.

In addition I would like to thank Elaine Werner and Deann Christianson for their work on the *Student's Solutions Manual* and the *Instructor's Solutions Manual*, Gayle Krzemien for creating the *Worksheets and Study Guide*, and Karen Sharp for developing the videos.

Again, very special thanks are due to my wife, Arlene, who spends numerous hours typing and proofreading manuscripts.

Jerome E. Kaufmann
Marble Falls, Texas

Some Basic Concepts of Arithmetic and Algebra

Karla started 1995 with $500 in her savings account and she planned to save an additional $15 per month for all of 1995. Without considering any accumulated interest, the numerical expression 500 + 12(15) represents the amount in her savings account at the end of 1995.

The numbers +2, −1, −3, +1, and −4 represent Woody's scores relative to par for five rounds of golf. The numerical expression 2 + (−1) + (−3) + 1 + (−4) can be used to determine how Woody stands relative to par at the end of the five rounds.

The temperature at 4 A.M. was −14°F. By noon the temperature had increased by 23°F. The numerical expression −14 + 23 can be used to determine the temperature at noon.

In the first two chapters of this text the concept of a numerical expression is used as a basis for reviewing addition, subtraction, multiplication, and division of various kinds of numbers. Then the concept of a variable allows us to move from numerical expressions to algebraic expressions; that is, to start the transition from arithmetic to algebra. Keep in mind that algebra is simply a generalized approach to arithmetic. Many algebraic concepts are extensions of arithmetic ideas; your knowledge of arithmetic will help you with your study of algebra.

1.1 Numerical and Algebraic Expressions

In arithmetic, we use symbols such as 4, 8, 17, and π to represent numbers. We indicate the basic operations of addition, subtraction, multiplication, and division by the symbols $+$, $-$, $\cdot$, and $\div$, respectively. Thus, we can formulate specific **numerical expressions**. For example, we can write the indicated sum of eight and four as $8 + 4$.

In algebra, **variables** allow us to generalize. By letting x and y represent *any* number, we can use the expression $x + y$ to represent the indicated sum of *any two* numbers. The x and y in such an expression are called variables and the phrase $x + y$ is called an **algebraic expression**. We commonly use letters of the alphabet such as x, y, z, and w as variables; the key idea is that they represent numbers. Our review of various operations and properties pertaining to numbers establishes the foundation for our study of algebra.

Many of the notational agreements made in arithmetic are extended to algebra with a few slight modifications. The following chart summarizes these notational agreements pertaining to the four basic operations. Notice the variety of ways to write a product by including parentheses to indicate multiplication. Actually the *ab* form is the simplest and probably the most used form; expressions such as *abc*, $6x$, and $7xyz$ all indicate multiplication. Also note the various forms for indicating division; the fractional form, $\dfrac{c}{d}$, is usually used in algebra although the other forms do serve a purpose at times.

Operation	Arithmetic	Algebra	Vocabulary
Addition	$4 + 6$	$x + y$	The *sum* of x and y
Subtraction	$7 - 2$	$w - z$	The *difference* of w and z
Multiplication	$9 \cdot 8$	$a \cdot b$, $a(b)$, $(a)b$, $(a)(b)$, or ab	The *product* of a and b
Division	$8 \div 2$, $\dfrac{8}{2}$, $2\overline{)8}$	$c \div d$, $\dfrac{c}{d}$, or $d\overline{)c}$	The *quotient* of c and d

As we review arithmetic ideas and introduce algebraic concepts, it is important to include some of the basic vocabulary and symbolism associated with sets. A **set** is a collection of objects, and the objects are called **elements** or **members** of the set. In arithmetic and algebra the elements of a set are often numbers. To communicate about sets, we use set braces, $\{\ \}$, to enclose the elements (or a description of the elements) and we use capital letters to name sets. For example, we can represent a set A, which consists of the vowels of the alphabet, as

$$A = \{\text{Vowels of the alphabet}\} \qquad \text{Word description, or}$$
$$A = \{a, e, i, o, u\} \qquad \text{List or roster description.}$$

We can modify the listing approach if the number of elements is large. For example, all of the letters of the alphabet can be listed as

$$\{a, b, c, \ldots, z\}.$$

We begin by simply writing enough elements to establish a pattern, then the three dots indicate that the set continues in that pattern. The final entry indicates the last element of the pattern. If we write

$$\{1, 2, 3, \ldots\}$$

the set begins with the counting numbers 1, 2, and 3. The three dots indicate that it continues in a like manner forever; there is no last element. A set that consists of no elements is called the **null set** (written $\varnothing$).

Two sets are said to be *equal* if they contain exactly the same elements. For example,

$$\{1, 2, 3\} = \{2, 1, 3\},$$

because both sets contain the same elements; the order in which the elements are written doesn't matter. The slash mark through the equality symbol denotes *not equal to*. Thus, if $A = \{1, 2, 3\}$ and $B = \{1, 2, 3, 4\}$, we can write $A \neq B$, which we read as "set A is not equal to set B."

Simplifying Numerical Expressions

Now let's simplify some numerical expressions that involve the set of **whole numbers**, that is, the set $\{0, 1, 2, 3, \ldots\}$.

Example 1

Simplify $8 + 7 - 4 + 12 - 7 + 14$.

Solution

The additions and subtractions should be performed from left to right in the order that they appear. Thus, $8 + 7 - 4 + 12 - 7 + 14$ simplifies to 30. ▲

Example 2

Simplify $7(9 + 5)$.

Solution

The parentheses indicate the product of 7 and the quantity $9 + 5$. Perform the addition inside the parentheses first and then multiply; $7(9 + 5)$ thus simplifies to $7(14)$, which becomes 98. ▲

Example 3

Simplify $(7 + 8) \div (4 - 1)$.

Solution

First, we perform the operations inside the parentheses. $(7 + 8) \div (4 - 1)$ thus becomes $15 \div 3$, which is 5. ▲

We frequently express a problem such as Example 3 in the form $\dfrac{7+8}{4-1}$. We don't need parentheses in this case because the fraction bar indicates that the sum of 7 and 8 is to be divided by the difference, $4-1$. A problem may, however, contain parentheses and fraction bars, as the next example illustrates.

Example 4

Simplify $\dfrac{(4+2)(7-1)}{9} + \dfrac{4}{7-3}$.

Solution

First, simplify above and below the fraction bars, and then proceed to evaluate as follows.

$$\frac{(4+2)(7-1)}{9} + \frac{4}{7-3} = \frac{(6)(6)}{9} + \frac{4}{4}$$

$$= \frac{36}{9} + 1 = 4 + 1 = 5$$

▲

Example 5

Simplify $7 \cdot 9 + 5$.

Solution

If there are no parentheses to indicate otherwise, multiplication takes precedence over addition. First perform the multiplication and then do the addition: $7 \cdot 9 + 5$ therefore simplifies to $63 + 5$, which is 68. ▲

REMARK Compare Example 2 and Example 5 and note the difference in meaning. △

Example 6

Simplify $8 + 4 \cdot 3 - 14 \div 2$.

Solution

The multiplication and division should be done first in the order that they appear, from left to right. Thus, $8 + 4 \cdot 3 - 14 \div 2$ simplifies to $8 + 12 - 7$. We perform the addition and subtraction in the order that they appear, which simplifies $8 + 12 - 7$ to 13. ▲

Example 7

Simplify $8 \cdot 5 \div 4 + 7 \cdot 3 - 32 \div 8 + 9 \div 3 \cdot 2$.

Solution

When we perform the multiplications and divisions first in the order that they appear and then do the additions and subtractions, our work takes on the following format.

$$8 \cdot 5 \div 4 + 7 \cdot 3 - 32 \div 8 + 9 \div 3 \cdot 2 = 10 + 21 - 4 + 6 = 33$$

▲

Example 8

Simplify $5 + 6[2(3+9)]$.

Solution

We use brackets for the same purpose as parentheses. In such a problem we need to simplify *from the inside out*; perform the operations in the innermost parentheses first.

$$5 + 6[2(3 + 9)] = 5 + 6[2(12)]$$
$$= 5 + 6[24]$$
$$= 5 + 144$$
$$= 149$$

Let us now summarize the ideas presented in the previous examples regarding **simplifying numerical expressions**. When simplifying a numerical expression, perform the operations in the following order.

Order of Operations

1. Perform the operations inside the symbols of inclusion (parentheses and brackets) and above and below each fraction bar. Start with the innermost inclusion symbol.
2. Perform all multiplications and divisions in the order that they appear from left to right.
3. Perform all additions and subtractions in the order that they appear from left to right.

Use of Variables

We can use the concept of a variable to generalize from numerical expressions to algebraic expressions. Each of the following is an example of an algebraic expression.

$$3x + 2y, \qquad 5a - 2b + c,$$

$$\frac{5d + 3e}{2c - d}, \qquad 2xy + 5yz,$$

$$7(w + z), \qquad (x + y)(x - y)$$

An algebraic expression takes on a numerical value whenever each variable in the expression is replaced by a specific number. For example, if x is replaced by 9 and z by 4, the algebraic expression $x - z$ becomes the numerical expression $9 - 4$, which simplifies to 5. We say that $x - z$ **has a value of 5** when x equals 9 and z equals 4. The value of $x - z$, when x equals 25 and z equals 12, is 13. The general algebraic expression $x - z$ has a specific value each time x and z are replaced by numbers.

Consider the following examples, which illustrate the process of finding a value of an algebraic expression. We call this process **evaluating algebraic expressions**.

Example 9

Find the value of $3x + 2y$ when x is replaced by 5 and y by 17.

Solution

The following format is convenient for such problems.

$$3x + 2y = 3(5) + 2(17) \quad \text{when } x = 5 \text{ and } y = 17$$
$$= 15 + 34$$
$$= 49 \qquad \blacktriangle$$

In the solution to Example 9, notice that parentheses were inserted to indicate multiplication as we switched from the algebraic expression to the numerical expression. We could also use the raised dot to indicate multiplication; that is, $3(5) + 2(17)$ could be written as $3 \cdot 5 + 2 \cdot 17$. Furthermore, notice that once we have a numerical expression, our previous agreements for simplifying numerical expressions are in effect.

Example 10

Find the value of $12a - 3b$ when $a = 5$ and $b = 9$.

Solution

$$12a - 3b = 12(5) - 3(9) \quad \text{when } a = 5 \text{ and } b = 9$$
$$= 60 - 27$$
$$= 33 \qquad \blacktriangle$$

Example 11

Evaluate $4xy + 2xz - 3yz$ when $x = 8$, $y = 6$, and $z = 2$.

Solution

$$4xy + 2xz - 3yz = 4(8)(6) + 2(8)(2) - 3(6)(2) \quad \text{when } x = 8, y = 6, \text{ and } z = 2$$
$$= 192 + 32 - 36$$
$$= 188 \qquad \blacktriangle$$

Example 12

Evaluate $\dfrac{5c + d}{3c - d}$ for $c = 12$ and $d = 4$.

Solution

$$\frac{5c + d}{3c - d} = \frac{5(12) + 4}{3(12) - 4} \quad \text{for } c = 12 \text{ and } d = 4$$
$$= \frac{60 + 4}{36 - 4} = \frac{64}{32} = 2 \qquad \blacktriangle$$

Example 13

Evaluate $(2x + 5y)(3x - 2y)$ when $x = 6$ and $y = 3$.

Solution

$$(2x + 5y)(3x - 2y) = (2 \cdot 6 + 5 \cdot 3)(3 \cdot 6 - 2 \cdot 3) \quad \text{when } x = 6 \text{ and } y = 3$$
$$= (12 + 15)(18 - 6)$$
$$= (27)(12)$$
$$= 324 \qquad \blacktriangle$$

Problem Set 1.1

For Problems 1–34, please simplify each numerical expression.

1. $9 + 14 - 7$
2. $32 - 14 + 6$
3. $7(14 - 9)$
4. $8(6 + 12)$
5. $16 + 5 \cdot 7$
6. $18 - 3(5)$
7. $4(12 + 9) - 3(8 - 4)$
8. $7(13 - 4) - 2(19 - 11)$
9. $4(7) + 6(9)$
10. $8(7) - 4(8)$
11. $6 \cdot 7 + 5 \cdot 8 - 3 \cdot 9$
12. $8(13) - 4(9) + 2(7)$
13. $(6 + 9)(8 - 4)$
14. $(15 - 6)(13 - 4)$
15. $6 + 4[3(9 - 4)]$
16. $92 - 3[2(5 - 2)]$
17. $16 \div 8 \cdot 4 + 36 \div 4 \cdot 2$
18. $7 \cdot 8 \div 4 - 72 \div 12$
19. $\dfrac{8 + 12}{4} - \dfrac{9 + 15}{8}$
20. $\dfrac{19 - 7}{6} + \dfrac{38 - 14}{3}$
21. $56 - [3(9 - 6)]$
22. $17 + 2[3(4 - 2)]$
23. $7 \cdot 4 \cdot 2 \div 8 + 14$
24. $14 \div 7 \cdot 8 - 35 \div 7 \cdot 2$
25. $32 \div 8 \cdot 2 + 24 \div 6 - 1$
26. $48 \div 12 + 7 \cdot 2 \div 2 - 1$
27. $4 \cdot 9 \div 12 + 18 \div 2 + 3$
28. $5 \cdot 8 \div 4 - 8 \div 4 \cdot 3 + 6$
29. $\dfrac{6(8 - 3)}{3} + \dfrac{12(7 - 4)}{9}$
30. $\dfrac{3(17 - 9)}{4} + \dfrac{9(16 - 7)}{3}$
31. $83 - \dfrac{4(12 - 7)}{5}$
32. $78 - \dfrac{6(21 - 9)}{4}$
33. $\dfrac{4 \cdot 6 + 5 \cdot 3}{7 + 2 \cdot 3} + \dfrac{7 \cdot 9 + 6 \cdot 5}{3 \cdot 5 + 8 \cdot 2}$
34. $\dfrac{7 \cdot 8 + 4}{5 \cdot 8 - 10} + \dfrac{9 \cdot 6 - 4}{6 \cdot 5 - 20}$

For Problems 35–54, evaluate each algebraic expression for the given values of the variables.

35. $7x + 4y$ for $x = 6$ and $y = 8$
36. $8x + 6y$ for $x = 9$ and $y = 5$
37. $16a - 9b$ for $a = 3$ and $b = 4$
38. $14a - 5b$ for $a = 7$ and $b = 9$
39. $4x + 7y + 3xy$ for $x = 4$ and $y = 9$
40. $x + 8y + 5xy$ for $x = 12$ and $y = 3$
41. $14xz + 6xy - 4yz$ for $x = 8$, $y = 5$, and $z = 7$
42. $9xy - 4xz + 3yz$ for $x = 7$, $y = 3$, and $z = 2$
43. $\dfrac{54}{n} + \dfrac{n}{3}$ for $n = 9$
44. $\dfrac{n}{4} + \dfrac{60}{n} - \dfrac{n}{6}$ for $n = 12$
45. $\dfrac{y + 16}{6} + \dfrac{50 - y}{3}$ for $y = 8$
46. $\dfrac{w + 57}{9} + \dfrac{90 - w}{7}$ for $w = 6$
47. $(x + y)(x - y)$ for $x = 8$ and $y = 3$
48. $(x + 2y)(2x - y)$ for $x = 7$ and $y = 4$
49. $(5x - 2y)(3x + 4y)$ for $x = 3$ and $y = 6$
50. $(3a + b)(7a - 2b)$ for $a = 5$ and $b = 7$
51. $6 + 3[2(x + 4)]$ for $x = 7$
52. $9 + 4[3(x + 3)]$ for $x = 6$
53. $81 - 2[5(n + 4)]$ for $n = 3$
54. $78 - 3[4(n - 2)]$ for $n = 4$

For Problems 55–60, find the value of $\dfrac{bh}{2}$ for each set of values for the variables b and h.

55. $b = 8$ and $h = 12$

56. $b = 6$ and $h = 14$

57. $b = 7$ and $h = 6$

58. $b = 9$ and $h = 4$

59. $b = 16$ and $h = 5$

60. $b = 18$ and $h = 13$

For Problems 61–66, find the value of $\dfrac{Bh}{3}$ for each set of values for the variables B and h.

61. $B = 27$ and $h = 9$

62. $B = 18$ and $h = 6$

63. $B = 25$ and $h = 12$

64. $B = 32$ and $h = 18$

65. $B = 36$ and $h = 7$

66. $B = 42$ and $h = 17$

For Problems 67–72, find the value of $\dfrac{h(b_1 + b_2)}{2}$ for each set of values for the variables h, b_1, and b_2.

(Subscripts are used to indicate that b_1 and b_2 are different variables.)

67. $h = 17$, $b_1 = 14$, and $b_2 = 6$

68. $h = 9$, $b_1 = 12$, and $b_2 = 16$

69. $h = 8$, $b_1 = 17$, and $b_2 = 24$

70. $h = 12$, $b_1 = 14$, and $b_2 = 5$

71. $h = 18$, $b_1 = 6$, and $b_2 = 11$

72. $h = 14$, $b_1 = 9$, and $b_2 = 7$

73. You should be able to do calculations like those in Problems 1–34 *with* and *without* a calculator. Be sure that you can do Problems 1–34 *with* your calculator.

THOUGHTS INTO WORDS

74. Explain the difference between a numerical expression and an algebraic expression.

75. Your friend keeps getting an answer of 45 when simplifying $3 + 2(9)$. What mistake is he making and how would you help him?

1.2   Prime and Composite Numbers

Occasionally, terms in mathematics are given a special meaning in the discussion of a particular topic. Such is the case with the term "divides" as it is used in this section.

We say that 6 *divides* 18 because 6 times the whole number 3 produces 18; but 6 *does not divide* 19 because there is no whole number such that 6 times the number produces 19. Likewise, 5 *divides* 35 because 5 times the whole number 7 produces 35; 5 *does not divide* 42 because there is no whole number such that 5 times the number produces 42. We can use the following general definition.

DEFINITION 1.1   Given that a and b are whole numbers, with a not equal to zero, a *divides* b if and only if there exists a whole number k such that $a \cdot k = b$.

REMARK   Notice the use of variables, a, b, and k, in the statement of a *general* definition. Also note that the definition merely generalizes the concept of *divides*, which was introduced in the specific examples prior to the definition.   △

The following statements further clarify Definition 1.1. Pay special attention to the italicized words for they indicate some of the terminology used for this topic.

1. 8 *divides* 56 because $8 \cdot 7 = 56$.

2. 7 *does not divide* 38 because there is no whole number, k, such that $7 \cdot k = 38$.

3. 3 is a *factor* of 27 because $3 \cdot 9 = 27$.

4. 4 is *not a factor* of 38 because there is no whole number, k, such that $4 \cdot k = 38$.

5. 35 is a *multiple* of 5 because $5 \cdot 7 = 35$.

6. 29 is *not a multiple* of 7 because there is no whole number, k, such that $7 \cdot k = 29$.

We use the *factor* terminology extensively. We say that 7 and 8 are factors of 56 because $7 \cdot 8 = 56$; 4 and 14 are also factors of 56 because $4 \cdot 14 = 56$. The factors of a number are also the divisors of the number.

Now consider two special kinds of whole numbers called **prime numbers** and **composite numbers** according to the following definition.

DEFINITION 1.2

A **prime number** is a whole number, greater than 1, that has no factors (divisors) other than itself and 1. Whole numbers, greater than 1, that are not prime numbers are called **composite numbers.**

The prime numbers less than 50 are 2, 3, 5, 7, 11, 13, 17, 19, 23, 29, 31, 37, 41, 43, and 47. Notice that each of these has no factors other than itself and 1. An interesting point is that the set of prime numbers is an infinite set; that is, the prime numbers go on forever, and there is no *largest* prime number.

We can express every composite number as the indicated product of prime numbers. Consider the following examples.

$$4 = 2 \cdot 2, \quad 6 = 2 \cdot 3, \quad 8 = 2 \cdot 2 \cdot 2, \quad 10 = 2 \cdot 5, \quad 12 = 2 \cdot 2 \cdot 3$$

In each case we expressed a composite number as the indicated product of prime numbers. The indicated product form is sometimes called the **prime factored form** of the number.

There are various procedures to find the prime factors of a given composite number. For our purposes, the simplest technique is to factor the given composite number into any two easily recognized factors and then to continue to factor each of these until we obtain only prime factors. Consider these examples.

$$18 = 2 \cdot 9 = 2 \cdot 3 \cdot 3, \qquad 27 = 3 \cdot 9 = 3 \cdot 3 \cdot 3,$$
$$24 = 4 \cdot 6 = 2 \cdot 2 \cdot 2 \cdot 3, \qquad 150 = 10 \cdot 15 = 2 \cdot 5 \cdot 3 \cdot 5$$

It does not matter which two factors we choose first. For example, we might start by expressing 18 as $3 \cdot 6$ and then factor 6 into $2 \cdot 3$, which would produce a final result of $18 = 3 \cdot 2 \cdot 3$. Either way, 18 contains two prime factors of 3 and one prime factor of 2. The order in which we write the prime factors is not important.

Greatest Common Factor

We can use the prime factorization form of two composite numbers to conveniently find their **greatest common factor.** Consider the following example.

$$42 = 2 \cdot 3 \cdot 7,$$
$$70 = 2 \cdot 5 \cdot 7$$

Notice that 2 is a factor of both, as is 7. Therefore, 14 (the product of 2 and 7) is the greatest common factor of 42 and 70. In other words, 14 is the largest whole number that divides both 42 and 70. The following examples should further clarify the process of finding the greatest common factor of two or more numbers.

Example 1

Find the greatest common factor of 48 and 60.

Solution

$$48 = 2 \cdot 2 \cdot 2 \cdot 2 \cdot 3$$
$$60 = 2 \cdot 2 \cdot 3 \cdot 5$$

Since two 2s and a 3 are common to both, the greatest common factor of 48 and 60 is $2 \cdot 2 \cdot 3 = \mathbf{12}$. ▲

Example 2

Find the greatest common factor of 21 and 75.

Solution

$$21 = 3 \cdot 7$$
$$75 = 3 \cdot 5 \cdot 5$$

Since only a 3 is common to both, the greatest common factor is **3**. ▲

Example 3

Find the greatest common factor of 24 and 35.

Solution

$$24 = 2 \cdot 2 \cdot 2 \cdot 3$$
$$35 = 5 \cdot 7$$

Since there are no common prime factors, the greatest common factor is **1**. ▲

The concept of greatest common factor can be extended to more than two numbers, as the next example demonstrates.

Example 4

Find the greatest common factor of 24, 56, and 120.

Solution

$$24 = 2 \cdot 2 \cdot 2 \cdot 3$$
$$56 = 2 \cdot 2 \cdot 2 \cdot 7$$
$$120 = 2 \cdot 2 \cdot 2 \cdot 3 \cdot 5$$

Since three 2s are common to the numbers, the greatest common factor of 24, 56, and 120 is $2 \cdot 2 \cdot 2 = 8$.   ▲

Least Common Multiple

We stated earlier in this section that 35 is a *multiple of* 5 because $5 \cdot 7 = 35$. The set of all whole numbers that are multiples of 5 consists of 0, 5, 10, 15, 20, 25, etc. In other words, 5 times each successive whole number ($5 \cdot 0 = 0, 5 \cdot 1 = 5$, $5 \cdot 2 = 10, 5 \cdot 3 = 15$, etc.) produces the multiples of 5. In a like manner, the set of multiples of 4 consists of 0, 4, 8, 12, 16, etc.

It is sometimes necessary to determine the smallest common *nonzero* multiple of two or more whole numbers. We use the phrase **least common multiple** to designate this nonzero number. For example, the least common multiple of 3 and 4 is 12, which means that 12 is the smallest nonzero multiple of both 3 and 4. Stated another way, 12 is the smallest nonzero whole number that is divisible by both 3 and 4. Likewise, we would say that the least common multiple of 6 and 8 is 24.

If we cannot determine the least common multiple by inspection, then the prime factorization form of composite numbers is helpful. Study the solutions to the following examples very carefully so that we can develop a systematic technique for finding the least common multiple of two or more numbers.

Example 5

Find the least common multiple of 24 and 36.

Solution

Let's first express each number as a product of prime factors.

$$24 = 2 \cdot 2 \cdot 2 \cdot 3,$$
$$36 = 2 \cdot 2 \cdot 3 \cdot 3$$

Since 24 contains three 2s, the least common multiple must have three 2s. Also, since 36 contains two 3s, we need to put two 3s in the least common multiple. The least common multiple of 24 and 36 is therefore $2 \cdot 2 \cdot 2 \cdot 3 \cdot 3 = 72$.   ▲

If the least common multiple is not obvious by inspection, then we can proceed as follows.

STEP 1 Express each number as a product of prime factors.

STEP 2 The least common multiple contains each different prime factor as many times as the *most* times it appears in any one of the factorizations from step 1.

Example 6

Solution

Find the least common multiple of 48 and 84.

$$48 = 2 \cdot 2 \cdot 2 \cdot 2 \cdot 3$$
$$84 = 2 \cdot 2 \cdot 3 \cdot 7$$

We need four 2s in the least common multiple because of the four 2s in 48. We need one 3 because of the 3 in each of the numbers and one 7 is needed because of the 7 in 84. The least common multiple of 48 and 84 is $2 \cdot 2 \cdot 2 \cdot 2 \cdot 3 \cdot 7 = 336$.

▲

Example 7

Solution

Find the least common multiple of 12, 18, and 28.

$$12 = 2 \cdot 2 \cdot 3$$
$$18 = 2 \cdot 3 \cdot 3$$
$$28 = 2 \cdot 2 \cdot 7$$

The least common multiple is $2 \cdot 2 \cdot 3 \cdot 3 \cdot 7 = 252$.

▲

Example 8

Solution

Find the least common multiple of 8 and 9.

$$8 = 2 \cdot 2 \cdot 2$$
$$9 = 3 \cdot 3$$

The least common multiple is $2 \cdot 2 \cdot 2 \cdot 3 \cdot 3 = 72$.

▲

Problem Set 1.2

For Problems 1–20, classify each statement as true or false.

1. 8 divides 56
2. 9 divides 54
3. 6 does not divide 54
4. 7 does not divide 42
5. 96 is a multiple of 8
6. 78 is a multiple of 6
7. 54 is not a multiple of 4
8. 64 is not a multiple of 6
9. 144 is divisible by 4
10. 261 is divisible by 9
11. 173 is divisible by 3
12. 149 is divisible by 7

13. 11 is a factor of 143
14. 11 is a factor of 187
15. 9 is a factor of 119
16. 8 is a factor of 98
17. 3 is a prime factor of 57
18. 7 is a prime factor of 91
19. 4 is a prime factor of 48
20. 6 is a prime factor of 72

For Problems 21–30, classify each number as prime or composite.

21. 53   22. 57
23. 59   24. 61
25. 91   26. 81

27. 89

28. 97

29. 111

30. 101

For Problems 31–42, factor each composite number into a product of prime numbers. For example, $18 = 2 \cdot 3 \cdot 3$.

31. 26

32. 16

33. 36

34. 80

35. 49

36. 92

37. 56

38. 144

39. 120

40. 84

41. 135

42. 98

For Problems 43–54, find the greatest common factor of the given numbers.

43. 12 and 16

44. 30 and 36

45. 56 and 64

46. 72 and 96

47. 63 and 81

48. 60 and 72

49. 84 and 96

50. 48 and 52

51. 36, 72, and 90

52. 27, 54, and 63

53. 48, 60, and 84

54. 32, 80, and 96

For Problems 55–66, find the least common multiple of the given numbers.

55. 6 and 8

56. 8 and 12

57. 12 and 16

58. 9 and 12

59. 28 and 35

60. 42 and 66

61. 49 and 56

62. 18 and 24

63. 8, 12, and 28

64. 6, 10, and 12

65. 9, 15, and 18

66. 8, 14, and 24

THOUGHTS INTO WORDS

67. How would you explain the concepts of *greatest common factor* and *least common multiple* to a friend who missed class during that discussion?

68. Is it always true that the greatest common factor of two numbers is less than the least common multiple of those same two numbers? Explain your answer.

Further Investigations

69. The numbers 0, 2, 4, 6, 8, etc. are multiples of 2. They are also called *even* numbers. Why is 2 the only even prime number?

70. Find the smallest nonzero whole number that is divisible by 2, 3, 4, 5, 6, 7, and 8.

71. Find the smallest whole number, greater than 1, that produces a remainder of 1 when divided by 2, 3, 4, 5, or 6.

72. What is the greatest common factor of x and y if x and y are both prime numbers and x does not equal y? Explain your answer.

73. What is the greatest common factor of x and y if x and y are nonzero whole numbers and y is a multiple of x? Explain your answer.

74. What is the least common multiple of x and y if they are both prime numbers and x does not equal y? Explain your answer.

75. What is the least common multiple of x and y if the greatest common factor of x and y is 1? Explain your answer.

For Problems 76–85, familiarity with a few basic divisibility rules will be helpful for determining the prime factors. For example, if you can quickly recognize that 51 is divisible by 3, then you can divide 51 by 3 to find another factor of 17. Since 3 and 17 are both prime numbers, we have $51 = 3 \cdot 17$. The divisibility rules for 2, 3, 5, and 9 are as follows.

Rule for 2

A whole number is divisible by 2 if and only if the units digit of its base-ten numeral is divisible by 2. (In other words, the units digit must be 0, 2, 4, 6, or 8.)

EXAMPLE   68 is divisible by 2 because 8 is divisible by 2.

EXAMPLE   57 is not divisible by 2 because 7 is not divisible by 2.

Rule for 3

A whole number is divisible by 3 if and only if the sum of the digits of its base-ten numeral is divisible by 3.

EXAMPLES   51 is divisible by 3 because $5 + 1 = 6$ and 6 is divisible by 3.
144 is divisible by 3 because $1 + 4 + 4 = 9$ and 9 is divisible by 3.
133 is not divisible by 3 because $1 + 3 + 3 = 7$ and 7 is not divisible by 3.

Rule for 5

A whole number is divisible by 5 if and only if the units digit of its base-ten numeral is divisible by 5. (In other words, the units digit must be 0 or 5.)

EXAMPLES   115 is divisible by 5 because 5 is divisible by 5.
172 is not divisible by 5 because 2 is not divisible by 5.

Rule for 9

A whole number is divisible by 9 if and only if the sum of the digits of its base-ten numeral is divisible by 9.

EXAMPLES   765 is divisible by 9 because $7 + 6 + 5 = 18$ and 18 is divisible by 9.

147 is not divisible by 9 because $1 + 4 + 7 = 12$ and 12 is not divisible by 9.

Use these divisibility rules to help determine the prime factorization of each of the following numbers.

76. 118		**77.** 76	
78. 201		**79.** 123	
80. 85		**81.** 115	
82. 117		**83.** 441	
84. 129		**85.** 153	

1.3   Integers: Addition and Subtraction

"A record temperature of 35° *below* zero was recorded on this date in 1904." "The PO stock closed *down* 3 points yesterday." "On a first-down sweep around left end, Moser *lost* 7 yards." "The Widget Manufacturing Company reported *assets* of 50 million dollars and *liabilities* of 53 million dollars for 1981." These examples illustrate our need for negative numbers.

    The number line becomes a helpful visual device for our work at this time. We can associate the set of whole numbers with evenly spaced points on a line as indicated in Figure 1.1. For each nonzero whole number we can associate its *negative* to the left of zero; with 1 we associate −1, with 2 we associate −2, etc. as indicated in Figure 1.2. The set of whole numbers along with −1, −2, −3,

FIGURE 1.1

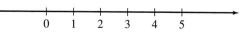

FIGURE 1.2

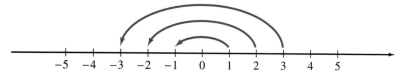

and so on, is called the set of **integers**. The following terminology is used with reference to the integers.

$\{\ldots, -3, -2, -1, 0, 1, 2, 3, \ldots\}$	Integers
$\{1, 2, 3, 4, \ldots\}$	Positive integers
$\{0, 1, 2, 3, 4, \ldots\}$	Nonnegative integers
$\{\ldots, -3, -2, -1\}$	Negative integers
$\{\ldots, -3, -2, -1, 0\}$	Nonpositive integers

The symbol −1 can be read as *negative one*, *opposite of one*, or *additive inverse of one*. The *opposite-of* and *additive-inverse-of* terminology is very helpful when working with variables. For example, the symbol −x, read as *opposite of x* or *additive inverse of x*, emphasizes an important issue. Since x can be any integer, −x (the opposite of x) can be zero, positive, or negative. If x is a positive integer, then −x is negative. If x is a negative integer, then −x is positive. If x is zero, then −x is zero. These statements can be written as follows and illustrated on the number lines in Figure 1.3.

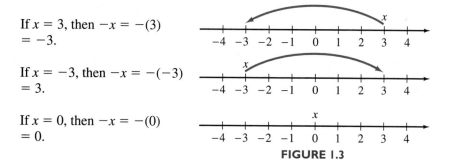

If $x = 3$, then $-x = -(3)$ = −3.

If $x = -3$, then $-x = -(-3)$ = 3.

If $x = 0$, then $-x = -(0)$ = 0.

FIGURE 1.3

From this discussion we also need to recognize the following general property.

PROPERTY 1.1

If *a* is any integer, then

$$-(-a) = a.$$

(The opposite of the opposite of any integer is the integer itself.)

Addition of Integers

The number line is also a convenient visual aid for interpreting **addition of integers.** In Figure 1.4 we see number line interpretations for the following examples.

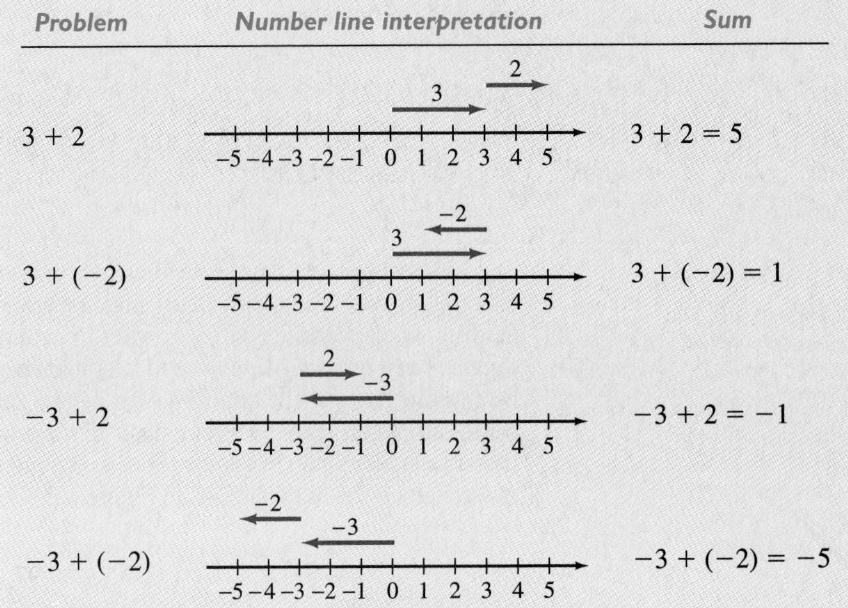

Problem	Number line interpretation	Sum
$3 + 2$		$3 + 2 = 5$
$3 + (-2)$		$3 + (-2) = 1$
$-3 + 2$		$-3 + 2 = -1$
$-3 + (-2)$		$-3 + (-2) = -5$

FIGURE 1.4

Once you acquire a feeling of movement on the number line, a mental image of this movement is sufficient. Consider the following addition problems and mentally picture the number line interpretation. Be sure that you agree with all of our answers.

$$5 + (-2) = 3, \qquad -6 + 4 = -2, \qquad -8 + 11 = 3,$$
$$-7 + (-4) = -11, \qquad -5 + 9 = 4, \qquad 9 + (-2) = 7,$$
$$14 + (-17) = -3, \qquad 0 + (-4) = -4, \qquad 6 + (-6) = 0$$

The last example illustrates a general property that should be noted: **Any integer plus its opposite produces zero.**

REMARK Profits and losses pertaining to investments also provide a good physical model for interpreting addition of integers. A loss of $25 on one investment along with a profit of $60 on a second investment produces an overall profit of $35. This can be expressed as $-25 + 60 = 35$. Perhaps it would be helpful for you to check the previous examples using a profit and loss interpretation.   △

Even though all problems involving addition of integers could be done by using the number line interpretation, it is sometimes convenient to give a more precise description of the addition process. For this purpose we need to briefly consider the concept of absolute value. The **absolute value** of a number is the distance between the number and zero on the number line. For example, the absolute value of 6 is 6. The absolute value of -6 is also 6. The absolute value of 0 is 0. Symbolically, absolute value is denoted with vertical bars. Thus, we write

$$|6| = 6, \qquad |-6| = 6, \qquad |0| = 0.$$

Notice that the absolute value of a positive number is the number itself, but the absolute value of a negative number is its opposite. Thus, the absolute value of any number, except 0, is positive, and the absolute value of 0 is 0.

We can describe the process of **adding integers** precisely by using the concept of absolute value as follows.

Two Positive Integers

The sum of two positive integers is the sum of their absolute values. (The sum of two positive integers is a positive integer.)

$$43 + 54 = |43| + |54| = 43 + 54 = 97$$

Two Negative Integers

The sum of two negative integers is the opposite of the sum of their absolute values. (The sum of two negative integers is a negative integer.)

$$(-67) + (-93) = -\big(|-67| + |-93|\big)$$
$$= -(67 + 93)$$
$$= -160$$

One Positive and One Negative Integer

We can find the sum of a positive and a negative integer by subtracting the smaller absolute value from the larger absolute value and giving the result the sign of the original number that has the larger absolute value. If the integers have the same absolute value, then their sum is 0.

$$82 + (-40) = |82| - |-40|$$
$$= 82 - 40$$
$$= 42$$
$$74 + (-90) = -\big(|-90| - |74|\big)$$
$$= -(90 - 74)$$
$$= -16$$
$$(-17) + 17 = |-17| - |17|$$
$$= 17 - 17$$
$$= 0$$

Zero and Another Integer

The sum of 0 and any integer is the integer itself.

$$0 + (-46) = -46$$
$$72 + 0 = 72$$

The following examples further demonstrate how to add integers. Be sure that you agree with each of the results.

$$-18 + (-56) = -\big(|-18| + |-56|\big) = -(18 + 56) = -74,$$
$$-71 + (-32) = -\big(|-71| + |-32|\big) = -(71 + 32) = -103,$$
$$64 + (-49) = |64| - |-49| = 64 - 49 = 15,$$
$$-56 + 93 = |93| - |-56| = 93 - 56 = 37,$$
$$-114 + 48 = -\big(|-114| - |48|\big) = -(114 - 48) = -66,$$
$$45 + (-73) = -\big(|-73| - |45|\big) = -(73 - 45) = -28,$$
$$46 + (-46) = 0, \qquad -48 + 0 = -48,$$
$$(-73) + 73 = 0, \qquad 0 + (-81) = -81$$

It is true that this *absolute value approach* does precisely describe the process of adding integers, but don't forget about the number line interpretation. Included in the next problem set are other physical models for interpreting the addition of integers. You may find these models helpful.

Subtraction of Integers

The following examples illustrate a relationship between addition and subtraction of *whole numbers*.

$$7 - 2 = 5 \quad \text{because } 2 + 5 = 7,$$

$$9 - 6 = 3 \quad \text{because } 6 + 3 = 9,$$

$$5 - 1 = 4 \quad \text{because } 1 + 4 = 5$$

This same relationship between addition and subtraction holds for *all integers*.

$$5 - 6 = -1 \quad \text{because } 6 + (-1) = 5,$$

$$-4 - 9 = -13 \quad \text{because } 9 + (-13) = -4,$$

$$-3 - (-7) = 4 \quad \text{because } -7 + 4 = -3,$$

$$8 - (-3) = 11 \quad \text{because } -3 + 11 = 8$$

Now consider a further observation:

$$5 - 6 = -1 \quad \text{and} \quad 5 + (-6) = -1,$$

$$-4 - 9 = -13 \quad \text{and} \quad -4 + (-9) = -13,$$

$$-3 - (-7) = 4 \quad \text{and} \quad -3 + 7 = 4,$$

$$8 - (-3) = 11 \quad \text{and} \quad 8 + 3 = 11.$$

The previous examples help us realize that we can state the subtraction of integers in terms of the addition of integers. More precisely, a general description for the subtraction of integers follows.

Subtraction of Integers

If a and b are integers, then $a - b = a + (-b)$.

It may be helpful for you to read $a - b = a + (-b)$ as "a minus b is equal to a plus the opposite of b." Every subtraction problem can be changed to an equivalent addition problem as illustrated by the following examples.

$$6 - 13 = 6 + (-13) = -7,$$

$$9 - (-12) = 9 + 12 = 21,$$

$$-8 - 13 = -8 + (-13) = -21,$$

$$-7 - (-8) = -7 + 8 = 1$$

It should be apparent that addition of integers is a key operation. The ability to effectively add integers is a necessary skill for further algebraic work.

Evaluating Algebraic Expressions

Let's conclude this section by evaluating some algebraic expressions using negative and positive integers.

Example 1

Evaluate each algebraic expression for the given values of the variables.

(a) $x - y$ for $x = -12$ and $y = 20$

(b) $-a + b$ for $a = -8$ and $b = -6$

(c) $-x - y$ for $x = 14$ and $y = -7$

Solution

(a) $x - y = -12 - 20$ when $x = -12$ and $y = 20$

$\qquad = -12 + (-20)$

$\qquad = -32$

(b) $-a + b = -(-8) + (-6)$ when $a = -8$ and $b = -6$

$\qquad = 8 + (-6)$

$\qquad = 2$

(c) $-x - y = -14 - (-7)$ when $x = 14$ and $y = -7$

$\qquad = -14 + 7$

$\qquad = -7$

Problem Set 1.3

For Problems 1–10, use the number line interpretation to find each sum.

1. $5 + (-3)$
2. $7 + (-4)$
3. $-6 + 2$
4. $-9 + 4$
5. $-3 + (-4)$
6. $-5 + (-6)$
7. $8 + (-2)$
8. $12 + (-7)$
9. $5 + (-11)$
10. $4 + (-13)$

For Problems 11–30, find each sum.

11. $17 + (-9)$
12. $16 + (-5)$
13. $8 + (-19)$
14. $9 + (-14)$
15. $-7 + (-8)$
16. $-6 + (-9)$
17. $-15 + 8$
18. $-22 + 14$
19. $-13 + (-18)$
20. $-15 + (-19)$
21. $-27 + 8$
22. $-29 + 12$
23. $32 + (-23)$
24. $27 + (-14)$
25. $-25 + (-36)$
26. $-34 + (-49)$
27. $54 + (-72)$
28. $48 + (-76)$
29. $-34 + (-58)$
30. $-27 + (-36)$

For Problems 31–50, subtract as indicated.

31. $3 - 8$
32. $5 - 11$
33. $-4 - 9$
34. $-7 - 8$
35. $5 - (-7)$
36. $9 - (-4)$
37. $-6 - (-12)$
38. $-7 - (-15)$
39. $-11 - (-10)$
40. $-14 - (-19)$
41. $-18 - 27$
42. $-16 - 25$
43. $34 - 63$
44. $25 - 58$
45. $45 - 18$
46. $52 - 38$
47. $-21 - 44$
48. $-26 - 54$
49. $-53 - (-24)$
50. $-76 - (-39)$

For Problems 51–66, add or subtract as indicated.

51. $6 - 8 - 9$
52. $5 - 9 - 4$
53. $-4 - (-6) + 5 - 8$
54. $-3 - 8 + 9 - (-6)$
55. $5 + 7 - 8 - 12$
56. $-7 + 9 - 4 - 12$
57. $-6 - 4 - (-2) + (-5)$
58. $-8 - 11 - (-6) + (-4)$
59. $-6 - 5 - 9 - 8 - 7$
60. $-4 - 3 - 7 - 8 - 6$
61. $7 - 12 + 14 - 15 - 9$
62. $8 - 13 + 17 - 15 - 19$
63. $-11 - (-14) + (-17) - 18$
64. $-15 + 20 - 14 - 18 + 9$
65. $16 - 21 + (-15) - (-22)$
66. $17 - 23 - 14 - (-18)$

The horizontal format is used extensively in algebra, but occasionally the vertical format shows up. Some exposure to the vertical format is therefore needed. Find the following *sums* for Problems 67–78.

| **67.** 5 | **68.** 8 | **69.** -13 |
| -9 | -13 | -18 |

| **70.** -14 | **71.** -18 | **72.** -17 |
| -28 | 9 | 9 |

| **73.** -21 | **74.** -15 | **75.** 27 |
| 39 | 32 | -19 |

| **76.** 31 | **77.** -53 | **78.** 47 |
| -18 | 24 | -28 |

For Problems 79–90, do the *subtraction* problems in vertical format.

| **79.** 5 | **80.** 8 | **81.** 6 |
| 12 | 19 | -9 |

| **82.** 13 | **83.** -7 | **84.** -6 |
| -7 | -8 | -5 |

| **85.** 17 | **86.** 18 | **87.** -23 |
| -19 | -14 | 16 |

| **88.** -27 | **89.** -12 | **90.** -13 |
| 15 | 12 | -13 |

For Problems 91–100, evaluate each algebraic expression for the given values of the variables.

91. $x - y$ for $x = -6$ and $y = -13$
92. $-x - y$ for $x = -7$ and $y = -9$
93. $-x + y - z$ for $x = 3$, $y = -4$, and $z = -6$
94. $x - y + z$ for $x = 5$, $y = 6$, and $z = -9$
95. $-x - y - z$ for $x = -2$, $y = 3$, and $z = -11$
96. $-x - y + z$ for $x = -8$, $y = -7$, and $z = -14$
97. $-x + y + z$ for $x = -11$, $y = 7$, and $z = -9$
98. $-x - y - z$ for $x = 12$, $y = -6$, and $z = -14$
99. $x - y - z$ for $x = -15$, $y = 12$, and $z = -10$
100. $x + y - z$ for $x = -18$, $y = 13$, and $z = 8$

A game such as football can also be used to interpret addition of integers. A gain of 3 yards on one play followed by a loss of 5 yards on the next play places the ball 2 yards *behind* the initial line of scrimmage and this could be expressed as $3 + (-5) = -2$. Use this *football interpretation* to find the following sums (Problems 101–110).

101. $4 + (-7)$ **102.** $3 + (-5)$

103. $-4 + (-6)$

104. $-2 + (-5)$

105. $-5 + 2$ **106.** $-10 + 6$

107. $-4 + 15$ **108.** $-3 + 22$

109. $-12 + 17$ **110.** $-9 + 21$

For Problems 111–120, refer to the Remark on page 19 and use the profit and loss interpretation for the addition of integers.

111. $60 + (-125)$

112. $50 + (-85)$

113. $-55 + (-45)$

114. $-120 + (-220)$

115. $-70 + 45$

116. $-125 + 45$

117. $-120 + 250$

118. $-75 + 165$

119. $145 + (-65)$

120. $275 + (-195)$

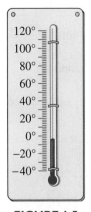

121. The temperature at 5 A.M. was $-17°$F. By noon the temperature had increased by $14°$F. Use the addition of integers to describe this situation and determine the temperature at noon (see Figure 1.5).

122. The temperature at 6 P.M. was $-6°$F and by 11 P.M. the temperature had dropped another $5°$F. Use the subtraction of integers to describe this situation and determine the temperature at 11 P.M. (see Figure 1.5).

123. Megan shot rounds of 3 over par, 2 under par, 3 under par, and 5 under par for a four-day golf tournament. Use the addition of integers to describe this situation and determine how much over or under par she was for the tournament.

124. The annual report of a company contained the following figures: a loss of $615,000 for 1990, a loss of $275,000 for 1991, a loss of $70,000 for 1992, and a profit of $115,000 for 1993. Use the addition of integers to describe this situation and determine the company's total loss or profit for the four-year period.

125. Use your calculator to check your answers for Problems 51–66.

THOUGHTS INTO WORDS

126. The statement $-6 - (-2) = -6 + 2 = -4$ can be read as "negative six minus negative two equals negative six plus two, which equals negative four." Express in words each of the following.

(a) $8 + (-10) = -2$

(b) $-7 - 4 = -7 + (-4) - 11$

(c) $9 - (-12) = 9 + 12 = 21$

(d) $-5 + (-6) = -11$

127. The algebraic expression $-x - y$ can be read as "the opposite of x minus y." Express in words each of the following.

(a) $-x + y$ (b) $x - y$ (c) $-x - y + z$

1.4 Integers: Multiplication and Division

Multiplication of whole numbers may be interpreted as repeated addition. For example, $3 \cdot 4$ means the sum of three 4s; thus, $3 \cdot 4 = 4 + 4 + 4 = 12$. Consider the following examples that use the repeated addition idea to find the product of a positive integer and a negative integer.

$$3(-2) = -2 + (-2) + (-2) = -6,$$
$$2(-4) = -4 + (-4) = -8,$$
$$4(-1) = -1 + (-1) + (-1) + (-1) = -4$$

Note the use of parentheses to indicate multiplication. Sometimes both numbers are enclosed in parentheses so that we have $(3)(-2)$.

When multiplying whole numbers, we realize that the order in which we multiply two factors does not change the product: $2(3) = 6$ and $3(2) = 6$. Using this idea we can now handle a negative integer times a positive integer as follows.

$$(-2)(3) = (3)(-2) = (-2) + (-2) + (-2) = -6,$$
$$(-3)(2) = (2)(-3) = (-3) + (-3) = -6,$$
$$(-4)(3) = (3)(-4) = (-4) + (-4) + (-4) = -12$$

Finally, let's consider the product of two negative integers. The following pattern helps us with the reasoning for this situation.

$$4(-3) = -12,$$
$$3(-3) = -9,$$
$$2(-3) = -6,$$
$$1(-3) = -3,$$
$$0(-3) = 0, \qquad \text{The product of zero and any integer is zero.}$$
$$(-1)(-3) = ?$$

Certainly, to continue this pattern, the product of -1 and -3 has to be 3. In general, this type of reasoning would help us to realize that the product of any two negative integers is a positive integer.

Using the concept of absolute value, we can precisely describe the **multiplication of integers**

1. The product of two positive integers or two negative integers is the product of their absolute values.

2. The product of a positive and a negative integer (either order) is the opposite of the product of their absolute values.

3. The product of zero and any integer is zero.

The following are examples of the multiplication of integers.

$$(-5)(-2) = |-5| \cdot |-2| = 5 \cdot 2 = 10,$$
$$(7)(-6) = -\big(|7| \cdot |-6|\big) = -(7 \cdot 6) = -42,$$
$$(-8)(9) = -\big(|-8| \cdot |9|\big) = -(8 \cdot 9) = -72,$$
$$(-14)(0) = 0,$$
$$(0)(-28) = 0$$

These examples show a step-by-step process for multiplying integers. In reality, however, the key issue is to remember whether the product is positive or negative. In other words, we need to remember that **the product of two positive integers or two negative integers is a positive integer; and the product of a positive integer and a negative integer (either order) is a negative integer.** Then we can avoid the step-by-step analysis and simply write the results as follows.

$$(7)(-9) = -63,$$
$$(8)(7) = 56,$$
$$(-5)(-6) = 30,$$
$$(-4)(12) = -48$$

Division of Integers

By looking back at our knowledge of whole numbers, we can get some guidance for our work with integers. We know, for example, that $\dfrac{8}{2} = 4$ because $2 \cdot 4 = 8$. In other words, we can find the quotient of two whole numbers by looking at a related multiplication problem. In the following examples we use this same link between multiplication and division to determine the quotients.

$$\frac{8}{-2} = -4 \quad \text{because } (-2)(-4) = 8,$$

$$\frac{-10}{5} = -2 \quad \text{because } (5)(-2) = -10,$$

$$\frac{-12}{-4} = 3 \quad \text{because } (-4)(3) = -12,$$

$$\frac{0}{-6} = 0 \quad \text{because } (-6)(0) = 0,$$

$$\frac{-9}{0} \quad \text{is undefined because no number times zero produces } -9,$$

$$\frac{0}{0} \quad \text{is undefined because any number times zero produces zero. } \textbf{Remember that division by zero is undefined!}$$

1. The quotient of two positive or two negative integers is the quotient of their absolute values.
2. The quotient of a positive integer and a negative integer (or a negative and a positive) is the opposite of the quotient of their absolute values.
3. The quotient of zero and any nonzero integer (zero divided by any nonzero integer) is zero.

The following are examples of the division of integers.

$$\frac{-8}{-4} = \frac{|-8|}{|-4|} = \frac{8}{4} = 2,$$

$$\frac{-14}{2} = -\left(\frac{|-14|}{|2|}\right) = -\left(\frac{14}{2}\right) = -7,$$

$$\frac{15}{-3} = -\left(\frac{|15|}{|-3|}\right) = -\left(\frac{15}{3}\right) = -5,$$

$$\frac{0}{-4} = 0$$

For practical purposes, the key is to remember whether the quotient is positive or negative. We need to remember that **the quotient of two positive integers or two negative integers is positive; and the quotient of a positive integer and a negative integer or a negative integer and a positive integer is negative.** We can then simply write the quotients as follows without showing all of the steps.

$$\frac{-18}{-6} = 3, \qquad \frac{-24}{12} = -2,$$

$$\frac{36}{-9} = -4$$

REMARK   Occasionally, people use the phrase "two negatives make a positive." We hope they realize that the reference is to multiplication and division only; in addition the sum of two negative integers is still a negative integer. It is probably best to avoid such imprecise statements. △

Simplifying Numerical Expressions

Now we can simplify numerical expressions involving any or all of the four basic operations with integers. Keep in mind the order of operations we stated in Section 1.1.

Example 1

Simplify $-4(-3) - 7(-8) + 3(-9)$.

Solution

$$-4(-3) - 7(-8) + 3(-9) = 12 - (-56) + (-27)$$
$$= 12 + 56 + (-27)$$
$$= 41$$

▲

Example 2

Simplify $\dfrac{-8 - 4(5)}{-4}$.

Solution

$$\frac{-8 - 4(5)}{-4} = \frac{-8 - 20}{-4}$$

$$= \frac{-28}{-4}$$

$$= 7$$

▲

Evaluating Algebraic Expressions

Evaluating algebraic expressions will often involve the use of two or more operations with integers. We use the final examples of this section to represent such situations.

Example 3

Find the value of $3x + 2y$ when $x = 5$ and $y = -9$.

Solution

$$3x + 2y = 3(5) + 2(-9) \quad \text{when } x = 5 \text{ and } y = -9$$
$$= 15 + (-18)$$
$$= -3$$

▲

Example 4

Evaluate $-2a + 9b$ for $a = 4$ and $b = -3$.

Solution

$$-2a + 9b = -2(4) + 9(-3) \quad \text{when } a = 4 \text{ and } b = -3$$
$$= -8 + (-27)$$
$$= -35$$

▲

Example 5

Find the value of $\dfrac{x - 2y}{4}$ when $x = -6$ and $y = 5$.

Solution

$$\frac{x - 2y}{4} = \frac{-6 - 2(5)}{4} \quad \text{when } x = -6 \text{ and } y = 5$$

$$= \frac{-6 - 10}{4}$$

$$= \frac{-16}{4}$$

$$= -4$$

▲

Problem Set 1.4

For Problems 1–40, find the product or quotient (multiply or divide) as indicated.

1. $5(-6)$

2. $7(-9)$

3. $\dfrac{-27}{3}$

4. $\dfrac{-35}{5}$

5. $\dfrac{-42}{-6}$

6. $\dfrac{-72}{-8}$

7. $(-7)(8)$

8. $(-6)(9)$

9. $(-5)(-12)$

10. $(-7)(-14)$

11. $\dfrac{96}{-8}$

12. $\dfrac{-91}{7}$

13. $14(-9)$

14. $17(-7)$

15. $(-11)(-14)$

16. $(-13)(-17)$

17. $\dfrac{135}{-15}$

18. $\dfrac{-144}{12}$

19. $\dfrac{-121}{-11}$

20. $\dfrac{-169}{-13}$

21. $(-15)(-15)$

22. $(-18)(-18)$

23. $\dfrac{112}{-8}$

24. $\dfrac{112}{-7}$

25. $\dfrac{0}{-8}$

26. $\dfrac{-8}{0}$

27. $\dfrac{-138}{-6}$

28. $\dfrac{-105}{-5}$

29. $\dfrac{76}{-4}$

30. $\dfrac{-114}{6}$

31. $(-6)(-15)$

32. $\dfrac{0}{-14}$

33. $(-56) \div (-4)$

34. $(-78) \div (-6)$

35. $(-19) \div 0$

36. $(-90) \div 15$

37. $(-72) \div 18$

38. $(-70) \div 5$

39. $(-36)(27)$

40. $(42)(-29)$

For Problems 41–60, simplify each numerical expression.

41. $3(-4) + 5(-7)$

42. $6(-3) + 5(-9)$

43. $7(-2) - 4(-8)$

44. $9(-3) - 8(-6)$

45. $(-3)(-8) + (-9)(-5)$

46. $(-7)(-6) + (-4)(-3)$

47. $5(-6) - 4(-7) + 3(2)$

48. $7(-4) - 8(-7) + 5(-8)$

49. $\dfrac{13 + (-25)}{-3}$

50. $\dfrac{15 + (-36)}{-7}$

51. $\dfrac{12 - 48}{6}$

52. $\dfrac{16 - 40}{8}$

53. $\dfrac{-7(10) + 6(-9)}{-4}$

54. $\dfrac{-6(8) + 4(-14)}{-8}$

55. $\dfrac{4(-7) - 8(-9)}{11}$

56. $\dfrac{5(-9) - 6(-7)}{3}$

57. $-2(3) - 3(-4) + 4(-5) - 6(-7)$

58. $2(-4) + 4(-5) - 7(-6) - 3(9)$

59. $-1(-6) - 4 + 6(-2) - 7(-3) - 18$

60. $-9(-2) + 16 - 4(-7) - 12 + 3(-8)$

For Problems 61–76, evaluate each algebraic expression for the given values of the variables.

61. $7x + 5y$ for $x = -5$ and $y = 9$

62. $4a + 6b$ for $a = -6$ and $b = -8$

63. $9a - 2b$ for $a = -5$ and $b = 7$

64. $8a - 3b$ for $a = -7$ and $b = 9$

65. $-6x - 7y$ for $x = -4$ and $y = -6$

66. $-5x - 12y$ for $x = -5$ and $y = -7$

67. $\dfrac{5x - 3y}{-6}$ for $x = -6$ and $y = 4$

68. $\dfrac{-7x + 4y}{-8}$ for $x = 8$ and $y = 6$

69. $3(2a - 5b)$ for $a = -1$ and $b = -5$

70. $4(3a - 7b)$ for $a = -2$ and $b = -4$

71. $-2x + 6y - xy$ for $x = 7$ and $y = -7$

72. $-3x + 7y - 2xy$ for $x = -6$ and $y = 4$

73. $-4ab - b$ for $a = 2$ and $b = -14$

74. $-5ab + b$ for $a = -1$ and $b = -13$

75. $(ab + c)(b - c)$ for $a = -2, b = -3$, and $c = 4$

76. $(ab - c)(a + c)$ for $a = -3, b = 2$, and $c = 5$

For Problems 77–82, find the value of $\dfrac{5(F - 32)}{9}$ for each of the given values for F.

77. $F = 59$ **78.** $F = 68$

79. $F = 14$ **80.** $F = -4$

81. $F = -13$ **82.** $F = -22$

For Problems 83–88, find the value of $\dfrac{9C}{5} + 32$ for each of the given values for C.

83. $C = 25$ **84.** $C = 35$

85. $C = 40$ **86.** $C = 0$

87. $C = -10$ **88.** $C = -30$

89. Monday morning Thad bought 800 shares of a stock at $19 per share. During that workweek the stock went up $2 per share on one day and dropped $1 per share on each of the other four days. Use multiplication and addition of integers to describe this situation and determine the value of the 800 shares by closing time on Friday.

90. In one workweek a small company showed a profit of $475 for one day and a loss of $65 for each of the other four days. Use multiplication and addition of integers to describe this situation and determine the company's profit or loss for the week.

91. Use a calculator to check your answers for Problems 41–60.

THOUGHTS INTO WORDS

92. Your friend keeps getting an answer of -7 when simplifying the expression $-6 + (-8) \div 2$. What mistake is she making and how would you help her?

93. Make up a problem that could be solved using $6(-4) = -24$.

94. Make up a problem that could be solved using $(-4)(-3) = 12$.

95. Explain why $\dfrac{0}{4} = 0$ but $\dfrac{4}{0}$ is undefined.

1.5 Use of Properties

We will begin this section by listing and briefly commenting on some of the basic properties of integers. We will then show how these properties facilitate manipulation with integers and also serve as a basis for some algebraic computation.

Commutative Property of Addition

If a and b are integers, then

$$a + b = b + a.$$

Commutative Property of Multiplication

If a and b are integers, then

$$ab = ba.$$

Addition and multiplication are said to be commutative operations. This means that the order in which you add or multiply two integers does not affect the result. For example, $3 + 5 = 5 + 3$ and $7(8) = 8(7)$. It is also important to realize that subtraction and division *are not* commutative operations; order does make a difference. For example, $8 - 7 \neq 7 - 8$ and $16 \div 4 \neq 4 \div 16$.

Associative Property of Addition

If a, b, and c are integers, then

$$(a + b) + c = a + (b + c).$$

Associative Property of Multiplication

If a, b, and c are integers, then

$$(ab)c = a(bc).$$

Addition and multiplication are associative operations. The associative properties can be thought of as grouping properties. For example $(-8 + 3) + 9 = -8 + (3 + 9)$. Changing the grouping of the numbers does not affect the final result. This is also true for multiplication as $[(-6)(5)](-4) = (-6)[(5)(-4)]$ illustrates. Subtraction and division *are not* associative operations. For example, $(8 - 4) - 7 = -3$ whereas $8 - (4 - 7) = 11$. An example, showing that division is not associative is $(8 \div 4) \div 2 = 1$; whereas $8 \div (4 \div 2) = 4$.

Identity Property of Addition

If a is an integer, then

$$a + 0 = 0 + a = a.$$

We refer to zero as the identity element for addition. This simply means that the sum of any integer and zero is exactly the same integer. For example, $-197 + 0 = 0 + (-197) = -197$.

Identity Property of Multiplication

If a is an integer, then

$$a(1) = 1(a) = a.$$

We call one the identity element for multiplication. The product of any integer and one is exactly the same integer. For example, $(-573)(1) = (1)(-573) = -573$.

Additive Inverse Property

For every integer a, there exists an integer $-a$, such that

$$a + (-a) = (-a) + a = 0.$$

The integer $-a$ is called the additive inverse of a or the opposite of a. Thus, 6 and -6 are additive inverses and their sum is 0. The additive inverse of 0 is 0.

Multiplication Property of Zero

If a is an integer, then

$$(a)(0) = (0)(a) = 0.$$

The product of zero and any integer is zero. For example, $(-873)(0) = (0)(-873) = 0$.

Multiplicative Property of Negative One

If a is an integer, then

$$(a)(-1) = (-1)(a) = -a.$$

The product of any integer and -1 is the opposite of the integer. For example, $(-1)(48) = (48)(-1) = -48$.

Distributive Property

If a, b, and c are integers, then

$$a(b + c) = ab + ac.$$

The distributive property involves both addition and multiplication. We say that **multiplication distributes over addition.** For example, $3(4 + 7) = 3(4) + 3(7)$.

Since $b - c = b + (-c)$, it follows that **multiplication also distributes over subtraction**. This could be stated as $a(b - c) = ab - ac$. For example, $7(8 - 2) = 7(8) - 7(2)$.

Let's now consider some examples that use the properties to help with certain types of manipulations.

Example 1

Find the sum $(43 + (-24)) + 24$.

Solution

In such a problem it is much more advantageous to group -24 and 24. Thus,

$$(43 + (-24)) + 24 = 43 + ((-24) + 24) \quad \text{Associative property for addition}$$
$$= 43 + 0.$$
$$= 43. \qquad \blacktriangle$$

Example 2

Find the product $[(-17)(25)](4)$.

Solution

In this problem it would be easier to group 25 and 4. Thus,

$$[(-17)(25)](4) = (-17)[(25)(4)] \quad \text{Associative property for multiplication}$$
$$= (-17)(100)$$
$$= -1700. \qquad \blacktriangle$$

Example 3

Find the sum $17 + (-24) + (-31) + 19 + (-14) + 29 + 43$.

Solution

Certainly we could add in the order that the numbers appear. However, since addition is *commutative* and *associative* we could change the order and group any convenient way. For example, we could add all of the positive integers and add all of the negative integers and then add these two results. It might be convenient to use the vertical format as follows.

$$
\begin{array}{rrr}
17 & & \\
19 & -24 & \\
29 & -31 & 108 \\
\underline{43} & -14 & -69 \\
108 & \underline{-69} & 39
\end{array}
\qquad \blacktriangle
$$

For a problem such as Example 3 it might be advisable to first add in the order that the numbers appear and then use the rearranging and regrouping idea as a check. Don't forget the link between addition and subtraction. A problem such as $18 - 43 + 52 - 17 - 23$ can be changed to $18 + (-43) + 52 + (-17) + (-23)$.

Example 4

Simplify $(-75)(-4 + 100)$.

Solution

For such a problem it might be convenient to apply the *distributive property* and then to simplify.

$$(-75)(-4 + 100) = (-75)(-4) + (-75)(100)$$
$$= 300 + (-7500)$$
$$= -7200 \qquad \blacktriangle$$

Example 5

Simplify $19(-26 + 25)$.

Solution

For this problem we are better off *not* to apply the distributive property, but simply to add the numbers inside the parentheses first and then to find the indicated product. Thus,

$$19(-26 + 25) = 19(-1) = -19. \qquad \blacktriangle$$

Example 6

Simplify $27(104) + 27(-4)$.

Solution

Keep in mind that the *distributive property* allows us to change from the form $a(b + c)$ to $ab + ac$ or from $ab + ac$ to $a(b + c)$. In this problem we want to use the latter change. Thus,

$$27(104) + 27(-4) = 27(104 + (-4))$$
$$= 27(100) = 2700. \qquad \blacktriangle$$

Examples 4, 5, and 6 demonstrate an important issue. Sometimes the form $a(b + c)$ is the most convenient, but at other times the form $ab + ac$ is better. A suggestion in regard to this issue—as well as to the use of the other properties —is to *think first*, and then decide whether or not the properties can be used to make the manipulations easier.

Combining Similar Terms

Algebraic expressions such as

$$3x, \quad 5y, \quad 7xy, \quad -4abc, \quad \text{and} \quad z$$

are called **terms**. A term is an indicated product and may have any number of factors. We call the variables in a term **literal factors**, and we call the numerical factor the **numerical coefficient**. Thus, in $7xy$, the x and y are literal factors, and 7 is the numerical coefficient. The numerical coefficient of the term, $-4abc$, is -4. Since $z = 1(z)$, the numerical coefficient of the term, z, is 1. Terms that have the same literal factors are called **like terms** or **similar terms**. Some examples of similar terms are

$$3x \quad \text{and} \quad 9x, \qquad 14abc \quad \text{and} \quad 29abc,$$
$$7xy \quad \text{and} \quad -15xy, \qquad 4z, \quad 9z, \quad \text{and} \quad -14z.$$

We can simplify algebraic expressions that contain similar terms by using a form of the distributive property. Consider the following examples.

$$3x + 5x = (3 + 5)x$$
$$= 8x$$
$$-9xy + 7xy = (-9 + 7)xy$$
$$= -2xy$$
$$18abc - 27abc = (18 - 27)abc$$
$$= (18 + (-27))abc$$
$$= -9abc$$
$$4x + x = (4 + 1)x \qquad \text{Don't forget that } x = 1(x).$$
$$= 5x$$

More complicated expressions might first require some rearranging of terms by using the commutative property.

$$7x + 3y + 9x + 5y = 7x + 9x + 3y + 5y$$
$$= (7 + 9)x + (3 + 5)y$$
$$= 16x + 8y$$
$$9a - 4 - 13a + 6 = 9a + (-4) + (-13a) + 6$$
$$= 9a + (-13a) + (-4) + 6$$
$$= (9 + (-13))a + 2$$
$$= -4a + 2$$

As you become more adept at handling the various simplifying steps, you may want to do the steps mentally and thereby go directly from the given expression to the simplified form as follows.

$$19x - 14y + 12x + 16y = 31x + 2y,$$

$$17ab + 13c - 19ab - 30c = -2ab - 17c,$$

$$9x + 5 - 11x + 4 + x - 6 = -x + 3$$

Simplifying some algebraic expressions requires repeated applications of the distributive property as the next examples demonstrate.

$$5(x - 2) + 3(x + 4) = 5(x) - 5(2) + 3(x) + 3(4)$$
$$= 5x - 10 + 3x + 12$$
$$= 5x + 3x - 10 + 12$$
$$= 8x + 2$$
$$-7(y + 1) - 4(y - 3) = -7(y) - 7(1) - 4(y) - 4(-3)$$
$$= -7y - 7 - 4y + 12 \qquad \text{Be careful with this sign.}$$
$$= -7y - 4y - 7 + 12$$
$$= -11y + 5$$

$$5(x + 2) - (x + 3) = 5(x + 2) - 1(x + 3) \qquad \text{Remember } -a = -1a.$$
$$= 5(x) + 5(2) - 1(x) - 1(3)$$
$$= 5x + 10 - x - 3$$
$$= 5x - x + 10 - 3$$
$$= 4x + 7$$

After you are sure of each step, you can use a more simplified format.

$$5(a + 4) - 7(a - 2) = 5a + 20 - 7a + 14$$
$$= -2a + 34$$

$$9(z - 7) + 11(z + 6) = 9z - 63 + 11z + 66$$
$$= 20z + 3$$

$$-(x - 2) + (x + 6) = -x + 2 + x + 6$$
$$= 8$$

Back to Evaluating Algebraic Expressions

To simplify by combining similar terms aids in the process of evaluating some algebraic expressions. The last examples of this section illustrate this idea.

Example 7

Evaluate $8x - 2y + 3x + 5y$ for $x = 3$ and $y = -4$.

Solution

Let's first simplify the given expression.

$$8x - 2y + 3x + 5y = 11x + 3y$$

Now, we can evaluate for $x = 3$ and $y = -4$.

$$11x + 3y = 11(3) + 3(-4)$$
$$= 33 + (-12) = 21 \qquad \blacktriangle$$

Example 8

Evaluate $2ab + 5c - 6ab + 12c$ for $a = 2$, $b = -3$, and $c = 7$.

Solution

$$2ab + 5c - 6ab + 12c = -4ab + 17c$$
$$= -4(2)(-3) + 17(7) \quad \text{when } a = 2, b = -3, \text{ and}$$
$$\qquad\qquad c = 7$$
$$= 24 + 119 = 143 \qquad \blacktriangle$$

Example 9

Evaluate $8(x - 4) + 7(x + 3)$ for $x = 6$.

Solution

$$8(x - 4) + 7(x + 3) = 8x - 32 + 7x + 21$$
$$= 15x - 11$$
$$= 15(6) - 11 \quad \text{when } x = 6$$
$$= 79 \qquad \blacktriangle$$

Problem Set 1.5

For Problems 1–12, state the property that justifies each statement. For example, $3 + (-4) = (-4) + 3$ because of the commutative property for addition.

1. $3(7 + 8) = 3(7) + 3(8)$
2. $(-9)(17) = 17(-9)$

3. $-2 + (5 + 7) = (-2 + 5) + 7$

4. $-19 + 0 = -19$
5. $143(-7) = -7(143)$

6. $5(9 + (-4)) = 5(9) + 5(-4)$

7. $-119 + 119 = 0$
8. $-4 + (6 + 9) = (-4 + 6) + 9$

9. $-56 + 0 = -56$
10. $5 + (-12) = -12 + 5$

11. $[5(-8)]4 = 5[-8(4)]$

12. $[6(-4)]8 = 6[-4(8)]$

For Problems 13–30, simplify each numerical expression. Don't forget to take advantage of the properties if they can be used to simplify the computation.

13. $(-18 + 56) + 18$
14. $-72 + [72 + (-14)]$
15. $36 - 48 - 22 + 41$
16. $-24 + 18 + 19 - 30$
17. $(25)(-18)(-4)$
18. $(2)(-71)(50)$
19. $(4)(-16)(-9)(-25)$
20. $(-2)(18)(-12)(-5)$
21. $37(-42 - 58)$
22. $-46(-73 - 27)$
23. $59(36) + 59(64)$
24. $-49(72) - 49(28)$

25. $15(-14) + 16(-8)$
26. $-9(14) - 7(-16)$
27. $17 + (-18) - 19 - 14 + 13 - 17$
28. $-16 - 14 + 18 + 21 + 14 - 17$
29. $-21 + 22 - 23 + 27 + 21 - 19$
30. $24 - 26 - 29 + 26 + 18 + 29 - 17 - 10$

For Problems 31–62, simplify each algebraic expression by combining similar terms.

31. $9x - 14x$
32. $12x - 14x + x$
33. $4m + m - 8m$
34. $-6m - m + 17m$
35. $-9y + 5y - 7y$
36. $14y - 17y - 19y$
37. $4x - 3y - 7x + y$
38. $9x + 5y - 4x - 8y$
39. $-7a - 7b - 9a + 3b$
40. $-12a + 14b - 3a - 9b$
41. $6xy - x - 13xy + 4x$
42. $-7xy - 2x - xy + x$
43. $5x - 4 + 7x - 2x + 9$
44. $8x + 9 + 14x - 3x - 14$
45. $-2xy + 12 + 8xy - 16$
46. $14xy - 7 - 19xy - 6$
47. $-2a + 3b - 7b - b + 5a - 9a$
48. $-9a - a + 6b - 3a - 4b - b + a$
49. $13ab + 2a - 7a - 9ab + ab - 6a$
50. $-ab - a + 4ab + 7ab - 3a - 11ab$
51. $3(x + 2) + 5(x + 6)$
52. $7(x + 8) + 9(x + 1)$
53. $5(x - 4) + 6(x + 8)$
54. $-3(x + 2) - 4(x - 10)$
55. $9(x + 4) - (x - 8)$
56. $-(x - 6) + 5(x - 9)$
57. $3(a - 1) - 2(a - 6) + 4(a + 5)$
58. $-4(a + 2) + 6(a + 8) - 3(a - 6)$

59. $-2(m + 3) - 3(m - 1) + 8(m + 4)$
60. $5(m - 10) + 6(m - 11) - 9(m - 12)$
61. $(y + 3) - (y - 2) - (y + 6) - 7(y - 1)$

62. $-(y - 2) - (y + 4) - (y + 7) - 2(y + 3)$

For Problems 63–80, simplify each algebraic expression and then evaluate the resulting expression for the given values for the variables.

63. $3x + 5y + 4x - 2y$ for $x = -2$ and $y = 3$
64. $5x - 7y - 9x - 3y$ for $x = -1$ and $y = -4$

65. $5(x - 2) + 8(x + 6)$ for $x = -6$
66. $4(x - 6) + 9(x + 2)$ for $x = 7$
67. $8(x + 4) - 10(x - 3)$ for $x = -5$
68. $-(n + 2) - 3(n - 6)$ for $n = 10$
69. $(x - 6) - (x + 12)$ for $x = -3$
70. $(x + 12) - (x - 14)$ for $x = -11$
71. $2(x + y) - 3(x - y)$ for $x = -2$ and $y = 7$
72. $5(x - y) - 9(x + y)$ for $x = 4$ and $y = -4$
73. $2xy + 6 + 7xy - 8$ for $x = 2$ and $y = -4$

74. $4xy - 5 - 8xy + 9$ for $x = -3$ and $y = -3$

75. $5x - 9xy + 3x + 2xy$ for $x = 12$ and $y = -1$

76. $-9x + xy - 4xy - x$ for $x = 10$ and $y = -11$

77. $(a - b) - (a + b)$ for $a = 19$ and $b = -17$

78. $(a + b) - (a - b)$ for $a = -16$ and $b = 14$

79. $-3x + 7x + 4x - 2x - x$ for $x = -13$
80. $5x - 6x + x - 7x - x - 2x$ for $x = -15$
81. Use a calculator to check your answers for Problems 13–30.

THOUGHTS INTO WORDS

82. State in your own words the associative property for addition of integers.
83. State in your own words the distributive property for multiplication over addition.
84. Is $2 \cdot 3 \cdot 5 \cdot 7 \cdot 11 + 7$ a prime or composite number? Defend your answer.

SUMMARY

(1.1) To simplify a numerical expression, perform the operations in the following order.

1. Perform the operations inside the symbols of inclusion (parentheses and brackets) and above and below each fraction bar. Start with the innermost inclusion symbol.

2. Perform all multiplications and divisions in the order that they appear from left to right.

3. Perform all additions and subtractions in the order that they appear from left to right.

To evaluate an algebraic expression, substitute the given values for the variables into the algebraic expression and simplify the resulting numerical expression.

(1.2) A prime number is a whole number greater than 1 that has no factors (divisors) other than itself and 1. Whole numbers greater than 1 that are not prime numbers are called composite numbers. Every composite number has one and only one prime factorization.

The greatest common factor of 6 and 8 is 2, which means that 2 is the largest whole number divisor of both 6 and 8.

The least common multiple of 6 and 8 is 24, which means that 24 is the smallest nonzero multiple of both 6 and 8.

(1.3) The number line is a convenient visual aid for interpreting addition of integers.

Subtraction of integers is defined in terms of addition: *a* − *b* means *a* + (−*b*).

(1.4) To multiply integers we must remember that *the product of two positives or two negatives is positive and the product of a positive and a negative (either order) is negative.*

To divide integers we must remember that *the quotient of two positives or two negatives is positive and the quotient of a positive and a negative (or a negative and a positive) is negative.*

(1.5) The following basic properties help with numerical manipulations and serve as a basis for algebraic computations.

Commutative Properties
$$a + b = b + a$$
$$ab = ba$$

Associative Properties
$$(a + b) + c = a + (b + c)$$
$$(ab)c = a(bc)$$

Identity Properties
$$a + 0 = 0 + a = a$$
$$a(1) = 1(a) = a$$

Additive Inverse Property
$$a + (-a) = (-a) + a = 0$$

Multiplication Property of Zero
$$a(0) = 0(a) = 0$$

Multiplication Property of Negative One
$$-1(a) = a(-1) = -a$$

Distributive Properties
$$a(b + c) = ab + ac$$
$$a(b - c) = ab - ac$$

Chapter 1 Review Problem Set

In Problems 1–10, do the indicated operations.

1. $7 + (-10)$ **2.** $(-12) + (-13)$

3. $8 - 13$ **4.** $-6 - 9$

5. $-12 - (-11)$ **6.** $-17 - (-19)$

7. $(13)(-12)$ **8.** $(-14)(-18)$

9. $(-72) \div (-12)$ **10.** $117 \div (-9)$

In Problems 11–15, classify each of the numbers as *prime* or *composite*.

11. 73   **12.** 87

13. 63   **14.** 81

15. 91

In Problems 16–20, express each of the numbers as the product of prime factors.

16. 24   **17.** 63

18. 57   **19.** 64

20. 84

21. Find the greatest common factor of 36 and 54.

22. Find the greatest common factor of 48, 60, and 84.

23. Find the least common multiple of 18 and 20.

24. Find the least common multiple of 15, 27, and 35.

In Problems 25–38, simplify each of the numerical expressions.

25. $(19 + 56) + (-9)$ **26.** $43 - 62 + 12$

27. $8 + (-9) + (-16) + (-14) + 17 + 12$

28. $19 - 23 - 14 + 21 + 14 - 13$

29. $3(-4) - 6$ **30.** $(-5)(-4) - 8$

31. $(5)(-2) + (6)(-4)$

32. $(-6)(8) + (-7)(-3)$

33. $(-6)(3) - (-4)(-5)$

34. $(-7)(9) - (6)(5)$

35. $\dfrac{4(-7) - (3)(-2)}{-11}$ **36.** $\dfrac{(-4)(9) + (5)(-3)}{1 - 18}$

37. $3 - 2[4(-3 - 1)]$ **38.** $-6 - [3(-4 - 7)]$

In Problems 39–50, simplify each algebraic expression by combining similar terms.

39. $12x + 3x - 7x$

40. $9y + 3 - 14y - 12$

41. $8x + 5y - 13x - y$

42. $9a + 11b + 4a - 17b$

43. $3ab - 4ab - 2a$

44. $5xy - 9xy + xy - y$

45. $3(x + 6) + 7(x + 8)$

46. $5(x - 4) - 3(x - 9)$

47. $-3(x - 2) - 4(x + 6)$

48. $-2x - 3(x - 4) + 2x$

49. $2(a - 1) - a - 3(a - 2)$

50. $-(a - 1) + 3(a - 2) - 4a + 1$

In Problems 51–64, evaluate each of the algebraic expressions for the given values of the variables.

51. $5x + 8y$ for $x = -7$ and $y = -3$

52. $7x - 9y$ for $x = -3$ and $y = 4$

53. $\dfrac{-5x - 2y}{-2x - 7}$ for $x = 6$ and $y = 4$

54. $\dfrac{-3x + 4y}{3x}$ for $x = -4$ and $y = -6$

55. $-2a + \dfrac{a - b}{a - 2}$ for $a = -5$ and $b = 9$

56. $\dfrac{2a + b}{b + 6} - 3b$ for $a = 3$ and $b = -4$

57. $5a + 6b - 7a - 2b$ for $a = -1$ and $b = 5$

58. $3x + 7y - 5x + y$ for $x = -4$ and $y = 3$

59. $2xy + 6 + 5xy - 8$ for $x = -1$ and $y = 1$

60. $7(x + 6) - 9(x + 1)$ for $x = -2$

61. $-3(x - 4) - 2(x + 8)$ for $x = 7$

62. $2(x - 1) - (x + 2) + 3(x - 4)$ for $x = -4$

63. $(a - b) - (a + b) - b$ for $a = -1$ and $b = -3$

64. $2ab - 3(a - b) + b + a$ for $a = 2$ and $b = -5$

CHAPTER 1 TEST

For Problems 1–10, simplify each of the numerical expressions.

1. $6 + (-7) - 4 + 12$

2. $7 + 4(9) + 2$

3. $-4(2 - 8) + 14$

4. $5(-7) - (-3)(8)$

5. $8 \div (-4) + (-6)(9) - 2$

6. $(-8)(-7) + (-6) - (9)(12)$

7. $\dfrac{6(-4) - (-8)(-5)}{-16}$

8. $-14 + 23 - 17 - 19 + 26$

9. $(-14)(4) \div 4 + (-6)$

10. $6(-9) - (-8) - (-7)(4) + 11$

For Problems 11–17, evaluate each of the algebraic expressions for the given values of the variables.

11. $7x - 9y$ for $x = -4$ and $y = -6$

12. $-4a - 6b$ for $a = -9$ and $b = 12$

13. $3xy - 8y + 5x$ for $x = 7$ and $y = -2$

14. $5(x - 4) - 6(x + 7)$ for $x = -5$

15. $3x - 2y - 4x - x + 7y$ for $x = 6$ and $y = -7$

16. $3(x - 2) - 5(x - 4) + 6(x - 1)$ for $x = -3$

17. $\dfrac{-x - y}{y - x}$ for $x = -9$ and $y = -6$

18. Classify 79 as a prime or composite number.

19. Express 360 as a product of prime factors.

20. Find the greatest common factor of 36, 60, and 84.

21. Find the least common multiple of 9 and 24.

22. State the property of integers demonstrated by $[-3 + (-4)] + (-6) = -3 + [(-4) + (-6)]$.

23. State the property of integers demonstrated by $8(25 + 37) = 8(25) + 8(37)$.

24. Simplify $-7x + 9y - y + x - 2y - 7x$ by combining similar terms.

25. Simplify $-2(x - 4) - 5(x + 7) - 6(x - 1)$ by applying the distributive property and combining similar terms.

Real Numbers

Leta bought 300 shares of one stock at $23\frac{3}{8}$ per share and 200 shares of another stock at $17\frac{1}{4}$ per share. We use the numerical expression $300(23\frac{3}{8}) + 200(17\frac{1}{4})$, which can be written as $300(23.375) + 200(17.25)$, to determine the total price of the two stocks.

When the market opened on Monday morning, Garth bought some shares of a stock at $13\frac{3}{4}$ per share. The rational numbers $+\frac{3}{4}$, $-1\frac{1}{8}$, $+\frac{3}{8}$, $-\frac{1}{4}$, and $-\frac{1}{2}$ represent the changes in the market for that stock for the week. We use the numerical expression $13\frac{3}{4} + \frac{3}{4} + (-1\frac{1}{8}) + \frac{3}{8} + (-\frac{1}{4}) + (-\frac{1}{2})$ to determine the value of one share of Garth's stock when the market closed on Friday.

The width of a rectangle is w feet and its length is four feet more than three times its width. The algebraic expression $2w + 2(3w + 4)$ represents the perimeter of the rectangle.

Again in this chapter we use the concepts of numerical and algebraic expressions to review some computational skills from arithmetic and to continue the transition from arithmetic to algebra. However, the set of rational numbers now becomes the primary focal point. We urge you to use this chapter to fine-tune your arithmetic skills so that the algebraic concepts in subsequent chapters can be built upon a solid foundation.

2.1 Rational Numbers: Multiplication and Division

Any number that can be written in the form $\dfrac{a}{b}$, where a and b are integers and b is not zero, we call a **rational number**. (We call the form $\dfrac{a}{b}$ a fraction or sometimes a common fraction.) The following are examples of rational numbers.

$$\frac{1}{2}, \quad \frac{7}{9}, \quad \frac{15}{7}, \quad \frac{-3}{4}, \quad \frac{5}{-7}, \quad \frac{-11}{-13}$$

All integers are rational numbers because every integer can be expressed as the indicated quotient of two integers. Some examples follow.

$$6 = \frac{6}{1} = \frac{12}{2} = \frac{18}{3}, \text{ etc.,}$$

$$27 = \frac{27}{1} = \frac{54}{2} = \frac{81}{3}, \text{ etc.,}$$

$$0 = \frac{0}{1} = \frac{0}{2} = \frac{0}{3}, \text{ etc.}$$

Our work with division involving negative integers in Chapter 1 helps with the next three examples.

$$-4 = \frac{-4}{1} = \frac{-8}{2} = \frac{-12}{3}, \text{ etc.,}$$

$$-6 = \frac{6}{-1} = \frac{12}{-2} = \frac{18}{-3}, \text{ etc.,}$$

$$10 = \frac{10}{1} = \frac{-10}{-1} = \frac{-20}{-2}, \text{ etc.}$$

Observe the following general properties.

PROPERTY 2.1

$$\frac{-a}{b} = \frac{a}{-b} = -\frac{a}{b} \qquad \text{and} \qquad \frac{-a}{-b} = \frac{a}{b}.$$

Therefore, a rational number such as $\dfrac{-2}{3}$ can also be written as $\dfrac{2}{-3}$ or $-\dfrac{2}{3}$. (However, we seldom express rational numbers with negative denominators.)

Multiplying Rational Numbers

We define multiplication of rational numbers in common fractional form as follows.

DEFINITION 2.1

If a, b, c, and d are integers with b and d not equal to zero, then

$$\frac{a}{b} \cdot \frac{c}{d} = \frac{a \cdot c}{b \cdot d}.$$

To multiply rational numbers in common fractional form we simply *multiply numerators and multiply denominators*. Furthermore, we see from the definition that the rational numbers are commutative and associative with respect to multiplication. We are free to rearrange and regroup factors as we do with integers. The following examples illustrate Definition 2.1.

$$\frac{1}{3} \cdot \frac{2}{5} = \frac{1 \cdot 2}{3 \cdot 5} = \frac{2}{15},$$

$$\frac{3}{4} \cdot \frac{5}{7} = \frac{3 \cdot 5}{4 \cdot 7} = \frac{15}{28},$$

$$\frac{-2}{3} \cdot \frac{7}{9} = \frac{-2 \cdot 7}{3 \cdot 9} = \frac{-14}{27} \quad \text{or} \quad -\frac{14}{27},$$

$$\frac{1}{5} \cdot \frac{9}{-11} = \frac{1 \cdot 9}{5(-11)} = \frac{9}{-55} \quad \text{or} \quad -\frac{9}{55},$$

$$-\frac{3}{4} \cdot \frac{7}{13} = \frac{-3}{4} \cdot \frac{7}{13} = \frac{-3 \cdot 7}{4 \cdot 13} = \frac{-21}{52} \quad \text{or} \quad -\frac{21}{52},$$

$$\frac{3}{5} \cdot \frac{5}{3} = \frac{3 \cdot 5}{5 \cdot 3} = \frac{15}{15} = 1$$

The last example is a very special case. **If the product of two numbers is 1, the numbers are said to be reciprocals of each other.**

We often use Property 2.2 when we work with rational numbers. We call this property the **fundamental principle of fractions.**

PROPERTY 2.2

If b and k are nonzero integers and a is any integer, then

$$\frac{a \cdot k}{b \cdot k} = \frac{a}{b}.$$

Property 2.2 provides the basis for what is often called *reducing fractions to lowest terms or expressing fractions in simplest or reduced form*. Let's apply the property to a few examples.

Example 1

Reduce $\dfrac{12}{18}$ to lowest terms.

Solution

$$\dfrac{12}{18} = \dfrac{2 \cdot 6}{3 \cdot 6} = \dfrac{2}{3}$$

▲

Example 2

Change $\dfrac{14}{35}$ to simplest form.

Solution

$$\dfrac{14}{35} = \dfrac{2 \cdot 7}{5 \cdot 7} = \dfrac{2}{5}$$ A common factor of 7 has been divided out of both the numerator and the denominator.

▲

Example 3

Express $\dfrac{-24}{32}$ in reduced form.

Solution

$$\dfrac{-24}{32} = -\dfrac{3 \cdot 8}{4 \cdot 8} = -\dfrac{3}{4} \cdot \dfrac{8}{8} = -\dfrac{3}{4} \cdot 1 = -\dfrac{3}{4}$$ The multiplication property of 1 is being used.

▲

Example 4

Reduce $-\dfrac{72}{90}$.

Solution

$$-\dfrac{72}{90} = -\dfrac{2 \cdot 2 \cdot 2 \cdot 3 \cdot 3}{2 \cdot 3 \cdot 3 \cdot 5} = -\dfrac{4}{5}$$ The prime factored forms of the numerator and denominator may be used to help recognize common factors.

▲

The fractions may contain variables in the numerator or denominator (or both), but this creates no great difficulty. Our thought processes remain the same, as these next examples illustrate. Variables appearing in denominators represent **nonzero** integers.

Example 5

Reduce $\dfrac{9x}{17x}$.

Solution

$$\dfrac{9x}{17x} = \dfrac{9 \cdot x}{17 \cdot x} = \dfrac{9}{17}$$

▲

Example 6

Simplify $\dfrac{8x}{36y}$.

Solution

$$\dfrac{8x}{36y} = \dfrac{2 \cdot 2 \cdot 2 \cdot x}{2 \cdot 2 \cdot 3 \cdot 3 \cdot y} = \dfrac{2x}{9y}$$

▲

Example 7

Express $\dfrac{-9xy}{30y}$ in reduced form.

Solution

$$\dfrac{-9xy}{30y} = -\dfrac{3 \cdot 3 \cdot x \cdot y}{2 \cdot 3 \cdot 5 \cdot y} = -\dfrac{3x}{10}$$

▲

Example 8

Reduce $\dfrac{-7abc}{-9ac}$.

Solution

$$\frac{-7abc}{-9ac} = \frac{7a\!\!\!/b\!\!\!/c}{9a\!\!\!/c} = \frac{7b}{9}$$ ▲

We are now ready to consider multiplication problems with the understanding that *the final answer should be expressed in reduced form.* Study the following examples carefully; we used different formats to handle such problems.

Example 9

Multiply $\dfrac{7}{9} \cdot \dfrac{5}{14}$.

Solution

$$\frac{7}{9} \cdot \frac{5}{14} = \frac{7 \cdot 5}{9 \cdot 14} = \frac{7\!\!\!/ \cdot 5}{3 \cdot 3 \cdot 2 \cdot 7\!\!\!/} = \frac{5}{18}$$ ▲

Example 10

Find the product of $\dfrac{8}{9}$ and $\dfrac{18}{24}$.

Solution

$$\frac{\overset{1}{8\!\!\!/}}{\underset{1}{9\!\!\!/}} \cdot \frac{\overset{2}{18\!\!\!/}}{\underset{3}{24\!\!\!/}} = \frac{2}{3}$$ A common factor of 8 has been divided out of 8 and 24, and a common factor of 9 has been divided out of 9 and 18. ▲

Example 11

Multiply $\left(-\dfrac{6}{8}\right)\left(\dfrac{14}{32}\right)$.

Solution

$$\left(-\frac{6}{8}\right)\left(\frac{14}{32}\right) = -\frac{\overset{3}{6\!\!\!/} \cdot \overset{7}{14\!\!\!/}}{\underset{4}{8\!\!\!/} \cdot \underset{16}{32\!\!\!/}} = -\frac{21}{64}$$ ▲

Example 12

Multiply $\left(-\dfrac{9}{4}\right)\left(-\dfrac{14}{15}\right)$.

Solution

$$\left(-\frac{9}{4}\right)\left(-\frac{14}{15}\right) = \frac{3\!\!\!/ \cdot 3 \cdot 2\!\!\!/ \cdot 7}{2 \cdot 2\!\!\!/ \cdot 3\!\!\!/ \cdot 5} = \frac{21}{10}$$ Immediately we recognize that *a negative times a negative is positive.* ▲

Example 13

Multiply $\dfrac{9x}{7y} \cdot \dfrac{14y}{45}$.

Solution

$$\frac{9x}{7y} \cdot \frac{14y}{45} = \frac{9\!\!\!/ \cdot x \cdot \overset{2}{14\!\!\!/} \cdot y\!\!\!/}{7\!\!\!/ \cdot y\!\!\!/ \cdot 45\!\!\!/} = \frac{2x}{5}$$ ▲

Example 14

Multiply $\dfrac{-6c}{7ab} \cdot \dfrac{14b}{5c}$.

Solution

$$\frac{-6c}{7ab} \cdot \frac{14b}{5c} = -\frac{2 \cdot 3 \cdot c\!\!\!/ \cdot 2 \cdot 7\!\!\!/ \cdot b\!\!\!/}{7\!\!\!/ \cdot a \cdot b\!\!\!/ \cdot 5 \cdot c\!\!\!/} = -\frac{12}{5a}$$ ▲

Dividing Rational Numbers

The following example motivates a definition for division of rational numbers in fractional form.

$$\frac{\frac{3}{4}}{\frac{2}{3}} = \left(\frac{\frac{3}{4}}{\frac{2}{3}}\right)\left(\frac{\frac{3}{2}}{\frac{3}{2}}\right) = \frac{\left(\frac{3}{4}\right)\left(\frac{3}{2}\right)}{1} = \left(\frac{3}{4}\right)\left(\frac{3}{2}\right) = \frac{9}{8}$$

Notice that this is a form of 1 and $\frac{3}{2}$ is the reciprocal of $\frac{2}{3}$.

In other words, $\frac{3}{4}$ divided by $\frac{2}{3}$ is equivalent to $\frac{3}{4}$ times $\frac{3}{2}$. The following definition for division should seem reasonable.

DEFINITION 2.2   If b, c, and d are nonzero integers and a is any integer, then

$$\frac{a}{b} \div \frac{c}{d} = \frac{a}{b} \cdot \frac{d}{c}.$$

Notice that to divide $\frac{a}{b}$ by $\frac{c}{d}$, we multiply $\frac{a}{b}$ times the reciprocal of $\frac{c}{d}$, which is $\frac{d}{c}$. The following examples demonstrate the important steps of a division problem.

$$\frac{2}{3} \div \frac{1}{2} = \frac{2}{3} \cdot \frac{2}{1} = \frac{4}{3},$$

$$\frac{5}{6} \div \frac{3}{4} = \frac{5}{6} \cdot \frac{4}{3} = \frac{5 \cdot 4}{6 \cdot 3} = \frac{5 \cdot \cancel{2} \cdot 2}{\cancel{2} \cdot 3 \cdot 3} = \frac{10}{9},$$

$$-\frac{9}{12} \div \frac{3}{6} = -\frac{\overset{3}{\cancel{9}}}{\underset{2}{\cancel{12}}} \cdot \frac{\overset{1}{\cancel{6}}}{\underset{1}{\cancel{3}}} = -\frac{3}{2},$$

$$\left(-\frac{27}{56}\right) \div \left(-\frac{33}{72}\right) = \left(-\frac{27}{56}\right)\left(-\frac{72}{33}\right) = \frac{\overset{9}{\cancel{27}} \cdot \overset{9}{\cancel{72}}}{\underset{7}{\cancel{56}} \cdot \underset{11}{\cancel{33}}} = \frac{81}{77},$$

$$\frac{\frac{6}{7}}{2} = \frac{\overset{3}{\cancel{6}}}{7} \cdot \frac{1}{\underset{1}{\cancel{2}}} = \frac{3}{7},$$

$$\frac{5x}{7y} \div \frac{10}{28y} = \frac{5x}{7y} \cdot \frac{28y}{10} = \frac{\cancel{5} \cdot x \cdot \overset{2}{\cancel{28}} \cdot \cancel{y}}{\cancel{7} \cdot \cancel{y} \cdot \underset{\cancel{2}}{\cancel{10}}} = 2x$$

Problem Set 2.1

For Problems 1–24, reduce each fraction to lowest terms.

1. $\dfrac{8}{12}$

2. $\dfrac{12}{16}$

3. $\dfrac{16}{24}$

4. $\dfrac{18}{32}$

5. $\dfrac{15}{9}$

6. $\dfrac{48}{36}$

7. $\dfrac{-8}{48}$

8. $\dfrac{-3}{15}$

9. $\dfrac{27}{-36}$

10. $\dfrac{9}{-51}$

11. $\dfrac{-54}{-56}$

12. $\dfrac{-24}{-80}$

13. $\dfrac{24x}{44x}$

14. $\dfrac{15y}{25y}$

15. $\dfrac{9x}{21y}$

16. $\dfrac{4y}{30x}$

17. $\dfrac{14xy}{35y}$

18. $\dfrac{55xy}{77x}$

19. $\dfrac{-20ab}{52bc}$

20. $\dfrac{-23ac}{41c}$

21. $\dfrac{-56yz}{-49xy}$

22. $\dfrac{-21xy}{-14ab}$

23. $\dfrac{65abc}{91ac}$

24. $\dfrac{68xyz}{85yz}$

35. $\dfrac{1}{4} \div \dfrac{-5}{6}$

36. $\dfrac{7}{8} \div \dfrac{14}{-16}$

37. $\left(-\dfrac{8}{10}\right)\left(-\dfrac{10}{32}\right)$

38. $\left(-\dfrac{6}{7}\right)\left(-\dfrac{21}{24}\right)$

39. $-9 \div \dfrac{1}{3}$

40. $-10 \div \dfrac{1}{4}$

41. $\dfrac{5x}{9y} \cdot \dfrac{7y}{3x}$

42. $\dfrac{4a}{11b} \cdot \dfrac{6b}{7a}$

43. $\dfrac{6a}{14b} \cdot \dfrac{16b}{18a}$

44. $\dfrac{5y}{8x} \cdot \dfrac{14z}{15y}$

45. $\dfrac{10x}{-9y} \cdot \dfrac{15}{20x}$

46. $\dfrac{3x}{4y} \cdot \dfrac{-8w}{9z}$

47. $ab \cdot \dfrac{2}{b}$

48. $3xy \cdot \dfrac{4}{x}$

49. $\left(-\dfrac{7x}{12y}\right)\left(-\dfrac{24y}{35x}\right)$

50. $\left(-\dfrac{10a}{15b}\right)\left(-\dfrac{45b}{65a}\right)$

51. $\dfrac{3}{x} \div \dfrac{6}{y}$

52. $\dfrac{6}{x} \div \dfrac{14}{y}$

53. $\dfrac{5x}{9y} \div \dfrac{13x}{36y}$

54. $\dfrac{3x}{5y} \div \dfrac{7x}{10y}$

55. $\dfrac{-7}{x} \div \dfrac{9}{x}$

56. $\dfrac{8}{y} \div \dfrac{28}{-y}$

57. $\dfrac{-4}{n} \div \dfrac{-18}{n}$

58. $\dfrac{-34}{n} \div \dfrac{-51}{n}$

For Problems 25–58, multiply or divide as indicated and express answers in reduced form.

25. $\dfrac{3}{4} \cdot \dfrac{5}{7}$

26. $\dfrac{4}{5} \cdot \dfrac{3}{11}$

27. $\dfrac{2}{7} \div \dfrac{3}{5}$

28. $\dfrac{5}{6} \div \dfrac{11}{13}$

29. $\dfrac{3}{8} \cdot \dfrac{12}{15}$

30. $\dfrac{4}{9} \cdot \dfrac{3}{2}$

31. $\dfrac{-6}{13} \cdot \dfrac{26}{9}$

32. $\dfrac{3}{4} \cdot \dfrac{-14}{12}$

33. $\dfrac{7}{9} \div \dfrac{5}{9}$

34. $\dfrac{3}{11} \div \dfrac{7}{11}$

For Problems 59–74, perform the operations as indicated and express answers in lowest terms.

59. $\dfrac{3}{4} \cdot \dfrac{8}{9} \cdot \dfrac{12}{20}$

60. $\dfrac{5}{6} \cdot \dfrac{9}{10} \cdot \dfrac{8}{7}$

61. $\left(-\dfrac{3}{8}\right)\left(\dfrac{13}{14}\right)\left(-\dfrac{12}{9}\right)$

62. $\left(-\dfrac{7}{9}\right)\left(\dfrac{5}{11}\right)\left(-\dfrac{18}{14}\right)$

63. $\left(\dfrac{3x}{4y}\right)\left(\dfrac{8}{9x}\right)\left(\dfrac{12y}{5}\right)$

64. $\left(\dfrac{2x}{3y}\right)\left(\dfrac{5y}{x}\right)\left(\dfrac{9}{4x}\right)$

65. $\left(-\dfrac{2}{3}\right)\left(\dfrac{3}{4}\right) \div \dfrac{1}{8}$

66. $\dfrac{3}{4} \cdot \dfrac{4}{5} \div \dfrac{1}{6}$

67. $\dfrac{5}{7} \div \left(-\dfrac{5}{6}\right)\left(-\dfrac{6}{7}\right)$

68. $\left(-\dfrac{3}{8}\right) \div \left(-\dfrac{4}{5}\right)\left(\dfrac{1}{2}\right)$

69. $\left(-\dfrac{6}{7}\right) \div \left(\dfrac{5}{7}\right)\left(-\dfrac{5}{6}\right)$

70. $\left(-\dfrac{4}{3}\right) \div \left(\dfrac{4}{5}\right)\left(\dfrac{3}{5}\right)$

71. $\left(\dfrac{4}{9}\right)\left(-\dfrac{9}{8}\right) \div \left(-\dfrac{3}{4}\right)$

72. $\left(-\dfrac{7}{8}\right)\left(\dfrac{4}{7}\right) \div \left(-\dfrac{3}{2}\right)$

73. $\left(\dfrac{5}{2}\right)\left(\dfrac{2}{3}\right) \div \left(-\dfrac{1}{4}\right) \div (-3)$

74. $\dfrac{1}{3} \div \left(\dfrac{3}{4}\right)\left(\dfrac{1}{2}\right) \div 2$

75. Maria's department has $\dfrac{3}{4}$ of all of the accounts within the ABC Advertising Agency. Maria is personally responsible for $\dfrac{1}{3}$ of all accounts in her department. For what portion of all of the accounts at ABC is Maria personally responsible?

76. Pablo has a board that is $4\dfrac{1}{2}$ feet long and he wants to cut it into three pieces of the same length (See Figure 2.1). Find the length of each of the three pieces.

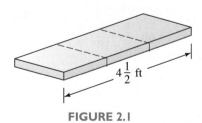

$4\dfrac{1}{2}$ ft

FIGURE 2.1

77. A certain recipe calls for $\dfrac{3}{4}$ cup of sugar. To make one-half of the recipe, how much sugar is needed?

78. Caleb left an estate valued at $750,000. His will states that three-fourths of the estate is to be divided equally among his three children. How much should each receive?

79. If your calculator is equipped to handle rational numbers in $\dfrac{a}{b}$ form, check your answers for Problems 1–12 and 59–74.

80. State in your own words the property

$$-\dfrac{a}{b} = \dfrac{-a}{b} = \dfrac{a}{-b}.$$

81. Explain how you would reduce $\dfrac{72}{117}$ to lowest terms.

82. What mistake was made in the following simplification process?

$$\dfrac{1}{2} \div \left(\dfrac{2}{3}\right)\left(\dfrac{3}{4}\right) \div 3 = \dfrac{1}{2} \div \dfrac{1}{2} \div 3 = \dfrac{1}{2} \cdot 2 \cdot \dfrac{1}{3} = \dfrac{1}{3}$$

How would you correct the error?

Further Investigations

83. The division problem $35 \div 7$ can be interpreted as "how many 7s are there in 35?" Likewise, a division problem such as $3 \div \dfrac{1}{2}$ can be interpreted as "how many one-halves in 3?" Use this how-many interpretation to do the following division problems.

(a) $4 \div \dfrac{1}{2}$ **(b)** $3 \div \dfrac{1}{4}$

(c) $5 \div \dfrac{1}{8}$ **(d)** $6 \div \dfrac{1}{7}$

(e) $\dfrac{5}{6} \div \dfrac{1}{6}$ **(f)** $\dfrac{7}{8} \div \dfrac{1}{8}$

84. Estimation is important in mathematics. In each of the following, estimate whether the answer is larger than 1 or smaller than 1 by using the how-many idea from Problem 83.

(a) $\dfrac{3}{4} \div \dfrac{1}{2}$

(b) $1 \div \dfrac{7}{8}$

(c) $\dfrac{1}{2} \div \dfrac{3}{4}$

(d) $\dfrac{8}{7} \div \dfrac{7}{8}$

(e) $\dfrac{2}{3} \div \dfrac{1}{4}$

(f) $\dfrac{3}{5} \div \dfrac{3}{4}$

85. Reduce each of the following to lowest terms. Don't forget that we reviewed some divisibility rules in Problem Set 1.2.

(a) $\dfrac{99}{117}$

(b) $\dfrac{175}{225}$

(c) $\dfrac{-111}{123}$

(d) $\dfrac{-234}{270}$

(e) $\dfrac{270}{495}$

(f) $\dfrac{324}{459}$

(g) $\dfrac{91}{143}$

(h) $\dfrac{187}{221}$

2.2 Addition and Subtraction of Rational Numbers

Suppose that it is one-fifth of a mile between your dorm and the union and two-fifths of a mile between the union and the library along a straight line as indicated in Figure 2.2. The total distance between your dorm and the library is three-fifths of a mile and we write $\dfrac{1}{5} + \dfrac{2}{5} = \dfrac{3}{5}$.

FIGURE 2.2

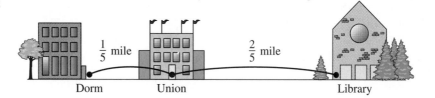

$\dfrac{1}{5}$ mile $\dfrac{2}{5}$ mile

Dorm   Union   Library

A pizza is cut into seven equal pieces and you eat two of the pieces. How much of the pizza remains (see Figure 2.3)? We represent the whole pizza by $\dfrac{7}{7}$ and then conclude that $\dfrac{7}{7} - \dfrac{2}{7} = \dfrac{5}{7}$ of the pizza remains.

FIGURE 2.3

These examples motivate the following definition for addition and subtraction of rational numbers in $\dfrac{a}{b}$ form.

DEFINITION 2.3   If a, b, and c are integers and b is not zero, then

$$\frac{a}{b} + \frac{c}{b} = \frac{a+c}{b}, \qquad \text{Addition}$$

$$\frac{a}{b} - \frac{c}{b} = \frac{a-c}{b}. \qquad \text{Subtraction}$$

We say that *rational numbers* **with common denominators** *can be added or subtracted by adding or subtracting the numerators and placing the results over the common denominator.* Consider the following examples.

$$\frac{3}{7} + \frac{2}{7} = \frac{3+2}{7} = \frac{5}{7},$$

$$\frac{7}{8} - \frac{2}{8} = \frac{7-2}{8} = \frac{5}{8},$$

$$\frac{2}{6} + \frac{1}{6} = \frac{2+1}{6} = \frac{3}{6} = \frac{1}{2}, \qquad \text{We agree to reduce final answer.}$$

$$\frac{3}{11} - \frac{5}{11} = \frac{3-5}{11} = \frac{-2}{11} \quad \left(\text{or } -\frac{2}{11} \right),$$

$$\frac{5}{x} + \frac{7}{x} = \frac{5+7}{x} = \frac{12}{x},$$

$$\frac{9}{y} - \frac{3}{y} = \frac{9-3}{y} = \frac{6}{y}$$

In the last two examples, the variables x and y cannot be equal to zero in order to exclude division by zero. It is always necessary to restrict denominators to nonzero values although we will not take the time or space to list such restrictions for every problem.

How do we add or subtract if the fractions do not have a common denominator? We use the fundamental principle of fractions, $\dfrac{a}{b} = \dfrac{a \cdot k}{b \cdot k}$, and obtain equivalent fractions that have a common denominator. **Equivalent fractions** are fractions that name the same number. Consider the following example, which shows the details.

Example 1

Add $\frac{1}{2} + \frac{1}{3}$.

Solution

$$\frac{1}{2} = \frac{1 \cdot 3}{2 \cdot 3} = \frac{3}{6} \qquad \frac{1}{2} \text{ and } \frac{3}{6} \text{ are equivalent fractions naming the same number.}$$

$$\frac{1}{3} = \frac{1 \cdot 2}{3 \cdot 2} = \frac{2}{6} \qquad \frac{1}{3} \text{ and } \frac{2}{6} \text{ are equivalent fractions naming the same number.}$$

$$\frac{5}{6} \qquad \frac{3}{6} + \frac{2}{6} = \frac{3+2}{6} = \frac{5}{6}$$

▲

Notice that we chose 6 as our common denominator and 6 is the least common multiple of the original denominators 2 and 3. (Recall that the least common multiple is the smallest nonzero whole number divisible by the given numbers.) In general, we use the least common multiple of the denominators of the fractions to be added or subtracted as a **least common denominator** (LCD).

Recall from Section 1.2 that the least common multiple may be found either by inspection or by using the prime factorization forms of the numbers. Let's consider some examples involving these procedures.

Example 2

Add $\frac{1}{4} + \frac{2}{5}$.

Solution

By inspection we see that the LCD is 20. Thus, both fractions can be changed to equivalent fractions that have a denominator of 20.

$$\frac{1}{4} + \frac{2}{5} = \frac{1 \cdot 5}{4 \cdot 5} + \frac{2 \cdot 4}{5 \cdot 4} = \frac{5}{20} + \frac{8}{20} = \frac{13}{20}$$

Use of fundamental
principle of fractions

▲

Example 3

Subtract $\frac{5}{8} - \frac{7}{12}$.

Solution

By inspection the LCD is 24.

$$\frac{5}{8} - \frac{7}{12} = \frac{5 \cdot 3}{8 \cdot 3} - \frac{7 \cdot 2}{12 \cdot 2} = \frac{15}{24} - \frac{14}{24} = \frac{1}{24}$$

▲

If the LCD is not obvious by inspection, then we can use the technique from Chapter 1 to find the least common multiple. So we proceed as follows.

STEP 1 Express each denominator as a product of prime factors.

STEP 2 The LCD contains each different prime factor as many times as the *most* times it appears in any one of the factorizations from step 1.

Example 4

Add $\dfrac{5}{18} + \dfrac{7}{24}$.

Solution

If we cannot find the LCD by inspection, then we can use the prime factorization forms.

$$\left.\begin{array}{l} 18 = 2 \cdot 3 \cdot 3 \\ 24 = 2 \cdot 2 \cdot 2 \cdot 3 \end{array}\right\} \longrightarrow \text{LCD} = 2 \cdot 2 \cdot 2 \cdot 3 \cdot 3 = 72$$

$$\frac{5}{18} + \frac{7}{24} = \frac{5 \cdot 4}{18 \cdot 4} + \frac{7 \cdot 3}{24 \cdot 3} = \frac{20}{72} + \frac{21}{72} = \frac{41}{72} \qquad \blacktriangle$$

Example 5

Subtract $\dfrac{3}{14} - \dfrac{8}{35}$.

Solution

$$\left.\begin{array}{l} 14 = 2 \cdot 7 \\ 35 = 5 \cdot 7 \end{array}\right\} \longrightarrow \text{LCD} = 2 \cdot 5 \cdot 7 = 70$$

$$\frac{3}{14} - \frac{8}{35} = \frac{3 \cdot 5}{14 \cdot 5} - \frac{8 \cdot 2}{35 \cdot 2} = \frac{15}{70} - \frac{16}{70} = \frac{-1}{70} \quad \left(\text{or} -\frac{1}{70}\right) \qquad \blacktriangle$$

Example 6

Add $\dfrac{-5}{8} + \dfrac{3}{14}$.

Solution

$$\left.\begin{array}{l} 8 = 2 \cdot 2 \cdot 2 \\ 14 = 2 \cdot 7 \end{array}\right\} \longrightarrow \text{LCD} = 2 \cdot 2 \cdot 2 \cdot 7 = 56$$

$$\frac{-5}{8} + \frac{3}{14} = \frac{-5 \cdot 7}{8 \cdot 7} + \frac{3 \cdot 4}{14 \cdot 4} = \frac{-35}{56} + \frac{12}{56} = \frac{-23}{56} \quad \left(\text{or} -\frac{23}{56}\right) \qquad \blacktriangle$$

Example 7

Add $-3 + \dfrac{2}{5}$.

Solution

$$-3 + \frac{2}{5} = \frac{-3 \cdot 5}{1 \cdot 5} + \frac{2}{5} = \frac{-15}{5} + \frac{2}{5} = \frac{-15 + 2}{5} = \frac{-13}{5} \quad \left(\text{or} -\frac{13}{5}\right) \qquad \blacktriangle$$

Denominators that contain variables do not complicate the situation very much, as the next examples illustrate.

Example 8

Add $\dfrac{2}{x} + \dfrac{3}{y}$.

Solution

By inspection, the LCD is xy.

$$\frac{2}{x} + \frac{3}{y} = \frac{2 \cdot y}{x \cdot y} + \frac{3 \cdot x}{y \cdot x} = \frac{2y}{xy} + \frac{3x}{xy} = \frac{2y + 3x}{xy}$$

Commutative property $\blacktriangle$

Example 9

Subtract $\dfrac{3}{8x} - \dfrac{5}{12y}$.

Solution

$$\left.\begin{array}{l} 8x = 2 \cdot 2 \cdot 2 \cdot x \\ 12y = 2 \cdot 2 \cdot 3 \cdot y \end{array}\right\} \rightarrow \text{LCD} = 2 \cdot 2 \cdot 2 \cdot 3 \cdot x \cdot y = 24xy$$

$$\frac{3}{8x} - \frac{5}{12y} = \frac{3 \cdot 3y}{8x \cdot 3y} - \frac{5 \cdot 2x}{12y \cdot 2x} = \frac{9y}{24xy} - \frac{10x}{24xy} = \frac{9y - 10x}{24xy}$$

▲

Example 10

Add $\dfrac{7}{4a} + \dfrac{-5}{6bc}$.

Solution

$$\left.\begin{array}{l} 4a = 2 \cdot 2 \cdot a \\ 6bc = 2 \cdot 3 \cdot b \cdot c \end{array}\right\} \rightarrow \text{LCD} = 2 \cdot 2 \cdot 3 \cdot a \cdot b \cdot c = 12abc$$

$$\frac{7}{4a} + \frac{-5}{6bc} = \frac{7 \cdot 3bc}{4a \cdot 3bc} + \frac{-5 \cdot 2a}{6bc \cdot 2a} = \frac{21bc}{12abc} + \frac{-10a}{12abc} = \frac{21bc - 10a}{12abc}$$

▲

Simplifying Numerical Expressions

Let's now consider simplifying numerical expressions that contain rational numbers. As with integers, multiplications and divisions are done first and then the additions and subtractions are performed. In these next examples only the major steps are shown, so be sure that you can fill in all of the details.

Example 11

Simplify $\dfrac{3}{4} + \dfrac{2}{3} \cdot \dfrac{3}{5} - \dfrac{1}{2} \cdot \dfrac{1}{5}$.

Solution

$$\frac{3}{4} + \frac{2}{3} \cdot \frac{3}{5} - \frac{1}{2} \cdot \frac{1}{5} = \frac{3}{4} + \frac{2}{5} - \frac{1}{10}$$

$$= \frac{15}{20} + \frac{8}{20} - \frac{2}{20} = \frac{15 + 8 - 2}{20} = \frac{21}{20}$$

▲

Example 12

Simplify $\dfrac{3}{5} \div \dfrac{8}{5} + \left(-\dfrac{1}{2}\right)\left(\dfrac{1}{3}\right) + \dfrac{5}{12}$.

Solution

$$\frac{3}{5} \div \frac{8}{5} + \left(-\frac{1}{2}\right)\left(\frac{1}{3}\right) + \frac{5}{12} = \frac{3}{5} \cdot \frac{5}{8} + \left(-\frac{1}{2}\right)\left(\frac{1}{3}\right) + \frac{5}{12}$$

$$= \frac{3}{8} + \frac{-1}{6} + \frac{5}{12}$$

$$= \frac{9}{24} + \frac{-4}{24} + \frac{10}{24}$$

$$= \frac{9 + (-4) + 10}{24}$$

$$= \frac{15}{24} = \frac{5}{8} \qquad \text{Reduce!}$$

▲

The distributive property, $a(b + c) = ab + ac$, holds true for rational numbers and (as with integers) can be used to facilitate manipulation.

Example 13

Simplify $12\left(\frac{1}{3} + \frac{1}{4}\right)$.

Solution

For help in this situation, let's change the form by applying the distributive property.

$$12\left(\frac{1}{3} + \frac{1}{4}\right) = 12\left(\frac{1}{3}\right) + 12\left(\frac{1}{4}\right)$$
$$= 4 + 3$$
$$= 7$$

▲

Example 14

Simplify $\frac{5}{8}\left(\frac{1}{2} + \frac{1}{3}\right)$.

Solution

In this case it may be easier not to apply the distributive property but to work with the expression in its given form.

$$\frac{5}{8}\left(\frac{1}{2} + \frac{1}{3}\right) = \frac{5}{8}\left(\frac{3}{6} + \frac{2}{6}\right)$$
$$= \frac{5}{8}\left(\frac{5}{6}\right)$$
$$= \frac{25}{48}$$

▲

Examples 13 and 14 emphasize a point we made in Chapter 1. *Think first*, and decide whether or not the properties can be used to make the manipulations easier. This section concludes with Example 15 which illustrates the combining of similar terms that have fractional coefficients.

Example 15

Simplify $\frac{1}{2}x + \frac{2}{3}x - \frac{3}{4}x$ by combining similar terms.

Solution

We can use the distributive property and our knowledge of adding and subtracting rational numbers to solve this type of problem.

$$\frac{1}{2}x + \frac{2}{3}x - \frac{3}{4}x = \left(\frac{1}{2} + \frac{2}{3} - \frac{3}{4}\right)x$$
$$= \left(\frac{6}{12} + \frac{8}{12} - \frac{9}{12}\right)x$$
$$= \frac{5}{12}x$$

▲

Problem Set 2.2

For Problems 1–64, add or subtract as indicated and express answers in lowest terms.

1. $\dfrac{2}{7} + \dfrac{3}{7}$

2. $\dfrac{3}{11} + \dfrac{5}{11}$

3. $\dfrac{7}{9} - \dfrac{2}{9}$

4. $\dfrac{11}{13} - \dfrac{6}{13}$

5. $\dfrac{3}{4} + \dfrac{9}{4}$

6. $\dfrac{5}{6} + \dfrac{7}{6}$

7. $\dfrac{11}{12} - \dfrac{3}{12}$

8. $\dfrac{13}{16} - \dfrac{7}{16}$

9. $\dfrac{1}{8} - \dfrac{5}{8}$

10. $\dfrac{2}{9} - \dfrac{5}{9}$

11. $\dfrac{5}{24} + \dfrac{11}{24}$

12. $\dfrac{7}{36} + \dfrac{13}{36}$

13. $\dfrac{8}{x} + \dfrac{7}{x}$

14. $\dfrac{17}{y} + \dfrac{12}{y}$

15. $\dfrac{5}{3y} + \dfrac{1}{3y}$

16. $\dfrac{3}{8x} + \dfrac{1}{8x}$

17. $\dfrac{1}{3} + \dfrac{1}{5}$

18. $\dfrac{1}{6} + \dfrac{1}{8}$

19. $\dfrac{15}{16} - \dfrac{3}{8}$

20. $\dfrac{13}{12} - \dfrac{1}{6}$

21. $\dfrac{7}{10} + \dfrac{8}{15}$

22. $\dfrac{7}{12} + \dfrac{5}{8}$

23. $\dfrac{11}{24} + \dfrac{5}{32}$

24. $\dfrac{5}{18} + \dfrac{8}{27}$

25. $\dfrac{5}{18} - \dfrac{13}{24}$

26. $\dfrac{1}{24} - \dfrac{7}{36}$

27. $\dfrac{5}{8} - \dfrac{2}{3}$

28. $\dfrac{3}{4} - \dfrac{5}{6}$

29. $-\dfrac{2}{13} - \dfrac{7}{39}$

30. $-\dfrac{3}{11} - \dfrac{13}{33}$

31. $-\dfrac{3}{14} + \dfrac{1}{21}$

32. $-\dfrac{3}{20} + \dfrac{14}{25}$

33. $-4 - \dfrac{3}{7}$

34. $-2 - \dfrac{5}{6}$

35. $\dfrac{3}{4} - 6$

36. $\dfrac{5}{8} - 7$

37. $\dfrac{3}{x} + \dfrac{4}{y}$

38. $\dfrac{5}{x} + \dfrac{8}{y}$

39. $\dfrac{7}{a} - \dfrac{2}{b}$

40. $\dfrac{13}{a} - \dfrac{4}{b}$

41. $\dfrac{2}{x} + \dfrac{7}{2x}$

42. $\dfrac{5}{2x} + \dfrac{7}{x}$

43. $\dfrac{10}{3x} - \dfrac{2}{x}$

44. $\dfrac{13}{4x} - \dfrac{3}{x}$

45. $\dfrac{1}{x} - \dfrac{7}{5x}$

46. $\dfrac{2}{x} - \dfrac{17}{6x}$

47. $\dfrac{3}{2y} + \dfrac{5}{3y}$

48. $\dfrac{7}{3y} + \dfrac{9}{4y}$

49. $\dfrac{5}{12y} - \dfrac{3}{8y}$

50. $\dfrac{9}{4y} - \dfrac{5}{9y}$

51. $\dfrac{1}{6n} - \dfrac{7}{8n}$

52. $\dfrac{3}{10n} - \dfrac{11}{15n}$

53. $\dfrac{5}{3x} + \dfrac{7}{3y}$

54. $\dfrac{3}{2x} + \dfrac{7}{2y}$

55. $\dfrac{8}{5x} + \dfrac{3}{4y}$

56. $\dfrac{1}{5x} + \dfrac{5}{6y}$

57. $\dfrac{7}{4x} - \dfrac{5}{9y}$

58. $\dfrac{2}{7x} - \dfrac{11}{14y}$

59. $-\dfrac{3}{2x} - \dfrac{5}{4y}$

60. $-\dfrac{13}{8a} - \dfrac{11}{10b}$

61. $3 + \dfrac{2}{x}$

62. $\dfrac{5}{x} + 4$

63. $2 - \dfrac{3}{2x}$

64. $-1 - \dfrac{1}{3x}$

For Problems 65–80, simplify each numerical expression expressing answers in reduced form.

65. $\frac{1}{4} - \frac{3}{8} + \frac{5}{12} - \frac{1}{24}$

66. $\frac{3}{4} + \frac{2}{3} - \frac{1}{6} + \frac{5}{12}$

67. $\frac{5}{6} + \frac{2}{3} \cdot \frac{3}{4} - \frac{1}{4} \cdot \frac{2}{5}$

68. $\frac{2}{3} + \frac{1}{2} \cdot \frac{2}{5} - \frac{1}{3} \cdot \frac{1}{5}$

69. $\frac{3}{4} \cdot \frac{6}{9} - \frac{5}{6} \cdot \frac{8}{10} + \frac{2}{3} \cdot \frac{6}{8}$

70. $\frac{3}{5} \cdot \frac{5}{7} + \frac{2}{3} \cdot \frac{3}{5} - \frac{1}{7} \cdot \frac{2}{5}$

71. $4 - \frac{2}{3} \cdot \frac{3}{5} - 6$

72. $3 + \frac{1}{2} \cdot \frac{1}{3} - 2$

73. $\frac{4}{5} - \frac{10}{12} - \frac{5}{6} \div \frac{14}{8} + \frac{10}{21}$

74. $\frac{3}{4} \div \frac{6}{5} + \frac{8}{12} \cdot \frac{6}{9} - \frac{5}{12}$

75. $24\left(\frac{3}{4} - \frac{1}{6}\right)$ Don't forget the distributive property!

76. $18\left(\frac{2}{3} + \frac{1}{9}\right)$

77. $64\left(\frac{3}{16} + \frac{5}{8} - \frac{1}{4} + \frac{1}{2}\right)$

78. $48\left(\frac{5}{12} - \frac{1}{6} + \frac{3}{8}\right)$

79. $\frac{7}{13}\left(\frac{2}{3} - \frac{1}{6}\right)$ **80.** $\frac{5}{9}\left(\frac{1}{2} + \frac{1}{4}\right)$

For Problems 81–96, simplify each algebraic expression by combining similar terms.

81. $\frac{1}{3}x + \frac{2}{5}x$ **82.** $\frac{1}{4}x + \frac{2}{3}x$

83. $\frac{1}{3}a - \frac{1}{8}a$ **84.** $\frac{2}{5}a - \frac{2}{7}a$

85. $\frac{1}{2}x + \frac{2}{3}x + \frac{1}{6}x$

86. $\frac{1}{3}x + \frac{2}{5}x + \frac{5}{6}x$

87. $\frac{3}{5}n - \frac{1}{4}n + \frac{3}{10}n$

88. $\frac{2}{5}n - \frac{7}{10}n + \frac{8}{15}n$

89. $n + \frac{4}{3}n - \frac{1}{9}n$

90. $2n - \frac{6}{7}n + \frac{5}{14}n$

91. $-n - \frac{7}{9}n - \frac{5}{12}n$

92. $-\frac{3}{8}n - n - \frac{3}{14}n$

93. $\frac{3}{7}x + \frac{1}{4}y + \frac{1}{2}x + \frac{7}{8}y$

94. $\frac{5}{6}x + \frac{3}{4}y + \frac{4}{9}x + \frac{7}{10}y$

95. $\frac{2}{9}x + \frac{5}{12}y - \frac{7}{15}x - \frac{13}{15}y$

96. $-\frac{9}{10}x - \frac{3}{14}y + \frac{2}{25}x + \frac{5}{21}y$

97. When the stock market opened on Monday morning Mona bought some shares of a stock at $11\frac{3}{4}$ per share. The numbers $+1\frac{1}{2}$, $-\frac{3}{8}$, $-\frac{1}{4}$, $+\frac{1}{2}$, and $-\frac{5}{8}$ represent the changes in the market for that stock for the week. Determine the value of one share of Mona's stock when the market closed on Friday.

98. Vinay has a board that is $6\frac{1}{2}$ feet long. If he cuts off a piece $2\frac{3}{4}$ feet long, how long is the remaining piece of board?

99. Mindy takes a daily walk of $2\frac{1}{2}$ miles. One day a thunderstorm forced her to stop her walk after $\frac{3}{4}$ of a mile. By how much was her walk shortened that day?

100. Blake Scott leaves $\frac{1}{4}$ of his estate to the Boy Scouts, $\frac{2}{5}$ to the local cancer fund, and the rest to his church. What fractional part of the estate does the church receive?

101. If your calculator handles rational numbers in $\frac{a}{b}$ form, check your answers for Problems 65–80.

THOUGHTS INTO WORDS

102. Give a step-by-step description of how to add the rational numbers $\frac{3}{8}$ and $\frac{5}{18}$.

103. Give a step-by-step description of how to add the fractions $\frac{5}{4x}$ and $\frac{7}{6x}$.

104. The will of a deceased collector of antique automobiles specified that his cars be left to his three children. Half were to go to his elder son, $\frac{1}{3}$ to his daughter, and $\frac{1}{9}$ to his younger son. At the time of his death, 17 cars were in the collection. The administrator of his estate borrowed a car to make 18. Then he distributed the cars as follows.

Elder son: $\quad\quad \frac{1}{2}(18) = 9$

Daughter: $\quad\quad \frac{1}{3}(18) = 6$

Younger son: $\quad\quad \frac{1}{9}(18) = 2$

This totaled 17 cars, so he then returned the borrowed car. Where is the error in this problem?

2.3 Real Numbers and Algebraic Expressions

We classify decimals—also called decimal fractions—as **terminating**, **repeating**, or **nonrepeating**. Some examples of each of these classifications follow.

$$\begin{bmatrix} .3 \\ .26 \\ .347 \\ .9865 \end{bmatrix} \quad \text{Terminating decimals}$$

Technically, a terminating decimal can be thought of as repeating zeros after the last digit. For example, $.3 = .30 = .300 = .3000$, etc.

$$\begin{bmatrix} .333333\ldots \\ .5466666\ldots \\ .14141414\ldots \\ .237237237\ldots \end{bmatrix} \quad \text{Repeating decimals}$$

A repeating decimal has a block of digits that repeats indefinitely. This repeating block of digits may contain any number of digits and may or may not begin repeating immediately after the decimal point.

$$
\begin{bmatrix}
.5918654279\ldots \\
.26224222722229\ldots \\
.145117211193111148\ldots
\end{bmatrix}
\quad \text{Nonrepeating decimals}
$$

In Section 2.1 we defined a rational number to be any number that can be written in the form $\dfrac{a}{b}$, where a and b are integers and b is not zero. **A rational number can also be defined as any number that has a terminating or repeating decimal representation.** Thus, rational numbers can be expressed in either common fractional form or decimal fraction form as the next examples illustrate.

$$
\begin{bmatrix}
\dfrac{3}{4} = .75 \\[2mm]
\dfrac{1}{8} = .125 \\[2mm]
\dfrac{5}{16} = .3125 \\[2mm]
\dfrac{7}{25} = .28
\end{bmatrix}
\quad \text{Terminating decimals}
$$

$$
\begin{bmatrix}
\dfrac{1}{3} = .33333\ldots \\[2mm]
\dfrac{2}{3} = .66666\ldots \\[2mm]
\dfrac{1}{6} = .166666\ldots \\[2mm]
\dfrac{1}{12} = .083333\ldots \\[2mm]
\dfrac{14}{99} = .14141414\ldots
\end{bmatrix}
\quad \text{Repeating decimals}
$$

The nonrepeating decimals are called **irrational numbers** and do appear in forms other than decimal form. For example, $\sqrt{2}$, $\sqrt{3}$, and π are irrational numbers; a partial decimal representation for each of these follows.

$$
\begin{bmatrix}
\sqrt{2} = 1.414213562373\ldots \\
\sqrt{3} = 1.73205080756887\ldots \\
\pi = 3.14159265358979\ldots
\end{bmatrix}
\quad \text{Nonrepeating decimals}
$$

(We will do more work with the irrationals in Chapter 9.)

The rational numbers together with the irrationals form the set of **real numbers.** The following tree diagram of the real number system is helpful for summarizing some basic ideas.

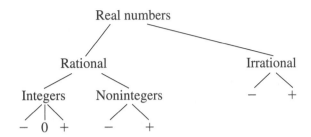

Any real number can be traced down through the diagram as follows.

  5 is real, rational, an integer, and positive,

  −4 is real, rational, an integer, and negative,

  $\frac{3}{4}$ is real, rational, a noninteger, and positive,

  .23 is real, rational, a noninteger, and positive,

  −.161616 . . . is real, rational, a noninteger, and negative,

  $\sqrt{7}$ is real, irrational, and positive,

  $-\sqrt{2}$ is real, irrational, and negative

  The properties we discussed in Section 1.5 pertaining to integers are true for all real numbers and are restated here for your convenience. The multiplicative inverse property was added to the list; a discussion follows.

Commutative Property of Addition

If a and b are real numbers, then

  $a + b = b + a$.

Commutative Property of Multiplication

If a and b are real numbers, then

  $ab = ba$.

Associative Property of Addition

If a, b, and c are real numbers, then

  $(a + b) + c = a + (b + c)$.

Associative Property of Multiplication

If a, b, and c are real numbers, then

$(ab)c = a(bc)$.

Identity Property of Addition

If a is any real number, then

$a + 0 = 0 + a = a$.

Identity Property of Multiplication

If a is any real number, then

$a(1) = 1(a) = a$.

Additive Inverse Property

For every real number a, there exists a real number $-a$, such that

$a + (-a) = (-a) + a = 0$.

Multiplication Property of Zero

If a is any real number, then

$a(0) = 0(a) = 0$.

Multiplicative Property of Negative One

If a is any real number, then

$a(-1) = -1(a) = -a$.

Multiplicative Inverse Property

For every nonzero real number a, there exists a real number $\dfrac{1}{a}$, such that

$a\left(\dfrac{1}{a}\right) = \dfrac{1}{a}(a) = 1$.

Distributive Property

If a, b, and c are real numbers, then

$a(b + c) = ab + ac$.

The number $\dfrac{1}{a}$ is called the **multiplicative inverse** or the **reciprocal of** a. For example, the reciprocal of 2 is $\dfrac{1}{2}$ and $2\left(\dfrac{1}{2}\right) = \dfrac{1}{2}(2) = 1$. Likewise, the reciprocal of $\dfrac{1}{2}$ is $\dfrac{1}{\frac{1}{2}} = 2$. Therefore, 2 and $\dfrac{1}{2}$ are said to be reciprocals (or multiplicative inverses) of each other. Also, $\dfrac{2}{5}$ and $\dfrac{5}{2}$ are multiplicative inverses and $\left(\dfrac{2}{5}\right)\left(\dfrac{5}{2}\right) = 1$. Since division by zero is undefined, zero does not have a reciprocal.

Basic Operations with Decimals

The basic operations with decimals may be related to the corresponding operation with common fractions. For example, $.3 + .4 = .7$ because $\dfrac{3}{10} + \dfrac{4}{10} = \dfrac{7}{10}$ and $.37 - .24 = .13$ because $\dfrac{37}{100} - \dfrac{24}{100} = \dfrac{13}{100}$. In general, to add or subtract decimals, we add or subtract the hundredths, the tenths, the ones, the tens, and so on. To keep place values aligned, we line up the decimal points.

	Addition		**Subtraction**
1	1 11	616	81113
2.14	5.214	7̸.6̸	9̸.2̸3̸5
3.12	3.162	4.9	6.781
5.16	7.218	2.7	2.454
10.42	8.914		
	24.508		

The following examples can be used to formulate a general rule for multiplying decimals.

Since $\dfrac{7}{10} \cdot \dfrac{3}{10} = \dfrac{21}{100}$, then $(.7)(.3) = .21$.

Since $\dfrac{9}{10} \cdot \dfrac{23}{100} = \dfrac{207}{1000}$, then $(.9)(.23) = .207$.

Since $\dfrac{11}{100} \cdot \dfrac{13}{100} = \dfrac{143}{10,000}$, then $(.11)(.13) = .0143$.

In general, to multiply decimals we (1) multiply the numbers and ignore the decimal points, and then (2) insert the decimal point in the product so that the number of digits to the right of the decimal point in the product is equal to the sum of the numbers of digits to the right of the decimal point in each factor.

(.7)		(.3)	=	.21
↑		↑		↑
One digit to right	+	One digit to right	=	Two digits to right

$$(.9) \quad + \quad (.23) \quad = \quad .207$$

One digit to right	+	Two digits to right	=	Three digits to right

$$(.11) \quad + \quad (.13) \quad = \quad .0143$$

Two digits to right	+	Two digits to right	=	Four digits to right

We frequently use the vertical format when multiplying decimals.

$$
\begin{array}{r}
41.2 \\
.13 \\
\hline
1236 \\
412 \\
\hline
5.356
\end{array}
$$

One digit to right
Two digits to right

Three digits to right

$$
\begin{array}{r}
.021 \\
.03 \\
\hline
.00063
\end{array}
$$

Three digits to right
Two digits to right

Five digits to right

Notice that in the last example we actually multiplied $3 \cdot 21$ and then inserted three 0s to the left so that there would be five digits to the right of the decimal point.

Once again let's look at some links between common fractions and decimals.

$$\text{Since } \frac{6}{10} \div 2 = \frac{\overset{3}{\cancel{6}}}{10} \cdot \frac{1}{\cancel{2}} = \frac{3}{10}, \quad \text{then } 2\overline{).6}^{.3};$$

$$\text{Since } \frac{39}{100} \div 13 = \frac{\overset{3}{\cancel{39}}}{100} \cdot \frac{1}{\cancel{13}} = \frac{3}{100}, \quad \text{then } 13\overline{).39}^{.03};$$

$$\text{Since } \frac{85}{100} \div 5 = \frac{\overset{17}{\cancel{85}}}{100} \cdot \frac{1}{\cancel{5}} = \frac{17}{100}, \quad \text{then } 5\overline{).85}^{.17}.$$

In general, to divide a decimal by a nonzero whole number we (1) place the decimal point in the quotient directly above the decimal point in the dividend

$$\left(\text{Divisor} \overline{)\text{Dividend}}^{\text{Quotient}} \right),$$

and then (2) divide as with whole numbers, except that in the division process, 0s are placed in the quotient immediately to the right of the decimal point in order to show the correct place value.

$$
4\overline{)\,.484}^{\,.121}
\qquad
\begin{array}{r}
.24 \\
32\overline{)7.68} \\
6\,4 \\
\hline
1\,28 \\
1\,28 \\
\hline
\end{array}
\qquad
\begin{array}{r}
.019 \\
12\overline{)\,.228} \\
12 \\
\hline
108 \\
108 \\
\hline
\end{array}
$$

Zero needed to show the correct place value

Don't forget that *division can be checked by multiplication*. For example, since $(12)(0.019) = 0.228$ we know that our last division example is correct.

We can easily handle problems involving division by a decimal by changing to an equivalent problem that has a whole number divisor. Consider the following examples in which the original division problem was changed to fractional form to show the reasoning involved in the procedure.

$$.6\overline{).24} \rightarrow \frac{.24}{.6} = \left(\frac{.24}{.6}\right)\left(\frac{10}{10}\right) = \frac{2.4}{6} \rightarrow 6\overline{)2.4,}^{\,.4}$$

$$.12\overline{).156} \rightarrow \frac{.156}{.12} = \left(\frac{.156}{.12}\right)\left(\frac{100}{100}\right) = \frac{15.6}{12} \rightarrow 12\overline{)15.6,}^{\,1.3}$$
$$\begin{array}{r} 12 \\ \hline 36 \\ 36 \end{array}$$

$$1.3\overline{).026} \rightarrow \frac{.026}{1.3} = \left(\frac{.026}{1.3}\right)\left(\frac{10}{10}\right) = \frac{.26}{13} \rightarrow 13\overline{).26}^{\,.02}$$
$$\begin{array}{r} 26 \\ \hline \end{array}$$

The format commonly used with such problems is as follows.

$$\begin{array}{r} 5.6 \\ 21.\overline{)1\,17.6} \\ \underline{1\,05} \\ 12\,6 \\ \underline{12\,6} \end{array}$$

The arrows indicate that the divisor and dividend were multiplied by 100, which changes the divisor to a whole number.

$$\begin{array}{r} .04 \\ 3.7.\overline{)1.48} \\ \underline{1\,48} \end{array}$$

The divisor and dividend were multiplied by 10.

Our agreements for operating with positive and negative integers extend to all real numbers. For example, the product of two negative real numbers is a positive real number. Make sure that you agree with the following results. (You may need to do some work on scratch paper since the steps are not shown.)

$$.24 + (-.18) = .06, \qquad (-.4)(.8) = -.32,$$
$$-7.2 + 5.1 = -2.1, \qquad (-.5)(-.13) = .065,$$
$$-.6 + (-.8) = -1.4, \qquad (1.4) \div (-.2) = -7,$$
$$2.4 - 6.1 = -3.7, \qquad (-.18) \div (.3) = -.6,$$
$$.31 - (-.52) = .83, \qquad (-.24) \div (-4) = .06$$
$$(.2)(-.3) = -.06,$$

Numerical and algebraic expressions may contain the decimal form as well as the fractional form of rational numbers. We continue to follow the agreement that multiplications and divisions are done *first* and then the additions and subtractions unless parentheses indicate otherwise. The following examples illustrate a variety of situations that involve both the decimal and fractional form of rational numbers.

Example 1

Simplify $6.3 \div 7 + (4)(2.1) - (.24) \div (-.4)$.

Solution

$$6.3 \div 7 + (4)(2.1) - (.24) \div (-.4) = .9 + 8.4 - (-.6)$$
$$= .9 + 8.4 + .6$$
$$= 9.9$$

▲

Example 2

Evaluate $\frac{3}{5}a - \frac{1}{7}b$ for $a = \frac{5}{2}$ and $b = -1$.

Solution

$$\frac{3}{5}a - \frac{1}{7}b = \frac{3}{5}\left(\frac{5}{2}\right) - \frac{1}{7}(-1) \quad \text{for } a = \frac{5}{2} \text{ and } b = -1$$
$$= \frac{3}{2} + \frac{1}{7}$$
$$= \frac{21}{14} + \frac{2}{14}$$
$$= \frac{23}{14}$$

▲

Example 3

Evaluate $\frac{1}{2}x + \frac{2}{3}x - \frac{1}{5}x$ for $x = -\frac{3}{4}$.

Solution

First, let's *combine similar terms* by using the distributive property.

$$\frac{1}{2}x + \frac{2}{3}x - \frac{1}{5}x = \left(\frac{1}{2} + \frac{2}{3} - \frac{1}{5}\right)x$$
$$= \left(\frac{15}{30} + \frac{20}{30} - \frac{6}{30}\right)x$$
$$= \frac{29}{30}x$$

Now we can evaluate.

$$\frac{29}{30}x = \frac{29}{30}\left(-\frac{3}{4}\right) \quad \text{when } x = -\frac{3}{4}$$
$$= \frac{29}{\overset{}{\underset{10}{30}}}\left(-\frac{\overset{1}{3}}{4}\right) = -\frac{29}{40}$$

▲

Example 4

Solution

Evaluate $2x + 3y$ for $x = 1.6$ and $y = 2.7$.

$$2x + 3y = 2(1.6) + 3(2.7) \quad \text{when } x = 1.6 \text{ and } y = 2.7$$
$$= 3.2 + 8.1$$
$$= 11.3$$

▲

Example 5

Solution

Evaluate $.9x + .7x - .4x + 1.3x$ for $x = .2$.

First, let's *combine similar terms* by using the distributive property.

$$.9x + .7x - .4x + 1.3x = (.9 + .7 - .4 + 1.3)x = 2.5x$$

Now we can evaluate.

$$2.5x = (2.5)(.2) \quad \text{for } x = .2$$
$$= .5$$

▲

▼ Problem Set 2.3

For Problems 1–32, perform the indicated operations.

1. $.37 + .25$
2. $7.2 + 4.9$
3. $2.93 - 1.48$
4. $14.36 - 5.89$
5. $(7.6) + (-3.8)$
6. $(6.2) + (-2.4)$
7. $(-4.7) + 1.4$
8. $(-14.1) + 9.5$
9. $-3.8 + 11.3$
10. $-2.5 + 14.8$
11. $6.6 - (-1.2)$
12. $18.3 - (-7.4)$
13. $-11.5 - (-10.6)$
14. $-14.6 - (-8.3)$
15. $-17.2 - (-9.4)$
16. $-21.4 - (-14.2)$
17. $(.4)(2.9)$
18. $(.3)(3.6)$
19. $(-.8)(.34)$
20. $(-.7)(.67)$
21. $(9)(-2.7)$
22. $(8)(-7.6)$
23. $(-.7)(-64)$
24. $(-.9)(-56)$
25. $(-.12)(-.13)$
26. $(-.11)(-.15)$
27. $1.56 \div 1.3$
28. $7.14 \div 2.1$
29. $5.92 \div (-.8)$
30. $-2.94 \div .6$
31. $-.266 \div (-.7)$
32. $-.126 \div (-.9)$

For Problems 33–46, simplify each of the numerical expressions.

33. $16.5 - 18.7 + 9.4$
34. $17.7 + 21.2 - 14.6$
35. $.34 - .21 - .74 + .19$
36. $-5.2 + 6.8 - 4.7 - 3.9 + 1.3$
37. $.76(.2 + .8)$ **38.** $9.8(1.8 - .8)$
39. $.6(4.1) + .7(3.2)$
40. $.5(74) - .9(87)$
41. $7(.6) + .9 - 3(.4) + .4$
42. $-5(.9) - .6 + 4.1(6) - .9$
43. $(.96) \div (-.8) + 6(-1.4) - 5.2$
44. $(-2.98) \div .4 - 5(-2.3) + 1.6$
45. $5(2.3) - 1.2 - 7.36 \div .8 + .2$
46. $.9(12) \div .4 - 1.36 \div 17 + 9.2$

For Problems 47–60, simplify each algebraic expression by combining similar terms.

47. $x - .4x - 1.8x$
48. $-2x + 1.7x - 4.6x$
49. $5.4n - .8n - 1.6n$

50. $6.2n - 7.8n - 1.3n$

51. $-3t + 4.2t - .9t + .2t$

52. $7.4t - 3.9t - .6t + 4.7t$

53. $3.6x - 7.4y - 9.4x + 10.2y$

54. $5.7x + 9.4y - 6.2x - 4.4y$

55. $.3(x - 4) + .4(x + 6) - .6x$

56. $.7(x + 7) - .9(x - 2) + .5x$

57. $6(x - 1.1) - 5(x - 2.3) - 4(x + 1.8)$

58. $4(x + .7) - 9(x + .2) - 3(x - .6)$

59. $5(x - .5) + .3(x - 2) - .7(x + 7)$

60. $-8(x - 1.2) + 6(x - 4.6) + 4(x + 1.7)$

For Problems 61–74, evaluate each algebraic expression for the given values of the variables. Don't forget that for some problems it might be helpful to combine similar terms first and then to evaluate.

61. $x + 2y + 3z$ for $x = \dfrac{3}{4}, y = \dfrac{1}{3}$, and $z = -\dfrac{1}{6}$

62. $2x - y - 3z$ for $x = -\dfrac{2}{5}, y = -\dfrac{3}{4}$, and $z = \dfrac{1}{2}$

63. $\dfrac{3}{5}y - \dfrac{2}{3}y - \dfrac{7}{15}y$ for $y = -\dfrac{5}{2}$

64. $\dfrac{1}{2}x + \dfrac{2}{3}x - \dfrac{3}{4}x$ for $x = \dfrac{7}{8}$

65. $-x - 2y + 4z$ for $x = 1.7, y = -2.3$, and $z = 3.6$

66. $-2x + y - 5z$ for $x = -2.9, y = 7.4$, and $z = -6.7$

67. $5x - 7y$ for $x = -7.8$ and $y = 8.4$

68. $8x - 9y$ for $x = -4.3$ and $y = 5.2$

69. $.7x + .6y$ for $x = -2$ and $y = 6$

70. $.8x + 2.1y$ for $x = 5$ and $y = -9$

71. $1.2x + 2.3x - 1.4x - 7.6x$ for $x = -2.5$

72. $3.4x - 1.9x + 5.2x$ for $x = .3$

73. $-3a - 1 + 7a - 2$ for $a = .9$

74. $5x - 2 + 6x + 4$ for $x = -1.1$

75. Tanya bought 400 shares of one stock at $\$14\dfrac{3}{4}$ per share and 250 shares of another stock at $\$16\dfrac{1}{8}$ per share. How much did she pay for the 650 shares?

76. On a trip Brent bought the following amounts of gasoline: 9.7 gallons, 12.3 gallons, 14.6 gallons, 12.2 gallons, 13.8 gallons, and 15.5 gallons. How many gallons of gasoline did he purchase on the trip?

77. Kathrin has a piece of copper tubing that is 76.4 centimeters long. She needs to cut it into four pieces of equal length. Find the length of each piece.

78. On a trip Biance filled the gasoline tank and noted that the odometer read 24,876.2 miles. After the next filling the odometer read 25,170.5 miles. It took 13.5 gallons of gasoline to fill the tank. How many miles per gallon did she get on that tank of gasoline?

79. Use a calculator to check your answers for Problems 33–46.

THOUGHTS INTO WORDS

80. At this time how would you describe the difference between arithmetic and algebra?

81. How have the properties of the real numbers been used thus far in your study of arithmetic and algebra?

82. Do you think that $2\sqrt{2}$ is a rational or an irrational number? Defend your answer.

Further Investigations

83. Without doing the actual dividing, defend the statement "$\dfrac{1}{7}$ produces a repeating decimal." [*Hint:* Think about the possible remainders when dividing by 7.]

84. Express each of the following in repeating decimal form.

(a) $\dfrac{1}{7}$ **(b)** $\dfrac{2}{7}$ **(c)** $\dfrac{4}{9}$

(d) $\dfrac{5}{6}$ **(e)** $\dfrac{3}{11}$ **(f)** $\dfrac{1}{12}$

85. (a) How can we tell that $\dfrac{5}{16}$ will produce a terminating decimal?

(b) How can we tell that $\dfrac{7}{15}$ will not produce a terminating decimal?

(c) Determine which of the following will produce a terminating decimal: $\dfrac{7}{8}, \dfrac{11}{16}, \dfrac{5}{12}, \dfrac{7}{24}, \dfrac{11}{75}, \dfrac{13}{32}, \dfrac{17}{40}, \dfrac{11}{30}, \dfrac{9}{20}, \dfrac{3}{64}.$

2.4 Exponents

We use exponents to indicate repeated multiplication. For example, we can write $5 \cdot 5 \cdot 5$ as 5^3 where the 3 indicates that 5 is to be used as a factor 3 times. The following general definition is helpful.

DEFINITION 2.4

If n is a positive integer and b is any real number, then

$$b^n = \underbrace{bbb \cdots b}_{n \text{ factors of } b}.$$

We refer to the b as the **base** and n as the **exponent**. The expression b^n can be read as "b to the nth **power**." We frequently associate the terms **squared** and **cubed** with exponents of 2 and 3, respectively. For example, b^2 is read as "b squared" and b^3 as "b cubed." An exponent of 1 is usually not written, so b^1 is written as b. The following examples further clarify the concept of an exponent.

$$2^3 = 2 \cdot 2 \cdot 2 = 8, \qquad (.6)^2 = (.6)(.6) = .36,$$

$$3^5 = 3 \cdot 3 \cdot 3 \cdot 3 \cdot 3 = 243, \qquad \left(\frac{1}{2}\right)^4 = \frac{1}{2} \cdot \frac{1}{2} \cdot \frac{1}{2} \cdot \frac{1}{2} = \frac{1}{16},$$

$$(-5)^2 = (-5)(-5) = 25, \qquad -5^2 = -(5 \cdot 5) = -25$$

We especially want to call your attention to the last two examples. Notice that $(-5)^2$ means that -5 is the base, which is to be used as a factor twice. However, -5^2 means that 5 is the base and after 5 is squared we take the opposite of that result.

Exponents provide a way of writing algebraic expressions in compact form. Sometimes we need to change from the compact form to an expanded form as these next examples demonstrate.

$$x^4 = x \cdot x \cdot x \cdot x, \qquad\qquad (2x)^3 = (2x)(2x)(2x),$$
$$2y^3 = 2 \cdot y \cdot y \cdot y, \qquad\qquad (-2x)^3 = (-2x)(-2x)(-2x),$$
$$-3x^5 = -3 \cdot x \cdot x \cdot x \cdot x \cdot x, \qquad -x^2 = -(x \cdot x),$$
$$a^2 + b^2 = a \cdot a + b \cdot b,$$

At other times we need to change from an expanded form to a more compact form using the exponent notation. The next examples illustrate this idea.

$$3 \cdot x \cdot x = 3x^2,$$
$$2 \cdot 5 \cdot x \cdot x \cdot x = 10x^3,$$
$$3 \cdot 4 \cdot x \cdot x \cdot y = 12x^2 y,$$
$$7 \cdot a \cdot a \cdot a \cdot b \cdot b = 7a^3 b^2,$$
$$(2x)(3y) = 2 \cdot x \cdot 3 \cdot y = 2 \cdot 3 \cdot x \cdot y = 6xy,$$
$$(3a^2)(4a) = 3 \cdot a \cdot a \cdot 4 \cdot a = 3 \cdot 4 \cdot a \cdot a \cdot a = 12a^3,$$
$$(-2x)(3x) = -2 \cdot x \cdot 3 \cdot x = -2 \cdot 3 \cdot x \cdot x = -6x^2$$

The commutative and associative properties for multiplication allowed us to rearrange and regroup factors in the last three examples above.

The concept of *exponent* can be used to extend our work with combining similar terms, operating with fractions, and evaluating algebraic expressions. Study the following examples very carefully; they will help you pull together many ideas.

Example 1

Simplify $4x^2 + 7x^2 - 2x^2$ by combining similar terms.

Solution

By applying the distributive property we obtain

$$4x^2 + 7x^2 - 2x^2 = (4 + 7 - 2)x^2$$
$$= 9x^2.$$

▲

Example 2

Simplify $-8x^3 + 9y^2 + 4x^3 - 11y^2$ by combining similar terms.

Solution

By rearranging terms and then applying the distributive property we obtain

$$-8x^3 + 9y^2 + 4x^3 - 11y^2 = -8x^3 + 4x^3 + 9y^2 - 11y^2$$
$$= (-8 + 4)x^3 + (9 - 11)y^2$$
$$= -4x^3 - 2y^2.$$

▲

Example 3

Simplify $-7x^2 + 4x + 3x^2 - 9x$.

Solution

$$-7x^2 + 4x + 3x^2 - 9x = -7x^2 + 3x^2 + 4x - 9x$$
$$= (-7 + 3)x^2 + (4 - 9)x$$
$$= -4x^2 - 5x$$

▲

As soon as you feel comfortable with this process of combining similar terms, you may want to do some of the steps mentally. Then your work may appear as follows.

$$9a^2 + 6a^2 - 12a^2 = 3a^2,$$

$$6x^2 + 7y^2 - 3x^2 - 11y^2 = 3x^2 - 4y^2,$$

$$7x^2y + 5xy^2 - 9x^2y + 10xy^2 = -2x^2y + 15xy^2,$$

$$2x^3 - 5x^2 - 10x - 7x^3 + 9x^2 - 4x = -5x^3 + 4x^2 - 14x$$

The next two examples illustrate the use of exponents when reducing fractions.

Example 4

Reduce $\dfrac{8x^2y}{12xy}$.

Solution

$$\frac{8x^2y}{12xy} = \frac{2 \cdot 2 \cdot 2 \cdot x \cdot x \cdot y}{2 \cdot 2 \cdot 3 \cdot x \cdot y} = \frac{2x}{3}$$

▲

Example 5

Reduce $\dfrac{15a^2b^3}{25a^3b}$.

Solution

$$\frac{15a^2b^3}{25a^3b} = \frac{3 \cdot 5 \cdot a \cdot a \cdot b \cdot b \cdot b}{5 \cdot 5 \cdot a \cdot a \cdot a \cdot b} = \frac{3b^2}{5a}$$

▲

The next three examples show how exponents may be used when multiplying and dividing fractions.

Example 6

Multiply $\left(\dfrac{4x}{6y}\right)\left(\dfrac{12y^2}{7x^2}\right)$ and express answer in reduced form.

Solution

$$\left(\frac{4x}{6y}\right)\left(\frac{12y^2}{7x^2}\right) = \frac{4 \cdot \overset{2}{12} \cdot x \cdot y \cdot y}{6 \cdot 7 \cdot y \cdot x \cdot x} = \frac{8y}{7x}$$

▲

Example 7

Multiply and simplify

$$\left(\frac{8a^3}{9b}\right)\left(\frac{12b^2}{16a}\right).$$

Solution

$$\left(\frac{8a^3}{9b}\right)\left(\frac{12b^2}{16a}\right) = \frac{\overset{4}{8} \cdot \overset{2}{12} \cdot a \cdot a \cdot a \cdot b \cdot b}{\underset{3}{9} \cdot \underset{2}{16} \cdot b \cdot a} = \frac{2a^2b}{3}$$

▲

Example 8

Divide and express in reduced form,

$$\frac{-2x^3}{3y^2} \div \frac{4}{9xy}.$$

Solution

$$\frac{-2x^3}{3y^2} \div \frac{4}{9xy} = -\frac{2x^3}{3y^2} \cdot \frac{9xy}{4} = -\frac{\overset{1}{\cancel{2}} \cdot \overset{3}{\cancel{9}} \cdot x \cdot x \cdot x \cdot x \cdot \cancel{y}}{\cancel{3} \cdot \cancel{4} \cdot \cancel{y} \cdot y} = -\frac{3x^4}{2y}$$
$$\underset{2}{}$$

▲

The next two examples demonstrate the use of exponents when adding and subtracting fractions.

Example 9

Add $\dfrac{4}{x^2} + \dfrac{7}{x}$.

Solution

The LCD is x^2. Thus,

$$\frac{4}{x^2} + \frac{7}{x} = \frac{4}{x^2} + \frac{7 \cdot x}{x \cdot x} = \frac{4}{x^2} + \frac{7x}{x^2} = \frac{4 + 7x}{x^2}.$$

▲

Example 10

Subtract $\dfrac{3}{xy} - \dfrac{4}{y^2}$.

Solution

$$\left.\begin{array}{l} xy = x \cdot y \\ y^2 = y \cdot y \end{array}\right\} \longrightarrow \text{ The LCD is } xy^2.$$

$$\frac{3}{xy} - \frac{4}{y^2} = \frac{3 \cdot y}{xy \cdot y} - \frac{4 \cdot x}{y^2 \cdot x} = \frac{3y}{xy^2} - \frac{4x}{xy^2}$$

$$= \frac{3y - 4x}{xy^2}$$

▲

Remember that exponents are used to indicate repeated multiplication. Therefore, to simplify numerical expressions containing exponents we proceed as follows.

1. Perform the operations inside the symbols of inclusion (parentheses and brackets) and above and below each fraction bar. Start with the innermost inclusion symbol.

2. Compute all indicated powers.

3. Perform all multiplications and divisions in the order that they appear from left to right.

4. Perform all additions and subtractions in the order that they appear from left to right.

Keep these steps in mind as we evaluate some algebraic expressions containing exponents.

Example 11

Solution

Evaluate $3x^2 - 4y^2$ for $x = -2$ and $y = 5$.

$$\begin{aligned} 3x^2 - 4y^2 &= 3(-2)^2 - 4(5)^2 \quad \text{when } x = -2 \text{ and } y = 5 \\ &= 3(-2)(-2) - 4(5)(5) \\ &= 12 - 100 \\ &= -88 \end{aligned}$$

▲

Example 12

Solution

Find the value of $a^2 - b^2$ when $a = \dfrac{1}{2}$ and $b = -\dfrac{1}{3}$.

$$\begin{aligned} a^2 - b^2 &= \left(\frac{1}{2}\right)^2 - \left(-\frac{1}{3}\right)^2 \quad \text{when } a = \frac{1}{2} \text{ and } b = -\frac{1}{3} \\ &= \frac{1}{4} - \frac{1}{9} \\ &= \frac{9}{36} - \frac{4}{36} \\ &= \frac{5}{36} \end{aligned}$$

▲

Example 13

Solution

Evaluate $5x^2 + 4xy$ for $x = .4$ and $y = -.3$.

$$\begin{aligned} 5x^2 + 4xy &= 5(.4)^2 + 4(.4)(-.3) \quad \text{when } x = .4 \text{ and } y = -.3 \\ &= 5(.16) + 4(-.12) \\ &= .80 + (-.48) \\ &= .32 \end{aligned}$$

▲

▼ Problem Set 2.4

For Problems 1–20, find the value of each numerical expression. For example, $2^4 = 2 \cdot 2 \cdot 2 \cdot 2 = 16$.

1. 2^6 **2.** 2^7 **3.** 3^4

4. 4^3 **5.** $(-2)^3$ **6.** $(-2)^5$

7. -3^2 **8.** -3^4 **9.** $(-4)^2$

10. $(-5)^4$ **11.** $\left(\dfrac{2}{3}\right)^4$

12. $\left(\dfrac{3}{4}\right)^3$ **13.** $-\left(\dfrac{1}{2}\right)^3$

14. $-\left(\dfrac{3}{2}\right)^3$ **15.** $\left(-\dfrac{3}{2}\right)^2$

16. $\left(-\dfrac{4}{3}\right)^2$ **17.** $(.3)^3$

18. $(.2)^4$ **19.** $-(1.2)^2$

20. $-(1.1)^2$

For Problems 21–34, simplify each numerical expression.

21. $3^2 + 2^3 - 4^3$

22. $2^4 - 3^3 + 5^2$

23. $(-2)^3 - 2^4 - 3^2$

24. $(-3)^3 - 3^2 - 6^2$

25. $5(2)^2 - 4(2) - 1$

26. $7(-2)^2 - 6(-2) - 8$

27. $-2(3)^3 - 3(3)^2 + 4(3) - 6$

28. $5(-3)^3 - 4(-3)^2 + 6(-3) + 1$

29. $-7^2 - 6^2 + 5^2$

30. $-8^2 + 3^4 - 4^3$

31. $-3(-4)^2 - 2(-3)^3 + (-5)^2$

32. $-4(-3)^3 + 5(-2)^3 - (4)^2$

33. $\dfrac{-3(2)^4}{12} + \dfrac{5(-3)^3}{15}$

34. $\dfrac{4(2)^3}{16} - \dfrac{2(3)^2}{6}$

For Problems 35–46, use exponents to help express each algebraic expression in a more compact form. For example, $3 \cdot 5 \cdot x \cdot x \cdot y = 15x^2y$ and $(3x)(2x^2) = 6x^3$.

35. $9 \cdot x \cdot x$

36. $8 \cdot x \cdot x \cdot x \cdot y$

37. $3 \cdot 4 \cdot x \cdot y \cdot y$

38. $7 \cdot 2 \cdot a \cdot a \cdot b \cdot b \cdot b$

39. $-2 \cdot 9 \cdot x \cdot x \cdot x \cdot x \cdot y$

40. $-3 \cdot 4 \cdot x \cdot y \cdot z \cdot z$

41. $(5x)(3y)$

42. $(3x^2)(2y)$

43. $(6x^2)(2x^2)$

44. $(-3xy)(6xy)$

45. $(-4a^2)(-2a^3)$

46. $(-7a^3)(-3a)$

For Problems 47–58, simplify each expression by combining similar terms.

47. $3x^2 - 7x^2 - 4x^2$

48. $-2x^3 + 7x^3 - 4x^3$

49. $-12y^3 + 17y^3 - y^3$

50. $-y^3 + 8y^3 - 13y^3$

51. $7x^2 - 2y^2 - 9x^2 + 8y^2$

52. $5x^3 + 9y^3 - 8x^3 - 14y^3$

53. $\dfrac{2}{3}n^2 - \dfrac{1}{4}n^2 - \dfrac{3}{5}n^2$

54. $-\dfrac{1}{2}n^2 + \dfrac{5}{6}n^2 - \dfrac{4}{9}n^2$

55. $5x^2 - 8x - 7x^2 + 2x$

56. $-10x^2 + 4x + 4x^2 - 8x$

57. $x^2 - 2x - 4 + 6x^2 - x + 12$

58. $-3x^3 - x^2 + 7x - 2x^3 + 7x^2 - 4x$

For Problems 59–68, reduce each fraction to simplest form.

59. $\dfrac{9xy}{15x}$

60. $\dfrac{8x^2y}{14x}$

61. $\dfrac{22xy^2}{6xy^3}$

62. $\dfrac{18x^3y}{12xy^4}$

63. $\dfrac{7a^2b^3}{17a^3b}$

64. $\dfrac{9a^3b^3}{22a^4b^2}$

65. $\dfrac{-24abc^2}{32bc}$

66. $\dfrac{4a^2c^3}{-22b^2c^4}$

67. $\dfrac{-5x^4y^3}{-20x^2y}$

68. $\dfrac{-32xy^2z^4}{-48x^3y^3z}$

For Problems 69–86, perform the indicated operations and express answers in reduced form.

69. $\left(\dfrac{7x^2}{9y}\right)\left(\dfrac{12y}{21x}\right)$

70. $\left(\dfrac{3x}{8y^2}\right)\left(\dfrac{14xy}{9y}\right)$

71. $\left(\dfrac{5c}{a^2b^2}\right) \div \left(\dfrac{12c}{ab}\right)$

72. $\left(\dfrac{13ab^2}{12c}\right) \div \left(\dfrac{26b}{14c}\right)$

73. $\dfrac{6}{x} + \dfrac{5}{y^2}$

74. $\dfrac{8}{y} - \dfrac{6}{x^2}$

75. $\dfrac{5}{x^4} - \dfrac{7}{x^2}$

76. $\dfrac{9}{x} - \dfrac{11}{x^3}$

77. $\dfrac{3}{2x^3} + \dfrac{6}{x}$

78. $\dfrac{5}{3x^2} + \dfrac{6}{x}$

79. $\dfrac{-5}{4x^2} + \dfrac{7}{3x^2}$

80. $\dfrac{-8}{5x^3} + \dfrac{10}{3x^3}$

81. $\dfrac{11}{a^2} - \dfrac{14}{b^2}$

82. $\dfrac{9}{x^2} + \dfrac{8}{y^2}$

83. $\dfrac{1}{2x^3} - \dfrac{4}{3x^2}$

84. $\dfrac{2}{3x^3} - \dfrac{5}{4x}$

85. $\dfrac{3}{x} - \dfrac{4}{y} - \dfrac{5}{xy}$

86. $\dfrac{5}{x} + \dfrac{7}{y} - \dfrac{1}{xy}$

For Problems 87–100, evaluate each algebraic expression for the given values of the variables.

87. $4x^2 + 7y^2$ for $x = -2$ and $y = -3$

88. $5x^2 + 2y^3$ for $x = -4$ and $y = -1$

89. $3x^2 - y^2$ for $x = \dfrac{1}{2}$ and $y = -\dfrac{1}{3}$

90. $x^2 - 2y^2$ for $x = -\dfrac{2}{3}$ and $y = \dfrac{3}{2}$

91. $x^2 - 2xy + y^2$ for $x = -\dfrac{1}{2}$ and $y = 2$

92. $x^2 + 2xy + y^2$ for $x = -\dfrac{3}{2}$ and $y = -2$

93. $-x^2$ for $x = -8$

94. $-x^3$ for $x = 5$

95. $-x^2 - y^2$ for $x = -3$ and $y = -4$

96. $-x^2 + y^2$ for $x = -2$ and $y = 6$

97. $-a^2 - 3b^3$ for $a = -6$ and $b = -1$

98. $-a^3 + 3b^2$ for $a = -3$ and $b = -5$

99. $y^2 - 3xy$ for $x = .4$ and $y = -.3$

100. $x^2 + 5xy$ for $x = -.2$ and $y = -.6$

101. Use a calculator to check your answers for Problems 1–34.

THOUGHTS INTO WORDS

102. Your friend keeps getting an answer of 16 when simplifying -2^4. What mistake is he making and how would you help him?

103. Explain how you would simplify $\dfrac{12x^2y}{18xy}$.

2.5 Translating from English to Algebra

In order to use the tools of algebra for solving problems, we must be able to translate back and forth between the English language and the language of algebra. In this section we want to translate algebraic expressions to English phrases (word phrases) and English phrases to algebraic expressions. Let's begin by translating from algebraic expressions to word phrases.

Algebraic expression	Word phrase
$x + y$	The sum of x and y
$x - y$	The difference of x and y
$y - x$	The difference of y and x
xy	The product of x and y
$\dfrac{x}{y}$	The quotient of x and y
$3x$	The product of 3 and x
$x^2 + y^2$	The sum of x squared and y squared
$2xy$	The product of 2, x, and y
$2(x + y)$	Two times the quantity x plus y
$x - 3$	Three less than x

Now let's consider the reverse process, translating from word phrases to algebraic expressions. Part of the difficulty in translating from English to algebra is the fact that different word phrases translate to the same algebraic expression. So we need to become familiar with *different ways of saying the same thing*, especially when referring to the four fundamental operations. The following examples should help to acquaint you with some of the phrases used in the basic operations.

$$\left.\begin{array}{l}\text{The sum of } x \text{ and 4} \\ x \text{ plus 4} \\ x \text{ increased by 4} \\ \text{4 added to } x \\ \text{4 more than } x\end{array}\right\} \rightarrow x + 4$$

$$\left.\begin{array}{l}\text{The difference of } n \text{ and 5} \\ n \text{ minus 5} \\ n \text{ less 5} \\ n \text{ decreased by 5} \\ \text{5 less than } n \\ \text{Subtract 5 from } n\end{array}\right\} \rightarrow n - 5$$

$$\left.\begin{array}{l}\text{The product of 4 and } y \\ \text{4 times } y \\ y \text{ multiplied by 4}\end{array}\right\} \rightarrow 4y$$

$$\begin{bmatrix} \text{The quotient of } n \text{ and } 6 \\ n \text{ divided by } 6 \\ 6 \text{ divided into } n \end{bmatrix} \rightarrow \frac{n}{6}$$

Often a word phrase indicates more than one operation. Furthermore, the standard vocabulary of sum, difference, product, and quotient may be replaced by other terminology. Study the following translations very carefully.

Word phrase	Algebraic expression
The sum of two times x and three times y	$2x + 3y$
The sum of the squares of a and b	$a^2 + b^2$
Five times x divided by y	$\dfrac{5x}{y}$
Two more than the square of x	$x^2 + 2$
Three less than the cube of b	$b^3 - 3$
Five less than the product of x and y	$xy - 5$
Nine minus the product of x and y	$9 - xy$
Four times the sum of x and 2	$4(x + 2)$
Six times the quantity w minus 4	$6(w - 4)$

Suppose you are told that the sum of two numbers is 12 and one of the numbers is 8. What is the other number? The other number is $12 - 8$, which equals 4. Now suppose that you are told that the product of two numbers is 56 and one of the numbers is 7. What is the other number? The other number is $56 \div 7$, which equals 8. The following examples illustrate the use of these addition-subtraction and multiplication-division relationships in a more general setting.

Example 1

The sum of two numbers is 83 and one of the numbers is x. What is the other number?

Solution

Using the addition and subtraction relationship we can represent the other number by $83 - x$. ▲

Example 2

The difference of two numbers is 14. The smaller number is n. What is the larger number?

Solution

Since the smaller number plus the difference must equal the larger number, we can represent the larger number by $n + 14$. ▲

Example 3

The product of two numbers is 39 and one of the numbers is *y*. Represent the other number.

Solution

Using the multiplication and division relationship we can represent the other number by $\frac{39}{y}$.

▲

The English statement may not contain key words such as sum, difference, product, or quotient; instead, the statement may describe a physical situation—from this description you need to deduce the operations involved. We make some suggestions for handling such situations in the following examples.

Example 4

Arlene can type 70 words per minute. How many words can she type in *m* minutes?

Solution

In 10 minutes she would type 70(10) = 700 words. In 50 minutes she would type 70(50) = 3500 words. Thus, in *m* minutes she would type 70*m* words.

▲

Notice the use of some specific examples (70(10) = 700 and 70(50) = 3500) to help formulate the general expression. This technique of first formulating some specific examples and then generalizing can be very effective.

Example 5

Lynn has *n* nickels and *d* dimes. Express, in cents, this amount of money.

Solution

Three nickels and 8 dimes would be 5(3) + 10(8) = 95 cents. Thus, *n* nickels and *d* dimes would be 5*n* + 10*d* cents.

▲

Example 6

A train travels at the rate of *r* miles per hour. How far will it travel in 8 hours?

Solution

Suppose that a train travels at 50 miles per hour. Using the formula *distance equals rate times time*, it would travel 50 · 8 = 400 miles. Therefore, at *r* miles per hour, it would travel *r* · 8 miles. We usually write the expression *r* · 8 as 8*r*.

▲

FIGURE 2.4

Example 7

The cost of a 5-pound box of candy (see Figure 2.4) is d dollars. How much is the cost per pound for the candy?

Solution

The price per pound is figured by dividing the total cost by the number of pounds. Therefore, the price per pound is represented by $\frac{d}{5}$. ▲

The English statement to be translated to algebra may contain some geometric ideas. For example, suppose that we want to express in inches the length of a line segment that is f feet long. Since 1 foot = 12 inches, we can represent f feet by 12 times f, written as $12f$ inches.

Tables 2.1 and 2.2 list some of the basic relationships pertaining to linear measurements in the English and metric systems, respectively. (Additional listings of both systems are located in the inside front cover of this book.)

TABLE 2.1 **English system**

12 inches = 1 foot
3 feet = 36 inches = 1 yard
5280 feet = 1760 yards = 1 mile

TABLE 2.2 **Metric system**

1 kilometer = 1000 meters
1 hectometer = 100 meters
1 dekameter = 10 meters
1 decimeter = 0.1 meter
1 centimeter = 0.01 meter
1 millimeter = 0.001 meter

Example 8

The distance between two cities is k kilometers. Express this distance in meters.

Solution

Since 1 kilometer equals 1000 meters, we need to multiply k by 1000. Therefore, the distance in meters is represented by $1000k$. ▲

Example 9

The length of a line segment is i inches. Express that length in yards.

Solution

To change from inches to yards, we must divide by 36. Therefore, $\frac{i}{36}$ represents in yards the length of the line segment. ▲

Example 10

The width of a rectangle is w centimeters and the length is 5 centimeters less than twice the width. What is the length of the rectangle? What is the perimeter of the rectangle?

Solution

The length of the rectangle can be represented by $2w - 5$. Now we can sketch a rectangle and record the given information. (Figure 2.5) The perimeter of a rectangle is the sum of the lengths of the four sides. Therefore, the perimeter is given by $2w + 2(2w - 5)$, which can be written as $2w + 4w - 10$ and then simplified to $6w - 10$. ▲

FIGURE 2.5

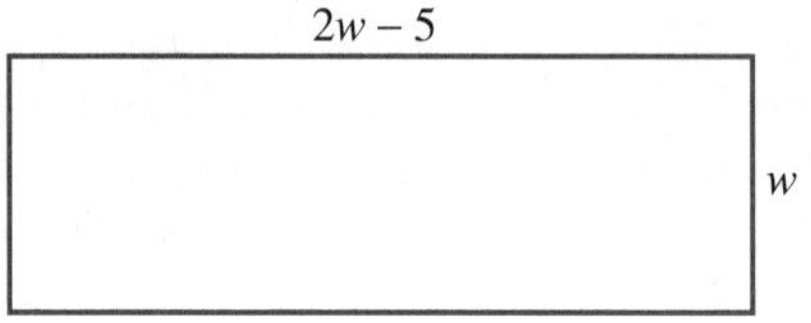

Problem Set 2.5

For Problems 1–12, write a word phrase for each of the algebraic expressions. For example, lw can be expressed as "the product of l and w."

1. $a - b$
2. $x + y$
3. $\frac{1}{3}Bh$
4. $\frac{1}{2}bh$
5. $2(l + w)$
6. πr^2
7. $\frac{A}{w}$
8. $\frac{C}{\pi}$
9. $\frac{a + b}{2}$
10. $\frac{a - b}{4}$
11. $3y + 2$
12. $3(x - y)$

For Problems 13–36, translate each word phrase into an algebraic expression. For example, "the sum of x and 14" translates into $x + 14$.

13. The sum of l and w
14. The difference of x and y
15. The product of a and b
16. The product of $\frac{1}{3}$, B, and h
17. The quotient of d and t
18. r divided into d
19. The product of l, w, and h
20. The product of π and the square of r
21. x subtracted from y
22. The difference "x subtract y"
23. Two larger than the product of x and y
24. Six plus the cube of x
25. Seven minus the square of y
26. The quantity, x minus 2, cubed
27. The quantity, x minus y, divided by four
28. Eight less than x

29. Ten less x

30. Nine times the quantity, n minus four

31. Ten times the quantity, n plus two

32. The sum of four times x and five times y

33. Seven subtracted from the product of x and y

34. Three times the sum n and 2

35. Twelve less than the product of x and y

36. Twelve less the product of x and y

For Problems 37–68, answer the question with an algebraic expression.

37. The sum of two numbers is 35 and one of the numbers is n. What is the other number?

38. The sum of two numbers is 100 and one of the numbers is x. What is the other number?

39. The difference of two numbers is 45 and the smaller number is n. What is the other number?

40. The product of two numbers is 25 and one of the numbers is x. What is the other number?

41. Janet is y years old. How old will she be in 10 years?

42. Hector is y years old. How old was he 5 years ago?

43. Debra is x years old and her mother is 3 years less than twice as old as Debra. How old is Debra's mother?

44. Jack is x years old and Dudley is 1 year more than three times as old as Jack. How old is Dudley?

45. Donna had d dimes and q quarters in her bank. How much money in cents does she have?

46. Andy has c cents that is all in dimes. How many dimes does he have?

47. A car travels d miles in t hours. What is the rate of the car?

48. If g gallons of gas cost d dollars, what is the price per gallon?

49. If p pounds of candy cost d dollars, what is the price per pound?

50. Sue can type x words per minute. How many words can she type in one hour?

51. Larry's annual salary is d dollars. What is his monthly salary?

52. Nancy's monthly salary is d dollars. What is her annual salary?

53. If n represents a whole number, what represents the next larger whole number?

54. If n represents an even number, what represents the next larger even number?

55. If n represents an odd number, what represents the next larger odd number?

56. Maria is y years old and her sister is twice as old. What represents the sum of their ages?

57. Willie is y years old and his father is 2 years less than twice Willie's age. What represents the sum of their ages?

58. Harriet has p pennies, n nickels, and d dimes. How much money in cents does she have?

59. The perimeter of a rectangle is y yards and f feet. What is the perimeter in inches?

60. The perimeter of a triangle is m meters and c centimeters. What is the perimeter in centimeters?

61. A rectangular plot of ground is f feet long. What is its length in yards?

62. The height of a telephone pole is f feet. What is the height in yards?

63. The width of a rectangle is w feet and its length is three times the width. What is the perimeter of the rectangle in feet?

64. The width of a rectangle is w feet and its length is 1 foot more than twice its width. What is the perimeter of the rectangle in feet?

65. The length of a rectangle is l inches and its width is 2 inches less than one-half of its length. What is the perimeter of the rectangle in inches?

66. The length of a rectangle is l inches and its width is 3 inches more than one-third of its length. What is the perimeter of the rectangle in inches?

67. The first side of a triangle is f feet long. The second side is 2 feet longer than the first side. The third side is twice as long as the second side. What is the perimeter of the triangle in inches?

68. The first side of a triangle is y yards long. The second side is 3 yards shorter than the first side. The third side is 3 times as long as the second side. What is the perimeter of the triangle in feet?

THOUGHTS INTO WORDS

69. What does the phrase "translating from English to algebra" mean to you?

70. Your friend is having trouble with Problems 61 and 62. For example, for Problem 61 she doesn't know if the answer should be $3f$ or $\dfrac{f}{3}$. What might you do to help her?

SUMMARY

(2.1) The property $\dfrac{a \cdot k}{b \cdot k} = \dfrac{a}{b}$ is used to express fractions in reduced form.

To **multiply** rational numbers in common fractional form we multiply numerators, multiply denominators, and express the result in reduced form.

To **divide** rational numbers in common fractional form, we multiply by the reciprocal of the divisor.

(2.2) **Addition** and **subtraction** of rational numbers in common fractional form are based on the following.

$$\frac{a}{b} + \frac{c}{b} = \frac{a+c}{b}, \qquad \text{Addition}$$

$$\frac{a}{b} - \frac{c}{b} = \frac{a-c}{b} \qquad \text{Subtraction}$$

To add or subtract fractions that do not have a common denominator, we use the fundamental principle of fractions, $\dfrac{a}{b} = \dfrac{a \cdot k}{b \cdot k}$, and obtain equivalent fractions that have a common denominator.

(2.3) To **add** or **subtract decimals**, we write the numbers in a column so that the decimal points are lined up, and then we add or subtract as we do with integers.

To **multiply decimals** we (1) multiply the numbers and ignore the decimal points, and then (2) insert the decimal point in the product so that the number of digits to the right of the decimal point in the product is equal to the sum of the numbers of digits to the right of the decimal point in each factor.

To **divide a decimal by a nonzero whole number** we (1) place the decimal point in the quotient directly above the decimal point in the dividend, and then (2) divide as with whole numbers, except that in the division process, we place zeros in the quotient immediately to the right of the decimal point (if necessary) to show the correct place value.

To **divide by a decimal** we change to an equivalent problem that has a whole number divisor.

(2.4) Expressions of the form b^n, where

$$b^n = b \cdot b \cdot b \cdots b, \qquad n \text{ factors of } b$$

are read as "b to the nth power"; b is the *base* and n is the *exponent*.

(2.5) To translate English phrases to algebraic expressions we must be familiar with the standard vocabulary of **sum**, **difference**, **product**, and **quotient** as well as other terms used to express the same ideas.

Chapter 2 Review Problem Set

For Problems 1–10, find the value of each of the following.

1. 2^6

2. $(-3)^3$

3. -4^2

4. $\left(\dfrac{3}{4}\right)^2$

5. $\left(\dfrac{1}{2} + \dfrac{2}{3}\right)^2$

6. $(.6)^3$

7. $(.12)^2$

8. $(.06)^2$

9. $\left(-\dfrac{2}{3}\right)^3$

10. $\left(-\dfrac{1}{2}\right)^4$

15. $\dfrac{5}{xy} - \dfrac{8}{x^2}$

16. $\left(\dfrac{7y}{8x}\right)\left(\dfrac{14x}{35}\right)$

17. $\left(\dfrac{6xy}{9y^2}\right) \div \left(\dfrac{15y}{18x^2}\right)$

18. $\left(\dfrac{-3x}{12y}\right)\left(\dfrac{8y}{-7x}\right)$

19. $\left(-\dfrac{4y}{3x}\right)\left(-\dfrac{3x}{4y}\right)$

20. $\left(\dfrac{6n}{7}\right)\left(\dfrac{9n}{8}\right)$

For Problems 11–20, perform the indicated operations and express answers in reduced form.

11. $\dfrac{3}{8} + \dfrac{5}{12}$

12. $\dfrac{9}{14} - \dfrac{3}{35}$

13. $\dfrac{2}{3} + \dfrac{-3}{5}$

14. $\dfrac{7}{x} + \dfrac{9}{2y}$

For Problems 21–30, simplify each of the following numerical expressions.

21. $\dfrac{1}{6} + \dfrac{2}{3} \cdot \dfrac{3}{4} - \dfrac{5}{6} \div \dfrac{8}{6}$

22. $\dfrac{3}{4} \cdot \dfrac{1}{2} - \dfrac{4}{3} \cdot \dfrac{3}{2}$

23. $\dfrac{7}{9} \cdot \dfrac{3}{5} + \dfrac{7}{9} \cdot \dfrac{2}{5}$

24. $\dfrac{4}{5} \div \dfrac{1}{5} \cdot \dfrac{2}{3} - \dfrac{1}{4}$

25. $\dfrac{2}{3} \cdot \dfrac{1}{4} \div \dfrac{1}{2} + \dfrac{2}{3} \cdot \dfrac{1}{4}$

26. $.48 + .72 - .35 - .18$

27. $.81 + (.6)(.4) - (.7)(.8)$

28. $1.28 \div .8 - .81 \div .9 + 1.7$

29. $(.3)^2 + (.4)^2 - (.6)^2$

30. $(1.76)(.8) + (1.76)(.2)$

For Problems 31–36, simplify each of the following algebraic expressions by combining similar terms. Express answers in reduced form when working with common fractions.

31. $\dfrac{3}{8}x^2 - \dfrac{2}{5}y^2 - \dfrac{2}{7}x^2 + \dfrac{3}{4}y^2$

32. $.24ab + .73bc - .82ab - .37bc$

33. $\dfrac{1}{2}x + \dfrac{3}{4}x - \dfrac{5}{6}x + \dfrac{1}{24}x$

34. $1.4a - 1.9b + .8a + 3.6b$

35. $\dfrac{2}{5}n + \dfrac{1}{3}n - \dfrac{5}{6}n$

36. $n - \dfrac{3}{4}n + 2n - \dfrac{1}{5}n$

For Problems 37–42, evaluate the following algebraic expressions for the given values of the variables.

37. $\dfrac{1}{4}x - \dfrac{2}{5}y$ for $x = \dfrac{2}{3}$ and $y = -\dfrac{5}{7}$

38. $a^3 + b^2$ for $a = -\dfrac{1}{2}$ and $b = \dfrac{1}{3}$

39. $2x^2 - 3y^2$ for $x = .6$ and $y = .7$

40. $.7w + .9z$ for $w = .4$ and $z = -.7$

41. $\dfrac{3}{5}x - \dfrac{1}{3}x + \dfrac{7}{15}x - \dfrac{2}{3}x$ for $x = \dfrac{15}{17}$

42. $\dfrac{1}{3}n + \dfrac{2}{7}n - n$ for $n = 21$

For Problems 43–50, answer each of the following questions with an algebraic expression.

43. The sum of two numbers is 72 and one of the number is n. What is the other number?

44. Joan has p pennies and d dimes. How much money in cents does she have?

45. Ellen types x words in an hour. What is her typing rate per minute?

46. Harry is y years old. His brother is 3 years less than twice as old as Harry. How old is Harry's brother?

47. Larry chose a number n. Cindy chose a number 3 more than 5 times the number chosen by Larry. What number did Cindy choose?

48. The height of a file cabinet is y yards and f feet. How tall is the file cabinet in inches?

49. The length of a rectangular room is m meters. How long in centimeters is the room?

50. Corinne has n nickels, d dimes, and q quarters. How much money in cents does she have?

For Problems 51–60, translate each word phrase into an algebraic expression.

51. Five less than n

52. Five less n

53. Ten times the quantity, x minus 2

54. Ten times x minus 2

55. x minus three

56. d divided by r

57. x squared plus nine

58. x plus nine, the quantity squared

59. The sum of the cubes of x and y

60. Four less than the product of x and y

CHAPTER 2 TEST

1. Find the value of $(-3)^4$ and also of -2^6.

2. Express $\dfrac{42}{54}$ in reduced form.

3. Simplify $\dfrac{18xy^2}{32y}$.

For Problems 4–7, simplify each numerical expression.

4. $5.7 - 3.8 + 4.6 - 9.1$

5. $.2(.4) - .6(.9) + .5(7)$

6. $-.4^2 + .3^2 - .7^2$

7. $4(.21) - 3(.17) - 6(.04)$

For Problems 8–11, perform the indicated operations and express answers in reduced form.

8. $\dfrac{5}{12} \div \dfrac{15}{8}$

9. $-\dfrac{2}{3} - \dfrac{1}{2}\left(\dfrac{3}{4}\right) + \dfrac{5}{6}$

10. $3\left(\dfrac{2}{5}\right) - 4\left(\dfrac{5}{6}\right) + 6\left(\dfrac{7}{8}\right)$

11. $4\left(\dfrac{1}{2}\right)^3 - 3\left(\dfrac{2}{3}\right)^2 + 9\left(\dfrac{1}{4}\right)^2$

For Problems 12–19, perform the indicated operations and express answers in reduced form.

12. $\dfrac{8x}{15y} \cdot \dfrac{9y^2}{6x}$

13. $\dfrac{6xy}{9} \div \dfrac{y}{3x}$

14. $\dfrac{4}{x} - \dfrac{5}{y^2}$

15. $\dfrac{3}{2x} + \dfrac{7}{6x}$

16. $\dfrac{5}{3y} + \dfrac{9}{7y^2}$

17. $\left(\dfrac{15a^2b}{12a}\right)\left(\dfrac{8ab}{9b}\right)$

18. $\dfrac{3}{x^2y} - \dfrac{4}{x} + \dfrac{6}{y}$

19. $\dfrac{3}{5x} - \dfrac{5}{4} + \dfrac{7}{10x}$

For Problems 20–23, evaluate each of the algebraic expressions for the given values of the variables.

20. $x^2 - xy + y^2$ for $x = \dfrac{1}{2}$ and $y = -\dfrac{2}{3}$

21. $.2x - .3y - xy$ for $x = .4$ and $y = .8$

22. $\dfrac{3}{4}x - \dfrac{2}{3}y$ for $x = -\dfrac{1}{2}$ and $y = \dfrac{3}{5}$

23. $3x - 2y + xy$ for $x = .5$ and $y = -.9$

24. David has n nickels, d dimes, and q quarters. How much money, in cents, does he have?

25. Hal chose a number n. Sheila chose a number 3 less than 4 times the number that Hal chose. Express the number that Sheila chose in terms of n.

Cumulative Review Problem Set

For Problems 1–12, simplify each of the numerical expressions.

1. $16 - 18 - 14 + 21 - 14 + 19$
2. $7(-6) - 8(-6) + 4(-9)$
3. $6 - [3 - (10 - 12)]$
4. $-9 - 2[4 - (-10 + 6)] - 1$
5. $\dfrac{-7(-4) - 5(-6)}{-2}$
6. $\dfrac{5(-3) + (-4)(6) - 3(4)}{-3}$
7. $\dfrac{3}{4} + \dfrac{1}{3} \div \dfrac{4}{3} - \dfrac{1}{2}$
8. $\left(\dfrac{2}{3}\right)\left(-\dfrac{3}{4}\right) - \left(\dfrac{5}{6}\right)\left(\dfrac{4}{5}\right)$
9. $\left(\dfrac{1}{2} - \dfrac{2}{3}\right)^2$
10. -4^3
11. $\left(-\dfrac{1}{2}\right)^3 - \left(-\dfrac{3}{4}\right)^2$
12. $(.2)^2 - (.3)^3 + (.4)^2$

For Problems 13–20, evaluate each algebraic expression for the given values of the variables.

13. $3xy - 2x - 4y$ for $x = -6$ and $y = 7$
14. $-4x^2y - 2xy^2 + xy$ for $x = -2$ and $y = -4$
15. $\dfrac{5x - 2y}{3x}$ for $x = \dfrac{1}{2}$ and $y = -\dfrac{1}{3}$
16. $.2x - .3y + 2xy$ for $x = .1$ and $y = .3$
17. $-7x + 4y + 6x - 9y + x - y$
   for $x = -.2$ and $y = .4$
18. $\dfrac{2}{3}x - \dfrac{3}{5}y + \dfrac{3}{4}x - \dfrac{1}{2}y$ for $x = \dfrac{6}{5}$ and $y = -\dfrac{1}{4}$
19. $\dfrac{1}{5}n - \dfrac{1}{3}n + n - \dfrac{1}{6}n$ for $n = \dfrac{1}{5}$
20. $-ab + \dfrac{1}{5}a - \dfrac{2}{3}b$ for $a = -2$ and $b = \dfrac{3}{4}$

For Problems 21–24, express each of the numbers as a product of prime factors.

21. 54
22. 78
23. 91
24. 153

For Problems 25–28, find the greatest common factor of the given numbers.

25. 42 and 70
26. 63 and 81
27. 28, 36, and 52
28. 48, 66, and 78

For Problems 29–32, find the least common multiple of the given numbers.

29. 20 and 28
30. 40 and 100
31. 12, 18, and 27
32. 16, 20, and 80

For Problems 33–38, simplify each algebraic expression by combining similar terms.

33. $\dfrac{2}{3}x - \dfrac{1}{4}y - \dfrac{3}{4}x - \dfrac{2}{3}y$
34. $-n - \dfrac{1}{2}n + \dfrac{3}{5}n + \dfrac{5}{6}n$
35. $3.2a - 1.4b - 6.2a + 3.3b$
36. $-(n - 1) + 2(n - 2) - 3(n - 3)$
37. $-x + 4(x - 1) - 3(x + 2) - (x + 5)$
38. $2a - 5(a + 3) - 2(a - 1) - 4a$

For Problems 39–46, perform the indicated operations and express answers in reduced form.

39. $\dfrac{5}{12} - \dfrac{3}{16}$
40. $\dfrac{3}{4} - \dfrac{5}{6} - \dfrac{7}{9}$
41. $\dfrac{5}{xy} - \dfrac{2}{x} + \dfrac{3}{y}$
42. $-\dfrac{7}{x^2} + \dfrac{9}{xy}$
43. $\left(\dfrac{7x}{9y}\right)\left(\dfrac{12y}{14}\right)$

44. $\left(-\dfrac{5a}{7b^2}\right)\left(-\dfrac{8ab}{15}\right)$

45. $\left(\dfrac{6x^2y}{11}\right) \div \left(\dfrac{9y^2}{22}\right)$

46. $\left(-\dfrac{9a}{8b}\right) \div \left(\dfrac{12a}{18b}\right)$

For Problems 47–50, answer each question with an algebraic expression.

47. Hector has p pennies, n nickels, and d dimes. How much money in cents does he have?

48. Ginny chose a number n. Penny chose a number 5 less than 4 times the number chosen by Ginny. What number did Penny choose?

49. The height of a flagpole is y yards, f feet, and i inches. How tall is the flagpole in inches?

50. A rectangular room is x meters by y meters. What is its perimeter in centimeters?

CHAPTER

First-Degree Equations and Inequalities of One Variable

Richard earned $305 last week for working 40 hours. Included in the $305 was a bonus of $25. How much was Richard's hourly pay? If we let h repesent Richard's hourly pay, then the equation $40h + 25 = 305$ can be used to determine that Richard makes $7 per hour.

Throughout this book we will develop some new skills; use the skills to help solve equations and inequalities; and use the equations and inequalities to solve applied problems. In this chapter we want to use the skills we developed in the first two chapters to help solve equations and inequalities, and then to begin our work with applied problems.

3.1 Solving First-Degree Equations

The following are examples of **numerical statements**.

$$3 + 4 = 7, \qquad 5 - 2 = 3, \qquad 7 + 1 = 12$$

The first two are true statements and the third one is a false statement.

When you use x as a variable, the statements

$$x + 3 = 4, \qquad 2x - 1 = 7, \qquad \text{and} \qquad x^2 = 4$$

are called **algebraic equations** in x. We call a number a a **solution** or **root** of an equation if a true numerical statement is formed when a is substituted for x. (We also say that a satisfies the equation.) For example, 1 is a solution of $x + 3 = 4$ because substituting 1 for x produces the true numerical statement $1 + 3 = 4$. We call the set of all solutions of an equation its **solution set**. Thus, the solution set of $x + 3 = 4$ is $\{1\}$. Likewise, the solution set of $2x - 1 = 7$ is $\{4\}$ and the solution set of $x^2 = 4$ is $\{-2, 2\}$. **Solving an equation** refers to the process of determining the solution set. Remember that a set that consists of no elements is called the **empty** or **null set** and is denoted by $\varnothing$. Thus, we say that the solution set of $x = x + 1$ is $\varnothing$; that is, there are no real numbers that satisfy $x = x + 1$.

In this chapter we will consider techniques for solving **first-degree equations of one variable**. This means that the equations contain only one variable and this variable has an exponent of one. The following are examples of first degree equations of one variable.

$$3x + 4 = 7, \qquad 8w + 7 = 5w - 4,$$

$$\frac{1}{2}y + 2 = 9, \qquad 7x + 2x - 1 = 4x - 1$$

Equivalent equations are equations that have the same solution set. For example,

$$5x - 4 = 3x + 8,$$
$$2x = 12,$$
$$x = 6,$$

are all equivalent equations (which can be verified by showing that 6 is the solution for all three equations).

As we work with equations we can use the following properties of equality.

PROPERTY 3.1   **Properties of Equality**

For all real numbers, a, b, and c,

1. $a = a$;   Reflexive property
2. If $a = b$ then $b = a$;   Symmetric property
3. If $a = b$ and $b = c$, then $a = c$;   Transitive property
4. If $a = b$, then a may be replaced by b, or b may be replaced by a, in any statement without changing the meaning of the statement. Substitution property

The general procedure for solving an equation is to continue replacing the given equation with equivalent but simpler equations until we obtain an equation of the form **variable = constant** or **constant = variable**. Thus, in the preceding example, $5x - 4 = 3x + 8$ was simplified to $2x = 12$, which was further simplified to $x = 6$, from which the solution of 6 is obvious. The exact procedure for simplifying equations becomes our next concern.

Two properties of equality play an important role in the process of solving equations. The first of these is the **addition-subtraction property of equality**, which we state as follows.

PROPERTY 3.2   **Addition-Subtraction Property of Equality**

For all real numbers a, b, and c,

1. $a = b$ if and only if $a + c = b + c$;
2. $a = b$ if and only if $a - c = b - c$.

Property 3.2 states that *any number can be added to or subtracted from both sides of an equation and an equivalent equation is produced.* Consider the use of this property in the next four examples.

Example 1   Solve $x - 8 = 3$.

Solution

$$x - 8 = 3$$
$$x - 8 + 8 = 3 + 8 \qquad \text{Add 8 to both sides.}$$
$$x = 11$$

The solution set is $\{11\}$.   ▲

REMARK   It is true that a simple equation such as Example 1 can be solved *by inspection*. That is to say, we could think "some number minus 8 produces 3." Obviously, the number is 11. However, as the equations become more complex, the technique of solving by inspection becomes ineffective. So it is necessary to develop more formal techniques for solving equations. Therefore, we will begin developing such techniques even with very simple types of equations.   ▲

Example 2

Solve $x + 14 = -8$

Solution

$$x + 14 = -8$$
$$x + 14 - 14 = -8 - 14 \qquad \text{Subtract 14 from both sides.}$$
$$x = -22$$

The solution set is $\{-22\}$.   ▲

Example 3

Solve $n - \dfrac{1}{3} = \dfrac{1}{4}$.

Solution

$$n - \frac{1}{3} = \frac{1}{4}$$
$$n - \frac{1}{3} + \frac{1}{3} = \frac{1}{4} + \frac{1}{3} \qquad \text{Add } \frac{1}{3} \text{ to both sides.}$$
$$n = \frac{3}{12} + \frac{4}{12}$$
$$n = \frac{7}{12}$$

The solution set is $\left\{\dfrac{7}{12}\right\}$.   ▲

Example 4

Solve $.72 = y + .35$.

Solution

$$.72 = y + .35$$
$$.72 - .35 = y + .35 - .35 \qquad \text{Subtract .35 from both sides.}$$
$$.37 = y$$

The solution set is $\{.37\}$.   ▲

Note in Example 4 the final equation is $.37 = y$ instead of $y = .37$. Technically, the **symmetric property of equality** (if $a = b$, then $b = a$) would permit us to change from $.37 = y$ to $y = .37$, but such a change is not necessary to determine that the solution is .37. You should also realize that the symmetric property could be applied to the original equation. Thus, $.72 = y + .35$ becomes $y + .35 = .72$ and to subtract .35 from both sides would produce $y = .37$.

One other comment that pertains to Property 3.2 should be made at this time. Because subtracting a number is equivalent to adding its opposite,

Property 3.2 could be stated only in terms of addition. Thus, to solve an equation such as Example 4 we could add $-.35$ to both sides rather than subtract $.35$ from both sides.

The other important property for solving equations is the **multiplication-division property of equality**.

PROPERTY 3.3

Multiplication-Division Property of Equality

For all real numbers, a, b, and c, where $c \neq 0$,

1. $a = b$ if and only if $ac = bc$;

2. $a = b$ if and only if $\dfrac{a}{c} = \dfrac{b}{c}$.

Property 3.3 states that *an equivalent equation is obtained whenever both sides of a given equation are multiplied or divided by the same nonzero real number.* The following examples illustrate the use of this property.

Example 5

Solve $\dfrac{3}{4}x = 6$.

Solution

$$\frac{3}{4}x = 6$$

$$\frac{4}{3}\left(\frac{3}{4}x\right) = \frac{4}{3}(6) \qquad \text{Multiply both sides by } \frac{4}{3} \text{ since } \left(\frac{4}{3}\right)\left(\frac{3}{4}\right) = 1.$$

$$x = 8$$

The solution set is $\{8\}$.

Example 6

Solve $5x = 27$.

Solution

$$5x = 27$$

$$\frac{5x}{5} = \frac{27}{5} \qquad \text{Divide both sides by 5.}$$

$$x = \frac{27}{5} \qquad \frac{27}{5} \text{ could be expressed as } 5\frac{2}{5} \text{ or 5.4.}$$

The solution set is $\left\{\dfrac{27}{5}\right\}$.

Example 7

Solve $-\dfrac{2}{3}p = \dfrac{1}{2}$.

Solution

$$-\frac{2}{3}p = \frac{1}{2}$$

$$-\frac{3}{2}\left(-\frac{2}{3}p\right) = \left(-\frac{3}{2}\right)\left(\frac{1}{2}\right) \qquad \text{Multiply both sides by } -\frac{3}{2}$$
$$\text{since } \left(-\frac{3}{2}\right)\left(-\frac{2}{3}\right) = 1.$$

$$p = -\frac{3}{4}$$

The solution set is $\left\{-\dfrac{3}{4}\right\}$.   ▲

Example 8

Solve $26 = -6x$.

Solution

$$26 = -6x$$

$$\frac{26}{-6} = \frac{-6x}{-6} \qquad \text{Divide both sides by } -6.$$

$$-\frac{26}{6} = x \qquad\qquad \frac{26}{-6} = -\frac{26}{6}$$

$$-\frac{13}{3} = x \qquad\qquad \text{Don't forget to reduce!}$$

The solution set is $\left\{-\dfrac{13}{3}\right\}$.   ▲

Look back at Examples 5–8 and you will notice that we divided both sides of the equation by the coefficient of the variable whenever the coefficient was an integer; otherwise we used the multiplication part of Property 3.3. Technically, because dividing by a number is equivalent to multiplying by its reciprocal, Property 3.3 could be stated only in terms of multiplication. Thus, to solve an equation such as $5x = 27$, we could multiply both sides by $\dfrac{1}{5}$ instead of dividing both sides by 5.

Example 9

Solve $.2n = 15$.

Solution

$$.2n = 15$$

$$\frac{.2n}{.2} = \frac{15}{.2} \qquad \text{Divide both sides by } .2.$$

$$n = 75$$

The solution set is $\{75\}$.   ▲

Problem Set 3.1

Use the properties of equality to help solve each of the following equations.

1. $x + 9 = 17$

2. $x + 7 = 21$

3. $x + 11 = 5$

4. $x + 13 = 2$

5. $-7 = x + 2$

6. $-12 = x + 4$

7. $8 = n + 14$

8. $6 = n + 19$

9. $21 + y = 34$

10. $17 + y = 26$

11. $x - 17 = 31$

12. $x - 22 = 14$

13. $14 = x - 9$

14. $17 = x - 28$

15. $-26 = n - 19$

16. $-34 = n - 15$

17. $y - \dfrac{2}{3} = \dfrac{3}{4}$

18. $y - \dfrac{2}{5} = \dfrac{1}{6}$

19. $x + \dfrac{3}{5} = \dfrac{1}{3}$

20. $x + \dfrac{5}{8} = \dfrac{2}{5}$

21. $b + .19 = .46$

22. $b + .27 = .74$

23. $n - 1.7 = -5.2$

24. $n - 3.6 = -7.3$

25. $15 - x = 32$

26. $13 - x = 47$

27. $-14 - n = 21$

28. $-9 - n = 61$

29. $7x = -56$

30. $9x = -108$

31. $-6x = 102$

32. $-5x = 90$

33. $5x = 37$

34. $7x = 62$

35. $-18 = 6n$

36. $-52 = 13n$

37. $-26 = -4n$

38. $-56 = -6n$

39. $\dfrac{t}{9} = 16$

40. $\dfrac{t}{12} = 8$

41. $\dfrac{n}{-8} = -3$

42. $\dfrac{n}{-9} = -5$

43. $-x = 15$

44. $-x = -17$

45. $\dfrac{3}{4}x = 18$

46. $\dfrac{2}{3}x = 32$

47. $-\dfrac{2}{5}n = 14$

48. $-\dfrac{3}{8}n = 33$

49. $\dfrac{2}{3}n = \dfrac{1}{5}$

50. $\dfrac{3}{4}n = \dfrac{1}{8}$

51. $\dfrac{5}{6}n = -\dfrac{3}{4}$

52. $\dfrac{6}{7}n = -\dfrac{3}{8}$

53. $\dfrac{3x}{10} = \dfrac{3}{20}$

54. $\dfrac{5x}{12} = \dfrac{5}{36}$

55. $\dfrac{-y}{2} = \dfrac{1}{6}$

56. $\dfrac{-y}{4} = \dfrac{1}{9}$

57. $-\dfrac{4}{3}x = -\dfrac{9}{8}$

58. $-\dfrac{6}{5}x = -\dfrac{10}{14}$

59. $-\dfrac{5}{12} = \dfrac{7}{6}x$

60. $-\dfrac{7}{24} = \dfrac{3}{8}x$

61. $-\dfrac{5}{7}x = 1$

62. $-\dfrac{11}{12}x = -1$

63. $-4n = \dfrac{1}{3}$

64. $-6n = \dfrac{3}{4}$

65. $-8n = \dfrac{6}{5}$

66. $-12n = \dfrac{8}{3}$

67. $1.2x = .36$

68. $2.5x = 17.5$

69. $30.6 = 3.4n$

70. $2.1 = 4.2n$

71. $-3.4x = 17$

72. $-4.2x = 50.4$

THOUGHTS INTO WORDS

73. Describe the difference between a numerical statement and an algebraic equation.

74. Are the equations $6 = 3x + 1$ and $1 + 3x = 6$ equivalent equations? Defend your answer.

3.2 Equations and Problem Solving

We often need more than one property of equality to help find the solution of an equation. Consider the following examples.

Example 1   Solve $3x + 1 = 7$.

Solution

$$3x + 1 = 7$$
$$3x + 1 - 1 = 7 - 1 \qquad \text{Subtract 1 from both sides.}$$
$$3x = 6$$
$$\frac{3x}{3} = \frac{6}{3} \qquad \text{Divide both sides by 3.}$$
$$x = 2$$

The potential solution can be *checked* by substituting it into the original equation to see if a true numerical statement is obtained.

 Check

$$3x + 1 = 7$$
$$3(2) + 1 \overset{?}{=} 7$$
$$6 + 1 \overset{?}{=} 7$$
$$7 = 7$$

Now we know that the solution set is $\{2\}$.   ▲

Example 2   Solve $5x - 6 = 14$.

Solution

$$5x - 6 = 14$$
$$5x - 6 + 6 = 14 + 6 \qquad \text{Add 6 to both sides.}$$
$$5x = 20$$
$$\frac{5x}{5} = \frac{20}{5} \qquad \text{Divide both sides by 5.}$$
$$x = 4$$

 Check

$$5x - 6 = 14$$
$$5(4) - 6 \overset{?}{=} 14$$
$$20 - 6 \overset{?}{=} 14$$
$$14 = 14$$

The solution set is $\{4\}$.   ▲

Example 3

Solve $4 - 3a = 22$.

Solution

$$4 - 3a = 22$$
$$4 - 3a - 4 = 22 - 4 \qquad \text{Subtract 4 from both sides.}$$
$$-3a = 18$$
$$\frac{-3a}{-3} = \frac{18}{-3} \qquad \text{Divide both sides by } -3.$$
$$a = -6$$

✔ **Check**

$$4 - 3a = 22$$
$$4 - 3(-6) \stackrel{?}{=} 22$$
$$4 + 18 \stackrel{?}{=} 22$$
$$22 = 22$$

The solution set is $\{-6\}$. ▲

Notice that in Examples 1, 2, and 3 we used the addition-subtraction property first and then we used the multiplication-division property. In general, this sequence of steps provides the easiest format for solving such equations. Perhaps you should convince yourself of that fact by doing Example 1 again, but this time use the multiplication-division property first and then the addition-subtraction property.

Example 4

Solve $19 = 2n + 4$.

Solution

$$19 = 2n + 4$$
$$19 - 4 = 2n + 4 - 4 \qquad \text{Subtract 4 from both sides.}$$
$$15 = 2n$$
$$\frac{15}{2} = \frac{2n}{2} \qquad \text{Divide both sides by 2.}$$
$$\frac{15}{2} = n$$

✔ **Check**

$$19 = 2n + 4$$
$$19 \stackrel{?}{=} 2\left(\frac{15}{2}\right) + 4$$
$$19 \stackrel{?}{=} 15 + 4$$
$$19 = 19$$

The solution set is $\left\{\frac{15}{2}\right\}$. ▲

Word Problems

In the last section of Chapter 2 we translated English phrases into algebraic expressions. We are now ready to extend that idea to the translation of English *sentences* into algebraic *equations*. Such translations allow us to use the concepts of algebra to solve word problems. Let's consider some examples.

Problem 1

A certain number added to 17 yields a sum of 29. What is the number?

Solution

Let n represent the number to be found. The sentence "A certain number added to 17 yields a sum of 29" translates to the algebraic equation $17 + n = 29$. To solve this equation, we obtain

$$17 + n = 29$$
$$17 + n - 17 = 29 - 17$$
$$n = 12.$$

The solution is 12, which is the number we asked for in the problem.   ▲

We often refer to the statement "let n represent the number to be found" as **declaring the variable**. We need to choose a letter to use as a variable and indicate what it represents for a specific problem—which may seem like an insignificant idea, but as the problems become more complex, the process of declaring the variable becomes even more important. We could solve a problem such as Problem 1 without setting up an algebraic equation; however, as problems increase in difficulty, the translation from English to an algebraic equation becomes a key issue. Therefore, even with these relatively simple problems we need to concentrate on the translation process.

Problem 2

Six years ago Bill was 13 years old. How old is he now?

Solution

Let y represent Bill's age now; therefore, $y - 6$ represents his age six years ago. Thus,

$$y - 6 = 13$$
$$y - 6 + 6 = 13 + 6$$
$$y = 19.$$

Bill is presently 19 years old.   ▲

Problem 3

Betty worked 8 hours Saturday and earned $32. How much did she earn per hour?

Solution A

Let x represent the amount Betty earned per hour. The number of hours worked times the wage per hour yields the total earnings. Thus,

$$8x = 32$$
$$\frac{8x}{8} = \frac{32}{8}$$
$$x = 4.$$

Betty earned $4 per hour.

Solution B

Let y represent the amount Betty earned per hour. The wage per hour equals the total wage divided by the number of hours. Thus,

$$y = \frac{32}{8}$$
$$y = 4.$$

Betty earned $4 per hour. ▲

Sometimes we can use more than one equation to solve a problem. In Solution A we set up the equation in terms of multiplication, whereas in Solution B we were thinking in terms of division.

Problem 4

If 2 is subtracted from five-times-a-certain-number, the result is 28. Find the number.

Solution

Let n represent the number to be found. Translating the first sentence in the problem into an algebraic equation, we obtain

$$5n - 2 = 28.$$

Solving this equation yields

$$5n - 2 + 2 = 28 + 2$$
$$5n = 30$$
$$\frac{5n}{5} = \frac{30}{5}$$
$$n = 6.$$

The number to be found is 6. ▲

Problem 5

A plumbing repair bill was $95. This bill included $17 for parts and an amount for 3 hours of labor. Find the hourly rate charged for labor.

Solution

Let l represent the hourly charge for labor; then $3l$ represents the total cost for labor. Thus, the cost for labor plus the cost for parts is the total bill of $95; so we can proceed as follows.

Cost for labor + Cost for parts = $95

$$3l \quad + \quad 17 \quad = 95$$

Solving this equation we obtain

$$3l = 78$$

$$\frac{3l}{3} = \frac{78}{3}$$

$$l = 26.$$

The charge for labor was $26 per hour. ▲

Problem Set 3.2

For Problems 1–40, solve each equation.

1. $2x + 5 = 13$ **2.** $3x + 4 = 19$

3. $5x + 2 = 32$ **4.** $7x + 3 = 24$

5. $3x - 1 = 23$ **6.** $2x - 5 = 21$

7. $4n - 3 = 41$ **8.** $5n - 6 = 19$

9. $6y - 1 = 16$ **10.** $4y - 3 = 14$

11. $2x + 3 = 22$ **12.** $3x + 1 = 21$

13. $10 = 3t - 8$ **14.** $17 = 2t + 5$

15. $5x + 14 = 9$ **16.** $4x + 17 = 9$

17. $18 - n = 23$ **18.** $17 - n = 29$

19. $-3x + 2 = 20$

20. $-6x + 1 = 43$

21. $7 + 4x = 29$ **22.** $9 + 6x = 23$

23. $16 = -2 - 9a$

24. $18 = -10 - 7a$

25. $-7x + 3 = -7$

26. $-9x + 5 = -18$

27. $17 - 2x = -19$

28. $18 - 3x = -24$

29. $-16 - 4x = 9$

30. $-14 - 6x = 7$

31. $-12t + 4 = 88$

32. $-16t + 3 = 67$

33. $14y + 15 = -33$

34. $12y + 13 = -15$

35. $32 - 16n = -8$

36. $-41 = 12n - 19$

37. $17x - 41 = -37$

38. $19y - 53 = -47$

39. $29 = -7 - 15x$

40. $49 = -5 - 14x$

For each of the following problems, (a) choose a variable and indicate what it represents in the problem, (b) set up an equation that represents the situation described, and (c) solve the equation.

41. Twelve added to a certain number is 21. What is the number?

42. A certain number added to 14 is 25. Find the number.

43. Nine subtracted from a certain number is 13. Find the number.

44. A certain number subtracted from 32 is 15. What is the number?

45. Suppose that two items cost $43. If one of the items costs $25, what is the cost of the other item?

46. Eight years ago Rosa was 22 years old. Find Rosa's present age.

47. Six years from now, Nora will be 41 years old. What is her present age?

48. The Chis bought 8 pizzas for a total of $50. What was the price per pizza?

49. Chad worked 6 hours Saturday for a total of $39. How much per hour did he earn?

50. Jill worked 8 hours Saturday at $5.50 per hour. How much did she earn?

51. If six is added to three times a certain number, the result is 24. Find the number.

52. If two is subtracted from five times a certain number, the result is 38. Find the number.

53. Nineteen is four larger than three times a certain number. Find the number.

54. If nine times a certain number is subtracted from seven, the result is 52. Find the number.

55. Forty-nine is equal to six less than five times a certain number. Find the number.

56. Seventy-one is equal to two more than three times a certain number. Find the number.

57. If one is subtracted from six times a certain number, the result is 47. Find the number.

58. Five less than four times a number equals 31. Find the number.

59. If eight times a certain number is subtracted from 27, the result is 3. Find the number.

60. Twenty is 22 less than six times a certain number. Find the number.

61. The price of a ring is $550. (see Figure 3.1.) This price represents $50 less than twice the cost of the ring. Find the cost of the ring.

FIGURE 3.1

62. Todd is following a 1750 calorie-per-day diet plan. This plan permits 650 calories less than twice the number of calories permitted by Lerae's diet plan. How many calories are permitted by Lerae's plan?

63. The length of a rectangular shaped floor is 18 meters (see Figure 3.2). This length represents 2 meters less than five times the width of the floor. Find the width of the floor.

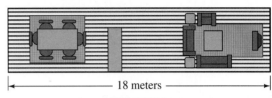

|← —————— 18 meters —————— →|

FIGURE 3.2

64. An executive is earning $45,000 per year. This represents $15,000 less than twice her salary four years ago. Find her salary four years ago.

65. The LM car dealership sold 32 cars during December of 1984. This number represents 4 more than twice the number of cars they sold during December of 1983. How many cars did they sell during December of 1983?

66. A car repair bill was $119, which included $42 for parts and $22 for each hour of labor. Find the number of hours of labor.

67. An electric golf cart repair bill was $156. This included $36 for parts and $24 per hour for labor. How many hours of labor were involved?

68. Richard earned $305 last week for working 40 hours. Included in the $305 was a bonus of $25. How much was Richard's hourly pay?

THOUGHTS INTO WORDS

69. Give a step-by-step description of how you would solve the equation $17 = -3x + 2$.

70. What does the phrase "declare a variable" mean when solving a word problem?

71. Suppose that you are helping a friend with his homework and he solves the equation $19 = 14 - x$ as follows.

$$19 = 14 - x$$
$$19 + x = 14 - x + x$$
$$19 + x = 14$$
$$19 + x - 19 = 14 - 19$$
$$x = -5$$

The solution set is $\{-5\}$.

Does he have a correct solution set? What would you say to him about his method of solving the equation?

72. Make up an equation whose solution set is the null set and explain why.

73. Make up an equation whose solution set is the set of all real numbers and explain why.

▼3.3 More on Solving Equations and Problem Solving

As equations become more complex, we need additional steps to solve them. So we need to carefully organize our work to minimize the chances for error. Let's begin this section with some suggestions for solving equations and then we will illustrate a *solution format* that is effective.

We can summarize the process of solving first-degree equations of one variable as follows.

STEP 1 Simplify both sides of the equation as much as possible.

STEP 2 Use the addition or subtraction property of equality to isolate a term that contains the variable on one side and a constant on the other side of the equation.

STEP 3 Use the multiplication or division property of equality to make the coefficient of the variable one.

The following examples illustrate this step-by-step process for solving equations. Study them carefully and be sure that you understand each step taken in the solving process.

Example 1

Solve $5y - 4 + 3y = 12$.

Solution

$$5y - 4 + 3y = 12$$
$$8y - 4 = 12 \qquad \text{Combine similar terms on the left side.}$$
$$8y - 4 + 4 = 12 + 4 \qquad \text{Add 4 to both sides.}$$
$$8y = 16$$
$$\frac{8y}{8} = \frac{16}{8} \qquad \text{Divide both sides by 8.}$$
$$y = 2$$

The solution set is $\{2\}$. You can do the check alone now!   ▲

Example 2

Solve $7x - 2 = 3x + 9$.

Solution

Notice that both sides of the equation are in simplified form; thus, we can begin by applying the subtraction property of equality.

$$7x - 2 = 3x + 9$$
$$7x - 2 - 3x = 3x + 9 - 3x \qquad \text{Subtract } 3x \text{ from both sides.}$$
$$4x - 2 = 9$$
$$4x - 2 + 2 = 9 + 2 \qquad \text{Add 2 to both sides.}$$
$$4x = 11$$
$$\frac{4x}{4} = \frac{11}{4} \qquad \text{Divide both sides by 4.}$$
$$x = \frac{11}{4}$$

The solution set is $\left\{\dfrac{11}{4}\right\}$. ▲

Example 3

Solve $5n + 12 = 9n - 16$.

Solution

$$5n + 12 = 9n - 16$$
$$5n + 12 - 9n = 9n - 16 - 9n \qquad \text{Subtract } 9n \text{ from both sides.}$$
$$-4n + 12 = -16$$
$$-4n + 12 - 12 = -16 - 12 \qquad \text{Subtract 12 from both sides.}$$
$$-4n = -28$$
$$\frac{-4n}{4} = \frac{-28}{-4} \qquad \text{Divide both sides by } -4.$$
$$n = 7$$

The solution set is $\{7\}$. ▲

Word Problems

As we expand our skills for solving equations, we also expand our capabilities for solving verbal problems. There is no one definite procedure that will insure success at solving word problems, but the following suggestions can be helpful.

Suggestions for Solving Word Problems

1. Read the problem carefully and make sure that you understand the meanings of all the words. Be especially alert for any technical terms in the statement of the problem.

2. Read the problem a second time (perhaps even a third time) to get an overview of the situation being described and to determine the known facts as well as what is to be found.

3. Sketch any figure, diagram, or chart that might be helpful in analyzing the problem.

4. Choose a meaningful variable to represent an unknown quantity in the problem (perhaps t, if time is an unknown quantity); represent any other unknowns in terms of that variable.

5. Look for a **guideline** that you can use to set up an equation. A guideline might be a formula such as "distance equals rate times time" or a statement of a relationship such as "the sum of the two numbers is 28." A guideline may also be indicated by a figure or diagram that you sketch for a particular problem.

6. Form an equation containing the variable that translates the conditions of the guideline from English to algebra.

7. Solve the equation and use the solution to determine all facts requested in the problem.

8. **Check** all answers back to the **original statement of the problem**.

If you decide not to check an answer, at least use the *reasonableness-of-answer* idea as a partial check. That is to say, ask yourself the question: Is this answer reasonable? For example, if the problem involves two investments that total $10,000, then an answer of $12,000 for one investment is certainly *not reasonable*.

Now let's consider some problems and use these suggestions.

Problem 1

Find two consecutive even numbers whose sum is 74.

Solution

To solve this problem we must know the meaning of the technical phrase "two consecutive even numbers." Two consecutive even numbers are two even numbers with one and only one whole number between them. For example, 2 and 4 are consecutive even numbers. Now we can proceed as follows. Let n represent the first even number; then $n + 2$ represents the next even number. Since their sum is 74, we can set up and solve the following equation.

$$n + n + 2 = 74$$
$$2n + 2 = 74$$
$$2n + 2 - 2 = 74 - 2$$
$$2n = 72$$
$$\frac{2n}{2} = \frac{72}{2}$$
$$n = 36$$

If $n = 36$, then $n + 2 = 38$; thus the numbers are 36 and 38.

✔ *Check*   To check your answers for Problem 1, determine whether they satisfy the conditions stated in the original problem. Since 36 and 38 are two consecutive even numbers and $36 + 38 = 74$ (their sum is 74), we know that the answers are correct.   ▲

Suggestion 5 in our list of problem solving suggestions was to "look for a *guideline* that can be used to set up an equation." The guideline may not be explicitly stated in the problem but may be implied by the nature of the problem. Consider the following example.

Problem 2

Barry sells bicycles on a salary-plus-commission basis. He receives a monthly salary of $300 and a commission of $15 for each bicycle that he sells. How many bicycles must he sell in a month to have a total monthly salary of $750?

Solution

Let b represent the number of bicycles to be sold in a month. Then $15b$ represents his commission for those bicycles. The *guideline* "fixed salary plus commission equals total monthly salary" generates the following equation.

Fixed salary + Commission = Total monthly salary
$$\$300 \quad + \quad 15b \quad = \quad \$750$$

Solving this equation yields

$$300 + 15b - 300 = 750 - 300$$
$$15b = 450$$
$$\frac{15b}{15} = \frac{450}{15}$$
$$b = 30.$$

He must sell 30 bicycles per month (Does this number check?)   ▲

Geometric Problems

Sometimes the guideline for setting up an equation to solve a problem is based on a geometric relationship. There are several basic geometric relationships that pertain to angle measure. Let's state some of these relationships and then consider some problems.

1. **Two angles** for which the sum of their measure is 90° (the symbol ° indicates degrees) are called **complementary angles**.
2. **Two angles** for which the sum of their measure is 180° are called **supplementary angles**.
3. The sum of the measures of the three angles of a triangle is 180°.

Problem 3

One of two complementary angles is 14° larger than the other. Find the measure of each of the angles.

Solution

If we let a represent the measure of the smaller angle, then $a + 14$ represents the measure of the larger angle. Since they are complementary angles their sum is 90° and we can proceed as follows.

$$a + a + 14 = 90$$
$$2a + 14 = 90$$
$$2a + 14 - 14 = 90 - 14$$
$$2a = 76$$
$$\frac{2a}{2} = \frac{76}{2}$$
$$a = 38$$

If $a = 38$, then $a + 14 = 52$, and the angles are of measure 38° and 52°.   ▲

Problem 4

Find the measures of the three angles of a triangle if the second is three times the first and the third is twice the second.

Solution

If we let a represent the measure of the smallest angle, then $3a$ and $2(3a)$ represent the measures of the other two angles. Therefore, we can set up and solve the following equation.

$$a + 3a + 2(3a) = 180$$
$$a + 3a + 6a = 180$$
$$10a = 180$$
$$\frac{10a}{10} = \frac{180}{10}$$
$$a = 18$$

If $a = 18$, then $3a = 54$ and $2(3a) = 108$. So the angles have measures of 18°, 54°, and 108°.   ▲

Problem Set 3.3

Solve each of the following equations.

1. $2x + 7 + 3x = 32$

2. $3x + 9 + 4x = 30$

3. $7x - 4 - 3x = -36$

4. $8x - 3 - 2x = -45$

5. $3y - 1 + 2y - 3 = 4$

6. $y + 3 + 2y - 4 = 6$

7. $5n - 2 - 8n = 31$

8. $6n - 1 - 10n = 51$

9. $-2n + 1 - 3n + n - 4 = 7$

10. $-n + 7 - 2n + 5n - 3 = -6$

11. $3x + 4 = 2x - 5$

12. $5x - 2 = 4x + 6$

13. $5x - 7 = 6x - 9$

14. $7x - 3 = 8x - 13$

15. $6x + 1 = 3x - 8$

16. $4x - 10 = x + 17$

17. $7y - 3 = 5y + 10$

18. $8y + 4 = 5y - 4$

19. $8n - 2 = 11n - 7$

20. $7n - 10 = 9n - 13$

21. $-2x - 7 = -3x + 10$

22. $-4x + 6 = -5x - 9$

23. $-3x + 5 = -5x - 8$

24. $-4x + 7 = -6x + 4$

25. $-7 - 6x = 9 - 9x$

26. $-10 - 7x = 14 - 12x$

27. $2x - 1 - x = 3x - 5$

28. $3x - 4 - 4x = 5 - 5x + 3x$

29. $5n - 4 - n = -3n - 6 + n$

30. $4x - 3 + 2x = 8x - 3 - x$

31. $-7 - 2n - 6 = 7n - 5n + 12$

32. $-3n + 6 + 5n = 7n - 8n - 9$

Solve each of the following word problems by setting up and solving an algebraic equation.

33. The sum of a number plus four times the number is 85. What is the number?

34. A number subtracted from three times the number yields 68. Find the number.

35. Find two consecutive odd numbers whose sum is 72.

36. Find two consecutive even numbers whose sum is 94.

37. Find three consecutive even numbers whose sum is 114.

38. Find three consecutive odd numbers whose sum is 159.

39. Two more than three times a certain number is the same as four less than seven times the number. Find the number.

40. One more than five times a certain number is equal to eleven less than nine times the number. What is the number?

41. The sum of a number and five times the number equals eighteen less than three times the number. Find the number.

42. One of two supplementary angles is five times as large as the other. Find the measure of each angle.

43. One of two complementary angles is six less than twice the other angle. Find the measure of each angle.

44. If two angles are complementary and the difference of their measures is 62°, find the measure of each angle.

45. If two angles are supplementary and the larger angle is 20° less than three times the smaller angle, find the measure of each angle.

46. Find the measures of the three angles of a triangle if the largest is 14° less than three times the smallest and the other angle is 4° more than the smallest.

47. One of the angles of a triangle has a measure of 40°. Find the measures of the other two angles if the difference of their measures is 10°.

48. Larry sold encyclopedias one summer on a salary-plus-commission basis. He was given a flat salary of $1800 for the summer plus a commission of $75 for each set of encyclopedias that he sold. His total earnings for the summer were $2550. How many sets of encyclopedias did he sell?

49. Daniel sold some stock at $35 per share. This was $17 a share less than twice what he paid for it. What price did he pay for each share of the stock?

50. Becky sold some stock at $37 per share. This was $14 a share less than three times what she paid for it. What price did she pay for the stock?

51. Suppose that Bob is paid two times his normal hourly rate for each hour he works over 40 hours in a week. Last week he earned $504 for 48 hours of work. What is his hourly wage?

52. Last week on an algebra test, the highest grade was 9 points less than three times the lowest grade. The sum of the two grades was 135. Find the lowest and highest grades on the test.

53. At a university-sponsored concert, there were three times as many females as males. A total of 600 people attended the concert. How many males and how many females attended?

54. Suppose that a triangular lot is enclosed with 135 yards of fencing (see Figure 3.3). The longest side of the lot is 5 yards more than twice the length of the shortest side. The other side is 10 yards longer than the shortest side. Find the lengths of the three sides of the lot.

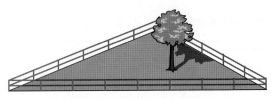

FIGURE 3.3

55. Melton and Sanchez were opposing candidates in a school board election. Sanchez received ten more than twice the number of votes that Melton received. If there was a total of 1030 votes cast, how many did Sanchez receive?

56. The plans for an apartment complex provided for 260 apartments, with each apartment having one, two, or three bedrooms. There were to be 25 more three-bedroom than one-bedroom apartments, and the number of two-bedroom apartments was to be 15 less than three times the number of one-bedroom apartments. How many apartments of each type were in the plans?

57. A board 20 feet long is to be cut into two pieces so that one piece is 8 feet longer than the other. How long should the shorter piece be?

THOUGHTS INTO WORDS

58. Give a step-by-step description of how you would solve the equation $3x + 4 = 5x - 2$.

59. Suppose that your friend solved the problem "find two consecutive odd integers whose sum is 28" as follows.

$$x + x + 1 = 28$$
$$2x = 27$$
$$x = \frac{27}{2} = 13\frac{1}{2}$$

She claims that $13\frac{1}{2}$ will check in the equation. Where has she gone wrong and how would you help her?

$\bigcirc$**Further Investigations**

60. Solve each of the following equations.
(a) $7x - 3 = 4x - 3$
(b) $-x - 4 + 3x = 2x - 7$
(c) $-3x + 9 - 2x = -5x + 9$
(d) $5x - 3 = 6x - 7 - x$
(e) $7x + 4 = -x + 4 + 8x$
(f) $3x - 2 - 5x = 7x - 2 - 5x$
(g) $-6x - 8 = 6x + 4$
(h) $-8x + 9 = -8x + 5$

▽ **3.4** # Equations Involving Parentheses and Fractional Forms

We will use the distributive property frequently in this section as we expand our techniques for solving equations. Recall that in symbolic form the distributive property states that $a(b + c) = ab + ac$. Consider the following examples, which illustrate the use of this property to *remove parentheses*. Pay special attention to the last two examples, which involve a negative number in front of the parentheses.

$$
\begin{array}{llll}
3(x + 2) = & 3 \cdot x + 3 \cdot 2 & = 3x + 6, & \\
5(y - 3) = & 5 \cdot y - 5 \cdot 3 & = 5y - 15, & (a(b - c) = ab - ac) \\
2(4x + 7) = & 2(4x) + 2(7) & = 8x + 14, & \\
-1(n + 4) = & (-1)(n) + (-1)(4) & = -n - 4, & \\
-6(x - 2) = & (-6)(x) - (-6)(2) & = -6x + 12 &
\end{array}
$$

$\downarrow$
Do this step
mentally!

It is often necessary to solve equations in which the variable is part of an expression enclosed in parentheses. The distributive property is used to remove the parentheses; we then proceed in the usual way. Consider the following examples. (Notice that we are beginning to show *only the major steps* when solving an equation.)

Example 1

Solve $3(x + 2) = 23$.

Solution

$$3(x + 2) = 23$$
$$3x + 6 = 23 \qquad \text{Applied distributive property to left side}$$
$$3x = 17 \qquad \text{Subtracted 6 from both sides}$$
$$x = \frac{17}{3} \qquad \text{Divided both sides by 3}$$

The solution set is $\left\{\dfrac{17}{3}\right\}$.

Example 2

Solve $4(x + 3) = 2(x - 6)$.

Solution

$$4(x + 3) = 2(x - 6)$$
$$4x + 12 = 2x - 12 \qquad \text{Applied distributive property on each side}$$
$$2x + 12 = -12 \qquad \text{Subtracted } 2x \text{ from both sides}$$
$$2x = -24 \qquad \text{Subtracted 12 from both sides}$$
$$x = -12 \qquad \text{Divided both sides by 2}$$

The solution set is $\{-12\}$.

It may be necessary to remove more than one set of parentheses and then to use the distributive property again to combine similar terms. Consider the following two examples.

Example 3

Solve $5(w + 3) + 3(w + 1) = 14$.

Solution

$$5(w + 3) + 3(w + 1) = 14$$
$$5w + 15 + 3w + 3 = 14 \qquad \text{Applied distributive property}$$
$$8w + 18 = 14 \qquad \text{Combined similar terms}$$
$$8w = -4 \qquad \text{Subtracted 18 from both sides}$$
$$w = -\frac{4}{8} \qquad \text{Divided both sides by 8}$$
$$w = -\frac{1}{2} \qquad \text{Reduced}$$

The solution set is $\left\{-\dfrac{1}{2}\right\}$.

Example 4

Solve $6(x - 7) - 2(x - 4) = 13$.

Solution

$$6(x - 7) - 2(x - 4) = 13 \qquad \text{Be careful with this sign!}$$

$$6x - 42 - 2x + 8 = 13 \qquad \text{Distributive property}$$

$$4x - 34 = 13 \qquad \text{Combined similar terms}$$

$$4x = 47 \qquad \text{Added 34 to both sides}$$

$$x = \frac{47}{4} \qquad \text{Divided both sides by 4}$$

The solution set is $\left\{\dfrac{47}{4}\right\}$.

In a previous section we solved equations like $x - \dfrac{2}{3} = \dfrac{3}{4}$ by adding $\dfrac{2}{3}$ to both sides. If an equation contains several fractions, then it is usually easier to *clear the equation of all fractions* by multiplying both sides by the least common denominator of all the denominators. Perhaps several examples will clarify this idea.

Example 5

Solve $\dfrac{1}{2}x + \dfrac{2}{3} = \dfrac{5}{6}$.

Solution

$$\frac{1}{2}x + \frac{2}{3} = \frac{5}{6}$$

$$6\left(\frac{1}{2}x + \frac{2}{3}\right) = 6\left(\frac{5}{6}\right) \qquad \text{6 is the LCD of 2, 3, and 6.}$$

$$6\left(\frac{1}{2}x\right) + 6\left(\frac{2}{3}\right) = 6\left(\frac{5}{6}\right) \qquad \text{Distributive property}$$

$$3x + 4 = 5 \qquad \text{Note how the equation has been } \textit{cleared of all fractions.}$$

$$3x = 1$$

$$x = \frac{1}{3}$$

The solution set is $\left\{\dfrac{1}{3}\right\}$.

Example 6

Solve $\dfrac{5n}{6} - \dfrac{1}{4} = \dfrac{3}{8}$.

Solution

$$\dfrac{5n}{6} - \dfrac{1}{4} = \dfrac{3}{8} \qquad \text{Remember } \dfrac{5n}{6} = \dfrac{5}{6}n.$$

$$24\left(\dfrac{5n}{6} - \dfrac{1}{4}\right) = 24\left(\dfrac{3}{8}\right) \qquad \text{24 is the LCD of 6, 4, and 8.}$$

$$24\left(\dfrac{5n}{6}\right) - 24\left(\dfrac{1}{4}\right) = 24\left(\dfrac{3}{8}\right) \qquad \text{Distributive property}$$

$$20n - 6 = 9$$

$$20n = 15$$

$$n = \dfrac{15}{20} = \dfrac{3}{4}$$

The solution set is $\left\{\dfrac{3}{4}\right\}$.

We use many of the ideas presented in this section to help solve the following equations. Study the solutions carefully and be sure that you can supply reasons for each step. It might be helpful to cover up the solutions and try to solve the equations on your own.

Example 7

Solve $\dfrac{x+3}{2} + \dfrac{x+4}{5} = \dfrac{3}{10}$.

Solution

$$\dfrac{x+3}{2} + \dfrac{x+4}{5} = \dfrac{3}{10}$$

$$10\left(\dfrac{x+3}{2} + \dfrac{x+4}{5}\right) = 10\left(\dfrac{3}{10}\right) \qquad \text{10 is the LCD of 2, 5, and 10.}$$

$$10\left(\dfrac{x+3}{2}\right) + 10\left(\dfrac{x+4}{5}\right) = 10\left(\dfrac{3}{10}\right) \qquad \text{Distributive property}$$

$$5(x+3) + 2(x+4) = 3$$

$$5x + 15 + 2x + 8 = 3$$

$$7x + 23 = 3$$

$$7x = -20$$

$$x = -\dfrac{20}{7}$$

The solution set is $\left\{-\dfrac{20}{7}\right\}$.

Example 8

Solve $\dfrac{x-1}{4} - \dfrac{x-2}{6} = \dfrac{2}{3}$.

Solution

$$\dfrac{x-1}{4} - \dfrac{x-2}{6} = \dfrac{2}{3}$$

$$12\left(\dfrac{x-1}{4} - \dfrac{x-2}{6}\right) = 12\left(\dfrac{2}{3}\right) \qquad \text{12 is the LCD of 4, 6, and 3.}$$

$$12\left(\dfrac{x-1}{4}\right) - 12\left(\dfrac{x-2}{6}\right) = 12\left(\dfrac{2}{3}\right) \qquad \text{Distributive property}$$

$$3(x-1) - 2(x-2) = 8$$

$$3x - 3 - 2x + 4 = 8 \qquad \text{Be careful with this sign.}$$

$$x + 1 = 8$$

$$x = 7$$

The solution set is $\{7\}$. ▲

Word Problems

We are now ready to solve some word problems using equations of the different types presented in this section. Again, it might be helpful for you to attempt to solve the problems on your own before looking at the book's approach.

Problem 1

Loretta has 19 coins (quarters and nickels) that amount to $2.35. How many coins of each kind does she have?

Solution

Let q represent the number of quarters. Then $19 - q$ represents the number of nickels. We can use the following guideline to help set up an equation.

Value of quarters in cents + Value of nickels in cents = Total value in cents

$$25q \qquad + \qquad 5(19-q) \qquad = \qquad 235$$

Solving the equation we obtain

$$25q + 95 - 5q = 235$$

$$20q + 95 = 235$$

$$20q = 140$$

$$q = 7.$$

If $q = 7$, then $19 - q = 12$; so she has 7 quarters and 12 nickels. ▲

Problem 2

Find a number such that 4-less-than-two-thirds the number is equal to one-sixth the number.

Solution

Let n represent the number. Then $\frac{2}{3}n - 4$ represents 4 less than two-thirds the number, and $\frac{1}{6}n$ represents one-sixth the number.

$$\frac{2}{3}n - 4 = \frac{1}{6}n$$
$$6\left(\frac{2}{3}n - 4\right) = 6\left(\frac{1}{6}n\right)$$
$$4n - 24 = n$$
$$3n - 24 = 0$$
$$3n = 24$$
$$n = 8$$

The number is 8.   ▲

Problem 3

Lance is paid $1\frac{1}{2}$ times his normal hourly rate for each hour he works over 40 hours in a week. Last week he worked 50 hours and earned $462. What is his normal hourly rate?

Solution

Let x represent his normal hourly rate. Then $\frac{3}{2}x$ represents $1\frac{1}{2}$ times his normal hourly rate. The following guideline can be used to help set up the equation.

Regular wages for first 40 hours + Wages for 10 hours of overtime = Total wages
$$\downarrow \qquad\qquad\qquad \downarrow \qquad\qquad\qquad \downarrow$$
$$40x \qquad + \qquad 10\left(\frac{3}{2}x\right) \qquad = \qquad 462$$

Solving this equation we obtain

$$40x + 15x = 462$$
$$55x = 462$$
$$x = 8.40.$$

His normal hourly rate is $8.40.   ▲

Problem 4

Find three consecutive whole numbers such that the sum of the first plus twice the second plus three times the third is 134.

Solution

Let n represent the first whole number. Then $n + 1$ represents the second whole number and $n + 2$ represents the third whole number.

$$n + 2(n + 1) + 3(n + 2) = 134$$
$$n + 2n + 2 + 3n + 6 = 134$$
$$6n + 8 = 134$$
$$6n = 126$$
$$n = 21$$

The numbers are 21, 22, and 23.

Keep in mind that the problem solving suggestions we offered in Section 3.3 simply outline a general algebraic approach to solving problems. You will add to this list throughout this course and in any subsequent mathematics courses that you take. Furthermore, you will be able to pick up additional problem solving ideas from your instructor and from fellow classmates as problems are discussed in class. Always be on the alert for any ideas that might help you become a better problem solver.

Problem Set 3.4

Solve each of the following equations.

1. $7(x + 2) = 21$

2. $4(x + 4) = 24$

3. $5(x - 3) = 35$

4. $6(x - 2) = 18$

5. $-3(x + 5) = 12$

6. $-5(x - 6) = -15$

7. $4(n - 6) = 5$

8. $3(n + 4) = 7$

9. $6(n + 7) = 8$

10. $8(n - 3) = 12$

11. $-10 = -5(t - 8)$

12. $-16 = -4(t + 7)$

13. $5(x - 4) = 4(x + 6)$

14. $7(x + 3) = 6(x - 8)$

15. $8(x + 1) = 9(x - 2)$

16. $4(x - 7) = 5(x + 2)$

17. $8(t + 5) = 6(t - 6)$

18. $7(t - 5) = 5(t + 3)$

19. $3(2t + 1) = 4(3t - 2)$

20. $2(3t - 1) = 3(3t - 2)$

21. $-2(x - 6) = -(x - 9)$

22. $-(x + 7) = -2(x + 10)$

23. $-3(t - 4) - 2(t + 4) = 9$

24. $5(t - 4) - 3(t - 2) = 12$

25. $3(n - 10) - 5(n + 12) = -86$

26. $4(n + 9) - 7(n - 8) = 83$

27. $3(x + 1) + 4(2x - 1) = 5(2x + 3)$

28. $4(x - 1) + 5(x + 2) = 3(x - 8)$

29. $-(x + 2) + 2(x - 3) = -2(x - 7)$

30. $-2(x + 6) + 3(3x - 2) = -3(x - 4)$

31. $5(2x - 1) - (3x + 4) = 4(x + 3) - 27$

32. $3(4x + 1) - 2(2x + 1) = -2(x - 1) - 1$

33. $-(a - 1) - (3a - 2) = 6 + 2(a - 1)$

34. $3(2a - 1) - 2(5a + 1) = 4(3a + 4)$

35. $3(x - 1) + 2(x - 3) = -4(x - 2) + 10(x + 4)$

36. $-2(x - 4) - (3x - 2) = -2 + (-6x + 2)$

37. $3 - 7(x - 1) = 9 - 6(2x + 1)$

38. $8 - 5(2x + 1) = 2 - 6(x - 3)$

39. $\dfrac{3}{4}x - \dfrac{2}{3} = \dfrac{5}{6}$

40. $\dfrac{1}{2}x - \dfrac{4}{3} = -\dfrac{5}{6}$

41. $\dfrac{5}{6}x + \dfrac{1}{4} = -\dfrac{9}{4}$

42. $\dfrac{3}{8}x + \dfrac{1}{6} = -\dfrac{7}{12}$

43. $\dfrac{1}{2}x - \dfrac{3}{5} = \dfrac{3}{4}$

44. $\dfrac{1}{4}x - \dfrac{2}{5} = \dfrac{5}{6}$

45. $\dfrac{n}{3} + \dfrac{5n}{6} = \dfrac{1}{8}$

46. $\dfrac{n}{6} + \dfrac{3n}{8} = \dfrac{5}{12}$

47. $\dfrac{5y}{6} - \dfrac{3}{5} = \dfrac{2y}{3}$

48. $\dfrac{3y}{7} + \dfrac{1}{2} = \dfrac{y}{4}$

49. $\dfrac{h}{6} + \dfrac{h}{8} = 1$

50. $\dfrac{h}{4} + \dfrac{h}{3} = 1$

51. $\dfrac{x + 2}{3} + \dfrac{x + 3}{4} = \dfrac{13}{3}$

52. $\dfrac{x - 1}{4} + \dfrac{x + 2}{5} = \dfrac{39}{20}$

53. $\dfrac{x - 1}{5} - \dfrac{x + 4}{6} = -\dfrac{13}{15}$

54. $\dfrac{x + 1}{7} - \dfrac{x - 3}{5} = \dfrac{4}{5}$

55. $\dfrac{x + 8}{2} - \dfrac{x + 10}{7} = \dfrac{3}{4}$

56. $\dfrac{x + 7}{3} - \dfrac{x + 9}{6} = \dfrac{5}{9}$

57. $\dfrac{x - 2}{8} - 1 = \dfrac{x + 1}{4}$

58. $\dfrac{x - 4}{2} + 3 = \dfrac{x - 2}{4}$

59. $\dfrac{x + 1}{4} = \dfrac{x - 3}{6} + 2$

60. $\dfrac{x + 3}{5} = \dfrac{x - 6}{2} + 1$

Solve each of the following problems by setting up and solving an appropriate algebraic equation.

61. Find two consecutive whole numbers such that the smaller number plus four times the larger number equals 39.

62. Find two consecutive whole numbers such that the smaller number subtracted from five times the larger number equals 57.

63. Find three consecutive whole numbers such that twice the sum of the two smallest numbers is ten more than three times the largest number.

64. Find four consecutive whole numbers such that the sum of the first three numbers equals the fourth number.

65. The sum of two numbers is 17. If twice the smaller number is one more than the larger number, find the numbers.

66. The sum of two numbers is 53. If three times the smaller number is one less than the larger number, find the numbers.

67. Find a number such that 20 more than one-third of the number equals three-fourths of the number.

68. The sum of three-eighths of a number and five-sixths of the same number is 29. Find the number.

69. The difference of two numbers is six. One-half of the larger number is five larger than one-third of the smaller. Find the numbers.

70. The difference of two numbers is 16. Three-fourths of the larger number is 14 larger than one-half of the smaller number. Find the numbers.

71. Suppose that a board 20 feet long is cut into two pieces. Four times the length of the shorter piece is 4 feet less than three times the length of the longer piece. Find the length of each piece.

72. Ellen is paid "time and a half" for each hour over 40 hours she works in a week. Last week she worked 44 hours and earned $299. What is her normal hourly rate?

73. Lucy has 69 coins consisting of nickels, dimes, and quarters. She has five more dimes than nickels, and the number of quarters is four more than three times the number of nickels. How many coins of each kind does she have?

74. Suppose that Julian has 44 coins consisting of pennies and nickels. If the number of nickels is two more than twice the number of pennies, find the number of coins of each kind.

75. Max has a collection of 210 coins consisting of nickels, dimes, and quarters. He has twice as many dimes as nickels, and ten more quarters than dimes. How many coins of each kind does he have?

76. Ginny has a collection of 425 coins consisting of pennies, nickels, and dimes. She has 50 more nickels than pennies and 25 more dimes than nickels. How many coins of each kind does she have?

77. Maida has 18 coins consisting of dimes and quarters amounting to $3.30. How many coins of each kind does she have?

78. Ike has some nickels and dimes amounting to $2.90. The number of dimes is one less than twice the number of nickels. How many coins of each kind does he have?

79. A collection of pennies, nickels, and dimes totals $30.75. There are twice as many nickels as pennies and three times as many dimes as pennies. How many coins of each kind are in the collection?

80. Tickets for a concert were priced at $3 for students and $5 for nonstudents. There were 1500 tickets sold for a total of $5000. How many student tickets were sold?

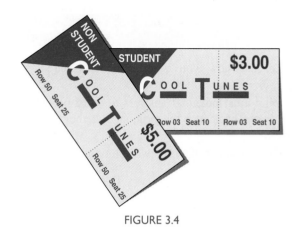

FIGURE 3.4

81. The supplement of an angle is 30° more than twice its complement. Find the measure of the angle.

82. The sum of the measure of an angle and three times its complement is 202°. Find the measure of the angle.

83. In the triangle ABC, the measure of angle A is 2° less than one-fifth of the measure of angle C. The measure of angle B is 5° less than one-half of the measure of angle C. Find the measures of the three angles of the triangle.

84. If one-fourth of the complement of an angle plus one-fifth of the supplement of the angle equals 36°, find the measure of the angle.

85. The supplement of an angle is 10° less than three times its complement. Find the size of the angle.

86. In triangle ABC, the measure of angle C is eight times the measure of angle A and the measure of angle B is 10° more than the measure of angle C. Find the measure of each angle of the triangle.

87. Discuss how you would solve the equation $3(x - 2) - 5(x + 3) = -4(x + 9)$.

88. Why must potential answers to word problems be checked back into the original statement of the problem?

89. Consider the following two solutions.

$$3(x + 2) = 9 \qquad\qquad 3(x - 4) = 7$$
$$\frac{3(x + 2)}{3} = \frac{9}{3} \qquad\qquad \frac{3(x - 4)}{3} = \frac{7}{3}$$
$$x + 2 = 3 \qquad\qquad x - 4 = \frac{7}{3}$$
$$x = 1 \qquad\qquad\qquad x = \frac{19}{3}$$

Are both of these solutions correct? Comment on the effectiveness of the two different approaches.

90. Solve each of the following equations.
(a) $-2(x - 1) = -2x + 2$
(b) $3(x + 4) = 3x - 4$
(c) $5(x - 1) = -5x - 5$
(d) $\dfrac{x - 3}{3} + 4 = 3$
(e) $\dfrac{x + 2}{3} + 1 = \dfrac{x - 2}{3}$
(f) $\dfrac{x - 1}{5} - 2 = \dfrac{x - 11}{5}$
(g) $4(x - 2) - 2(x + 3) = 2(x + 6)$
(h) $5(x + 3) - 3(x - 5) = 2(x + 15)$
(i) $7(x - 1) + 4(x - 2) = 15(x - 1)$

91. Find three consecutive integers such that the sum of the smallest integer and the largest integer is equal to twice the middle integer.

3.5   Inequalities

Just as we use the symbol $=$ to represent *is equal to*, we also use the symbols $<$ and $>$ to represent *is less than* and *is greater than*, respectively. The following are examples of **statements of inequality**. Notice that the first four are true statements and the last two are false.

$$6 + 4 > 7,$$
$$8 - 2 < 14,$$
$$4 \cdot 8 > 4 \cdot 6,$$
$$5 \cdot 2 < 5 \cdot 7,$$
$$5 + 8 > 19,$$
$$9 - 2 < 3.$$

Algebraic inequalities contain one or more variables. The following are examples of algebraic inequalities.

$$x + 3 > 4,$$
$$2x - 1 < 6,$$
$$x^2 + 2x - 1 > 0,$$
$$2x + 3y < 7,$$
$$7ab < 9$$

An algebraic inequality such as $x + 1 > 2$ is neither true nor false as it stands; it is called an **open sentence**. Each time a number is substituted for x, the algebraic inequality $x + 1 > 2$ becomes a numerical statement that is either true or false. For example, if $x = 0$, then $x + 1 > 2$ becomes $0 + 1 > 2$, which is false. If $x = 2$, then $x + 1 > 2$ becomes $2 + 1 > 2$, which is true. **Solving an inequality** refers to the process of finding the numbers that make an algebraic inequality a true numerical statement. We say that such numbers, called the **solutions of the inequality**, satisfy the inequality. The set of all solutions of an inequality is called its **solution set**. We often state solution sets for inequalities with **set builder notation**. For example, the solution set for $x + 1 > 2$ is the set of real numbers greater than 1, expressed as $\{x \mid x > 1\}$. The set builder notation $\{x \mid x > 1\}$ is read as "the set of all x such that x is greater than 1." We sometimes graph solution sets for inequalities on a number line; the solution set for $\{x \mid x > 1\}$ is pictured in Figure 3.5.

FIGURE 3.5

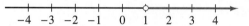

The open circle at 1 indicates that 1 is not a solution; the shaded portion indicates that all numbers to the right of (greater than) 1 are solutions. We refer to the shaded portion of the number line as the *graph* of the solution set $\{x \mid x > 1\}$.

The following examples contain some simple algebraic inequalities, their solution sets, and the graphs of the solution sets in Figure 3.6. Look them over very carefully because they also introduce some additional symbolism.

Algebraic inequality	Solution set	Graph of solution set
$x < 2$	$\{x \mid x < 2\}$	⟵─┼─┼─┼─┼─┼─┼─○─┼─┼─⟶ −5 −4 −3 −2 −1 0 1 2 3 4 5
$x > -1$	$\{x \mid x > -1\}$	─┼─┼─┼─○─┼─┼─┼─┼─┼─⟶ −5 −4 −3 −2 −1 0 1 2 3 4 5
$1 > x$	$\{x \mid x < 1\}$	⟵─┼─┼─┼─┼─┼─○─┼─┼─┼─⟶ −5 −4 −3 −2 −1 0 1 2 3 4 5
$3 < x$	$\{x \mid x > 3\}$	─┼─┼─┼─┼─┼─┼─┼─○─┼─⟶ −5 −4 −3 −2 −1 0 1 2 3 4 5
$x \geq 1$ ($\geq$ is read "greater than or equal to")	$\{x \mid x \geq 1\}$	─┼─┼─┼─┼─┼─●─┼─┼─┼─⟶ −5 −4 −3 −2 −1 0 1 2 3 4 5
$x \leq 2$ ($\leq$ is read "less than or equal to")	$\{x \mid x \leq 2\}$	⟵─┼─┼─┼─┼─┼─┼─●─┼─┼─⟶ −5 −4 −3 −2 −1 0 1 2 3 4 5

FIGURE 3.6

The general process for solving inequalities closely parallels that for solving equations. We continue to replace the given inequality with equivalent, but simpler inequalities. For example,

$$2x + 1 > 9, \tag{1}$$
$$2x > 8, \tag{2}$$
$$x > 4 \tag{3}$$

are all equivalent inequalities; that is, they have the same solutions. Thus, to solve (1) we can solve (3), which is obviously all numbers greater than 4. The exact procedure for simplifying inequalities is primarily based on two properties and they become our topics of discussion at this time. The first of these is the **addition-subtraction property of inequality**.

PROPERTY 3.4

For all real numbers a, b, and c,
1. $a > b$ if and only if $a + c > b + c$,
2. $a > b$ if and only if $a - c > b - c$.

Property 3.4 states that any number can be added to or subtracted from both sides of an inequality and an equivalent inequality is produced. The property is stated in terms of $>$, but analogous properties exist for $<$, $\geq$, and $\leq$. Consider the use of this property in the next three examples.

Example 1

Solve $x - 3 > -1$ and graph the solutions.

Solution

$$x - 3 > -1$$
$$x - 3 + 3 > -1 + 3 \qquad \text{Add 3 to both sides.}$$
$$x > 2$$

The solution set is $\{x \mid x > 2\}$ and it can be graphed as follows (Figure 3.7).

FIGURE 3.7

Example 2

Solve $x + 4 \leq 5$ and graph the solutions.

Solution

$$x + 4 \leq 5$$
$$x + 4 - 4 \leq 5 - 4 \qquad \text{Subtract 4 from both sides.}$$
$$x \leq 1$$

The solution set is $\{x \mid x \leq 1\}$ and it can be graphed as follows (Figure 3.8).

FIGURE 3.8

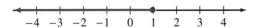

Example 3

Solution

Solve $5 > 6 + x$ and graph the solutions.

$$5 > 6 + x$$
$$5 - 6 > 6 + x - 6 \qquad \text{Subtract 6 from both sides.}$$
$$-1 > x$$

Since $-1 > x$ is equivalent to $x < -1$, the solution set is $\{x \mid x < -1\}$ and it can be graphed as follows (Figure 3.9).

FIGURE 3.9

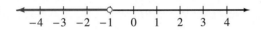

Now let's look at some numerical examples to see what happens when both sides of an inequality are multiplied or divided by some number.

$$4 > 3 \quad \longrightarrow \quad \underline{5}(4) > \underline{5}(3),$$
$$-2 > -3 \quad \longrightarrow \quad \underline{4}(-2) > \underline{4}(-3),$$
$$6 > 4 \quad \longrightarrow \quad \frac{6}{\underline{2}} > \frac{4}{\underline{2}},$$
$$8 > -2 \quad \longrightarrow \quad \frac{8}{\underline{4}} > \frac{-2}{\underline{4}}$$

Notice that multiplying or dividing both sides of an inequality by a positive number produces an inequality of the same sense. This means that if the original inequality is *greater than*, then the new inequality is *greater than*, and if the original is *less than*, then the resulting inequality is *less than*.

Now note what happens when we multiply or divide both sides by a negative number.

$$3 < 5 \quad \text{but } \underline{-2}(3) > \underline{-2}(5),$$
$$-4 < 1 \quad \text{but } \underline{-5}(-4) > \underline{-5}(1),$$
$$4 > 2 \quad \text{but } \frac{4}{\underline{-2}} < \frac{2}{\underline{-2}},$$
$$-3 > -6 \quad \text{but } \frac{-3}{\underline{-3}} < \frac{-6}{\underline{-3}}$$

Multiplying or dividing both sides of an inequality *by a negative number reverses the sense of the inequality*. Property 3.5 summarizes these ideas.

PROPERTY 3.5

(a) For all real numbers a, b, and c, with $c > 0$,
   1. $a > b$ if and only if $ac > bc$,
   2. $a > b$ if and only if $\dfrac{a}{c} > \dfrac{b}{c}$.

(b) For all real numbers a, b, and c, with $c < 0$,
   1. $a > b$ if and only if $ac < bc$,
   2. $a > b$ if and only if $\dfrac{a}{c} < \dfrac{b}{c}$.

Similar properties hold if each inequality is reversed or if $>$ is replaced with $\geq$ and $<$ with $\leq$. For example, if $a \leq b$ and $c < 0$, then $ac \geq bc$ and $\dfrac{a}{c} \geq \dfrac{b}{c}$.
Observe the use of Property 3.5 in the next three examples.

Example 4

Solve $2x > 4$.

Solution

$$2x > 4$$
$$\frac{2x}{2} > \frac{4}{2} \qquad \text{Divide both sides by 2.}$$
$$x > 2$$

The solution set is $\{x \mid x > 2\}$. ▲

Example 5

Solve $\dfrac{3}{4}x \leq \dfrac{1}{5}$.

Solution

$$\frac{3}{4}x \leq \frac{1}{5}$$
$$\frac{4}{3}\left(\frac{3}{4}x\right) \leq \frac{4}{3}\left(\frac{1}{5}\right) \qquad \text{Multiply both sides by } \frac{4}{3}.$$
$$x \leq \frac{4}{15}$$

The solution set is $\left\{x \mid x \leq \dfrac{4}{15}\right\}$. ▲

Example 6

Solve $-3x > 9$.

$-3(-3) > 9$

Solution

$$-3x > 9$$

$$\frac{-3x}{-3} < \frac{9}{-3} \qquad \text{Divide both sides by } -3, \text{ which reverses the inequality.}$$

$$x < -3$$

The solution set is $\{x \mid x < -3\}$. ▲

As we mentioned earlier, many of the same techniques used to solve equations may be used to solve inequalities. However, we must be extremely careful when we apply Property 3.5. Study the following examples and notice the similarities between solving equations and solving inequalities.

Example 7

Solve $4x - 3 > 9$.

Solution

$$4x - 3 > 9$$

$$4x - 3 + 3 > 9 + 3 \qquad \text{Add 3 to both sides.}$$

$$4x > 12$$

$$\frac{4x}{4} > \frac{12}{4} \qquad \text{Divide both sides by 4.}$$

$$x > 3$$

The solution set is $\{x \mid x > 3\}$. ▲

Example 8

Solve $-3n + 5 < 11$.

Solution

$$-3n + 5 < 11$$

$$-3n + 5 - 5 < 11 - 5 \qquad \text{Subtract 5 from both sides.}$$

$$-3n < 6$$

$$\frac{-3n}{-3} > \frac{6}{-3} \qquad \text{Divide both sides by } -3, \text{ which reverses the inequality.}$$

$$n > -2$$

The solution set is $\{n \mid n > -2\}$. ▲

Problem Set 3.5

For Problems 1–10, determine whether each numerical inequality is *true* or *false*.

1. $2(3) - 4(5) < 5(3) - 2(-1) + 4$

2. $5 + 6(-3) - 8(-4) > 17$

3. $\frac{2}{3} - \frac{3}{4} + \frac{1}{6} > \frac{1}{5} + \frac{3}{4} - \frac{7}{10}$

4. $\frac{1}{2} + \frac{1}{3} < \frac{1}{3} + \frac{1}{4}$

5. $\left(-\frac{1}{2}\right)\left(\frac{4}{9}\right) > \left(\frac{3}{5}\right)\left(-\frac{1}{3}\right)$

6. $\left(\frac{5}{6}\right)\left(\frac{8}{12}\right) < \left(\frac{3}{7}\right)\left(\frac{14}{15}\right)$

7. $\frac{3}{4} + \frac{2}{3} \div \frac{1}{5} > \frac{2}{3} + \frac{1}{2} \div \frac{3}{4}$

8. $1.9 - 2.6 - 3.4 < 2.5 - 1.6 - 4.2$

9. $.16 + .34 > .23 + .17$

10. $(.6)(1.4) > (.9)(1.2)$

For Problems 11–22, state the solution set and graph it on a number line.

11. $x > -2$

12. $x > -4$

13. $x \le 3$

14. $x \le 0$

15. $2 < x$

16. $-3 \le x$

17. $-2 \ge x$

18. $1 > x$

19. $-x > 1$

20. $-x < 2$

21. $-2 < -x$

22. $-1 > -x$

For Problems 23–60, solve each of the inequalities.

23. $x + 6 < -14$

24. $x + 7 > -15$

25. $x - 4 \ge -13$

26. $x - 3 \le -12$

27. $4x > 36$

28. $3x < 51$

29. $6x < 20$

30. $8x > 28$

31. $-5x > 40$

32. $-4x < 24$

33. $-7n \le -56$

34. $-9n \ge -63$

35. $48 > -14n$

36. $36 < -8n$

37. $16 < 9 + n$

38. $19 > 27 + n$

39. $3x + 2 > 17$

40. $2x + 5 < 19$

41. $4x - 3 \le 21$

42. $5x - 2 \ge 28$

43. $-2x - 1 \ge 41$

44. $-3x - 1 \le 35$

45. $6x + 2 < 18$

46. $8x + 3 > 25$

47. $3 > 4x - 2$

48. $7 < 6x - 3$

49. $-2 < -3x + 1$

50. $-6 > -2x + 4$

51. $-38 \ge -9t - 2$

52. $36 \ge -7t + 1$

53. $5x - 4 - 3x > 24$

54. $7x - 8 - 5x < 38$

55. $4x + 2 - 6x < -1$

56. $6x + 3 - 8x > -3$

57. $-5 \ge 3t - 4 - 7t$

58. $6 \le 4t - 7t - 10$

59. $-x - 4 - 3x > 5$

60. $-3 - x - 3x < 10$

61. Do the *greater than* and *less than* relations possess the symmetric property? Explain your answer.

62. Is the solution set for $x < 3$ the same as for $3 > x$? Explain your answer.

63. How would you convince someone that it is necessary to reverse the inequality symbol when multiplying both sides of an inequality by a negative number?

Further Investigations

Solve each of the following inequalities.

64. $x + 3 < x - 4$

65. $x - 4 < x + 6$

66. $2x + 4 > 2x - 7$

67. $5x + 2 > 5x + 7$

68. $3x - 4 - 3x > 6$

69. $-2x + 7 + 2x > 1$

70. $-5 \leq -4x - 1 + 4x$

71. $-7 \geq 5x - 2 - 5x$

3.6 Inequalities, Compound Inequalities, and Problem Solving

Let's begin this section by solving three inequalities with the same basic steps we used with equations. Again, be careful when applying the multiplication and division properties of inequality.

Example 1

Solve $5x + 8 \leq 3x - 10$.

Solution

$$5x + 8 \leq 3x - 10$$
$$5x + 8 - 3x \leq 3x - 10 - 3x \qquad \text{Subtract } 3x \text{ from both sides.}$$
$$2x + 8 \leq -10$$
$$2x + 8 - 8 \leq -10 - 8 \qquad \text{Subtract 8 from both sides.}$$
$$2x \leq -18$$
$$\frac{2x}{2} \leq \frac{-18}{2} \qquad \text{Divide both sides by 2.}$$
$$x \leq -9$$

The solution set is $\{x \mid x \leq -9\}$. ▲

Example 2

Solve $4(x + 3) + 3(x - 4) \geq 2(x - 1)$.

Solution

$$4(x + 3) + 3(x - 4) \geq 2(x - 1)$$
$$4x + 12 + 3x - 12 \geq 2x - 2 \qquad \text{Distributive property}$$
$$7x \geq 2x - 2 \qquad \text{Combined similar terms}$$
$$7x - 2x \geq 2x - 2 - 2x \qquad \text{Subtract } 2x \text{ from both sides.}$$

$$5x \geq -2$$

$$\frac{5x}{5} \geq \frac{-2}{5} \qquad \text{Divide both sides by 5.}$$

$$x \geq -\frac{2}{5}$$

The solution set is $\left\{ x \mid x \geq -\frac{2}{5} \right\}$. ▲

Example 3

Solve $-\frac{3}{2}n + \frac{1}{6}n < \frac{3}{4}$.

Solution

$$-\frac{3}{2}n + \frac{1}{6}n < \frac{3}{4}$$

$$12\left(-\frac{3}{2}n + \frac{1}{6}n\right) < 12\left(\frac{3}{4}\right) \qquad \begin{array}{l}\text{Multiply both sides by 12, which is the LCD}\\ \text{of all denominators.}\end{array}$$

$$12\left(-\frac{3}{2}n\right) + 12\left(\frac{1}{6}n\right) < 12\left(\frac{3}{4}\right) \qquad \text{Distributive property}$$

$$-18n + 2n < 9$$

$$-16n < 9$$

$$\frac{-16n}{-16} > \frac{9}{-16} \qquad \begin{array}{l}\text{Divide both sides by } -16, \text{ which reverses}\\ \text{the inequality.}\end{array}$$

$$n > -\frac{9}{16}$$

The solution set is $\left\{ n \mid n > -\frac{9}{16} \right\}$. ▲

Checking solutions for an inequality presents a problem. Obviously, we cannot check all of the infinitely many solutions for a particular inequality. However, by checking at least one solution, especially when the multiplication property was used, we could make sure that we did not forget to reverse an inequality sign. In Example 3 we are claiming that "all numbers greater than $-\frac{9}{16}$ will satisfy the original inequality." Let's check one number, say 0.

$$-\frac{3}{2}n + \frac{1}{6}n < \frac{3}{4}$$

$$-\frac{3}{2}(0) + \frac{1}{6}(0) \stackrel{?}{<} \frac{3}{4}$$

$$0 < \frac{3}{4}$$

Therefore, 0 satisfies the original inequality. Had we forgotten to reverse the inequality sign when both sides were divided by -16, then our answer would have been $n < -\dfrac{9}{16}$ and we would have detected such an error by the check.

Compound Statements

The words *and* and *or* are used in mathematics to form **compound statements**. The following are examples of some compound numerical statements using *and*. Such statements are called **conjunctions**. We agree to call a conjunction true only if all of its component parts are true.

1. $3 + 6 = 9$ and $4 < 7$ True

2. $4 > 2$ and $0 > -2$ True

3. $8 + 7 = 15$ and $2 < 1$ False

4. $9 > 14$ and $13 < 18$ False

5. $5 > 8$ and $-6 > 0$ False

Compound statements using *or* are called **disjunctions**. A disjunction is true if at least one of its component parts is true. Said another way, a disjunction is false only if all of its component parts are false. Consider the following disjunctions.

1. $17 > 14$ or $18 < 25$ True

2. $19 > 12$ or $13 > 17$ True

3. $4 + 6 = 8$ or $0 < 1$ True

4. $7 < -4$ or $9 < 2$ False

Now let's consider finding solutions for some compound statements that involve algebraic inequalities. Keep in mind that the previous discussion regarding labeling conjunctions and disjunctions true or false forms the basis for our reasoning.

Example 4

Solution

Graph the solutions for $x > 2$ and $x < 5$.

The key word is *and*, so we need to satisfy both inequalities. Thus, all numbers between 2 and 5 are solutions and we can graph them as in Figure 3.10.

FIGURE 3.10

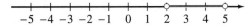

Example 5

Solution

Graph the solutions for $x < -1$ or $x > 3$.

The key word is *or*, so any numbers satisfying either (or both) inequalities are solutions. Thus, all numbers greater than 3 along with all numbers less than -1 are solutions and we can graph them as in Figure 3.11.

FIGURE 3.11

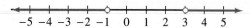

Example 6

Solution

Graph the solutions for $x \geq -2$ and $x < 1$.

Since it is an *and* statement, we need to satisfy both inequalities. Thus, all numbers between -2 and 1, including -2 (but not including 1), are solutions. Notice on the graph that there is a solid circle at -2, but an open circle at 1 (Figure 3.12).

FIGURE 3.12

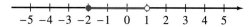

Example 7

Solution

Graph the solutions for $x > 1$ or $x > 4$.

Since it is an *or* statement, numbers that satisfy either inequality (or both) are solutions. Thus, all numbers greater than 1 are included and the graph is shown in Figure 3.13.

FIGURE 3.13

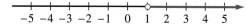

Back to Problem Solving

Let's consider some word problems that translate into inequality statements. We gave *suggestions for solving word problems* in Section 3.3 that apply here except that the situation being described in a problem will translate into an inequality instead of an equation.

Problem 1

Solution

Debbie had scores of 95, 82, 93, and 84 on her first four exams of the semester. What score must she obtain on the fifth exam to have an average of 90 or better for the five exams?

Let s represent the score needed on the fifth exam. Since the average is computed by adding all five scores and dividing by 5 (the number of scores), we have the following inequality to solve.

$$\frac{95 + 82 + 93 + 84 + s}{5} \geq 90$$

Solving this inequality, we obtain

$$\frac{354 + s}{5} \geq 90 \qquad \text{Simplified numerator of left side}$$

$$5\left(\frac{354 + s}{5}\right) \geq 5(90) \qquad \text{Multiply both sides by 5.}$$

$$354 + s \geq 450$$

$$354 + s - 354 \geq 450 - 354 \qquad \text{Subtract 354 from both sides.}$$

$$s \geq 96.$$

She must receive a score of 96 or better on the fifth exam.   ▲

Problem 2

The Cubs have won 40 baseball games and have lost 62 games. They have 60 more games to play. To win more than 50% of all their games, how many of the 60 games remaining must they win?

Solution

Let w represent the number of games they must win out of the 60 games remaining. Since they are playing a total of $40 + 62 + 60 = 162$ games, to win more than 50% of their games they would need to win more than 81 games. Thus, we have the following inequality.

$$w + 40 > 81$$

Solving this yields

$$w > 41.$$

They need to win at least 42 of the 60 games remaining.   ▲

Problem Set 3.6

Solve each of the following inequalities.

1. $3x + 4 > x + 8$

2. $5x + 3 < 3x + 11$

3. $7x - 2 < 3x - 6$

4. $8x - 1 > 4x - 21$

5. $6x + 7 > 3x - 3$

6. $7x + 5 < 4x - 12$

7. $5n - 2 \leq 6n + 9$

8. $4n - 3 \geq 5n + 6$

9. $2t + 9 \geq 4t - 13$

10. $6t + 14 \leq 8t - 16$

11. $-3x - 4 < 2x + 7$

12. $-x - 2 > 3x - 7$

13. $-4x + 6 > -2x + 1$

14. $-6x + 8 < -4x + 5$

15. $5(x - 2) \leq 30$

16. $4(x + 1) \geq 16$

17. $2(n + 3) > 9$

18. $3(n - 2) < 7$

19. $-3(y - 1) < 12$

20. $-2(y + 4) > 18$

21. $-2(x + 6) > -17$

22. $-3(x - 5) < -14$

23. $3(x - 2) < 2(x + 1)$

24. $5(x + 3) > 4(x - 2)$

25. $4(x + 3) > 6(x - 5)$

26. $6(x - 1) < 8(x + 5)$

27. $3(x - 4) + 2(x + 3) < 24$

28. $2(x + 1) + 3(x + 2) > -12$

29. $5(n + 1) - 3(n - 1) > -9$

30. $4(n - 5) - 2(n - 1) < 13$

31. $\dfrac{1}{2}n - \dfrac{2}{3}n \geq -7$

32. $\dfrac{3}{4}n + \dfrac{1}{6}n \leq 1$

33. $\dfrac{3}{4}n - \dfrac{5}{6}n < \dfrac{3}{8}$

34. $\dfrac{2}{3}n - \dfrac{1}{2}n > \dfrac{1}{4}$

35. $\dfrac{3x}{5} - \dfrac{2}{3} > \dfrac{x}{10}$

36. $\dfrac{5x}{4} + \dfrac{3}{8} < \dfrac{7x}{12}$

37. $n \geq 3.4 + .15n$

38. $x \geq 2.1 + .3x$

39. $.09t + .1(t + 200) > 77$

40. $.07t + .08(t + 100) > 38$

41. $.06x + .08(250 - x) \geq 19$

42. $.08x + .09(2x) \leq 130$

43. $\dfrac{x - 1}{2} + \dfrac{x + 3}{5} > \dfrac{1}{10}$

44. $\dfrac{x + 3}{4} + \dfrac{x - 5}{7} < \dfrac{1}{28}$

45. $\dfrac{x + 2}{6} - \dfrac{x + 1}{5} < -2$

46. $\dfrac{x - 6}{8} - \dfrac{x + 2}{7} > -1$

47. $\dfrac{n + 3}{3} + \dfrac{n - 7}{2} > 3$

48. $\dfrac{n - 4}{4} + \dfrac{n - 2}{3} < 4$

49. $\dfrac{x - 3}{7} - \dfrac{x - 2}{4} \geq \dfrac{9}{14}$

50. $\dfrac{x - 1}{5} - \dfrac{x + 2}{6} \geq \dfrac{7}{15}$

Graph the solutions for each of the following compound inequalities.

51. $x > -1$ and $x < 2$

52. $x > 1$ and $x < 4$

53. $x < -2$ or $x > 1$

54. $x < 0$ or $x > 3$

55. $x > -2$ and $x \leq 2$

56. $x \geq -1$ and $x < 3$

57. $x > -1$ and $x > 2$

58. $x < -2$ and $x < 3$

59. $x > -4$ or $x > 0$

60. $x < 2$ or $x < 4$

61. $x > 3$ and $x < -1$

62. $x < -3$ and $x > 6$

63. $x \leq 0$ or $x \geq 2$

64. $x \leq -2$ or $x \geq 1$

65. $x > -4$ or $x < 3$

66. $x > -1$ or $x < 2$

Solve each of the following problems by setting up and solving an appropriate inequality.

67. Five more than three times a number is greater than 26. Find all of the numbers that satisfy this relationship.

68. Fourteen increased by twice a number is less than or equal to three times the number. Find the numbers that satisfy this relationship.

69. Suppose that the perimeter of a rectangle is to be no greater than 70 inches and the length of the rectangle must be 20 inches. Find the largest possible value for the width of the rectangle.

70. One side of a triangle is three times as long as another side. The third side is 15 centimeters long. If the perimeter of the triangle is to be no greater than 75 centimeters, find the largest lengths that the other two sides can be.

71. Sue bowled 132 and 160 in her first two games. What must she bowl in the third game to have an average of at least 150 for the three games?

72. Mike has scores of 87, 81, and 74 on his first three algebra tests. What score must he make on the fourth test to have an average of 85 or better for the four tests?

73. This semester Lance has scores of 96, 90, and 94 on his first three algebra exams. What must he average on the last two exams to have an average of better than 92 for all five exams?

74. The Mets have won 45 baseball games and lost 55 games. They have 62 more games to play. To win more than 50% of all their games, how many of the 62 games remaining must they win?

75. Mona has $1000 to invest. If she invests $500 at 8% interest, at what rate must she invest the other $500 so that the two investments together yield more than $100 of yearly interest?

76. The average height of the two forwards and the center of a basketball team is 6 feet and 8 inches. What must the average height of the two guards be so that the team average is at least 6 feet and 4 inches?

77. Scott shot rounds of 82, 84, 78, and 79 on the first four days of the golf tournament. What must he shoot on the fifth day of the tournament to average 80 or less for the five days?

78. Terrell has $500 to invest. If he invests $300 at 9% interest, at what rate must he invest the other $200 so that the two investments together yield more than $47 in yearly interest?

THOUGHTS INTO WORDS

79. Explain the difference between a conjunction and a disjunction. Give an example of each, outside the field of mathematics.

80. Give a step-by-step description of how you would solve the inequality $3x - 2 > 4(x + 6)$.

81. Find the solution set for each of the following compound statements and in each case explain your reasoning.
(a) $x > 2$ and $5 > 4$
(b) $x > 2$ or $5 > 4$
(c) $x > 2$ and $4 > 10$
(d) $x > 2$ or $4 > 10$

SUMMARY

(3.1) **Numerical equations** may be true or false. **Algebraic equations** (open sentences) contain one or more variables. **Solving an equation** refers to the process of finding the number (or numbers) that makes an algebraic equation a true statement. A **first-degree equation of one variable** is an equation that contains only one variable and this variable has an exponent of one.

Properties 3.1 and 3.2 provide the basis for solving equations. Be sure that you can use these properties to solve the variety of equations presented in this chapter.

(3.2) It is often necessary to use both the addition and multiplication properties of equality to solve an equation.

Be sure to *declare your variable* as you translate English sentences into algebraic equations.

(3.3) Keep the following suggestions in mind as you solve word problems.

    1. Read the problem carefully.

    2. Sketch any figure, diagram, or chart that might be helpful.

    3. Choose a meaningful variable.

    4. Look for a *guideline*.

    5. Form an equation or inequality.

    6. Solve the equation or inequality.

    7. Check your answers.

(3.4) The **distributive property** is used to *remove parentheses*.

If an equation contains several fractions, then it is usually advisable to *clear the equation of all fractions* by multiplying both sides by the least common denominator of all of the denominators in the equation.

(3.5) Properties 3.4 and 3.5 provide the basis for solving inequalities. Be sure that you can use these properties to solve the variety of inequalities presented in this chapter.

We can use many of the same techniques used to solve equations to solve inequalities *but* we must be very careful when multiplying or dividing both sides of an inequality by the same number. *Don't forget* that when multiplying or dividing both sides of an inequality by a *negative number*, the resulting inequality *reverses sense*.

(3.6) The words **and** and **or** are used to form compound inequalities.

To solve inequalities involving **and** we must satisfy all of the conditions. Thus, the compound inequality $x > 1$ **and** $x < 3$ is satisfied by all numbers between 1 and 3.

To solve inequalities involving **or** we must satisfy one or more of the conditions. Thus, the compound inequality $x < -1$ **or** $x > 2$ is satisfied by (*a*) all numbers less than -1, or (*b*) all numbers greater than 2, or (*c*) both (*a*) and (*b*).

Chapter 3 Review Problem Set

For Problems 1–20, solve each of the equations.

1. $9x - 2 = -29$
2. $-3 = -4y + 1$
3. $7 - 4x = 10$
4. $6y - 5 = 4y + 13$
5. $4n - 3 = 7n + 9$
6. $7(y - 4) = 4(y + 3)$
7. $2(x + 1) + 5(x - 3) = 11(x - 2)$
8. $-3(x + 6) = 5x - 3$
9. $\frac{2}{5}n - \frac{1}{2}n = \frac{7}{10}$
10. $\frac{3n}{4} + \frac{5n}{7} = \frac{1}{14}$
11. $\frac{x - 3}{6} + \frac{x + 5}{8} = \frac{11}{12}$
12. $\frac{n}{2} - \frac{n - 1}{4} = \frac{3}{8}$
13. $-2(x - 4) = -3(x + 8)$
14. $3x - 4x - 2 = 7x - 14 - 9x$
15. $5(n - 1) - 4(n + 2) = -3(n - 1) + 3n + 5$
16. $\frac{x - 3}{9} = \frac{x + 4}{8}$
17. $\frac{x - 1}{-3} = \frac{x + 2}{-4}$
18. $-(t - 3) - (2t + 1) = 3(t + 5) - 2(t + 1)$
19. $\frac{2x - 1}{3} = \frac{3x + 2}{2}$
20. $3(2t - 4) + 2(3t + 1) = -2(4t + 3) - (t - 1)$

For Problems 21–36, solve each of the inequalities.

21. $3x - 2 > 10$
22. $-2x - 5 < 3$
23. $2x - 9 \geq x + 4$
24. $3x + 1 \leq 5x - 10$
25. $6(x - 3) > 4(x + 13)$
26. $2(x + 3) + 3(x - 6) < 14$
27. $\frac{2n}{5} - \frac{n}{4} < \frac{3}{10}$
28. $\frac{n + 4}{5} + \frac{n - 3}{6} > \frac{7}{15}$
29. $-16 < 8 + 2y - 3y$
30. $-24 > 5x - 4 - 7x$
31. $-3(n - 4) > 5(n + 2) + 3n$
32. $-4(n - 2) - (n - 1) < -4(n + 6)$
33. $\frac{3}{4}n - 6 \leq \frac{2}{3}n + 4$
34. $\frac{1}{2}n - \frac{1}{3}n - 4 \geq \frac{3}{5}n + 2$
35. $-12 > -4(x - 1) + 2$
36. $36 < -3(x + 2) - 1$

For Problems 37–40, graph the solutions for each of the compound inequalities.

37. $x > -3$ and $x < 2$
38. $x < -1$ or $x > 4$
39. $x < 2$ or $x > 0$
40. $x > 1$ and $x > 0$

Set up an equation or an inequality and solve each of the following problems.

41. Three-fourths of a number equals 18. Find the number.

42. Nineteen is two less than three times a certain number. Find the number.

43. The difference of two numbers is 21. If 12 is the smaller number, find the other number.

44. One subtracted from nine times a certain number is the same as 15 added to seven times the number. Find the number.

45. Monica has scores of 83, 89, 78, and 86 on her first four exams. What score must she receive on the fifth exam so that her average for all five exams is 85 or better?

46. The sum of two numbers is 40. Six times the smaller number equals four times the larger. Find the numbers.

47. Find a number such that two less than two-thirds of the number is one more than one-half of the number.

48. Ameya's average score for her first three psychology exams is 84. What must she get on the fourth exam so that her average for the four exams is 85 or better?

49. Miriam has 30 coins (nickels and dimes) that amount to $2.60. How many coins of each kind does she have?

50. Suppose that Khoa has a bunch of nickels, dimes, and quarters amounting to $15.40. The number of dimes is one more than three times the number of nickels and the number of quarters is twice the number of dimes. How many coins of each kind does he have?

51. The supplement of an angle is 14° more than three times the complement of the angle. Find the measure of the angle.

52. Pam rented a car from a rental agency that charges $25 a day and $0.20 per mile. She kept the car for 3 days and her bill was $215. How many miles did she drive during that 3-day period?

CHAPTER 3 TEST

For Problems 1–12, solve each of the equations.

1. $7x - 3 = 11$

2. $-7 = -3x + 2$

3. $4n + 3 = 2n - 15$

4. $3n - 5 = 8n + 20$

5. $4(x - 2) = 5(x + 9)$

6. $9(x + 4) = 6(x - 3)$

7. $5(y - 2) + 2(y + 1) = 3(y - 6)$

8. $\dfrac{3}{5}x - \dfrac{2}{3} = \dfrac{1}{2}$

9. $\dfrac{x - 2}{4} = \dfrac{x + 3}{6}$

10. $\dfrac{x + 2}{3} + \dfrac{x - 1}{2} = 2$

11. $\dfrac{x - 3}{6} - \dfrac{x - 1}{8} = \dfrac{13}{24}$

12. $-5(n - 2) = -3(n + 7)$

For Problems 13–18, solve each of the inequalities.

13. $3x - 2 < 13$

14. $-2x + 5 \geq 3$

15. $3(x - 1) \leq 5(x + 3)$

16. $-4 > 7(x - 1) + 3$

17. $-2(x - 1) + 5(x - 2) < 5(x + 3)$

18. $\dfrac{1}{2}n + 2 \leq \dfrac{3}{4}n - 1$

For Problems 19 and 20, graph the solutions for each of the compound inequalities.

19. $x \geq -2$ and $x \leq 4$

20. $x < 1$ or $x > 3$

For Problems 21–25, set up an equation or an inequality and solve each problem.

21. A car repair bill without the tax was $127. This included $53 for parts and four hours of labor. Find the hourly rate that was charged for labor.

22. Suppose that a triangular plot of ground is enclosed with 70 meters of fencing. The longest side of the lot is two times the length of the shortest side and the third side is ten meters longer than the shortest side. Find the length of each side of the plot.

23. Tina had scores of 86, 88, 89, and 91 on her first four history exams. What score must she get on the fifth exam to have an average of 90 or better for the five exams?

24. Sean has 103 coins consisting of nickels, dimes, and quarters. The number of dimes is one less than twice the number of nickels and the number of quarters is two more than three times the number of nickels. How many coins of each kind does he have?

25. The complement of an angle is 10° less than one-half of the supplement of the angle. Find the measure of the angle.

Formulas and Problem Solving

Kirk starts jogging at 5 miles per hour. One-half hour later Consuela starts jogging on the same route at 7 miles per hour. How long will it take Consuela to catch Kirk? If we let t represent the time that Consuela jogs, then $t + \frac{1}{2}$ represents Kirk's time. We can use the equation $7t = 5\left(t + \frac{1}{2}\right)$ to determine that Consuela should catch Kirk in $1\frac{1}{4}$ hours.

We used the *formula distance equals rate times time*, which is usually expressed as $d = rt$, to set up the equation $7t = t\left(t + \frac{1}{2}\right)$. Throughout this chapter we will use a variety of formulas to solve problems that connect algebraic and geometric concepts.

4.1 Ratio, Proportion, and Percent

In Figure 4.1, as gear *A* revolves 4 times, gear *B* will revolve 3 times. We say that the *gear ratio* of *A* to *B* is 4 to 3 or the gear ratio of *B* to *A* is 3 to 4. Mathematically, a **ratio** *is the comparison of two numbers by division.* We can write the gear ratio of *A* to *B* as

$$4 \text{ to } 3 \quad \text{or} \quad 4{:}3 \quad \text{or} \quad \frac{4}{3}.$$

FIGURE 4.1

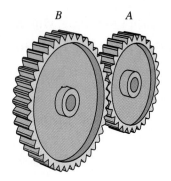

B   *A*

We express ratios as fractions in reduced form. For example, if there are 7500 females and 5000 males at a certain university, then the ratio of females to males is $\dfrac{7500}{5000} = \dfrac{3}{2}$.

A statement of equality between two ratios is called a **proportion**. For example,

$$\frac{2}{3} = \frac{8}{12}$$

is a proportion that states that the ratios $\dfrac{2}{3}$ and $\dfrac{8}{12}$ are equal. In the general proportion

$$\frac{a}{b} = \frac{c}{d}, \qquad b \neq 0 \text{ and } d \neq 0,$$

if we multiply both sides of the equation by the common denominator, *bd*, we obtain

$$(bd)\left(\frac{a}{b}\right) = (bd)\left(\frac{c}{d}\right)$$

$$ad = bc.$$

Let's state this as a property of proportions.

$$\frac{a}{b} = \frac{c}{d} \quad \text{if and only if } ad = bc, \quad \text{where } b \neq 0 \text{ and } d \neq 0$$

Example 1

Solve $\frac{x}{20} = \frac{3}{4}$.

Solution

$\frac{x}{20} = \frac{3}{4}$

$4x = 60$ Cross products are equal.

$x = 15$

The solution set is $\{15\}$. ▲

Example 2

Solve $\frac{x-3}{5} = \frac{x+2}{4}$.

Solution

$\frac{x-3}{5} = \frac{x+2}{4}$

$4(x-3) = 5(x+2)$ Cross products are equal.

$4x - 12 = 5x + 10$ Distributive property

$-12 = x + 10$ Subtracted $4x$ from both sides

$-22 = x$ Subtracted 10 from both sides

The solution set is $\{-22\}$. ▲

If a variable appears in one or both of the denominators, then a proper restriction should be made to avoid division by zero, as the next example illustrates.

Example 3

Solve $\frac{7}{a-2} = \frac{4}{a+3}$.

Solution

$\frac{7}{a-2} = \frac{4}{a+3}, \quad a \neq 2 \text{ and } a \neq -3$

$7(a+3) = 4(a-2)$ Cross products are equal.

$7a + 21 = 4a - 8$ Distributive property

$3a + 21 = -8$ Subtracted $4a$ from both sides

$3a = -29$ Subtracted 21 from both sides

$a = -\frac{29}{3}$ Divided both sides by 3

The solution set is $\left\{-\frac{29}{3}\right\}$. ▲

Example 4

Solve $\frac{x}{4} + 3 = \frac{x}{5}$.

Solution

This is *not* a proportion, so let's multiply both sides by 20 to clear the equation of all fractions.

$$\frac{x}{4} + 3 = \frac{x}{5}$$

$$20\left(\frac{x}{4} + 3\right) = 20\left(\frac{x}{5}\right) \qquad \text{Multiply both sides by 20.}$$

$$20\left(\frac{x}{4}\right) + 20(3) = 20\left(\frac{x}{5}\right) \qquad \text{Applied the distributive property}$$

$$5x + 60 = 4x$$

$$x + 60 = 0 \qquad \text{Subtracted } 4x \text{ from both sides}$$

$$x = -60 \qquad \text{Subtracted 60 from both sides}$$

The solution set is $\{-60\}$.   ▲

REMARK   Example 4 demonstrates the importance of *thinking first before pushing the pencil*. Since the equation was not in the form of a proportion, we needed to revert back to a previous technique for solving such equations.   △

Problem Solving Using Proportions

Some word problems can be conveniently set up and solved using the concepts of ratio and proportion. Consider the following examples.

Problem I

On the map in Figure 4.2 (see page 141), 1 inch represents 20 miles. If two cities are $6\frac{1}{2}$ inches apart on the map, find the number of miles between the cities.

Solution

Let m represent the number of miles between the two cities. Now let's set up a proportion where one ratio compares distances in inches on the map, and the other ratio compares *corresponding* distances in miles on land.

$$\frac{1}{6\frac{1}{2}} = \frac{20}{m}$$

To solve this equation, equate the cross products.

$$m(1) = \left(6\frac{1}{2}\right)(20)$$

$$m = \left(\frac{13}{2}\right)(20) = 130$$

The distance between the two cities is 130 miles.   ▲

FIGURE 4.2

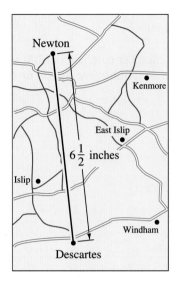

Problem 2

A sum of $1750 is to be divided between two people in the ratio of 3 to 4. How much does each person receive?

Solution

Let d represent the amount of money to be received by one person. Then $1750 - d$ represents the amount for the other person. To solve Problem 2, we set up the following proportion.

$$\frac{d}{1750 - d} = \frac{3}{4}$$

$$4d = 3(1750 - d)$$
$$4d = 5250 - 3d$$
$$7d = 5250$$
$$d = 750$$

If $d = 750$, then $1750 - d = 1000$; therefore, one person receives $750 and the other person receives $1000. ▲

Percent

The word **percent** means *per one hundred* and we use the symbol % to express it. For example, we write 7 percent as 7%, which means $\frac{7}{100}$ or .07. In other words, percent is a special kind of ratio, namely, one in which the denominator is always 100. Proportions provide a convenient basis for changing common fractions to percents. Consider the following examples.

Example 5

Express $\dfrac{7}{20}$ as a percent.

Solution

We are asking "what number compares to 100 as 7 compares to 20?" Therefore, if we let n represent that number, we can set up the following proportion.

$$\frac{n}{100} = \frac{7}{20}$$

$$20n = 700$$

$$n = 35$$

Thus, $\dfrac{7}{20} = \dfrac{35}{100} = 35\%$.   ▲

Example 6

Express $\dfrac{5}{6}$ as a percent.

Solution

$$\frac{n}{100} = \frac{5}{6}$$

$$6n = 500$$

$$n = \frac{500}{6} = \frac{250}{3} = 83\frac{1}{3}$$

Therefore, $\dfrac{5}{6} = 83\dfrac{1}{3}\%$.   ▲

Some Basic Percent Problems

What is 8% of 35? Fifteen percent of what number is 24? Twenty-one is what percent of 70? These are often referred to as the three basic types of percent problems. Each of these problems can be easily solved by translating into and solving a simple algebraic equation.

Problem 3

What is 8% of 35?

Solution

Let n represent the number to be found. The "is" refers to equality and "of" means multiplication. Thus, the question translates into

$$n = (8\%)(35),$$

which can be solved as follows.

$$n = (.08)(35)$$

$$= 2.8$$

Therefore, 2.8 is 8% of 35.   ▲

Problem 4

Solution

Fifteen percent of what number is 24?

Let n represent the number to be found.

$$(15\%)(n) = (24)$$
$$.15n = 24$$
$$15n = 2400 \qquad \text{Multiplied both sides by 100}$$
$$n = 160$$

Therefore, 15% of 160 is 24. ▲

Problem 5

Solution

Twenty-one is what percent of 70?

Let r represent the percent to be found.

$$21 = r(70)$$
$$\frac{21}{70} = r$$
$$\frac{3}{10} = r \qquad \text{Reduce!}$$
$$\frac{30}{100} = r \qquad \text{Changed } \frac{3}{10} \text{ to } \frac{30}{100}$$
$$30\% = r$$

Therefore, 21 is 30% of 70. ▲

Problem 6

Solution

Seventy-two is what percent of 60?

Let r represent the percent to be found.

$$72 = r(60)$$
$$\frac{72}{60} = r$$
$$\frac{6}{5} = r$$
$$\frac{120}{100} = r$$
$$120\% = r$$

Therefore, 72 is 120% of 60. ▲

Again, it is helpful to get into the habit of checking answers for *reasonableness*. We also suggest that you alert yourself to a potential computational error by *estimating* the answer before you actually do the problem. For example, prior to doing Problem 6 you may have estimated as follows: Since 72 is larger than

60, you know that the answer has to be greater than 100%. Furthermore, 1.5 (or 150%) times 60 equals 90. Therefore, you can estimated the answer to be somewhere between 100% and 150%. That may seem like a rather rough estimate, but many times such an estimate will detect a computational error.

Problem Set 4.1

For Problems 1–32, solve each of the equations.

1. $\dfrac{x}{6} = \dfrac{3}{2}$

2. $\dfrac{x}{9} = \dfrac{5}{3}$

3. $\dfrac{5}{12} = \dfrac{n}{24}$

4. $\dfrac{7}{8} = \dfrac{n}{16}$

5. $\dfrac{x}{3} = \dfrac{5}{2}$

6. $\dfrac{x}{7} = \dfrac{4}{3}$

7. $\dfrac{x-2}{4} = \dfrac{x+4}{3}$

8. $\dfrac{x-6}{7} = \dfrac{x+9}{8}$

9. $\dfrac{x+1}{6} = \dfrac{x+2}{4}$

10. $\dfrac{x-2}{6} = \dfrac{x-6}{8}$

11. $\dfrac{h}{2} - \dfrac{h}{3} = 1$

12. $\dfrac{h}{5} + \dfrac{h}{4} = 2$

13. $\dfrac{x+1}{3} - \dfrac{x+2}{2} = 4$

14. $\dfrac{x-2}{5} - \dfrac{x+3}{6} = -4$

15. $\dfrac{-4}{x+2} = \dfrac{-3}{x-7}$

16. $\dfrac{-9}{x+1} = \dfrac{-8}{x+5}$

17. $\dfrac{-1}{x-7} = \dfrac{5}{x-1}$

18. $\dfrac{3}{x-10} = \dfrac{-2}{x+6}$

19. $\dfrac{3}{2x-1} = \dfrac{2}{3x+2}$

20. $\dfrac{1}{4x+3} = \dfrac{2}{5x-3}$

21. $\dfrac{n+1}{n} = \dfrac{8}{7}$

22. $\dfrac{5}{6} = \dfrac{n}{n+1}$

23. $\dfrac{x-1}{2} - 1 = \dfrac{3}{4}$

24. $-2 + \dfrac{x+3}{4} = \dfrac{5}{6}$

25. $-3 - \dfrac{x+4}{5} = \dfrac{3}{2}$

26. $\dfrac{x-5}{3} + 2 = \dfrac{5}{9}$

27. $\dfrac{n}{150-n} = \dfrac{1}{2}$

28. $\dfrac{n}{200-n} = \dfrac{3}{5}$

29. $\dfrac{300-n}{n} = \dfrac{3}{2}$

30. $\dfrac{80-n}{n} = \dfrac{7}{9}$

31. $\dfrac{-1}{5x-1} = \dfrac{-2}{3x+7}$

32. $\dfrac{-3}{2x-5} = \dfrac{-4}{x-3}$

For Problems 33–44, use proportions to change each common fraction to a percent.

33. $\dfrac{11}{20}$

34. $\dfrac{17}{20}$

35. $\dfrac{3}{5}$

36. $\dfrac{7}{25}$

37. $\frac{1}{6}$ **38.** $\frac{5}{7}$

39. $\frac{3}{8}$ **40.** $\frac{1}{16}$

41. $\frac{3}{2}$ **42.** $\frac{5}{4}$

43. $\frac{12}{5}$ **44.** $\frac{13}{6}$

For Problems 45–56, answer the question by setting up and solving an appropriate equation.

45. What is 7% of 38?

46. What is 35% of 52?

47. 15% of what number is 6.3?

48. 55% of what number is 38.5?

49. 76 is what percent of 95?

50. 72 is what percent of 120?

51. What is 120% of 50?

52. What is 160% of 70?

53. 46 is what percent of 40?

54. 26 is what percent of 20?

55. 160% of what number is 144?

56. 220% of what number is 66?

For Problems 57–73, solve each problem using a proportion.

57. A blueprint has a scale where 1 inch represents 6 feet. Find the dimensions of a rectangular room that measures $2\frac{1}{2}$ inches by $3\frac{1}{4}$ inches on the blueprint.

58. On a certain map, 1 inch represents 15 miles. If two cities are 7 inches apart on the map, find the number of miles between the cities.

59. Suppose that a car can travel 264 miles using 12 gallons of gasoline. How far will it go on 15 gallons?

60. Jesse used 10 gallons of gasoline to drive 170 miles. How much gasoline will he need to travel 238 miles?

61. If the ratio of the length of a rectangle to its width is $\frac{5}{2}$ and the width is 24 centimeters, find its length.

62. If the ratio of the width of a rectangle to its length is $\frac{4}{5}$ and the length is 45 centimeters, find the width.

63. A saltwater solution is made by dissolving 3 pounds of salt in 10 gallons of water. At this rate, how many pounds of salt are needed for 25 gallons of water (see Figure 4.3)?

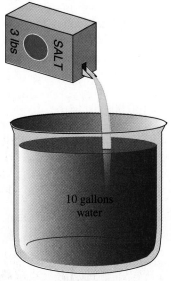

FIGURE 4.3

64. A home valued at $50,000 is assessed $900 in real estate taxes. At the same rate, how much are the taxes on a home assessed at $60,000?

65. If 20 pounds of fertilizer will cover 1500 square feet of lawn, how many pounds are needed for 2500 square feet?

66. It was reported that a flu epidemic is affecting 6 out of every 10 college students in a certain part of the country. At this rate, how many students will be affected at a university of 15,000 students?

67. A preelection poll indicated that 3 out of every 7 eligible voters were going to vote in an upcoming election. At this rate, how many people would be expected to vote in a city of 210,000?

68. A board 28 feet long is cut into two pieces whose lengths are in the ratio of 2 to 5. Find the lengths of the two pieces.

69. The perimeter of a rectangle is 50 inches. If the ratio of its length to its width is 3 to 2, find the dimensions of the rectangle.

70. The ratio of male students to female students at a certain university is 5 to 4. If there is a total of 6975 students, find the number of male students and the number of female students.

71. An investment of $500 earns $45 in a year. At the same rate, how much additional money must be invested to raise the earnings to $72 per year?

72. A sum of $1250 is to be divided between two people in the ratio of 2 to 3. How much does each person receive?

73. An inheritance of $180,000 is to be divided between a child and the local cancer fund in the ratio of 5 to 1. How much money will the child receive?

74. Explain the difference between a ratio and a proportion.

75. What is wrong with the following solution? Explain how it should be done.

$$\frac{x}{2} + 4 = \frac{x}{6}$$

$$6\left(\frac{x}{2} + 4\right) = 2(x)$$

$$3x + 24 = 2x$$

$$x = -24$$

76. Estimate an answer for each of the following problems. Also explain how you arrived at your estimate. Then work out the problem to see how well you estimated.

(a) The ratio of female students to male students at a small private college is 5 to 3. If there is a total of 1096 students, find the number of male students.

(b) If 15 pounds of fertilizer will cover 1200 square feet of lawn, how many pounds are needed for 3000 square feet?

(c) An investment of $5000 earns $300 interest in a year. At the same rate, how much money must be invested to earn $450?

(d) If the ratio of the length of a rectangle to its width is 5 to 3 and the length is 70 centimeters, find its width.

 Further Investigations

Solve each of the following equations. Don't forget that division by zero is undefined.

77. $\dfrac{3}{x - 2} = \dfrac{6}{2x - 4}$

78. $\dfrac{8}{2x + 1} = \dfrac{4}{x - 3}$

79. $\dfrac{5}{x - 3} = \dfrac{10}{x - 6}$

80. $\dfrac{6}{x - 1} = \dfrac{5}{x - 1}$

81. $\dfrac{x - 2}{2} = \dfrac{x}{2} - 1$

82. $\dfrac{x + 3}{x} = 1 + \dfrac{3}{x}$

4.2   More on Percents and Problem Solving

The equation $x + .35 = .72$ can be solved by subtracting .35 from both sides of the equation. Another technique for solving equations that contain decimals is to *clear the equation of all decimals* by multiplying both sides by an appropriate power of 10. The following examples demonstrate the use of the clear-equation-of-all-decimals strategy in a variety of situations.

Example 1

Solution

Solve $.5x = 14$.

$$.5x = 14$$
$$5x = 140 \qquad \text{Multiplied both sides by 10}$$
$$x = 28 \qquad \text{Divided both sides by 5}$$

The solution set is $\{28\}$.

▲

Example 2

Solution

Solve $x + .07x = .13$.

$$x + .07x = .13$$
$$100(x + .07x) = 100(.13) \qquad \text{Multiply both sides by 100}$$
$$100(x) + 100(.07x) = 100(.13) \qquad \text{Distributive property}$$
$$100x + 7x = 13$$
$$107x = 13$$
$$x = \frac{13}{107}$$

The solution set is $\left\{\frac{13}{107}\right\}$.

▲

Example 3

Solution

Solve $.08y + .09y = 3.4$.

$$.08y + .09y = 3.4$$
$$8y + 9y = 340 \qquad \text{Multiplied both sides by 100}$$
$$17y = 340$$
$$y = 20$$

The solution set is $\{20\}$.

▲

Example 4

Solution

Solve $.10t = 560 - .12(t + 1000)$.

$$.10t = 560 - .12(t + 1000)$$
$$10t = 56{,}000 - 12(t + 1000) \qquad \text{Multiplied both sides by 100}$$
$$10t = 56{,}000 - 12t - 12{,}000 \qquad \text{Distributive property}$$
$$22t = 44{,}000$$
$$t = 2000$$

The solution set is $\{2000\}$.

▲

Problems Involving Percents

Many consumer problems can be solved with an equation approach. For example, the following is a general guideline regarding discount sales.

> Original selling price − Discount = Discount sale price

Let's consider some examples using our algebraic techniques along with this basic guideline.

Problem 1

Amy bought a dress at a 30% discount sale for $35. What was the original price of the dress?

Solution

Let p represent the original price of the dress. We can use the basic discount guideline to set up an algebraic equation.

Original selling price −   Discount   =Discount sale price

$$(100\%)(p) \quad -(30\%)(p) = \quad \$35$$

Solving this equation we obtain

$$(100\%)(p) - (30\%)(p) = 35$$
$$(70\%)(p) = 35$$
$$.7p = 35$$
$$7p = 350$$
$$p = 50.$$

The original price of the dress was $50.   ▲

Don't forget that if an item is on sale for 30% off, then you are going to pay $100\% - 30\% = 70\%$ of the original price. So at a 30% discount sale a $50 dress can be purchased for (70%)($50) = $35. (Note that we just checked our answer for Problem 1.)

Problem 2

Find the cost of a $60 pair of jogging shoes on sale for 20% off (see Figure 4.4 on page 149).

Solution

Let x represent the discount sale price. Since the shoes are on sale for 20% off, we must pay 80% of the original price.

$$x = (80\%)(60)$$
$$= (.8)(60) = 48$$

The sale price is $48.   ▲

FIGURE 4.4

Another equation that we can use in consumer problems is the following.

| Selling price = Cost + Profit |

Profit (also called *markup*, *markon*, *margin*, and *margin of profit*) may be stated in different ways. It may be stated as a percent of the selling price, a percent of the cost, or simply in terms of dollars and cents. Let's consider some problems where the profit is either a percent of the selling price or a percent of the cost.

Problem 3

A retailer has some shirts that cost him $20 each. He wants to sell them at a profit of 60% of the cost. What selling price should be marked on the shirts?

Solution

Let *s* represent the selling price. The basic relationship *selling price equals cost plus profit* can be used as a guideline.

Selling price = Cost + Profit (% of cost)
$$s \qquad = \$20 + \quad (60\%)(20)$$

Solving this equation we obtain

$$s = 20 + (60\%)(20)$$
$$s = 20 + (.6)(20)$$
$$s = 20 + 12$$
$$s = 32.$$

The selling price should be $32. ▲

Problem 4

Kathrin bought a painting for $120 and later decided to resell it. She made a profit of 40% of the selling price. What did she receive for the painting?

Solution

We can use the same basic relationship as a guideline except this time the profit is a percent of the selling price. Let *s* represent the selling price.

$$\text{Selling price} = \text{Cost} + \text{Profit (\% of selling price)}$$

$$\downarrow \qquad\qquad \downarrow \qquad\qquad\qquad \downarrow$$

$$s \qquad = 120 + \qquad (40\%)(s)$$

Solving this equation we obtain

$$s = 120 + (40\%)(s)$$
$$s = 120 + .4s$$
$$.6s = 120 \qquad \text{We subtracted .4s from both sides.}$$
$$s = \frac{120}{.6} = 200.$$

She received $200 for the painting.   ▲

Certain types of investment problems can also be translated into algebraic equations. In some of these problems, we use the simple interest formula $i = Prt$, where i represents the amount of interest earned by investing P dollars at a yearly rate of r percent for t years.

Problem 5

A woman invests a total of $5000. Part of it is invested at 10% and the remainder at 12%. Her total yearly interest from the two investments is $560. How much did she invest at each rate?

Solution

Let x represent the amount invested at 12%. Then $5000 - x$ represents the amount invested at 10%. Use the following guideline.

$$\begin{array}{ccc} \text{Interest earned from} & + & \text{Interest earned from} & = & \text{Total interest} \\ \text{12\% investment} & & \text{10\% investment} & & \text{earned} \end{array}$$

$$\downarrow \qquad\qquad\qquad \downarrow \qquad\qquad\qquad \downarrow$$

$$(12\%)(x) \quad + (10\%)(\$5000 - x) = \quad \$560$$

Solving this equation yields

$$(12\%)(x) + (10\%)(5000 - x) = 560$$
$$.12x + .10(5000 - x) = 560$$
$$12x + 10(5000 - x) = 56{,}000$$
$$12x + 50{,}000 - 10x = 56{,}000$$
$$2x + 50{,}000 = 56{,}000$$
$$2x = 6000$$
$$x = 3000.$$
$$\text{Therefore } 5000 - x = 2000.$$

She invested $3000 at 12% and $2000 at 10%.   ▲

Problem 6

An investor invests a certain amount of money at 9%. Then he finds a better deal and invests $5000 more than that amount at 12%. His yearly income from the two investments was $1650. How much did he invest at each rate?

Solution

Let x represent the amount invested at 9%. Then $x + 5000$ represents the amount invested at 12%.

$$(9\%)(x) + (12\%)(x + 5000) = 1650$$
$$.09x + .12(x + 5000) = 1650$$
$$9x + 12(x + 5000) = 165{,}000$$
$$9x + 12x + 60{,}000 = 165{,}000$$
$$21x + 60{,}000 = 165{,}000$$
$$21x = 105{,}000$$
$$x = 5000$$
$$\text{Therefore } x + 5000 = 10{,}000.$$

He invested $5000 at 9% and $10,000 at 12%.   ▲

Problem Set 4.2

For Problems 1–22, solve each of the equations.

1. $x - .36 = .75$

2. $x - .15 = .42$

3. $x + 7.6 = 14.2$

4. $x + 11.8 = 17.1$

5. $.62 - y = .14$ **6.** $7.4 - y = 2.2$

7. $.7t = 56$ **8.** $1.3t = 39$

9. $x = 3.36 - .12x$

10. $x = 5.3 - .06x$ **11.** $s = 35 + .3s$

12. $s = 40 + .5s$ **13.** $s = 42 + .4s$

14. $s = 24 + .6s$

15. $.07x + .08(x + 600) = 78$

16. $.06x + .09(x + 200) = 63$

17. $.09x + .1(2x) = 130.5$

18. $.11x + .12(3x) = 188$

19. $.08x + .11(500 - x) = 50.5$

20. $.07x + .09(2000 - x) = 164$

21. $.09x = 550 - .11(5400 - x)$

22. $.08x = 580 - .1(6000 - x)$

For Problems 23–48, set up an equation and solve each problem.

23. Tom bought a pair of trousers at a 30% discount sale for $35. What was the original price of the trousers?

24. Magda bought a dress for $140, which represents a 20% discount of the original price. What was the original price of the dress?

25. Find the cost of a $48 sweater that is on sale for 25% off.

26. Byron purchased a bicycle at a 10% discount sale for $121.50. What was the original price of the bicycle?

27. Suppose that Jack bought a $32 putter on sale for 35% off (see Figure 4.5). How much did he pay for the putter?

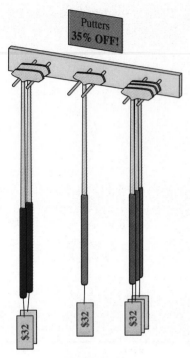

Putters
35% OFF!

$32 $32 $32

FIGURE 4.5

28. Swati bought a 13-inch portable color TV for 20% off of the list price. The list price was $349.95. What did she pay for the TV?

29. Pierre bought a coat for $126 that was listed for $180. What rate of discount did he receive?

30. Phoebe paid $32 for a pair of golf shoes that were listed for $40. What rate of discount did she receive?

31. A retailer has some ties that cost him $5 each. He wants to sell them at a profit of 70% of the cost. What should be the selling price of the ties?

32. A retailer has some blouses that cost her $25 each. She wants to sell them at a profit of 80% of the cost. What price should she charge for the blouses?

33. The owner of a pizza parlor wants to make a profit of 55% of the cost for each pizza sold. If it costs $3 to make a pizza, at what price should it be sold?

34. Produce in a food market usually has a high markup because of loss due to spoilage. If a head of lettuce costs a retailer $.40, at what price should it be sold to realize a profit of 130% of the cost?

35. Jewelry has a very high markup rate. If a ring costs a jeweler $400, at what price should it be sold to gain a profit of 60% of the selling price?

36. If a box of candy costs a retailer $2.50 and he wants to make a profit of 50% based on the selling price, what price should he charge for the candy?

37. If the cost of a pair of shoes for a retailer is $32 and he sells them for $44.80, what is his rate of profit based on the cost?

38. A retailer has some skirts that cost her $24. If she sells them for $31.20, find her rate of profit based on the cost.

39. Suppose that Lou invested a certain amount of money at 9% interest and $250 more than that amount at 10%. His total yearly interest was $101. How much did he invest at each rate?

40. Nina received an inheritance of $12,000 from her grandmother. She invested part of it at 12% interest and the remainder at 14%. If the total yearly interest from both investments was $1580, how much did she invest at each rate?

41. Udit received $1200 from his parents as a graduation present. He invested part of it at 9% interest and the remainder at 12%. If the total yearly interest amounted to $129, how much did he invest at each rate?

42. Sally invested a certain sum of money at 9%, twice that sum at 10%, and three times that sum at 11%. Her total yearly interest from all three investments was $310. How much did she invest at each rate?

43. If $2000 is invested at 8% interest, how much money must be invested at 11% interest so that the total return for both investments averages 10%?

44. Fawn invested a certain amount of money at 10% interest and $250 more than that amount at 11%. Her total yearly interest was $153.50. How much did she invest at each rate?

45. A sum of $2300 is invested, part of it at 10% interest and the remainder at 12%. If the interest earned by the 12% investment is $100 more than the interest earned by the 10% investment, find the amount invested at each rate.

46. If $3000 is invested at 9% interest, how much money must be invested at 12% so that the total return for both investments averages 11%?

47. How can $5400 be invested, part of it at 8% and the remainder at 10%, so that the two investments will produce the same amount of interest?

48. A sum of $6000 is invested, part of it at 9% interest and the remainder at 11%. If the interest earned by the 9% investment is $160 less than the interest earned by the 11% investment, find the amount invested at each rate.

THOUGHTS INTO WORDS

49. What is wrong with the following solution and how should it be changed?

$$1.2x + 2 = 3.8$$
$$10(1.2x) + 2 = 10(3.8)$$
$$12x + 2 = 38$$
$$12x = 36$$
$$x = 3$$

50. From a consumer's viewpoint, would you prefer that a retailer figure his profit based on the cost or the selling price of an item? Explain your answer.

Further Investigations

51. A retailer buys an item for $40, resells it for $50, and claims that she is only making a 20% profit. Is her claim correct?

52. A store has a special discount sale of 40% off on all items. It also advertises an additional 10% off on items bought in quantities of a dozen or more. How much will it cost to buy a dozen items of some particular kind that regularly sell for $5 per item? (Be careful, a 40% discount followed by a 10% discount is not equal to a 50% discount.)

53. Is a 10% discount followed by a 40% discount the same as a 40% discount followed by a 10% discount? Justify your answer.

54. Some people use the following formula for determining the selling price of an item when the profit is based on a percent of the selling price.

$$\text{Selling price} = \frac{\text{Cost}}{100\% - \text{Percent of profit}}$$

Show how to develop this formula.

Solve each of the following equations and express the solutions in decimal form. Your calculator might be of some help.

55. $2.4x + 5.7 = 9.6$

56. $-3.2x - 1.6 = 5.8$

57. $.08x + .09(800 - x) = 68.5$

58. $.10x + .12(720 - x) = 80$

59. $7x - .39 = .03$

60. $9x - .37 = .35$

61. $.2(t + 1.6) = 3.4$

62. $.4(t - 3.8) = 2.2$

4.3   Formulas

To find the distance traveled in 3 hours at a rate of 50 miles per hour, we multiply the rate by the time. Thus, the distance is $50(3) = 150$ miles. We usually state the rule *distance equals rate times time* as a **formula**: $d = rt$. Formulas are simply rules we state in symbolic language and express as equations. Thus, the formula $d = rt$ is an equation that involves three variables, d, r, and t.

As we work with formulas it is often necessary to solve for a specific variable when numerical values for the remaining variables are known. Consider the following examples.

Example 1

Solve $d = rt$ for r if $d = 330$ and $t = 6$.

Solution

Substitute 330 for d and 6 for t in the given formula to obtain

$$330 = r(6).$$

Solving this equation yields

$$330 = 6r$$
$$55 = r. \qquad \blacktriangle$$

Example 2

Solve $C = \dfrac{5}{9}(F - 32)$ for F if $C = 10$. (This formula expresses the relationship between the Fahrenheit and Celsius temperature scales.)

Solution

Substitute 10 for C to obtain

$$10 = \frac{5}{9}(F - 32).$$

Solving this equation produces

$$\frac{9}{5}(10) = \frac{9}{5}\left(\frac{5}{9}\right)(F - 32) \qquad \text{Multiply both sides by } \frac{9}{5}.$$
$$18 = F - 32$$
$$50 = F. \qquad \blacktriangle$$

Sometimes it may be convenient to change a formula's form by using the properties of equality. For example, the formula, $d = rt$, can be changed as follows.

$$d = rt$$
$$\frac{d}{r} = \frac{rt}{r} \qquad \text{Divide both sides by } r.$$
$$\frac{d}{r} = t$$

We say that the formula $d = rt$ has been **solved for the variable** t. The formula can also be **solved for** r as follows.

$$d = rt$$

$$\frac{d}{t} = \frac{rt}{t} \qquad \text{Divide both sides by } t.$$

$$\frac{d}{t} = r$$

Geometric Formulas

There are several formulas in geometry that we use quite often. Let's briefly review them at this time; they will be used periodically throughout the remainder of the text. These formulas (along with some others) and Figures 4.6 through 4.16 are also listed in the inside back cover of this text.

Rectangle

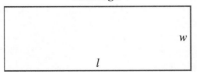

$$A = lw \qquad P = 2l + 2w$$

A area
P perimeter
l length
w width

FIGURE 4.6

Triangle

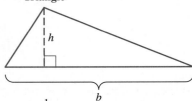

$$A = \frac{1}{2}bh$$

A area
b base
h altitude (height)

FIGURE 4.7

Trapezoid

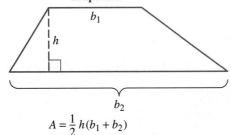

$$A = \frac{1}{2}h(b_1 + b_2)$$

A area
b_1, b_2 bases
h altitude

FIGURE 4.8

Parallelogram

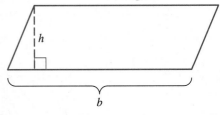

$$A = bh$$

A area
b base
h altitude (height)

FIGURE 4.9

Circle

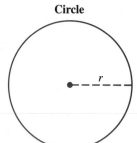

$$A = \pi r^2 \qquad C = 2\pi r$$

A area
C circumference
r radius

FIGURE 4.10

Sphere

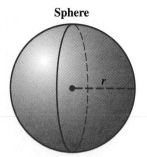

$$V = \frac{4}{3}\pi r^3 \qquad S = 4\pi r^2$$

S surface area
V volume
r radius

FIGURE 4.11

Prism

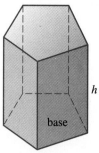

$$V = Bh$$

V volume
B area of base
h altitude (height)

FIGURE 4.12

Rectangular Prism

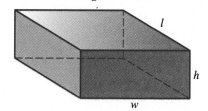

$$V = lwh \qquad S = 2hw + 2hl + 2lw$$

V volume
S total surface area
w width
l length
h altitude (height)

FIGURE 4.13

Pyramid

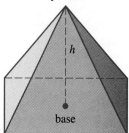

$$V = \frac{1}{3}Bh$$

V volume
B area of base
h altitude (height)

FIGURE 4.14

Right Circular Cylinder

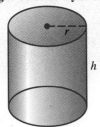

$$V = \pi r^2 h \qquad S = 2\pi r^2 + 2\pi rh$$

V volume
S total surface area
r radius
h altitude (height)

FIGURE 4.15

Right Circular Cone

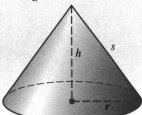

$$V = \frac{1}{3}\pi r^2 h \qquad S = \pi r^2 + \pi rs$$

V volume
S total surface area
r radius
h altitude (height)
s slant height

FIGURE 4.16

Example 3

Solve $C = 2\pi r$ for r.

Solution

$$C = 2\pi r$$

$$\frac{C}{2\pi} = \frac{2\pi r}{2\pi} \qquad \text{Divide both sides by } 2\pi.$$

$$\frac{C}{2\pi} = r$$

▲

Example 4

Solve $V = \frac{1}{3} Bh$ for h.

Solution

$$V = \frac{1}{3} Bh$$

$$3(V) = 3\left(\frac{1}{3}Bh\right) \qquad \text{Multiply both sides by 3.}$$

$$3V = Bh$$

$$\frac{3V}{B} = \frac{Bh}{B} \qquad \text{Divide both sides by B.}$$

$$\frac{3V}{B} = h$$

▲

Example 5

Solve $P = 2l + 2w$ for w.

Solution

$$P = 2l + 2w$$

$$P - 2l = 2l + 2w - 2l \qquad \text{Subtract } 2l \text{ from both sides.}$$

$$P - 2l = 2w$$

$$\frac{P - 2l}{2} = \frac{2w}{2} \qquad \text{Divide both sides by 2.}$$

$$\frac{P - 2l}{2} = w$$

▲

Example 6

Find the total surface area of a right circular cylinder that has a radius of 10 inches and a height of 14 inches.

Solution

Let's sketch a right circular cylinder and record the given information (see Figure 4.17).

FIGURE 4.17

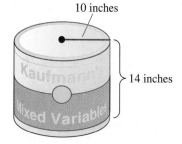

10 inches

14 inches

Substitute 10 for r and 14 for h in the formula for finding total surface area of a right circular cylinder to obtain

$$S = 2\pi r^2 + 2\pi rh$$
$$= 2\pi(10)^2 + 2\pi(10)(14)$$
$$= 200\pi + 280\pi$$
$$= 480\pi.$$

The total surface area is 480π square inches.   ▲

    In Example 6 we used the figure to record the given information and it also served as a reminder of the geometric figure under consideration. Now let's consider an example where a figure is very useful in the analysis of the problem.

Example 7

A sidewalk 3 feet wide surrounds a rectangular plot of ground that measures 75 feet by 100 feet. Find the area of the sidewalk.

Solution

Let's make a sketch and record the given information (see Figure 4.18).

FIGURE 4.18

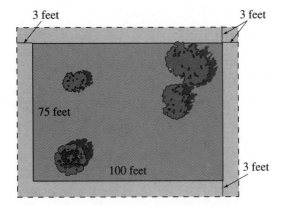

3 feet   3 feet

75 feet

100 feet   3 feet

The area of the sidewalk can be found by subtracting the area of the rectangular plot from the area of the plot plus the sidewalk (the large dashed rectangle). The width of the large rectangle is $75 + 3 + 3 = 81$ feet and its length is $100 + 3 + 3 = 106$ feet.

$$A = (81)(106) - (75)(100)$$
$$= 8586 - 7500 = 1086$$

The area of the sidewalk is 1086 square feet. ▲

    In Chapter 8 you will be working with equations that contain two variables. At times you will need to solve for one variable in terms of the other variable; that is, change the form of the equation as we have been doing with formulas. The following examples illustrate once again how we can use the properties of equality for such situations.

Example 8

Solution

Solve $3x + y = 4$ for x.

$$3x + y = 4$$
$$3x + y - y = 4 - y \qquad \text{Subtract } y \text{ from both sides.}$$
$$3x = 4 - y$$
$$\frac{3x}{3} = \frac{4 - y}{3} \qquad \text{Divide both sides by 3.}$$
$$x = \frac{4 - y}{3}$$

▲

Example 9

Solution

Solve $4x - 5y = 7$ for y.

$$4x - 5y = 7$$
$$4x - 5y - 4x = 7 - 4x \qquad \text{Subtract } 4x \text{ from both sides.}$$
$$-5y = 7 - 4x$$

$$\frac{-5y}{-5} = \frac{7 - 4x}{-5}$$ Divide both sides by −5.

$$y = \frac{7 - 4x}{-5}\left(\frac{-1}{-1}\right)$$ Multiply numerator and denominator of fraction on the right by −1.

$$y = \frac{4x - 7}{5}$$ We commonly do this so that the denominator is positive. ▲

Example 10 Solve $y = mx + b$ for m.

Solution

$$y = mx + b$$
$$y - b = mx + b - b$$ Subtract b from both sides.
$$y - b = mx$$
$$\frac{y - b}{x} = \frac{mx}{x}$$ Divide both sides by x.
$$\frac{y - b}{x} = m$$ ▲

▼ Problem Set 4.3

For Problems 1–10, solve for the specified variable using the given facts.

1. Solve $d = rt$ for t if $d = 336$ and $r = 48$.

2. Solve $d = rt$ for r if $d = 486$ and $t = 9$.

3. Solve $i = Prt$ for P if $i = 200$, $r = .08$, and $t = 5$.

4. Solve $i = Prt$ for t if $i = 540$, $P = 750$, and $r = .09$.

5. Solve $F = \frac{9}{5}C + 32$ for C if F = 68.

6. Solve $C = \frac{5}{9}(F - 32)$ for F if C = 15.

7. Solve $V = \frac{1}{3}Bh$ for B if $V = 112$ and $h = 7$.

8. Solve $V = \frac{1}{3}Bh$ for h if $V = 216$ and $B = 54$.

9. Solve $A = P + Prt$ for t if $A = 652$, $P = 400$, and $r = .07$.

10. Solve $A = P + Prt$ for P if $A = 1032$, $r = .06$, and $t = 12$.

For Problems 11–32, use the geometric formulas given in this section to help solve the problems.

11. Find the perimeter of a rectangle that is 14 centimeters long, and 9 centimeters wide.

12. If the perimeter of a rectangle is 80 centimeters and its length is 24 centimeters, find its width.

13. If the perimeter of a rectangle is 108 inches and its length is $3\frac{1}{4}$ feet, find its width in inches.

14. How many yards of fencing would it take to enclose a rectangular plot of ground that is 69 feet long and 42 feet wide?

15. A dirt path 4 feet wide surrounds a rectangular garden that is 38 feet long and 17 feet wide. Find the area of the dirt path.

16. Find the area of a cement walk 3 feet wide that surrounds a rectangular plot of ground 86 feet long and 42 feet wide.

17. Suppose that paint costs $2.00 per liter and that one liter will cover 9 square meters of surface. We are going to paint (on one side only) 50 pieces of wood of the same size, which are rectangular in shape, and have a length of 60 centimeters and a width of 30 centimeters. What will the cost for the paint be?

18. A lawn is in the shape of a triangle with one side 130 feet long and the altitude to that side 60 feet long. Will one sack of fertilizer, which covers 4000 square feet, be enough to fertilize the lawn?

19. Find the length of an altitude of a trapezoid with bases of 8 inches and 20 inches and an area of 98 square inches.

20. A flower garden is in the shape of a trapezoid with bases of 6 yards and 10 yards. The distance between the bases is 4 yards. Find the area of the garden.

21. In Figure 4.19, you'll notice that the diameter of a metal washer is 4 centimeters. The diameter of the hole is 2 centimeters. How many square centimeters of metal are there in 50 washers? Express the answer in terms of π.

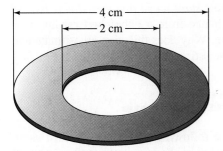

FIGURE 4.19

22. Find the area of a circular plot of ground that has a radius of length 14 meters. Use $3\frac{1}{7}$ as an approximation for π.

23. Find the area of a circular region that has a diameter of 1 yard. Express the answer in terms of π.

24. Find the area of a circular region if the circumference is 12π units. Express the answer in terms of π.

25. Find the total surface area and volume of a sphere that has a radius 9 inches long. Express the answers in terms of π.

26. A circular pool is 34 feet in diameter and has a flagstone walk around it that is 3 feet wide (see Figure 4.20). Find the area of the walk. Express the answer in terms of π.

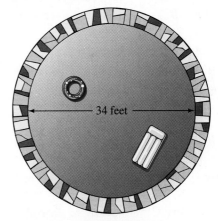

FIGURE 4.20

27. Find the volume and total surface area of a right circular cylinder that has a radius of 8 feet and a height of 18 feet. Express answers in terms of π.

28. Find the total surface area and volume of a sphere that has a diameter 12 centimeters long. Express the answers in terms of π.

29. If the volume of a right circular cone is 324π cubic inches and a radius of the base is 9 inches long, find the height of the cone.

30. Find the volume and total surface area of a tin can if the radius of the base is 3 centimeters, and the height of the can is 10 centimeters. Express answers in terms of π.

31. If the total surface area of a right circular cone is 65π square feet and a radius of the base is 5 feet long, find the slant height of the cone.

32. If the total surface area of a right circular cylinder is 104π square meters and a radius of the base is 4 meters long, find the height of the cylinder.

For Problems 33–44, solve each of the formulas for the indicated variable. (Before doing these problems, cover the right-hand column and see how many of these formulas you recognize!)

33. $V = Bh$ for h Volume of a prism

34. $A = lw$ for l Area of a rectangle

35. $V = \dfrac{1}{3}Bh$ for B Volume of a pyramid

36. $A = \dfrac{1}{2}bh$ for h Area of a triangle

37. $P = 2l + 2w$ for w Perimeter of a rectangle

38. $V = \pi r^2 h$ for h Volume of a cylinder

39. $V = \dfrac{1}{3}\pi r^2 h$ for h Volume of a cone

40. $i = Prt$ for t Simple interest formula

41. $F = \dfrac{9}{5}C + 32$ for C Celsius to Fahrenheit

42. $A = P + Prt$ for t Simple interest formula

43. $A = 2\pi r^2 + 2\pi rh$ for h Surface area of a cylinder

44. $C = \dfrac{5}{9}(F - 32)$ for F Fahrenheit to Celsius

For Problems 45–60, solve each equation for the indicated variable.

45. $3x + 7y = 9$ for x

46. $5x + 2y = 12$ for x

47. $9x - 6y = 13$ for y

48. $3x - 5y = 19$ for y

49. $-2x + 11y = 14$ for x

50. $-x + 14y = 17$ for x

51. $y = -3x - 4$ for x

52. $y = -7x + 10$ for x

53. $\dfrac{x - 2}{4} = \dfrac{y - 3}{6}$ for y

54. $\dfrac{x + 1}{3} = \dfrac{y - 5}{2}$ for y

55. $ax - by - c = 0$ for y

56. $ax + by = c$ for y

57. $\dfrac{x + 6}{2} = \dfrac{y + 4}{5}$ for x

58. $\dfrac{x - 3}{6} = \dfrac{y - 4}{8}$ for x

59. $m = \dfrac{y - b}{x}$ for y

60. $y = mx + b$ for x

THOUGHTS INTO WORDS

61. Suppose that both the length and width of a rectangle are doubled. How does this affect the perimeter of the rectangle? Defend your answer.

62. Suppose that the length of a radius of a circle is doubled. How does this affect the area of the circle? Defend your answer.

63. Some people *subtract 32 and then divide by 2* to estimate the change from a Fahrenheit reading to a Celsius reading. Why does this give an estimate and how good is the estimate?

Further Investigations

For each of the following problems, use 3.14 as an approximation for π. Your calculator should be of some help with these problems.

64. Find the area of a circular plot of ground that has a radius 16.3 meters long. Express your answer to the nearest tenth of a square meter.

65. Find the area, to the nearest tenth of a square centimeter, of the "shaded ring" in Figure 4.21.

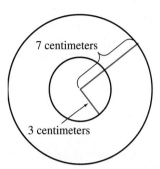

FIGURE 4.21

66. Find the area, to the nearest square inch, of each of the following pizzas.

10-inch diameter _____

12-inch diameter _____

14-inch diameter _____

67. Find the total surface area, to the nearest square centimeter, of the accompanying tin can (Figure 4.22).

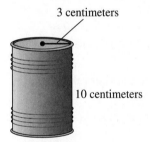

FIGURE 4.22

68. Find the total surface area, to the nearest square centimeter, of a baseball that has a radius of 4 centimeters (Figure 4.23).

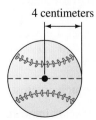

FIGURE 4.23

69. Find the volume, to the nearest cubic inch, of a softball that has a diameter of 5 inches (Figure 4.24).

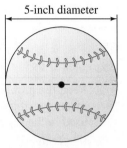

FIGURE 4.24

70. Find the volume, to the nearest cubic meter, of the accompanying figure (Figure 4.25).

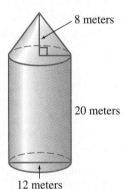

FIGURE 4.25

4.4   Problem Solving

Let's begin this section by restating the suggestions for solving word problems we offered in Section 3.3.

Suggestions for Solving Word Problems

---

1. Read the problem carefully and make certain that you understand the meanings of all the words. Be especially alert for any technical terms used in the statement of the problem.

2. Read the problem a second time (perhaps even a third time) to get an overview of the situation being described and to determine the known facts as well as what is to be found.

3. Sketch any figure, diagram, or chart that might be helpful in analyzing the problem.

4. Choose a meaningful variable to represent an unknown quantity in the problem (perhaps t, if time is the unknown quantity); represent any other unknowns in terms of that variable.

5. Look for a guideline that can be used to set up an equation. A guideline might be a formula such as *selling price equals cost plus profit*, or a relationship such as *interest earned from a 9% investment plus interest earned from a 10% investment equals total amount of interest earned*. A guideline may also be indicated by a figure or diagram that you sketch for a particular problem.

6. Form an equation that contains the variable which translates the conditions of the guideline from English to algebra.

7. Solve the equation and use the solution to determine all facts requested in the problem.

8. **Check all answers back into the original statement of the problem.**

Again we emphasize the importance of suggestion number 5. Determining the guideline to follow when setting up the equation is a vital part of the analysis of a problem. Sometimes the guideline is a formula, such as one of the formulas we presented in the previous section and accompanying problem set. Let's consider a problem of that type.

Problem 1   How long will it take $500 to double itself if it is invested at 8% simple interest?

Solution   Let's use the basic simple interest formula, $i = Prt$, where i represents interest, P is the principal (money invested), r is the rate (percent), and t is the time in years. For $500 to "double itself" means that we want $500 to earn another $500 in interest. Thus, using $i = Prt$ as a guideline we can proceed as follows.

$$i = Prt$$
$$500 = 500(8\%)(t)$$

Now let's solve this equation.

$$500 = 500(.08)(t)$$
$$1 = .08t$$
$$100 = 8t$$
$$\frac{100}{8} = t$$
$$12\frac{1}{2} = t$$

It will take $12\frac{1}{2}$ years. ▲

If the problem involves a geometric formula, then a sketch of the figure is helpful for recording the given information and analyzing the problem. The next example illustrates this idea.

Problem 2

The length of a football field is 40 feet more than twice its width and the perimeter of the field is 1040 feet. Find the length and width of the field.

Solution

Since the length is stated in terms of the width, we can let w represent the width and then $2w + 40$ represents the length (Figure 4.26).

FIGURE 4.26

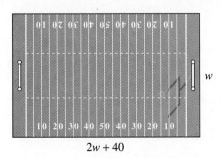

$$2w + 40$$

A guideline for this problem is the perimeter formula $P = 2l + 2w$. Thus, the following equation can be set up and solved.

$$P = 2l + 2w$$
$$1040 = 2(2w + 40) + 2w$$

$$1040 = 4w + 80 + 2w$$
$$1040 = 6w + 80$$
$$960 = 6w$$
$$160 = w$$

If $w = 160$, then $2w + 40 = 2(160) + 40 = 360$. Thus, the football field is 360 feet long and 160 feet wide. ▲

Sometimes the formulas we use when we are analyzing a problem are different than those we use as a guideline for setting up the equation. For example, uniform motion problems involve the formula $d = rt$, but the main guideline for setting up an equation for such problems is usually a statement about either *times*, *rates*, or *distances*. Let's consider an example.

Problem 3

Pablo leaves city A on a moped traveling toward city B at 18 miles per hour. At the same time, Cindy leaves city B on a moped traveling toward city A at 23 miles per hour. The distance between the two cities is 123 miles. How long will it take before Pablo and Cindy meet on their mopeds?

Solution

First, let's sketch a diagram (see Figure 4.27).

FIGURE 4.27

Pablo traveling at 18 mph   Cindy traveling at 23 mph

A   B

total of 123 miles

Let t represent the time that Pablo travels. Then t also represents the time that Cindy travels.

Distance Pablo travels + Distance Cindy travels = Total distance

$$18t \qquad + \qquad 23t \qquad = \qquad 123$$

Solving this equation yields

$$18t + 23t = 123$$
$$41t = 123$$
$$t = 3.$$

They both travel for 3 hours. ▲

Some people find it helpful to use a chart to organize the known and unknown facts in a uniform motion problem We will illustrate with an example.

Problem 4

A car leaves a town at 60 kilometers per hour. How long will it take a second car traveling at 75 kilometers per hour to catch the first car if the second car leaves 1 hour later?

Solution

Let t represent the time of the second car. Then $t + 1$ represents the time of the first car since it travels one hour longer. We can now record the information of the problem in a chart as follows.

	Rate	Time	Distance
First car	60	$t + 1$	$60(t + 1)$
Second car	75	t	$75t$

$d = rt$

Since the second car is to overtake the first car, the distances must be equal.

Distance of second car   Equals   Distance of first car
$$75t \qquad = \qquad 60(t + 1)$$

Solving this equation yields

$$75t = 60(t + 1)$$
$$75t = 60t + 60$$
$$15t = 60$$
$$t = 4.$$

The second car should overtake the first car in 4 hours. (Check the answer!)

We would like to offer one bit of advice at this time. Don't become discouraged if solving word problems is giving you trouble. Problem solving is not a skill that can be developed overnight. It takes time, patience, hard work, and an open mind. Keep giving it your best shot and gradually you should become more confident in your approach to such problems. Furthermore, we realize that some (perhaps many) of these problems may not seem "practical" to you. However, keep in mind that the real goal here is to develop problem solving techniques. Finding and using a guideline, sketching a figure to record information and help in the analysis, estimating an answer before attempting to solve the problem, and using a chart to record information are the key issues.

Problem Set 4.4

For Problems 1–12, solve each of the equations. These equations are of the types you will be using in Problems 13–40.

1. $950(.12)t = 950$

2. $1200(.09)t = 1200$

3. $l + \frac{1}{4}l - 1 = 19$

4. $l + \frac{2}{3}l + 1 = 41$

5. $500(.08)t = 1000$

6. $800(.11)t = 1600$

7. $s + (2s - 1) + (3s - 4) = 37$

8. $s + (3s - 2) + (4s - 4) = 42$

9. $\frac{5}{2}r + \frac{5}{2}(r + 6) = 135$

10. $\frac{10}{3}r + \frac{10}{3}(r - 3) = 90$

11. $24\left(t - \frac{2}{3}\right) = 18t + 8$

12. $16t + 8\left(\frac{9}{2} - t\right) = 60$

Set up an equation and solve each of the following problems. Keep in mind the suggestions we offered in this section.

13. How long will it take $750 to double itself if it is invested at 8% simple interest?

14. How many years will it take $1000 to double itself if it is invested at 10% simple interest?

15. How long will it take $800 to triple itself if it is invested at 10% simple interest?

16. How many years will it take $500 to earn $750 in interest if it is invested at 6% simple interest?

17. The length of a rectangle is three times its width. If the perimeter of the rectangle is 112 inches, find its length and width.

18. The width of a rectangle is one-half of its length. If the perimeter of the rectangle is 54 feet, find its length and width.

19. Suppose that the length of a rectangle is 2 centimeters less than three times its width. The perimeter of the rectangle is 92 centimeters. Find the length and width of the rectangle.

20. Suppose that the length of a certain rectangle is 1 meter more than five times its width. The perimeter of the rectangle is 98 meters. Find the length and width of the rectangle.

21. The width of a rectangle is 3 inches less than one-half of its length. If the perimeter of the rectangle is 42 inches, find the area of the rectangle.

22. The width of a rectangle is 1 foot more than one-third of its length. If the perimeter of the rectangle is 74 feet, find the area of the rectangle.

23. The perimeter of a triangle is 100 feet. The longest side is 3 feet less than twice the shortest side, and the third side is 7 feet longer than the shortest side. Find the lengths of the sides of the triangle.

24. A triangular plot of ground has a perimeter of 54 yards. The longest side is twice the shortest side and the third side is two yards longer than the shortest side. Find the lengths of the sides of the triangle.

25. The second side of a triangle is 1 centimeter more than three times the first side. The third side is 2 centimeters longer than the second side. If the perimeter is 46 centimeters, find the length of each side of the triangle.

26. The second side of a triangle is 3 meters less than twice the first side. The third side is 4 meters longer than the second side. If the perimeter is 58 meters, find the length of each side of the triangle.

27. The perimeter of an equilateral triangle is 4 centimeters more than the perimeter of a square and the length of a side of the triangle is 4 centimeters more than the length of a side of the square. Find the length of a side of the equilateral triangle. (An equilateral triangle has three sides of the same length.)

28. Suppose that a square and an equilateral triangle have the same perimeter. Each side of the equilateral triangle is 6 centimeters longer than each side of the square. Find the length of each side of the square. (An equilateral triangle has three sides of the same length).

29. Suppose that the length of a radius of a circle is the same as the length of a side of a square. If the circumference of the circle is 15.96 centimeters longer than the perimeter of the square, find the length of a radius of the circle. (Use 3.14 as an approximation for π.)

30. The circumference of a circle is 2.24 centimeters more than six times the length of a radius. Find the radius of the circle. (Use 3.14 as an approximation for π.)

31. Sandy leaves a town traveling in her car at a rate of 45 miles per hour. One hour later, Monica leaves the same town traveling the same route at a rate of 50 miles per hour. How long will it take Monica to overtake Sandy?

32. Two cars start from the same place traveling in opposite directions. One car travels 4 miles per hour faster than the other car. Find their speeds if after 5 hours they are 520 miles apart.

33. The distance between city A and city B is 325 miles. A freight train leaves city A and travels toward city B at 40 miles per hour. At the same time, a passenger train leaves city B and travels toward city A at 90 miles per hour. How long will it take the two trains to meet?

34. Kirk starts jogging at 5 miles per hour. One-half hour later Nancy starts jogging on the same route at

7 miles per hour. How long will it take Nancy to catch Kirk?

35. A car leaves a town at 40 miles per hour. Two hours later a second car leaves the town traveling the same route and overtakes the first car in 5 hours and 20 minutes. How fast was the second car traveling?

36. Two airplanes leave St. Louis at the same time and fly in opposite directions (see Figure 4.28). If one travels at 500 kilometers per hour and the other at 600 kilometers per hour, how long will it take for them to be 1925 kilometers apart?

FIGURE 4.28

37. Two trains leave at the same time, one traveling east and the other traveling west. At the end of $9\frac{1}{2}$ hours they are 1292 miles apart. If the rate of the train traveling east is 8 miles per hour faster than the other train, find their rates.

38. Dawn starts on a 58-mile trip on her moped at 20 miles per hour. After a while the motor stopped and she pedals the remainder of the trip at 12 miles per hour. The entire trip took $3\frac{1}{2}$ hours. How far had Dawn traveled when the motor on the moped quit running?

39. Jeff leaves home and rides his bicycle out into the country for 3 hours. On his return trip along the same route, it takes him three-quarters of an hour longer. If his rate on the return trip was 2 miles per hour slower than on the trip out into the country, find the total roundtrip distance.

40. In $1\frac{1}{4}$ hours more time, Rita, riding her bicycle at 12 miles per hour, rode 2 miles further than Sonya, who was riding her bicycle at 16 miles per hour. How long did each girl ride?

41. Suppose that your friend Koen analyzes Problem 31 as follows: Sandy has traveled 45 miles before Monica starts. Since Monica travels 5 miles per hour faster than Sandy, it will take her $\frac{45}{5} = 9$ hours to catch Sandy. How would you react to this analysis of the problem?

42. Summarize the new ideas relative to problem solving that you have acquired thus far in this course.

4.5   More about Problem Solving

Let's begin this section with an important but often overlooked facet of problem solving: The importance of *looking back* over your solution and considering some of the following questions.

1. Is your answer to the problem a reasonable answer? Does it agree with the estimated answer you arrived at before doing the problem?

2. Have you checked your answer by substituting it back into the conditions stated in the problem?

3. Do you now see another plan that could be used to solve the problem? Perhaps there is even another guideline that could be used.

4. Do you now see that this problem is closely related to another problem that you have previously solved?

5. Have you "tucked away for future reference" the technique used to solve this problem?

Looking back over the solution of a newly solved problem can often lay important groundwork for solving problems in the future.

Now let's consider three problems that we often refer to as mixture problems. There is no basic formula that applies for all of these problems, but the suggestion that *you think in terms of a pure substance* is often helpful in setting up a guideline. For example, a phrase such as "30% solution of acid" means that 30% of the amount of solution is acid and the remaining 70% is water.

Problem 1

How many milliliters of pure acid must be added to 150 milliliters of a 30% solution of acid to obtain a 40% solution?

REMARK   If a guideline is not apparent from reading the problem, it might help to guess an answer and then check that guess. Suppose that we guess that 30 milliliters of pure acid need to be added. To check we must determine if the final solution is 40% acid. Since we started with .30(150) = 45 milliliters of pure acid and added our guess of 30 milliliters, the final solution will have 45 + 30 = 75 milliliters of pure acid. The final amount of solution is $150 + 30 = 180$ milliliters. Thus, the final solution is $\dfrac{75}{180} = 41\dfrac{2}{3}\%$ pure acid (see Figure 4.29).

FIGURE 4.29

150 millileters
30% solution

Solution

We hope that by guessing and checking our guess the following guideline becomes apparent.

| Amount of pure acid in original solution | $+$ | Amount of pure acid to be added | $=$ | Amount of pure acid in final solution |

Let p represent the amount of pure acid to be added. Then using the guideline, we can form the following equation.

$$(30\%)(150) + p = 40\%(150 + p)$$

Now let's solve this equation to determine the amount of pure acid to be added.

$$(.30)(150) + p = .40(150 + p)$$
$$45 + p = 60 + .4p$$
$$.6p = 15$$
$$p = \frac{15}{.6} = 25$$

We must add 25 milliliters of pure acid. (Perhaps you should check this answer.)

Problem 2

Suppose that you have a supply of a 30% solution of alcohol and a 70% solution. How many quarts of each should be mixed to produce a 20-quart solution that is 40% alcohol?

Solution

We can use a guideline similar to the one in Problem 1.

$$\boxed{\begin{array}{c}\text{Pure alcohol in}\\\text{30\% solution}\end{array}} + \boxed{\begin{array}{c}\text{Pure alcohol in}\\\text{70\% solution}\end{array}} = \boxed{\begin{array}{c}\text{Pure alcohol in}\\\text{40\% solution}\end{array}}$$

Let x represent the amount of 30% solution. Then $20 - x$ represents the amount of 70% solution. Now using the guideline, we translate to

$$(30\%)(x) + (70\%)(20 - x) = (40\%)(20).$$

Solving this equation we obtain

$$.30x + .70(20 - x) = 8$$
$$30x + 70(20 - x) = 800$$
$$30x + 1400 - 70x = 800$$
$$-40x = -600$$
$$x = 15.$$

Therefore, $20 - x = 5$. We should mix 15 quarts of the 30% solution with 5 quarts of the 70% solution.   ▲

Problem 3

A 4-gallon radiator is full and contains a 40% solution of antifreeze. How much needs to be drained out and replaced with pure antifreeze to obtain a 70% solution?

Solution

The following guideline can be used.

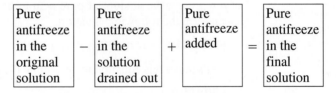

Let x represent the amount of pure antifreeze to be added. Then x also represents the amount of the 40% solution to be drained out. Thus, the guideline translates into the following equation.

$$(40\%)(4) - (40\%)(x) + x = (70\%)(4)$$

Solving this equation we obtain

$$.4(4) - .4x + x = .7(4)$$
$$1.6 + .6x = 2.8$$
$$.6x = 1.2$$
$$x = 2.$$

Therefore, we must drain out 2 gallons of the 40% solution and then add 2 gallons of pure antifreeze. (Checking this answer would be a worthwhile exercise for you!)   ▲

Now let's consider a problem where the process of representing the various unknown quantities in terms of one variable is the key to solving the problem.

Problem 4

Jody is 6 years younger than her sister Cathy, and in 7 years Jody will be three-fourths as old as Cathy. Find their present ages.

Solution

By letting c represent Cathy's present age we can represent all of the unknown quantities as follows.

c:	Cathy's present age
$c - 6$:	Jody's present age
$c + 7$:	Cathy's age in 7 years
$c - 6 + 7$ or $c + 1$:	Jody's age in 7 years

The statement that Jody's age in 7 years will be three-fourths of Cathy's age at that time serves as the guideline. So we can set up and solve the following equation.

$$c + 1 = \frac{3}{4}(c + 7)$$
$$4c + 4 = 3(c + 7)$$
$$4c + 4 = 3c + 21$$
$$c = 17$$

Therefore, Cathy's present age is 17 and Jody's present age is $17 - 6 = 11$.

Problem Set 4.5

For Problems 1–12, solve each of the equations. You will be using these types of equations in Problems 13–37.

1. $.3x + .7(20 - x) = .4(20)$

2. $.4x + .6(50 - x) = .5(50)$

3. $.2(20) + x = .3(20 + x)$

4. $.3(32) + x = .4(32 + x)$

5. $.7(15) - x = .6(15 - x)$

6. $.8(25) - x = .7(25 - x)$

7. $.4(10) - .4x + x = .5(10)$

8. $.2(15) - .2x + x = .4(15)$

9. $20x + 12\left(4\frac{1}{2} - x\right) = 70$

10. $30x + 14\left(3\frac{1}{2} - x\right) = 97$

11. $3t = \frac{11}{2}\left(t - \frac{3}{2}\right)$

12. $5t = \frac{7}{3}\left(t + \frac{1}{2}\right)$

Set up an equation and solve each of the following problems.

13. How many milliliters of pure acid must be added to 100 milliliters of a 10% acid solution to obtain a 20% solution?

14. How many liters of pure alcohol must be added to 20 liters of a 40% solution to obtain a 60% solution?

15. How many centiliters of distilled water must be added to 10 centiliters of a 50% acid solution to obtain a 20% acid solution?

16. How many milliliters of distilled water must be added to 50 milliliters of a 40% acid solution to reduce it to a 10% acid solution?

17. Suppose that we want to mix some 30% alcohol solution with some 50% alcohol solution to obtain 10 quarts of a 35% solution. How many quarts of each kind should we use?

18. We have a 20% alcohol solution and a 50% solution. How many pints must be used from each to obtain 8 pints of a 30% solution?

19. How much water needs to be removed from 20 gallons of a 30% salt solution to change it to a 40% salt solution?

20. How much water needs to be removed from 30 liters of a 20% salt solution to change it to a 50% salt solution?

21. Suppose that a 12-quart radiator contains a 20% solution of antifreeze. How much solution needs to be drained out and replaced with pure antifreeze to obtain a 40% solution of antifreeze?

22. A tank contains 50 gallons of a 40% solution of antifreeze. How much solution needs to be drained out and replaced with pure antifreeze to obtain a 50% solution?

23. How many gallons of a 15% salt solution must be mixed with 8 gallons of a 20% salt solution to obtain a 17% salt solution?

24. How many liters of a 10% salt solution must be mixed with 15 liters of a 40% salt solution to obtain a 20% salt solution?

25. Thirty ounces of a punch that contains 10% grapefruit juice is added to 50 ounces of punch that contains 20% grapefruit juice. Find the percent of grapefruit juice in the resulting mixture.

26. Suppose that 20 gallons of a 20% salt solution is mixed with 30 gallons of a 25% salt solution. What is the percent of salt in the resulting solution?

27. Suppose that the perimeter of a square equals the perimeter of a rectangle. The width of the rectangle is 9 inches less than twice the side of the square and the length of the rectangle is 3 inches less than twice the side of the square. Find the dimensions of the square and the rectangle.

28. The perimeter of a triangle is 40 centimeters. The longest side is 1 centimeter more than twice the shortest side. The other side is 2 centimeters shorter than the longest side. Find the lengths of the three sides.

29. Butch starts walking from point A at 2 miles per hour. One-half hour later, Dick starts walking from point A at $3\frac{1}{2}$ miles per hour and follows the same route. How long will it take Dick to catch up with Butch?

30. Suppose that Karen, riding her bicycle at 15 miles per hour, rode 10 miles farther than Michelle, who was riding her bicycle at 14 miles per hour. Karen rode for 30 minutes longer than Michelle. How long did Michelle and Karen each ride their bicycles?

31. Pam is half as old as her brother Bill. Six years ago Bill was four times older than Pam. How old is each now?

32. Suppose that the sum of the present ages of Tom and his father is 100 years. Ten years ago Tom's father was three times as old as Tom was at that time. Find their present ages.

33. The difference of the present ages of Abby and her mother is 21 years. In ten years Abby's mother's age will be three years less than twice Abby's age at that time. Find their present ages.

34. Amina has a coin collection in which a penny is twice as old as a dime. In 10 years, the age of the penny will be $1\frac{1}{2}$ times the age of the dime at that time. Find the present ages of the penny and the dime.

35. The sum of the ages of our two cars is 11 years. In five years, the age of the older car will be three years more than twice the age of the newer car at that time. Find the present ages of the two cars.

36. Mitch is three-fourths as old as his sister Candy. In 8 years he will be five-sixths as old as Candy. Find their present ages.

37. Ramon rode his bicycle from one town to another at 14 miles per hour. If he had increased his speed to 16 miles per hour he could have made the trip in one-half hour less time. How far apart are the towns?

THOUGHTS INTO WORDS

38. Explain how a *trial and error* approach could be used to solve Problem 35.

39. Now try a *trial and error* approach on Problem 33. Are you having more difficulty than you did with Problem 35? What is causing the difficulty?

SUMMARY

(4.1) A **ratio** is the comparison of two numbers by division.

A statement of equality between two ratios is a **proportion**. In a proportion, the *cross products* are equal. That is to say,

$$\text{if } \frac{a}{b} = \frac{c}{d}, \quad \text{then } ad = bc, \quad b \neq 0 \text{ and } d \neq 0.$$

The cross product property can be used to solve equations that are in the form of a proportion.

A variety of word problems can be conveniently set up and solved using proportions.

The concept of **percent** means "per one hundred" and is therefore a special ratio, namely, one with a denominator of 100. Proportions provide a convenient way to change common fractions to percents.

(4.2) To solve equations that contain decimals, you can clear the equation of all decimals by multiplying both sides by an appropriate power of ten.

Many consumer problems involve the concept of *percent* and can be solved with an equation approach. The basic relationships *selling price equals cost plus profit* and *original selling price minus discount equals discount sale price* are used frequently.

(4.3) **Formulas** are rules stated in symbolic form. A formula such as $P = 2l + 2w$ can be solved for l ($l = \frac{P - 2w}{2}$) or for w ($w = \frac{P - 2l}{2}$) by applying the properties of equality. Many of the formulas used in this section connected algebra, geometry, and the real world.

(4.4) and **(4.5)** Don't forget the following suggestions for solving word problems.

1. Read the problem carefully.
2. Sketch a figure, diagram, or chart that might be helpful to organize the facts.
3. Choose a meaningful variable.
4. Look for a guideline.
5. Use the guideline to help set up an equation.
6. Solve the equation and determine the facts requested in the problem.
7. Check your answers back into the statement of the problem.

Determining a guideline is often a key issue when solving a problem. Many times formulas are used as guidelines or as part of the analysis within the framework of a guideline. A review of the examples of these two sections should help you better understand the role of formulas in problem solving.

▼ Chapter 4 Review Problem Set

For Problems 1–5, solve each of the equations.

1. $.5x + .7x = 1.7$
2. $.07t + .12(t - 3) = .59$
3. $.1x + .12(1700 - x) = 188$
4. $x - .25x = 12$
5. $.2(x - 3) = 14$
6. Solve $P = 2l + 2w$ for w if $P = 50$ and $l = 19$.
7. Solve $F = \frac{9}{5}C + 32$ for C if $F = 77$.
8. Solve $A = P + Prt$ for t.
9. Solve $2x - 3y = 13$ for x.
10. Find the area of a trapezoid that has one base 8 inches long and the other base 14 inches long, if the altitude between the two bases is 7 inches.

11. If the area of a triangle is 27 square centimeters and the length of one side is 9 centimeters, find the length of the altitude to that side.
12. If the total surface area of a right circular cylinder is 152π square feet and a radius of a base is 4 feet long, find the height of the cylinder.

Set up an equation and solve each of the following problems.

13. Eighteen is what percent of 30?
14. The sum of two numbers is 96 and their ratio is 5 to 7. Find the numbers.
15. Fifteen percent of a certain number is six. Find the number.
16. Suppose that the length of a certain rectangle is 5 meters more than twice the width. The perimeter of

the rectangle is 46 meters. Find the length and width of the rectangle.

17. Two airplanes leave Chicago at the same time and fly in opposite directions. If one travels at 350 miles per hour and the other at 400 miles per hour, how long will it take them to be 1125 miles apart?

18. How many liters of pure alcohol must be added to 10 liters of a 70% solution to obtain a 90% solution?

19. A copper wire 110 centimeters long was bent in the shape of a rectangle. The length of the rectangle was 10 centimeters more than twice the width. Find the dimensions of the rectangle.

20. Seventy-eight yards of fencing was purchased to enclose a rectangular garden. The length of the garden is 1 yard less than three times its width. Find the length and width of the garden.

21. The ratio of the complement of an angle to the supplement of the angle is 7 to 16. Find the measure of the angle.

22. If a car uses 18 gallons of gasoline for a 369-mile trip, at the same rate of consumption how many gallons will it use on a 615-mile trip?

23. A sum of $2100 is invested, part of it at 9% interest and the remainder at 11%. If the interest earned by the 11% investment is $51 more than the interest from the 9% investment, find the amount invested at each rate.

24. A retailer has some sweaters that cost $28 each. At what price should the sweaters be sold to obtain a profit of 30% of the selling price?

25. Anastasia bought a dress on sale for $39 and the original price of the dress was $60. What percent of discount did she receive?

26. One angle of a triangle has a measure of 47°. Of the other two angles, one of them is 3° less than three times the other angle. Find the measures of the two remaining angles.

27. Connie rides out into the country on her bicycle at a rate of 10 miles per hour. An hour later Zak leaves from the same place that Connie did and rides his bicycle along the same route at 12 miles per hour. How long will it take Zak to catch Connie?

28. How many gallons of a 10% salt solution must be mixed with 12 gallons of a 15% salt solution to obtain a 12% salt solution?

29. Suppose that 20 ounces of a punch containing 20% orange juice is added to 30 ounces of punch containing 30% orange juice. Find the percent of orange juice in the resulting mixture.

30. The present ages of Bill and Dan are 20 years and 30 years, respectively. Will there be or was there ever a time when Bill's age is (was) one-half of Dan's age?

CHAPTER 4 TEST

For Problems 1–10, solve each of the equations.

1. $\dfrac{x + 2}{4} = \dfrac{x - 3}{5}$

2. $\dfrac{-4}{2x - 1} = \dfrac{3}{3x + 5}$

3. $\dfrac{x - 1}{6} - \dfrac{x + 2}{5} = 2$

4. $\dfrac{x + 8}{7} - 2 = \dfrac{x - 4}{4}$

5. $\dfrac{n}{20 - n} = \dfrac{7}{3}$

6. $\dfrac{h}{4} + \dfrac{h}{6} = 1$

7. $.05n + .06(400 - n) = 23$

8. $s = 35 + .5s$

9. $.07n = 45.5 - .08(600 - n)$

10. $12t + 8\left(\dfrac{7}{2} - t\right) = 50$

11. Solve $F = \dfrac{9C + 160}{5}$ for C.

12. Solve $y = 2(x - 4)$ for x.

13. Solve $\dfrac{x + 3}{4} = \dfrac{y - 5}{9}$ for y.

For Problems 14–16, use the geometric formulas given in this chapter to help solve the problems.

14. Find the area of a circular region if the circumference is 16π centimeters. Express the answer in terms of π.

15. If the perimeter of a rectangle is 100 inches and its length is 32 inches, find the area of the rectangle.

16. The area of a triangular plot of ground is 133 square yards. If the length of one side of the plot is 19 yards, find the length of the altitude to that side.

For Problems 17–25, set up an equation and solve each problem.

17. Express $\dfrac{5}{4}$ as a percent.

18. Thirty five percent of what number is 24.5?

19. Cora bought a blouse for $28, which represented a 30% discount of the original price. What was the original price of the blouse?

20. A retailer has some skirts that cost her $40 each. She wants to sell them at a profit of 30% of the cost. What price should she charge for the skirts?

21. Hugh bought a pair of golf shoes for $48 that was listed for $80. What rate of discount did he receive?

(continued on next page)

CHAPTER 4 TEST *(continued)*

22. The election results of a certain precinct indicated that the ratio of female voters to male voters was 7 to 5. If a total of 1500 people voted, how many females voted?

23. A car leaves a city at 50 miles per hour. One hour later a second car leaves the same city traveling the same route at 55 miles per hour. How long will it take the second car to overtake the first car?

24. How many centiliters of pure acid must be added to 6 centiliters of a 50% acid solution to obtain a 70% acid solution?

25. At the present time, Bianca is 20 years younger than her mother. In 4 years, the ratio of Bianca's age to her mother's age will be 4 to 9. Find Bianca's present age.

Cumulative Review Problem Set

For Problems 1–10, simplify each algebraic expression by combining similar terms.

1. $7x - 9x - 14x$
2. $-10a - 4 + 13a + a - 2$
3. $5(x - 3) + 7(x + 6)$
4. $3(x - 1) - 4(2x - 1)$
5. $-3n - 2(n - 1) + 5(3n - 2) - n$
6. $6n + 3(4n - 2) - 2(2n - 3) - 5$
7. $\frac{1}{2}x - \frac{3}{4}x + \frac{2}{3}x - \frac{1}{6}x$
8. $\frac{1}{3}n - \frac{4}{15}n + \frac{5}{6}n - n$
9. $.4x - .7x - .8x + x$
10. $.5(x - 2) + .4(x + 3) - .2x$

For Problems 11–20, evaluate each of the algebraic expressions for the given values of the variables.

11. $5x - 7y + 2xy$ for $x = -2$ and $y = 5$
12. $2ab - a + 6b$ for $a = 3$ and $b = -4$
13. $-3(x - 1) + 2(x + 6)$ for $x = -5$
14. $5(n + 3) - (n + 4) - n$ for $n = 7$

15. $\frac{3x - 2y}{2x - 3y}$ for $x = 3$ and $y = -6$
16. $\frac{3}{4}n - \frac{1}{3}n + \frac{5}{6}n$ for $n = -\frac{2}{3}$
17. $2a^2 - 4b^2$ for $a = .2$ and $b = -.3$
18. $x^2 - 3xy - 2y^2$ for $x = \frac{1}{2}$ and $y = \frac{1}{4}$
19. $5x - 7y - 8x + 3y$ for $x = 9$ and $y = -8$
20. $\frac{3a - b - 4a + 3b}{a - 6b - 4b - 3a}$ for $a = -1$ and $b = 3$

For Problems 21–26, evaluate each of the expressions.

21. 3^4
22. -2^6
23. $\left(\frac{2}{3}\right)^3$
24. $\left(-\frac{1}{2}\right)^5$
25. $\left(\frac{1}{2} + \frac{1}{3}\right)^2$
26. $\left(\frac{3}{4} - \frac{7}{8}\right)^3$

For Problems 27–38, solve each of the equations.

27. $-5x + 2 = 22$

28. $3x - 4 = 7x + 4$

29. $7(n - 3) = 5(n + 7)$

30. $2(x - 1) - 3(x - 2) = 12$

31. $\dfrac{2}{5}x - \dfrac{1}{3} = \dfrac{1}{3}x + \dfrac{1}{2}$

32. $\dfrac{t - 2}{4} + \dfrac{t + 3}{3} = \dfrac{1}{6}$

33. $\dfrac{2n - 1}{5} - \dfrac{n + 2}{4} = 1$

34. $.09x + .12(500 - x) = 54$

35. $-5(n - 1) - (n - 2) = 3(n - 1) - 2n$

36. $\dfrac{-2}{x - 1} = \dfrac{-3}{x + 4}$

37. $.2x + .1(x - 4) = .7x - 1$

38. $-(t - 2) + (t - 4) = 2\left(t - \dfrac{1}{2}\right) - 3\left(t + \dfrac{1}{3}\right)$

For Problems 39–46, solve each of the inequalities.

39. $4x - 6 > 3x + 1$

40. $-3x - 6 < 12$

41. $-2(n - 1) \le 3(n - 2) + 1$

42. $\dfrac{2}{7}x - \dfrac{1}{4} \ge \dfrac{1}{4}x + \dfrac{1}{2}$

43. $.08t + .1(300 - t) > 28$

44. $-4 > 5x - 2 - 3x$

45. $\dfrac{2}{3}n - 2 \ge \dfrac{1}{2}n + 1$

46. $-3 < -2(x - 1) - x$

For Problems 47–54, set up an equation or an inequality and solve each problem.

47. Erin's salary this year is $32,000. This represents $2000 more than twice her salary five years ago. Find her salary five years ago.

48. One of two supplementary angles is 45° less than four times the other angle. Find the measure of each angle.

49. Jaamal has 25 coins (nickels and dimes) that amount to $2.10. How many coins of each kind does he have?

50. Hana bowled 144 and 176 in her first two games. What must she bowl in the third game to have an average of at least 150 for the three games?

51. A board 30 feet long is cut into two pieces whose lengths are in the ratio of 2 to 3. Find the lengths of the two pieces.

52. A retailer has some shoes that cost him $32 per pair. He wants to sell them at a profit of 20% of the selling price. What price should he charge for the shoes?

53. Two cars start from the same place traveling in opposite directions. One car travels 5 miles per hour faster than the other car. Find their speeds if after six hours they are 570 miles apart.

54. How many liters of pure alcohol must be added to 15 liters of a 20% solution to obtain a 40% solution?

Exponents and Polynomials

A strip of uniform width is shaded along both sides and both ends of a rectangular poster that measures 12 inches by 16 inches. How wide is the strip if one-half of the poster is shaded? **If we let *x* represent the width of the strip, then we can use the equation $16(12) - (16 - 2x)(12 - 2x) = 96$ to determine that the width of the strip is 2 inches.**

The equation we used to solve this problem is called a quadratic equation. **Quadratic equations belong to a larger classification called polynomial equations. To solve problems involving polynomial equations, we need to develop some basic skills that pertain to polynomials. That is to say, we need to be able to add, subtract, multiply, divide, and factor polynomials. These next two chapters are designed to help develop those skills and to present problems that involve quadratic equations.**

5.1   Addition and Subtraction of Polynomials

In the previous chapters, algebraic expressions such as $4x$, $5y$, $-6ab$, $7x^2$, and $-9xy^2z^3$ are called *terms*. Recall that a term is an indicated product that may contain any number of factors. The variables in a term are called *literal factors* and the numerical factor is called the *numerical coefficient* of the term. Thus, in $-6ab$, the a and b are literal factors and the numerical coefficient is -6. Terms that have the same literal factors are similar, or like, terms.

Terms that contain variables with only whole numbers as exponents are called **monomials**. The previously listed terms, $4x$, $5y$, $-6ab$, $7x^2$, and $-9xy^2z^3$ are all monomials. (We will work with some algebraic expressions later, such as $7x^{-1}y^{-1}$ and $4a^{-2}b^{-3}$, which are not monomials.) The **degree of a monomial** is the sum of the exponents of the literal factors.

$4xy$ is of degree 2;

$5x$ is of degree 1;

$14a^2b$ is of degree 3;

$-17xy^2z^3$ is of degree 6;

$-9y^4$ is of degree 4.

If the monomial contains only one variable, then the exponent of the variable is the degree of the monomial. Any nonzero constant term is said to be of degree zero.

A **polynomial** is a monomial or a finite sum (or difference) of monomials. The **degree of a polynomial** is the degree of the term with the highest degree in the polynomial. Some special classifications of polynomials are made according to the number of terms. We call a one-term polynomial a **monomial**, a two-term polynomial a **binomial**, and a three-term polynomial a **trinomial**. The following examples illustrate some of this terminology.

The polynomial $5x^3y^4$ is a monomial of degree 7.

The polynomial $4x^2y - 3xy$ is a binomial of degree 3.

The polynomial $5x^2 - 6x + 4$ is a trinomial of degree 2.

The polynomial $9x^4 - 7x^3 + 6x^2 + x - 2$ is given no special name, but is of degree 4.

Adding Polynomials

In the preceding chapters you have worked many problems involving the addition and subtraction of polynomials. For example, simplifying $4x^2 + 6x + 7x^2 - 2x$ to $11x^2 + 4x$ by combining similar terms can actually be considered the addition problem $(4x^2 + 6x) + (7x^2 - 2x)$. At this time we simply want to review and extend some of those ideas.

Example 1

Add $5x^2 + 7x - 2$ and $9x^2 - 12x + 13$.

Solution

We commonly use the horizontal format for such work. Thus,

$$(5x^2 + 7x - 2) + (9x^2 - 12x + 13) = (5x^2 + 9x^2) + (7x - 12x) + (-2 + 13)$$
$$= 14x^2 - 5x + 11. \quad \blacktriangle$$

The commutative, associative, and distributive properties provide the basis for rearranging, regrouping, and combining similar terms.

Example 2

Add $5x - 1$, $3x + 4$, and $9x - 7$.

Solution

$$(5x - 1) + (3x + 4) + (9x - 7) = (5x + 3x + 9x) + (-1 + 4 + (-7))$$
$$= 17x - 4 \quad \blacktriangle$$

Example 3

Add $-x^2 + 2x - 1$, $2x^3 - x + 4$, and $-5x + 6$.

Solution

$$(-x^2 + 2x - 1) + (2x^3 - x + 4) + (-5x + 6)$$
$$= 2x^3 - x^2 + 2x - x - 5x - 1 + 4 + 6$$
$$= 2x^3 - x^2 - 4x + 9 \quad \blacktriangle$$

Subtracting Polynomials

Recall from Chapter 1 that $a - b = a + (-b)$. We define subtraction as *adding the opposite*. This same idea extends to polynomials in general. The opposite of a polynomial can be formed by taking the opposite of each term. For example, the opposite of $(2x^2 - 7x + 3)$ is $-2x^2 + 7x - 3$. Symbolically, we express this as

$$-(2x^2 - 7x + 3) = -2x^2 + 7x - 3.$$

Now consider the following subtraction problems.

Example 4

Subtract $2x^2 + 9x - 3$ from $5x^2 - 7x - 1$.

Solution

Use the horizontal format to get

$$(5x^2 - 7x - 1) - (2x^2 + 9x - 3) = (5x^2 - 7x - 1) + (-2x^2 - 9x + 3)$$
$$= (5x^2 - 2x^2) + (-7x - 9x) + (-1 + 3)$$
$$= 3x^2 - 16x + 2. \quad \blacktriangle$$

Example 5

Subtract $-8y^2 - y + 5$ from $2y^2 + 9$.

Solution

$$(2y^2 + 9) - (-8y^2 - y + 5) = (2y^2 + 9) + (8y^2 + y - 5)$$
$$= (2y^2 + 8y^2) + (y) + (9 - 5)$$
$$= 10y^2 + y + 4 \quad \blacktriangle$$

Later, when dividing polynomials, we will need to use a vertical format to subtract polynomials. Therefore, let's consider two such examples.

Example 6

Subtract $3x^2 + 5x - 2$ from $9x^2 - 7x - 1$.

Solution

$$9x^2 - 7x - 1$$
$$\underline{3x^2 + 5x - 2}$$

Notice which polynomial goes on the bottom and the alignment of similar terms in columns.

Now we can *mentally form the opposite of the bottom polynomial* and add.

$$9x^2 - 7x - 1$$
$$\underline{3x^2 + 5x - 2}$$
$$6x^2 - 12x + 1$$

The opposite of $3x^2 + 5x - 2$ is $-3x^2 - 5x + 2$.

▲

Example 7

Subtract $15y^3 + 5y^2 + 3$ from $13y^3 + 7y - 1$.

Solution

$$13y^3 + 7y - 1$$
$$\underline{15y^3 + 5y^2 + 3}$$
$$-2y^3 - 5y^2 + 7y - 4$$

The similar terms are arranged in columns.

We mentally formed the opposite of the bottom polynomial and added.

▲

We can use the distributive property along with the properties $a = 1(a)$ and $-a = -1(a)$ when adding and subtracting polynomials. The next examples illustrate this approach.

Example 8

Perform the indicated operations.

$$(3x - 4) + (2x - 5) - (7x - 1)$$

Solution

$$(3x - 4) + (2x - 5) - (7x - 1)$$
$$= 1(3x - 4) + 1(2x - 5) - 1(7x - 1)$$
$$= 1(3x) - 1(4) + 1(2x) - 1(5) - 1(7x) - 1(-1)$$
$$= 3x - 4 + 2x - 5 - 7x + 1$$
$$= 3x + 2x - 7x - 4 - 5 + 1$$
$$= -2x - 8$$

▲

Certainly we can do some of the steps mentally; Example 9 gives a possible format.

Example 9

Perform the indicated operations.

$$(-y^2 + 5y - 2) - (-2y^2 + 8y + 6) + (4y^2 - 2y - 5)$$

Solution

$$(-y^2 + 5y - 2) - (-2y^2 + 8y + 6) + (4y^2 - 2y - 5)$$
$$= -y^2 + 5y - 2 + 2y^2 - 8y - 6 + 4y^2 - 2y - 5$$
$$= -y^2 + 2y^2 + 4y^2 + 5y - 8y - 2y - 2 - 6 - 5$$
$$= 5y^2 - 5y - 13 \qquad \blacktriangle$$

When we use the horizontal format, as in Examples 8 and 9, we use parentheses to indicate a quantity. In Example 8 the quantities $(3x - 4)$ and $(2x - 5)$ are to be added; from this result we are to subtract the quantity $(7x - 1)$. Brackets, [ ], are also sometimes used as grouping symbols, especially if there is a need to indicate quantities within quantities. To remove the grouping symbols, perform the indicated operations, starting with the innermost set of symbols. Let's consider two examples of this type.

Example 10

Perform the indicated operations.

$$3x - [2x + (3x - 1)]$$

Solution

First, we need to add the quantities $2x$ and $(3x - 1)$.

$$3x - [2x + (3x - 1)] = 3x - [2x + 3x - 1]$$
$$= 3x - [5x - 1]$$

Now we need to subtract the quantity $[5x - 1]$ from $3x$.

$$3x - [5x - 1] = 3x - 5x + 1$$
$$= -2x + 1 \qquad \blacktriangle$$

Example 11

Perform the indicated operations.

$$8 - [7x - [2 + (x - 1)] + 4x]$$

Solution

Start with the innermost set of grouping symbols (the parentheses) and proceed as follows.

$$8 - [7x - [2 + (x - 1)] + 4x] = 8 - [7x - [2 + x - 1] + 4x]$$
$$= 8 - [7x - [x + 1] + 4x]$$
$$= 8 - [7x - x - 1 + 4x]$$
$$= 8 - [10x - 1]$$
$$= 8 - 10x + 1$$
$$= -10x + 9 \qquad \blacktriangle$$

As a final example in this section, let's look at polynomials in a geometric setting.

Example 12

Suppose that a parallelogram and a rectangle have dimensions as indicated in Figure 5.1. Find a polynomial that represents the sum of the areas of the two figures.

FIGURE 5.1

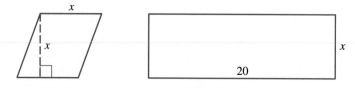

Solution

Using the area formulas $A = bh$ and $A = lw$ for parallelograms and rectangles, respectively, we can represent the sum of the areas of the two figures with the polynomial $x^2 + 20x$.   ▲

Problem Set 5.1

For Problems 1–8, determine the degree of each polynomial.

1. $7x^2y + 6xy$ **2.** $4xy - 7x$

3. $5x^2 - 9$

4. $8x^2 y^2 - 2xy^2 - x$

5. $5x^3 - x^2 - x + 3$

6. $8x^4 - 2x^2 + 6$ **7.** $5xy$

8. $-7x + 4$

For Problems 9–22, add the polynomials.

9. $3x + 4$ and $5x + 7$

10. $3x - 5$ and $2x - 9$

11. $-5y - 3$ and $9y + 13$

12. $x^2 - 2x - 1$ and $-2x^2 + x + 4$

13. $-2x^2 + 7x - 9$ and $4x^2 - 9x - 14$

14. $3a^2 + 4a - 7$ and $-3a^2 - 7a + 10$

15. $5x - 2, 3x - 7$, and $9x - 10$

16. $-x - 4, 8x + 9$, and $-7x - 6$

17. $2x^2 - x + 4, -5x^2 - 7x - 2$, and $9x^2 + 3x - 6$

18. $-3x^2 + 2x - 6, 6x^2 + 7x + 3$, and $-4x^2 - 9$

19. $-4n^2 - n - 1$ and $4n^2 + 6n - 5$

20. $-5n^2 + 7n - 9$ and $-5n - 4$

21. $2x^2 - 7x - 10, -6x - 2$, and $-9x^2 + 5$

22. $7x - 11, -x^2 - 5x + 9$,   and $-4x + 5$

For Problems 23–34, subtract the polynomials using a horizontal format.

23. $7x + 1$ from $12x + 6$

24. $10x + 3$ from $14x + 13$

25. $5x - 2$ from $3x - 7$

26. $7x - 2$ from $2x + 3$

27. $-x - 1$ from $-4x + 6$

28. $-3x + 2$ from $-x - 9$

29. $x^2 - 7x + 2$ from $3x^2 + 8x - 4$

30. $2x^2 + 6x - 1$ from $8x^2 - 2x + 6$

31. $-2n^2 - 3n + 4$ from $3n^2 - n + 7$

32. $3n^2 - 7n - 9$ from $-4n^2 + 6n + 10$

33. $-4x^3 - x^2 + 6x - 1$ from $-7x^3 + x^2 + 6x - 12$

34. $-4x^2 + 6x - 2$ from $-3x^3 + 2x^2 + 7x - 1$

For Problems 35–44, subtract the polynomials using a vertical format.

35. $3x - 2$ from $12x - 4$

36. $-4x + 6$ from $7x - 3$

37. $-5a - 6$ from $-3a + 9$

38. $7a - 11$ from $-2a - 1$

39. $8x^2 - x + 6$ from $6x^2 - x + 11$

40. $3x^2 - 2$ from $-2x^2 + 6x - 4$

41. $-2x^3 - 6x^2 + 7x - 9$ from $4x^3 + 6x^2 + 7x - 14$

42. $4x^3 + x - 10$ from $3x^2 - 6$

43. $2x^2 - 6x - 14$ from $4x^3 - 6x^2 + 7x - 2$

44. $3x - 7$ from $7x^3 + 6x^2 - 5x - 4$

For Problems 45–64, perform the indicated operations.

45. $(5x + 3) - (7x - 2) + (3x + 6)$

46. $(3x - 4) + (9x - 1) - (14x - 7)$

47. $(-x - 1) - (-2x + 6) + (-4x - 7)$

48. $(-3x + 6) + (-x - 8) - (-7x + 10)$

49. $(x^2 - 7x - 4) + (2x^2 - 8x - 9) - (4x^2 - 2x - 1)$

50. $(3x^2 + x - 6) - (8x^2 - 9x + 1) - (7x^2 + 2x - 6)$

51. $(-x^2 - 3x + 4) + (-2x^2 - x - 2) - (-4x^2 + 7x + 10)$

52. $(-3x^2 - 2) + (7x^2 - 8) - (9x^2 - 2x - 4)$

53. $(3a - 2b) - (7a + 4b) - (6a - 3b)$

54. $(5a + 7b) + (-8a - 2b) - (5a + 6b)$

55. $(n - 6) - (2n^2 - n + 4) + (n^2 - 7)$

56. $(3n + 4) - (n^2 - 9n + 10) - (-2n + 4)$

57. $7x + [3x - (2x - 1)]$

58. $-6x + [-2x - (5x + 2)]$

59. $-7n - [4n - (6n - 1)]$

60. $9n - [3n - (5n + 4)]$

61. $(5a - 1) - [3a + (4a - 7)]$

62. $(-3a + 4) - [-7a + (9a - 1)]$

63. $13x - [5x - [4x - (x - 6)]]$

64. $-10x - [7x - [3x - (2x - 3)]]$

65. Subtract $5x - 3$ from the sum of $4x - 2$ and $7x + 6$.

66. Subtract $7x + 5$ from the sum of $9x - 4$ and $-3x - 2$.

67. Subtract the sum of $-2n - 5$ and $-n + 7$ from $-8n + 9$.

68. Subtract the sum of $7n - 11$ and $-4n - 3$ from $13n - 4$.

69. Find a polynomial that represents the perimeter of the rectangle in Figure 5.2.

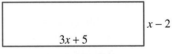

$x - 2$

$3x + 5$

FIGURE 5.2

70. Find a polynomial that represents the area of the shaded region in Figure 5.3. The length of a radius of the larger circle is r units and the length of a radius of the smaller circle is 4 units.

FIGURE 5.3

71. Find a polynomial that represents the sum of the areas of the rectangles and squares in Figure 5.4.

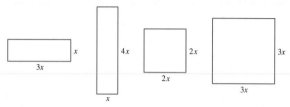

FIGURE 5.4

72. Find a polynomial that represents the total surface area of the rectangular solid in Figure 5.5.

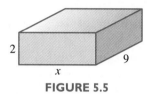

FIGURE 5.5

THOUGHTS INTO WORDS

73. Explain how to subtract the polynomial $3x^2 + 6x - 2$ from $4x^2 + 7$.

74. Is the sum of two binomials always another binomial? Defend your answer.

75. Is the sum of two binomials ever a trinomial? Defend your answer.

5.2 Multiplying Monomials

In Section 2.4 we used exponents and some of the basic properties of real numbers to simplify algebraic expressions into a more compact form. For example,

$$(3x)(4xy) = 3 \cdot 4 \cdot x \cdot x \cdot y = 12x^2y.$$

Actually, we were **multiplying monomials**, and it is this topic that we will pursue further at this time. Multiplying monomials is made easier by using some basic properties of exponents. These properties are the direct result of the definition of an exponent. The following examples lead to the first property.

$$x^2 \cdot x^3 = (x \cdot x)(x \cdot x \cdot x) = x^5$$
$$a^3 \cdot a^4 = (a \cdot a \cdot a)(a \cdot a \cdot a \cdot a) = a^7$$
$$b \cdot b^2 = (b)(b \cdot b) = b^3$$

In general,

$$b^n \cdot b^m = \underbrace{(b \cdot b \cdot b \cdots b)}_{n \text{ factors of } b}\underbrace{(b \cdot b \cdot b \cdots b)}_{m \text{ factors of } b}$$

$$= \underbrace{b \cdot b \cdot b \cdots b}_{\substack{(n + m) \text{ factors} \\ \text{of } b}}$$

$$= b^{n+m}.$$

PROPERTY 5.1

If b is any real number and n and m are positive integers, then

$$b^n \cdot b^m = b^{n+m}.$$

Property 5.1 states that when multiplying powers with the same base, add exponents.

Example 1

Multiply each of the following.

(a) $x^4 \cdot x^3$ (b) $a^8 \cdot a^7$

Solution

(a) $x^4 \cdot x^3 = x^{4+3} = x^7$ (b) $a^8 \cdot a^7 = a^{8+7} = a^{15}$ ▲

Another property of exponents is demonstrated by the following examples.

$$(x^2)^3 = x^2 \cdot x^2 \cdot x^2 = x^{2+2+2} = x^6$$
$$(a^3)^2 = a^3 \cdot a^3 = a^{3+3} = a^6$$
$$(b^3)^4 = b^3 \cdot b^3 \cdot b^3 \cdot b^3 = b^{3+3+3+3} = b^{12}$$

In general,

$$(b^n)^m = \underbrace{b^n \cdot b^n \cdot b^n \cdot \cdots \cdot b^n}_{m \text{ factors of } b^n}$$

$$= b^{\overbrace{n+n+n+ \cdots +n}^{m \text{ of these } ns}}$$

$$= b^{mn}.$$

PROPERTY 5.2

If b is any real number and m and n are positive integers, then

$$(b^n)^m = b^{mn}.$$

Property 5.2 states that when raising a power to a power, multiply exponents.

Example 2

Raise each of the following to the indicated power.

(a) $(x^4)^3$ (b) $(a^5)^6$

Solution

(a) $(x^4)^3 = x^{3 \cdot 4} = x^{12}$ (b) $(a^5)^6 = a^{6 \cdot 5} = a^{30}$ ▲

The third property of exponents we will use in this section raises a monomial to a power.

$$(2x)^3 = (2x)(2x)(2x) = 2 \cdot 2 \cdot 2 \cdot x \cdot x \cdot x = 2^3 \cdot x^3$$
$$(3a^4)^2 = (3a^4)(3a^4) = 3 \cdot 3 \cdot a^4 \cdot a^4 = (3)^2(a^4)^2$$
$$(-2xy^5)^2 = (-2xy^5)(-2xy^5) = (-2)(-2)(x)(x)(y^5)(y^5) = (-2)^2(x)^2(y^5)^2$$

In general,

$$(ab)^n = \underbrace{ab \cdot ab \cdot ab \cdots ab}_{n \text{ factors of } ab}$$

$$= \underbrace{(a \cdot a \cdot a \cdots a)}_{n \text{ factors of } a}\underbrace{(b \cdot b \cdot b \cdots b)}_{n \text{ factors of } b}$$

$$= a^n b^n.$$

PROPERTY 5.3

If a and b are real numbers and n is a positive integer, then

$$(ab)^n = a^n b^n.$$

Property 5.3 states that when raising a monomial to a power, raise each factor to that power.

Example 3

Raise each of the following to the indicated power.

(a) $(2x^2y^3)^4$ (b) $(-3ab^5)^3$

Solution

(a) $(2x^2y^3)^4 = (2)^4(x^2)^4(y^3)^4 = 16x^8y^{12}$
(b) $(-3ab^5)^3 = (-3)^3(a^1)^3(b^5)^3 = -27a^3b^{15}$ ▲

Consider the following examples in which we use the properties of exponents to help simplify the process of multiplying monomials.

1. $(3x^3)(5x^4) = 3 \cdot 5 \cdot x^3 \cdot x^4$
$$= 15x^7 \qquad x^3 \cdot x^4 = x^{3+4} = x^7$$

2. $(-4a^2b^3)(6ab^2) = -4 \cdot 6 \cdot a^2 \cdot a \cdot b^3 \cdot b^2$
$$= -24a^3b^5$$

3. $(xy)(7xy^5) = 1 \cdot 7 \cdot x \cdot x \cdot y \cdot y^5$ The numerical coefficient of xy is 1.
$$= 7x^2y^6$$

4. $\left(\frac{3}{4}x^2y^3\right)\left(\frac{1}{2}x^3y^5\right) = \frac{3}{4} \cdot \frac{1}{2} \cdot x^2 \cdot x^3 \cdot y^3 \cdot y^5$
$$= \frac{3}{8}x^5y^8$$

It is a simple process to raise a monomial to a power when using the properties of exponents. Study the following examples.

5. $(2x^3)^4 = (2)^4(x^3)^4$ by using $(ab)^n = a^n b^b$

$\qquad\qquad = (2)^4(x^{12})$ by using $(b^n)^m = b^{mn}$

$\qquad\qquad = 16x^{12}$

6. $(-2a^4)^5 = (-2)^5(a^4)^5$

$\qquad\qquad\; = -32a^{20}$

7. $\left(\dfrac{2}{5}x^2y^3\right)^3 = \left(\dfrac{2}{5}\right)^3(x^2)^3(y^3)^3$

$\qquad\qquad\quad = \dfrac{8}{125}x^6y^9$

8. $(.2a^6b^7)^2 = (.2)^2(a^6)^2(b^7)^2$

$\qquad\qquad\; = .04a^{12}b^{14}$

Sometimes problems involve first raising monomials to a power and then multiplying the resulting monomials, as in the following examples.

9. $(3x^2)^3(2x^3)^2 = (3)^3(x^2)^3(2)^2(x^3)^2$

$\qquad\qquad\qquad = (27)(x^6)(4)(x^6)$

$\qquad\qquad\qquad = 108x^{12}$

10. $(-x^2y^3)^5(-2x^2y)^2 = (-1)^5(x^2)^5(y^3)^5(-2)^2(x^2)^2(y)^2$

$\qquad\qquad\qquad\qquad = (-1)(x^{10})(y^{15})(4)(x^4)(y^2)$

$\qquad\qquad\qquad\qquad = -4x^{14}y^{17}$

The distributive property along with the properties of exponents forms a basis for finding the product of a monomial and a polynomial. The following examples illustrate these ideas.

11. $(3x)(2x^2 + 6x + 1) = (3x)(2x^2) + (3x)(6x) + (3x)(1)$

$\qquad\qquad\qquad\qquad = 6x^3 + 18x^2 + 3x$

12. $(5a^2)(a^3 - 2a^2 - 1) = (5a^2)(a^3) - (5a^2)(2a^2) - (5a^2)(1)$

$\qquad\qquad\qquad\qquad = 5a^5 - 10a^4 - 5a^2$

13. $(-2xy)(6x^2y - 3xy^2 - 4y^3)$

$\qquad = (-2xy)(6x^2y) - (-2xy)(3xy^2) - (-2xy)(4y^3)$

$\qquad = -12x^3y^2 + 6x^2y^3 + 8xy^4$

Once you feel comfortable with this process, you may want to perform most of the work mentally and simply write down the final result. See if you understand the following examples.

14. $3x(2x + 3) = 6x^2 + 9x$

15. $-4x(2x^2 - 3x - 1) = -8x^3 + 12x^2 + 4x$

16. $ab(3a^2b - 2ab^2 - b^3) = 3a^3b^2 - 2a^2b^3 - ab^4$

Let's conclude this section with a connection between algebra and geometry.

Example 4

Suppose that the dimensions of a rectangular solid are represented by x, $2x$, and $3x$ as shown in Figure 5.6. Express the volume and total surface area of the figure.

FIGURE 5.6

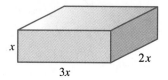

Solution

Using the formula $V = lwh$, the volume of the rectangular solid can be expressed as $(2x)(3x)(x)$, which equals $6x^3$. The total surface area can be described as follows:

Area of front and back rectangles	$2(x)(3x) = 6x^2$
Area of left-side and right-side	$2(2x)(x) = 4x^2$
Area of top and bottom	$2(2x)(3x) = 12x^2$

The total surface area is represented by $6x^2 + 4x^2 + 12x^2$ or $22x^2$.   ▲

Problem Set 5.2

For Problems 1–30, multiply using the properties of exponents to help with the manipulation.

1. $(5x)(9x)$

2. $(7x)(8x)$

3. $(3x^2)(7x)$

4. $(9x)(4x^3)$

5. $(-3xy)(2xy)$

6. $(6xy)(-3xy)$

7. $(-2x^2y)(-7x)$

8. $(-5xy^2)(-4y)$

9. $(4a^2b^2)(-12ab)$

10. $(-3a^3b)(13ab^2)$

11. $(-xy)(-5x^3)$

12. $(-7y^2)(-x^2y)$

13. $(8ab^2c)(13a^2c)$

14. $(9abc^3)(14bc^2)$

15. $(5x^2)(2x)(3x^3)$

16. $(4x)(2x^2)(6x^4)$

17. $(4xy)(-2x)(7y^2)$

18. $(5y^2)(-3xy)(5x^2)$

19. $(-2ab)(-ab)(-3b)$

20. $(-7ab)(-4a)(-ab)$

21. $(6cd)(-3c^2d)(-4d)$

22. $(2c^3d)(-6d^3)(-5cd)$

23. $\left(\frac{2}{3}xy\right)\left(\frac{3}{5}x^2y^4\right)$

24. $\left(-\dfrac{5}{6}x\right)\left(\dfrac{8}{3}x^2y\right)$

25. $\left(-\dfrac{7}{12}a^2b\right)\left(\dfrac{8}{21}b^4\right)$

26. $\left(-\dfrac{9}{5}a^3b^4\right)\left(-\dfrac{15}{6}ab^2\right)$

27. $(.4x^5)(.7x^3)$

28. $(-1.2x^4)(.3x^2)$

29. $(-4ab)(1.6a^3b)$

30. $(-6a^2b)(-1.4a^2b^4)$

For Problems 31–46, raise each monomial to the indicated power. Use the properties of exponents to help with the manipulation.

31. $(2x^4)^2$ **32.** $(3x^3)^2$

33. $(-3a^2b^3)^2$ **34.** $(-8a^4b^5)^2$

35. $(3x^2)^3$ **36.** $(2x^4)^3$

37. $(-4x^4)^3$ **38.** $(-3x^3)^3$

39. $(9x^4y^5)^2$ **40.** $(8x^6y^4)^2$

41. $(2x^2y)^4$ **42.** $(2x^2y^3)^5$

43. $(-3a^3b^2)^4$ **44.** $(-2a^4b^2)^4$

45. $(-x^2y)^6$ **46.** $(-x^2y^3)^7$

For Problems 47–60, multiply by using the distributive property.

47. $5x(3x + 2)$

48. $7x(2x + 5)$

49. $3x^2(6x - 2)$

50. $4x^2(7x - 2)$

51. $-4x(7x^2 - 4)$

52. $-6x(9x^2 - 5)$

53. $2x(x^2 - 4x + 6)$

54. $3x(2x^2 - x + 5)$

55. $-6a(3a^2 - 5a - 7)$

56. $-8a(4a^2 - 9a - 6)$

57. $7xy(4x^2 - x + 5)$

58. $5x^2y(3x^2 + 7x - 9)$

59. $-xy(9x^2 - 2x - 6)$

60. $xy^2(6x^2 - x - 1)$

For Problems 61–70, remove parentheses by multiplying and then simplify by combining similar terms. For example,

$$3(x - y) + 2(x - 3y) = 3x - 3y + 2x - 6y$$
$$= 5x - 9y.$$

61. $5(x + 2y) + 4(2x + 3y)$

62. $3(2x + 5y) + 2(4x + y)$

63. $4(x - 3y) - 3(2x - y)$

64. $2(5x - 3y) - 5(x + 4y)$

65. $2x(x^2 - 3x - 4) + x(2x^2 + 3x - 6)$

66. $3x(2x^2 - x + 5) - 2x(x^2 + 4x + 7)$

67. $3[2x - (x - 2)] - 4(x - 2)$

68. $2[3x - (2x + 1)] - 2(3x - 4)$

69. $-4(3x + 2) - 5[2x - (3x + 4)]$

70. $-5(2x - 1) - 3[x - (4x - 3)]$

For Problems 71–80, perform the indicated operations and simplify.

71. $(3x)^2(2x^3)$

72. $(-2x)^3(4x^5)$

73. $(-3x)^3(-4x)^2$

74. $(3xy)^2(2x^2y)^4$

75. $(5x^2y)^2(xy^2)^3$

76. $(-x^2y)^3(6xy)^2$

77. $(-a^2bc^3)^3(a^3b)^2$

78. $(ab^2c^3)^4(-a^2b)^3$

79. $(-2x^2y^2)^4(-xy^3)^3$

80. $(-3xy)^3(-x^2y^3)^4$

81. Express, in simplified form, the sum of the areas of the two rectangles shown in Figure 5.7.

FIGURE 5.7

82. Express, in simplified form, the volume and the total surface area of the rectangular solid in Figure 5.8.

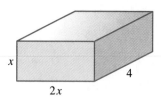

2x

FIGURE 5.8

83. Represent the area of the shaded region in Figure 5.9. The length of a radius of the smaller circle is x and the length of a radius of the larger circle is $2x$.

FIGURE 5.9

84. Represent the area of the shaded region in Figure 5.10.

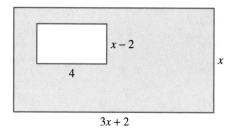

FIGURE 5.10

85. How would you explain to someone why the product of x^3 and x^4 is x^7 and not x^{12}?

86. Suppose your friend was absent from class the day that this section was discussed. How would you help her understand why the property $(b^n)^m = b^{mn}$ is true?

87. How can Figure 5.11 be used to geometrically demonstrate that $x(x + 2) = x^2 + 2x$?

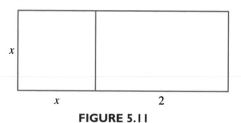

FIGURE 5.11

Further Investigations

For Problems 88–97, find each of the indicated products. Assume that the variables in the exponents represent positive integers. For example,

$$(x^{2n})(x^{4n}) = x^{2n+4n} = x^{6n}.$$

88. $(x^n)(x^{3n})$ **89.** $(x^{2n})(x^{5n})$

90. $(x^{2n-1})(x^{3n+2})$

91. $(x^{5n+2})(x^{n-1})$ **92.** $(x^3)(x^{4n-5})$

93. $(x^{6n-1})(x^4)$ **94.** $(2x^n)(3x^{2n})$

95. $(4x^{3n})(-5x^{7n})$

96. $(-6x^{2n+4})(5x^{3n-4})$

97. $(-3x^{5n-2})(-4x^{2n+2})$

5.3 Multiplying Polynomials

In general, to go from multiplying a monomial and a polynomial to multiplying two polynomials requires the use of the distributive property. Consider the following examples.

Example 1

Find the product of $(x + 3)$ and $(y + 4)$.

Solution

$$(x + 3)(y + 4) = x(y + 4) + 3(y + 4)$$
$$= x(y) + x(4) + 3(y) + 3(4)$$
$$= xy + 4x + 3y + 12 \qquad \blacktriangle$$

Notice that each term of the first polynomial is multiplied times each term of the second polynomial.

Example 2

Find the product of $(x - 2)$ and $(y + z + 5)$.

Solution

$$(x - 2)(y + z + 5) = x(y) + x(z) + x(5) - 2(y) - 2(z) - 2(5)$$
$$= xy + xz + 5x - 2y - 2z - 10 \qquad \blacktriangle$$

Frequently, multiplying polynomials will produce similar terms that can be combined to simplify the resulting polynomial.

Example 3

Multiply $(x + 3)(x + 2)$.

Solution

$$(x + 3)(x + 2) = x(x + 2) + 3(x + 2)$$
$$= x^2 + 2x + 3x + 6$$
$$= x^2 + 5x + 6 \qquad \blacktriangle$$

Example 4

Multiply $(x - 4)(x + 9)$.

Solution

$$(x - 4)(x + 9) = x(x + 9) - 4(x + 9)$$
$$= x^2 + 9x - 4x - 36$$
$$= x^2 + 5x - 36 \qquad \blacktriangle$$

Example 5

Multiply $(x + 4)(x^2 + 3x + 2)$.

Solution

$$(x + 4)(x^2 + 3x + 2) = x(x^2 + 3x + 2) + 4(x^2 + 3x + 2)$$
$$= x^3 + 3x^2 + 2x + 4x^2 + 12x + 8$$
$$= x^3 + 7x^2 + 14x + 8 \qquad \blacktriangle$$

Example 6

Multiply $(2x - y)(3x^2 - 2xy + 4y^2)$.

Solution

$$(2x - y)(3x^2 - 2xy + 4y^2) = 2x(3x^2 - 2xy + 4y^2) - y(3x^2 - 2xy + 4y^2)$$
$$= 6x^3 - 4x^2y + 8xy^2 - 3x^2y + 2xy^2 - 4y^3$$
$$= 6x^3 - 7x^2y + 10xy^2 - 4y^3 \qquad \blacktriangle$$

Perhaps the most frequently used type of multiplication problem is the product of two binomials. It will be a big help later if you can become proficient at multiplying binomials without showing all of the intermediate steps. This is quite easy to do if you use a three-step shortcut pattern demonstrated by the following examples.

Example 7

Multiply $(x + 5)(x + 7)$.

Solution

FIGURE 5.12

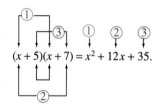

STEP 1   Multiply $x \cdot x$.
STEP 2   Multiply $5 \cdot x$ and $7 \cdot x$ and combine.
STEP 3   Multiply $5 \cdot 7$. $\qquad \blacktriangle$

Example 8

Multiply $(x - 8)(x + 3)$.

Solution

FIGURE 5.13

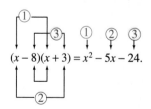

$\qquad \blacktriangle$

Example 9

Multiply $(3x + 2)(2x - 5)$.

Solution

FIGURE 5.14

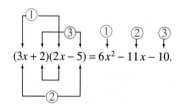

$\qquad \blacktriangle$

Now see if *you* can use the pattern to find the following products.

$(x + 3)(x + 7) =$ _____

$(3x + 1)(2x + 5) =$ _____

$(x - 2)(x - 3) =$ _____

$(4x + 5)(x - 2) =$ _____

Your answers should be $x^2 + 10x + 21$, $6x^2 + 17x + 5$, $x^2 - 5x + 6$, and $4x^2 - 3x - 10$.

Keep in mind that the shortcut pattern discussed above applies only to finding the product of two binomials. For other situations, such as finding the product of a binomial and a trinomial, we would suggest showing the intermediate steps as follows.

$$(x + 3)(x^2 + 6x - 7) = x(x^2) + x(6x) - x(7) + 3(x^2) + 3(6x) - 3(7)$$
$$= x^3 + 6x^2 - 7x + 3x^2 + 18x - 21$$
$$= x^3 + 9x^2 + 11x - 21$$

Perhaps we could omit the first step and shorten the form as follows.

$$(x - 4)(x^2 - 5x - 6) = x^3 - 5x^2 - 6x - 4x^2 + 20x + 24$$
$$= x^3 - 9x^2 + 14x + 24$$

Remember that you are multiplying each term of the first polynomial times each term of the second polynomial and combining similar terms.

Exponents are also used to indicate repeated multiplication of polynomials. For example, $(x + 4)(x + 4)$ can be written as $(x + 4)^2$. Thus, to square a binomial we simply write it as the product of two equal binomials and apply the shortcut pattern.

$(x + 4)^2 = (x + 4)(x + 4) = x^2 + 8x + 16$;

$(x - 5)^2 = (x - 5)(x - 5) = x^2 - 10x + 25$;

$(2x + 3)^2 = (2x + 3)(2x + 3) = 4x^2 + 12x + 9$

When squaring binomials, be careful not to forget the middle term. That is to say, $(x + 3)^2 \neq x^2 + 3^2$; instead, $(x + 3)^2 = (x + 3)(x + 3) = x^2 + 6x + 9$.

The following example suggests a format to use when cubing a binomial.

$$(x + 4)^3 = (x + 4)(x + 4)(x + 4)$$
$$= (x + 4)(x^2 + 8x + 16)$$
$$= x(x^2 + 8x + 16) + 4(x^2 + 8x + 16)$$
$$= x^3 + 8x^2 + 16x + 4x^2 + 32x + 64$$
$$= x^3 + 12x^2 + 48x + 64$$

Special Product Patterns

When multiplying binomials, there are some special patterns that occur which you should recognize. We can use these patterns to find products and later to factor polynomials. Each of the patterns will be stated in general terms followed by examples to illustrate the use of each pattern.

Pattern

$$(a + b)^2 = (a + b)(a + b) = a^2 + \quad 2ab \quad + \quad b^2$$

| Square of first term of binomial | + | Twice the product of the two terms of binomial | + | Square of second term of binomial |

Examples

$$(x + 4)^2 = x^2 + 8x + 16.$$

$$(2x + 3y)^2 = 4x^2 + 12xy + 9y^2.$$

$$(5a + 7b)^2 = 25a^2 + 70ab + 49b^2$$

▲

Pattern

$$(a - b)^2 = (a - b)(a - b) = a^2 - \quad 2ab \quad + \quad b^2$$

| Square of first term of binomial | − | Twice the product of the two terms of binomial | + | Square of second term of binomial |

Examples

$$(x - 8)^2 = x^2 - 16x + 64.$$

$$(3x - 4y)^2 = 9x^2 - 24xy + 16y^2.$$

$$(4a - 9b)^2 = 16a^2 - 72ab + 81b^2$$

▲

Pattern

$$(a + b)(a - b) = a^2 - \quad b^2$$

| Square of first term of binomial | − | Square of second term of binomial |

Examples

$$(x + 7)(x - 7) = x^2 - 49$$

$$(2x + y)(2x - y) = 4x^2 - y^2$$

$$(3a - 2b)(3a + 2b) = 9a^2 - 4b^2$$

▲

As you might expect, there are geometric interpretations for many of the algebraic concepts presented in this section. We will give you the opportunity to make some of these connections between algebra and geometry in the next problem set. Let's conclude this section with a problem that allows us to use some algebra and geometry.

Example 10

A rectangular piece of tin is 16 inches long and 12 inches wide as shown in Figure 5.15. From each corner a square piece x inches on a side is cut out. The flaps are then turned up to form an open box. Find polynomials that represent the volume and outside surface area of the box.

FIGURE 5.15

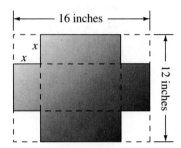

Solution

The length of the box will be $16 - 2x$, the width $12 - 2x$, and the height x. Using the volume formula $V = lwh$, the polynomial $(16 - 2x)(12 - 2x)(x)$, which simplifies to $4x^3 - 56x^2 + 192x$, represents the volume.

The outside surface area of the box is the area of the original piece of tin minus the four corners that were cut off. Therefore, the polynomial $16(12) - 4x^2$ or $192 - 4x^2$ represents the outside surface area of the box. ▲

Problem Set 5.3

For Problems 1–10, find the indicated products by applying the distributive property. For example,

$$(x + 1)(y + 5) = x(y) + x(5) + 1(y) + 1(5)$$
$$= xy + 5x + y + 5.$$

1. $(x + 2)(y + 3)$

2. $(x + 3)(y + 6)$

3. $(x - 4)(y + 1)$

4. $(x - 5)(y + 7)$

5. $(x - 5)(y - 6)$

6. $(x - 7)(y - 9)$

7. $(x + 2)(y + z + 1)$

8. $(x + 4)(y - z + 4)$

9. $(2x + 3)(3y + 1)$

10. $(3x - 2)(2y - 5)$

For Problems 11–36, find the indicated products by applying the distributive property and combining similar terms. Use the following format to show your work.

$$(x + 3)(x + 8) = x(x) + x(8) + 3(x) + 3(8)$$
$$= x^2 + 8x + 3x + 24$$
$$= x^2 + 11x + 24$$

11. $(x + 3)(x + 7)$

12. $(x + 4)(x + 2)$

13. $(x + 8)(x - 3)$

14. $(x + 9)(x - 6)$

15. $(x - 7)(x + 1)$

16. $(x - 10)(x + 8)$

17. $(n - 4)(n - 6)$

18. $(n - 3)(n - 7)$

19. $(3n + 1)(n + 6)$

20. $(4n + 3)(n + 6)$

21. $(5x - 2)(3x + 7)$

22. $(3x - 4)(7x + 1)$

23. $(x + 3)(x^2 + 4x + 9)$

24. $(x + 2)(x^2 + 6x + 2)$

25. $(x + 4)(x^2 - x - 6)$

26. $(x + 5)(x^2 - 2x - 7)$

27. $(x - 5)(2x^2 + 3x - 7)$

28. $(x - 4)(3x^2 + 4x - 6)$

29. $(2a - 1)(4a^2 - 5a + 9)$

30. $(3a - 2)(2a^2 - 3a - 5)$

31. $(3a + 5)(a^2 - a - 1)$

32. $(5a + 2)(a^2 + a - 3)$

33. $(x^2 + 2x + 3)(x^2 + 5x + 4)$

34. $(x^2 - 3x + 4)(x^2 + 5x - 2)$

35. $(x^2 - 6x - 7)(x^2 + 3x - 9)$

36. $(x^2 - 5x - 4)(x^2 + 7x - 8)$

For Problems 37–80, find the indicated products by using the shortcut pattern for multiplying binomials.

37. $(x + 2)(x + 9)$

38. $(x + 3)(x + 8)$

39. $(x + 6)(x - 2)$

40. $(x + 8)(x - 6)$

41. $(x + 3)(x - 11)$

42. $(x + 4)(x - 10)$

43. $(n - 4)(n - 3)$

44. $(n - 5)(n - 9)$

45. $(n + 6)(n + 12)$

46. $(n + 8)(n + 13)$

47. $(y + 3)(y - 7)$

48. $(y + 2)(y - 12)$

49. $(y - 7)(y - 12)$

50. $(y - 4)(y - 13)$

51. $(x - 5)(x + 7)$

52. $(x - 1)(x + 9)$

53. $(x - 14)(x + 8)$

54. $(x - 15)(x + 6)$

55. $(a + 10)(a - 9)$

56. $(a + 7)(a - 6)$

57. $(2a + 1)(a + 6)$

58. $(3a + 2)(a + 4)$

59. $(5x - 2)(x + 7)$

60. $(2x - 3)(x + 8)$

61. $(3x - 7)(2x + 1)$

62. $(5x - 6)(4x + 3)$

63. $(4a + 3)(3a - 4)$

64. $(5a + 4)(4a - 5)$

65. $(6n - 5)(2n - 3)$

66. $(4n - 3)(6n - 7)$

67. $(7x - 4)(2x + 3)$

68. $(8x - 5)(3x + 7)$

69. $(5 - x)(9 - 2x)$

70. $(4 - 3x)(2 + x)$

71. $(-2x + 3)(4x - 5)$

72. $(-3x + 1)(9x - 2)$

73. $(-3x - 1)(3x - 4)$

74. $(-2x - 5)(4x + 1)$

75. $(8n + 3)(9n - 4)$

76. $(6n + 5)(9n - 7)$

77. $(3 - 2x)(9 - x)$

78. $(5 - 4x)(4 - 5x)$

79. $(-4x + 3)(-5x - 2)$

80. $(-2x + 7)(-7x - 3)$

For Problems 81–110, use one of the appropriate patterns $(a + b)^2 = a^2 + 2ab + b^2$, $(a - b)^2 = a^2 - 2ab + b^2$, or $(a + b)(a - b) = a^2 - b^2$ to help find the indicated products.

81. $(x + 7)^2$

82. $(x + 9)^2$

83. $(5x - 2)(5x + 2)$

84. $(6x + 1)(6x - 1)$

85. $(x - 1)^2$

86. $(x - 4)^2$

87. $(3x + 7)^2$

88. $(2x + 9)^2$

89. $(2x - 3)^2$

90. $(4x - 5)^2$

91. $(2x + 3y)(2x - 3y)$

92. $(3a - b)(3a + b)$

93. $(1 - 5n)^2$

94. $(2 - 3n)^2$

95. $(3x + 4y)^2$

96. $(2x + 5y)^2$

97. $(3 + 4y)^2$

98. $(7 + 6y)^2$

99. $(1 + 7n)(1 - 7n)$

100. $(2 + 9n)(2 - 9n)$

101. $(4a - 7b)^2$

102. $(6a - b)^2$

103. $(x + 8y)^2$

104. $(x + 6y)^2$

105. $(5x - 11y)(5x + 11y)$

106. $(7x - 9y)(7x + 9y)$

107. $x(8x + 1)(8x - 1)$

108. $3x(5x + 7)(5x - 7)$

109. $-2x(4x + y)(4x - y)$

110. $-4x(2 - 3x)(2 + 3x)$

For Problems 111–118, find the indicated products. Don't forget that $(x + 2)^3$ means $(x + 2)(x + 2)(x + 2)$.

111. $(x + 2)^3$

112. $(x + 4)^3$

113. $(x - 3)^3$

114. $(x - 1)^3$

115. $(2n + 1)^3$

116. $(3n + 2)^3$

117. $(3n - 2)^3$

118. $(4n - 3)^3$

119. Explain how Figure 5.16 can be used to demonstrate geometrically that $(x + 3)(x + 5) = x^2 + 8x + 15$.

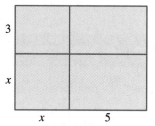

FIGURE 5.16

120. Explain how Figure 5.17 can be used to demonstrate geometrically that $(x + 5)(x - 3) = x^2 + 2x - 15$.

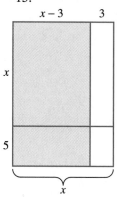

FIGURE 5.17

121. A square piece of cardboard is 14 inches on a side. From each corner a square piece x inches on a side is cut out as shown in Figure 5.18. The flaps are then turned up to form an open box. Find polynomials that represent the volume and outside surface area of the box.

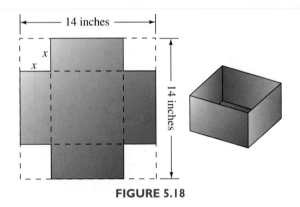

14 inches

14 inches

FIGURE 5.18

122. Describe the process of multiplying two polynomials.

123. Illustrate as many uses of the distributive property as you can.

124. Determine the number of terms in the product of $(x + y + z)$ and $(a + b + c)$ without doing the multiplication. Explain how you arrived at your answer.

Further Investigations

125. The following two patterns result from cubing binomials.

$$(a + b)^3 = a^3 + 3a^2b + 3ab^2 + b^3$$

$$(a - b)^3 = a^3 - 3a^2b + 3ab^2 - b^3$$

Use these patterns and redo Problems 111–118.

126. Find a pattern for the expansion of $(a + b)^4$. Then use the pattern to expand $(x + 2)^4$, $(x + 3)^4$, and $(2x + 1)^4$.

127. Some of the product patterns can be used to do arithmetic computations mentally. For example, let's use the pattern $(a + b)^2 = a^2 + 2ab + b^2$ to mentally compute 31^2. Your thought process should be "$31^2 = (30 + 1)^2 = 30^2 + 2(30)(1) + 1^2 = 961$." Compute each of the following numbers mentally and then check your answers.

(a) 21^2 (b) 41^2 (c) 71^2
(d) 32^2 (e) 52^2 (f) 82^2

128. Use the pattern $(a - b)^2 = a^2 - 2ab + b^2$ to compute each of the following numbers mentally and then check your answers.

(a) 19^2 (b) 29^2 (c) 49^2
(d) 79^2 (e) 38^2 (f) 58^2

129. Every whole number with a units digit of 5 can be represented by the expression $10x + 5$, where x is a whole number. For example, $35 = 10(3) + 5$ and $145 = 10(14) + 5$. Now let's observe the following pattern when squaring such a number.

$$(10x + 5)^2 = 100x^2 + 100x + 25$$
$$= \boxed{100x(x + 1) + 25}$$

The pattern inside the dashed box can be stated as "add 25 to the product of x, $x + 1$, and 100." Thus, to mentally compute 35^2 we can think "$35^2 = 3(4)(100) + 25 = 1225$." Compute each of the following numbers mentally and then check your answers.

(a) 15^2 (b) 25^2 (c) 45^2
(d) 55^2 (e) 65^2 (f) 75^2
(g) 85^2 (h) 95^2 (i) 105^2

5.4 Dividing by Monomials

To develop an effective process for dividing by a monomial we must rely on yet another property of exponents. This property is also a direct consequence of the definition of exponent and is illustrated by the following examples.

$$\frac{x^5}{x^2} = \frac{\cancel{x} \cdot \cancel{x} \cdot x \cdot x \cdot x}{\cancel{x} \cdot \cancel{x}} = x^3$$

$$\frac{a^4}{a^3} = \frac{\cancel{a} \cdot \cancel{a} \cdot \cancel{a} \cdot a}{\cancel{a} \cdot \cancel{a} \cdot \cancel{a}} = a$$

$$\frac{y^7}{y^3} = \frac{y \cdot y \cdot y \cdot y \cdot y \cdot y \cdot y}{\cancel{y} \cdot \cancel{y} \cdot \cancel{y}} = y^4$$

$$\frac{x^4}{x^4} = \frac{\cancel{x} \cdot \cancel{x} \cdot \cancel{x} \cdot \cancel{x}}{\cancel{x} \cdot \cancel{x} \cdot \cancel{x} \cdot \cancel{x}} = 1$$

$$\frac{y^3}{y^3} = \frac{\cancel{y} \cdot \cancel{y} \cdot \cancel{y}}{\cancel{y} \cdot \cancel{y} \cdot \cancel{y}} = 1$$

PROPERTY 5.4

If b is any nonzero real number and n and m are positive integers, then

1. $\dfrac{b^n}{b^m} = b^{n-m}$ when $n > m$;

2. $\dfrac{b^n}{b^m} = 1$ when $n = m$.

(The situation when $n < m$ will be discussed in a later section.)

Applying Property 5.4 to the previous examples yields

$$\frac{x^5}{x^2} = x^{5-2} = x^3$$

$$\frac{a^4}{a^3} = a^{4-3} = a^1 \qquad \text{Usually written as } a$$

$$\frac{y^7}{y^3} = y^{7-3} = y^4$$

$$\frac{x^4}{x^4} = 1$$

$$\frac{y^3}{y^3} = 1$$

Property 5.4 along with our knowledge of dividing integers provides the basis for dividing a monomial by another monomial. Consider the following examples.

$$\frac{16x^5}{2x^3} = 8x^{5-3} = 8x^2 \qquad \frac{-81a^{12}}{-9a^4} = 9a^{12-4} = 9a^8$$

$$\frac{-35x^9}{5x^4} = -7x^{9-4} = -7x^5 \qquad \frac{45x^4}{9x^4} = 5 \qquad \frac{x^4}{x^4} = 1$$

$$\frac{56y^6}{-7y^2} = -8y^{6-2} = -8y^4 \qquad \frac{54x^3y^7}{-6xy^5} = -9x^{3-1}y^{7-5} = -9x^2y^2$$

Recall that $\dfrac{a}{c} + \dfrac{b}{c} = \dfrac{a+b}{c}$; this same property $\left(\text{except viewed as} \right.$ $\left. \dfrac{a+b}{c} = \dfrac{a}{c} + \dfrac{b}{c} \right)$ serves as the basis for dividing a polynomial by a monomial. Consider the following examples.

$$\frac{25x^3 + 10x^2}{5x} = \frac{25x^3}{5x} + \frac{10x^2}{5x} = 5x^2 + 2x$$

$$\frac{-35x^8 - 28x^6}{7x^3} = \frac{-35x^8}{7x^3} - \frac{28x^6}{7x^3} = -5x^5 - 4x^3. \qquad \frac{a-b}{c} = \frac{a}{c} - \frac{b}{c}$$

To divide a polynomial by a monomial we simply divide each term of the polynomial by the monomial. Here are some additional examples.

$$\frac{12x^3y^2 - 14x^2y^5}{-2xy} = \frac{12x^3y^2}{-2xy} - \frac{14x^2y^5}{-2xy} = -6x^2y + 7xy^4$$

$$\frac{48ab^5 + 64a^2b}{-16ab} = \frac{48ab^5}{-16ab} + \frac{64a^2b}{-16ab} = -3b^4 - 4a$$

$$\frac{33x^6 - 24x^5 - 18x^4}{3x} = \frac{33x^6}{3x} - \frac{24x^5}{3x} - \frac{18x^4}{3x}$$

$$= 11x^5 - 8x^4 - 6x^3$$

As with many skills, once you feel comfortable with the process, you may want to mentally perform some of the steps. Your work could take on the following format.

$$\frac{24x^4y^5 - 56x^3y^9}{8x^2y^3} = 3x^2y^2 - 7xy^6$$

$$\frac{13a^2b - 12ab^2}{-ab} = -13a + 12b$$

Problem Set 5.4

For Problems 1–24, divide the monomials.

1. $\dfrac{x^{10}}{x^2}$

2. $\dfrac{x^{12}}{x^5}$

3. $\dfrac{4x^3}{2x}$

4. $\dfrac{8x^5}{4x^3}$

5. $\dfrac{-16n^6}{2n^2}$

6. $\dfrac{-54n^8}{6n^4}$

7. $\dfrac{72x^3}{-9x^3}$

8. $\dfrac{84x^5}{-7x^5}$

9. $\dfrac{65\,x^2y^3}{5\,xy}$

10. $\dfrac{70x^3y^4}{5x^2y}$

11. $\dfrac{-91a^4b^6}{-13a^3b^4}$

12. $\dfrac{-72a^5b^4}{-12ab^2}$

13. $\dfrac{18\,x^2y^6}{xy^2}$

14. $\dfrac{24\,x^3y^4}{x^2y^2}$

15. $\dfrac{32x^6y^2}{-x}$

16. $\dfrac{54x^5y^3}{-y^2}$

17. $\dfrac{-96\,x^5y^7}{12y^3}$

18. $\dfrac{-84x^4y^9}{14x^4}$

19. $\dfrac{-ab}{ab}$

20. $\dfrac{6ab}{-ab}$

21. $\dfrac{56a^2b^3c^5}{4abc}$

22. $\dfrac{60a^3b^2c}{15a^2c}$

23. $\dfrac{-80xy^2z^6}{-5xyz^2}$

24. $\dfrac{-90\,x^3y^2z^8}{-6xy^2z^4}$

For Problems 25–50, perform the following divisions of polynomials by monomials.

25. $\dfrac{8x^4 + 12x^5}{2x^2}$

26. $\dfrac{12x^3 + 16x^6}{4x}$

27. $\dfrac{9x^6 - 24x^4}{3x^3}$

28. $\dfrac{35x^8 - 45x^6}{5x^4}$

29. $\dfrac{-28n^5 + 36n^2}{4n^2}$

30. $\dfrac{-42n^6 + 54n^4}{6n^4}$

31. $\dfrac{35x^6 - 56x^5 - 84x^3}{7x^2}$

32. $\dfrac{27x^7 - 36x^5 - 45x^3}{3x}$

33. $\dfrac{-24n^8 + 48n^5 - 78n^3}{-6n^3}$

34. $\dfrac{-56n^9 + 84n^6 - 91n^2}{-7n^2}$

35. $\dfrac{-60a^7 - 96a^3}{-12a}$

36. $\dfrac{-65a^8 - 78a^4}{-13a^2}$

37. $\dfrac{27x^2y^4 - 45\,xy^4}{-9xy^3}$

38. $\dfrac{-40\,x^4y^7 + 64\,x^5y^8}{-8x^3y^4}$

39. $\dfrac{48a^2b^2 + 60a^3b^4}{-6ab}$

40. $\dfrac{45a^3b^4 - 63a^2b^6}{-9ab^2}$

41. $\dfrac{12a^2b^2c^2 - 52a^2b^3c^5}{-4a^2bc}$

42. $\dfrac{48a^3b^2c + 72a^2b^4c^5}{-12ab^2c}$

43. $\dfrac{9x^2y^3 - 12x^3y^4}{-xy}$

44. $\dfrac{-15x^3y + 27x^2y^4}{xy}$

45. $\dfrac{-42x^6 - 70x^4 + 98x^2}{14x^2}$

46. $\dfrac{-48x^8 - 80x^6 + 96x^4}{16x^4}$

47. $\dfrac{15a^3b - 35a^2b - 65ab^2}{-5ab}$

48. $\dfrac{-24a^4b^2 + 36a^3b - 48a^2b}{-6ab}$

49. $\dfrac{-xy + 5x^2y^3 - 7x^2y^6}{xy}$

50. $\dfrac{-9x^2y^3 - xy + 14xy^4}{-xy}$

THOUGHTS INTO WORDS

51. How would you explain to someone why the quotient of x^8 and x^2 is x^6 and not x^4?

52. Your friend is having difficulty with problems such as $\dfrac{12x^2y}{xy}$ and $\dfrac{36x^3y^2}{-xy}$ where there appears to be no numerical coefficient in the denominator. What can you tell him that might help?

5.5 Dividing by Binomials

Perhaps the easiest way to explain the process of dividing a polynomial by a binomial is to work a few examples and describe the step-by-step procedure as we go along.

Example 1

Divide $x^2 + 5x + 6$ by $x + 2$.

Solution

STEP 1 Use the conventional long division format from arithmetic and arrange both the dividend and the divisor in descending powers of the variable.

$$x + 2\overline{)x^2 + 5x + 6}$$

STEP 2 Find the first term of the quotient by dividing the first term of the dividend by the first term of the divisor.

$$x + 2\overline{)x^2 + 5x + 6}^{\,x} \qquad \dfrac{x^2}{x} = x$$

STEP 3 Multiply the entire divisor by the term of the quotient found in step 2 and position this product to be subtracted from the dividend.

$$\begin{array}{r} x \\ x + 2\overline{)x^2 + 5x + 6} \\ \underline{x^2 + 2x} \end{array} \qquad (x(x+2) = x^2 + 2x)$$

STEP 4 Subtract.
Remember to add the opposite! ⟶

$$\begin{array}{r} x \\ x + 2\overline{)x^2 + 5x + 6} \\ \underline{x^2 + 2x} \\ 3x + 6 \end{array}$$

STEP 5 Repeat the process beginning with step 2; use the polynomial that resulted from the subtraction in step 4 as a new dividend.

$$
\begin{array}{r}
x + 3 \\
x + 2{\overline{\smash{\big)}\,x^2 + 5x + 6}} \\
\underline{x^2 + 2x} \\
3x + 6 \\
\underline{3x + 6}
\end{array}
$$

$\dfrac{3x}{x} = 3$

$3(x + 2) = 3x + 6$

Thus, $(x^2 + 5x + 6) \div (x + 2) = x + 3$, which can be checked by multiplying $(x + 2)$ and $(x + 3)$.

$$(x + 2)(x + 3) = x^2 + 5x + 6$$

▲

A division problem such as $(x^2 + 5x + 6) \div (x + 2)$ can also be written as $\dfrac{x^2 + 5x + 6}{x + 2}$. Using this format the final result for Example 1 could be expressed as $\dfrac{x^2 + 5x + 6}{x + 2} = x + 3$. (Technically, the restriction $x \neq -2$ should be made to avoid division by zero.)

In general, to check a division problem we can multiply the divisor times the quotient and add the remainder, which can be expressed as

Dividend = (Divisor)(Quotient) + Remainder.

Sometimes the remainder is expressed as a fractional part of the divisor. The relationship then becomes

$$\frac{\text{Dividend}}{\text{Divisor}} = \text{Quotient} + \frac{\text{Remainder}}{\text{Divisor}}.$$

Example 2

Divide $2x^2 - 3x - 20$ by $x - 4$.

Solution

STEP 1 $x - 4{\overline{\smash{\big)}\,2x^2 - 3x - 20}}$

STEP 2

$$
\begin{array}{r}
2x \\
x - 4{\overline{\smash{\big)}\,2x^2 - 3x - 20}}
\end{array}
$$

$\dfrac{2x^2}{x} = 2x$

STEP 3

$$
\begin{array}{r}
2x \\
x - 4{\overline{\smash{\big)}\,2x^2 - 3x - 20}} \\
\underline{2x^2 - 8x}
\end{array}
$$

$2x(x - 4) = 2x^2 - 8x$

STEP 4

$$
\begin{array}{r}
2x \\
x - 4{\overline{\smash{\big)}\,2x^2 - 3x - 20}} \\
\underline{2x^2 - 8x} \\
5x - 20
\end{array}
$$

STEP 5

$$
\begin{array}{r}
2x + 5 \\
x - 4{\overline{\smash{\big)}\,2x^2 - 3x - 20}} \\
\underline{2x^2 - 8x} \\
5x - 20 \\
\underline{5x - 20}
\end{array}
$$

$\dfrac{5x}{x} = 5$

$5(x - 4) = 5x - 20$

✔ *Check* $(x - 4)(2x + 5) = 2x^2 - 3x - 20$

Therefore, $\dfrac{2x^2 - 3x - 20}{x - 4} = 2x + 5$.   ▲

Now let's continue to think in terms of the step-by-step division process but we'll organize our work in the typical long division format.

Example 3   Divide $12x^2 + x - 6$ by $3x - 2$.

Solution

$$
\begin{array}{r}
4x + 3 \\
3x - 2 \overline{)12x^2 + x - 6} \\
\underline{12x^2 - 8x } \\
9x - 6 \\
\underline{9x - 6}
\end{array}
$$

✔ *Check* $(3x - 2)(4x + 3) = 12x^2 + x - 6$

Therefore, $\dfrac{12x^2 + x - 6}{3x - 2} = 4x + 3$.   ▲

Each of the next three examples illustrates another aspect of the division process. Study them carefully; then you should be ready to work the exercises in the next problem set.

Example 4   Perform the division $(7x^2 - 3x - 4) \div (x - 2)$.

Solution

$$
\begin{array}{r}
7x + 11 \\
x - 2 \overline{)7x^2 - 3x - 4} \\
\underline{7x^2 - 14x } \\
11x - 4 \\
\underline{11x - 22} \\
18 \leftarrow \text{A remainder of 18}
\end{array}
$$

✔ *Check*   Just as in arithmetic, we check by *adding* the remainder to the product of the divisor and quotient.

$$(x - 2)(7x + 11) + 18 \stackrel{?}{=} 7x^2 - 3x - 4$$
$$7x^2 - 3x - 22 + 18 \stackrel{?}{=} 7x^2 - 3x - 4$$
$$7x^2 - 3x - 4 = 7x^2 - 3x - 4$$

Therefore, $\dfrac{7x^2 - 3x - 4}{x - 2} = 7x + 11 + \dfrac{18}{x - 2}$. ▲

Example 5

Perform the division $\dfrac{x^3 - 8}{x - 2}$.

Solution

$$
\begin{array}{r}
x^2 + 2x\ + 4 \\
x - 2\overline{)x^3 + 0x^2 + 0x - 8} \\
\underline{x^3 - 2x^2} \\
2x^2 + 0x - 8 \\
\underline{2x^2 - 4x} \\
4x - 8 \\
\underline{4x - 8}
\end{array}
$$

⟵ Notice the inserting of x squared and x terms with zero coefficients.

✔ **Check**

$$(x - 2)(x^2 + 2x + 4) \overset{?}{=} x^3 - 8$$
$$x^3 + 2x^2 + 4x - 2x^2 - 4x - 8 \overset{?}{=} x^3 - 8$$
$$x^3 - 8 = x^3 - 8$$

Therefore, $\dfrac{x^3 - 8}{x - 2} = x^2 + 2x + 4$. ▲

Example 6

Perform the division $\dfrac{x^3 + 5x^2 - 3x - 4}{x^2 + 2x}$.

Solution

$$
\begin{array}{r}
x\ + 3 \\
x^2 + 2x\overline{)x^3 + 5x^2 - 3x - 4} \\
\underline{x^3 + 2x^2} \\
3x^2 - 3x - 4 \\
\underline{3x^2 + 6x} \\
- 9x - 4
\end{array}
$$

⟵ A remainder of $-9x - 4$

The division process is stopped when the degree of the remainder is less than the degree of the divisor.

✔ **Check**

$$(x^2 + 2x)(x + 3) + (-9x - 4) \overset{?}{=} x^3 + 5x^2 - 3x - 4$$
$$x^3 + 3x^2 + 2x^2 + 6x - 9x - 4 \overset{?}{=} x^3 + 5x^2 - 3x - 4$$
$$x^3 + 5x^2 - 3x - 4 = x^3 + 5x^2 - 3x - 4.$$

Therefore, $\dfrac{x^3 + 5x^2 - 3x - 4}{x^2 + 2x} = x + 3 + \dfrac{-9x - 4}{x^2 + 2x}$. ▲

Problem Set 5.5

Perform the following divisions.

1. $(x^2 + 16x + 48) \div (x + 4)$

2. $(x^2 + 15x + 54) \div (x + 6)$

3. $(x^2 - 5x - 14) \div (x - 7)$

4. $(x^2 + 8x - 65) \div (x - 5)$

5. $(x^2 + 11x + 28) \div (x + 3)$

6. $(x^2 + 11x + 15) \div (x + 2)$

7. $(x^2 - 4x - 39) \div (x - 8)$

8. $(x^2 - 9x - 30) \div (x - 12)$

9. $(5n^2 - n - 4) \div (n - 1)$

10. $(7n^2 - 61n - 90) \div (n - 10)$

11. $(8y^2 + 53y - 19) \div (y + 7)$

12. $(6y^2 + 47y - 72) \div (y + 9)$

13. $(20x^2 - 31x - 7) \div (5x + 1)$

14. $(27x^2 + 21x - 20) \div (3x + 4)$

15. $(6x^2 + 25x + 8) \div (2x + 7)$

16. $(12x^2 + 28x + 27) \div (6x + 5)$

17. $(2x^3 - x^2 - 2x - 8) \div (x - 2)$

18. $(3x^3 - 7x^2 - 26x + 24) \div (x - 4)$

19. $(5n^3 + 11n^2 - 15n - 9) \div (n + 3)$

20. $(6n^3 + 29n^2 - 6n - 5) \div (n + 5)$

21. $(n^3 - 40n + 24) \div (n - 6)$

22. $(n^3 - 67n - 24) \div (n + 8)$

23. $(x^3 - 27) \div (x - 3)$

24. $(x^3 + 8) \div (x + 2)$

25. $\dfrac{27x^3 - 64}{3x - 4}$

26. $\dfrac{8x^3 + 27}{2x + 3}$

27. $\dfrac{1 + 3n^2 - 2n}{n + 2}$

28. $\dfrac{x + 5 + 12x^2}{3x - 2}$

29. $\dfrac{9t^2 + 3t + 4}{-1 + 3t}$

30. $\dfrac{4n^2 + 6n - 1}{4 + 2n}$

31. $\dfrac{6n^3 - 5n^2 - 7n + 4}{2n - 1}$

32. $\dfrac{21n^3 + 23n^2 - 9n - 10}{3n + 2}$

33. $\dfrac{4x^3 + 23x^2 - 30x + 32}{x + 7}$

34. $\dfrac{5x^3 - 12x^2 + 13x - 14}{x - 1}$

35. $(x^3 + 2x^2 - 3x - 1) \div (x^2 - 2x)$

36. $(x^3 - 6x^2 - 2x + 1) \div (x^2 + 3x)$

37. $(2x^3 - 4x^2 + x - 5) \div (x^2 + 4x)$

38. $(2x^3 - x^2 - 3x + 5) \div (x^2 + x)$

39. $(x^4 - 16) \div (x + 2)$

40. $(x^4 - 81) \div (x - 3)$

THOUGHTS INTO WORDS

41. Give a step-by-step description of how you would do the division problem $(2x^3 + 8x^2 - 29x - 30) \div (x + 6)$.

42. How do you know by inspection that the answer to the following division problem is incorrect?

$$(3x^3 - 7x^2 - 22x + 8) \div (x - 4) = 3x^2 + 5x + 1$$

5.6 Zero and Negative Integers as Exponents

Thus far in this text we have used only positive integers as exponents. The following definition and properties serve as a basis for our work with exponents.

DEFINITION 5.1

If n is a positive integer and b is any real number, then

$$b^n = \underbrace{bbb \cdots b}_{n \text{ factors of } b}.$$

PROPERTY 5.5

If m and n are positive integers and a and b are real numbers, except $b \neq 0$ whenever it appears in a denominator, then

1. $b^n \cdot b^m = b^{n+m}$
2. $(b^n)^m = b^{mn}$
3. $(ab)^n = a^n b^n$
4. $\left(\dfrac{a}{b}\right)^n = \dfrac{a^n}{b^n}$ Part 4 has not been previously stated.
5. $\dfrac{b^n}{b^m} = b^{n-m}$ When $n > m$

   $\dfrac{b^n}{b^m} = 1$ When $n = m$

Zero and Negative Integers as Exponents

Property 5.5 pertains to the use of positive integers as exponents. Zero and the negative integers can also be used as exponents. First, let's consider the use of 0 as an exponent. We want to use 0 as an exponent in such a way that the basic properties of exponents will continue to hold. So consider the example $x^4 \cdot x^0$.

If part 1 of Property 5.5 is to hold, then

$$x^4 \cdot x^0 = x^{4+0} = x^4.$$

Note that x^0 *acts like* 1 since $x^4 \cdot x^0 = x^4$. This suggests the following definition.

DEFINITION 5.2

If b is a nonzero real number, then

$$b^0 = 1.$$

According to Definition 5.2 the following statements are all true.

$$4^0 = 1$$

$$(-628)^0 = 1$$

$$\left(\frac{4}{7}\right)^0 = 1$$

$$n^0 = 1, \qquad n \neq 0$$

$$(x^2y^5)^0 = 1, \qquad x \neq 0 \text{ and } y \neq 0$$

A similar line of reasoning indicates how negative integers should be used as exponents. Consider the example $x^3 \cdot x^{-3}$. If part 1 of Property 5.5 is to hold, then

$$x^3 \cdot x^{-3} = x^{3+(-3)} = x^0 = 1.$$

Thus, x^{-3} must be the reciprocal of x^3 since their product is 1; that is,

$$x^{-3} = \frac{1}{x^3}.$$

This process suggests the following definition.

DEFINITION 5.3

If n is a positive integer and b is a nonzero real number, then

$$b^{-n} = \frac{1}{b^n}.$$

According to Definition 5.3 the following statements are all true.

$$x^{-6} = \frac{1}{x^6}$$

$$2^{-3} = \frac{1}{2^3} = \frac{1}{8}$$

$$10^{-2} = \frac{1}{10^2} = \frac{1}{100} \text{ or } .01$$

$$\frac{1}{x^{-4}} = \frac{1}{\dfrac{1}{x^4}} = x^4$$

$$\left(\frac{2}{3}\right)^{-2} = \frac{1}{\left(\dfrac{2}{3}\right)^2} = \frac{1}{\dfrac{4}{9}} = \frac{9}{4}$$

REMARK Note in the last example that $\left(\frac{2}{3}\right)^{-2} = \left(\frac{3}{2}\right)^2$. In other words, to raise a fraction to a negative power, we can invert the fraction and raise it to the corresponding positive power. △

We can verify (we will not do so in this text) that all parts of Property 5.5 hold for *all integers*. In fact, part 5 can be replaced with the following statement.

Replacement for part 5 of Property 5.5

$$\frac{b^n}{b^m} = b^{n-m} \quad \text{for all integers } n \text{ and } m.$$

The following examples illustrate the use of this new property. In each example we have simplified the original expression and used only positive exponents in the final result.

$$\frac{x^2}{x^5} = x^{2-5} = x^{-3} = \frac{1}{x^3}$$

$$\frac{a^{-3}}{a^{-7}} = a^{-3-(-7)} = a^{-3+7} = a^4$$

$$\frac{y^{-5}}{y^{-2}} = y^{-5-(-2)} = y^{-5+2} = y^{-3} = \frac{1}{y^3}$$

$$\frac{x^{-6}}{x^{-6}} = x^{-6-(-6)} = x^{-6+6} = x^0 = 1$$

The properties of exponents provide a basis for simplifying certain types of numerical expressions, as the following examples illustrate.

$$2^{-4} \cdot 2^6 = 2^{-4+6} = 2^2 = 4$$

$$10^5 \cdot 10^{-6} = 10^{5+(-6)} = 10^{-1} = \frac{1}{10} \text{ or } .1$$

$$\frac{10^2}{10^{-2}} = 10^{2-(-2)} = 10^{2+2} = 10^4 = 10,000$$

$$(2^{-3})^{-2} = 2^{-3(-2)} = 2^6 = 64$$

Having the use of all integers as exponents also expands the type of work that we can do with algebraic expressions. In each of the following examples we have simplified a given expression and used only positive exponents in the final result.

$$x^8 x^{-2} = x^{8+(-2)} = x^6$$

$$a^{-4} a^{-3} = a^{-4+(-3)} = a^{-7} = \frac{1}{a^7}$$

$$(y^{-3})^4 = y^{-3(4)} = y^{-12} = \frac{1}{y^{12}}$$

$$(x^{-2}y^4)^{-3} = (x^{-2})^{-3}(y^4)^{-3} = x^6y^{-12} = \frac{x^6}{y^{12}}$$

$$\left(\frac{x^{-1}}{y^2}\right)^{-2} = \frac{(x^{-1})^{-2}}{(y^2)^{-2}} = \frac{x^2}{y^{-4}} = x^2y^4$$

$$(4x^{-2})(3x^{-1}) = 12x^{-2+(-1)} = 12x^{-3} = \frac{12}{x^3}$$

$$\left(\frac{12x^{-6}}{6x^{-2}}\right)^{-2} = (2x^{-6-(-2)})^{-2} = (2x^{-4})^{-2}$$

$$= (2)^{-2}(x^{-4})^{-2}$$

$$= \left(\frac{1}{2^2}\right)(x^8) = \frac{x^8}{4}$$

Scientific Notation

Many scientific applications of mathematics involve the use of very large and very small numbers. For example:

1. The speed of light is approximately 29,979,200,000 centimeters per second.

2. A light year—the distance light travels in one year—is approximately 5,865,696,000,000 miles.

3. A millimicron equals .000000001 of a meter.

Working with numbers of this type in standard form is quite cumbersome. It is much more convenient to represent very small and very large numbers in **scientific notation**, sometimes called scientific form. A number is in scientific notation when it is written as a product of a number between 1 and 10 (including 1) and an integral power of 10. Symbolically, a number in scientific notation has the form $(N)(10^k)$, where $1 \le N < 10$ and k is an integer. For example, 621 can be written as $(6.21)(10^2)$ and .0023 can be written $(2.3)(10^{-3})$.

To switch from ordinary notation to scientific notation, you can use the following procedure.

> Write the given number as the product of a number between 1 and 10 and an integral power of 10. To determine the exponent of 10, count the number of places that the decimal point moved when going from the original number to the number between 1 and 10. This exponent is (a) negative if the original number is less than 1, (b) positive if the original number is greater than 10, and (c) zero if the original number itself is between 1 and 10.

Thus, we can write

$$.000179 = (1.79)(10^{-4}) \qquad \text{According to part (a) of the rule}$$

$$8175 = (8.175)(10^3) \qquad \text{According to part (b)}$$
$$3.14 = (3.14)(10^0) \qquad \text{According to part (c)}$$

We can express the applications given earlier in scientific notation as follows.

Speed of light $29,979,200,000 = (2.99792)(10^{10})$ centimeters per second.

Light year $5,865,696,000,000 = (5.865696)(10^{12})$ miles.

Metric units   A millimicron is $.000000001 = (1)(10^{-9})$ of a meter.

To switch from scientific notation to ordinary decimal notation you can use the following procedure.

Move the decimal point the number of places indicated by the exponent of 10. The decimal point is moved to the right if the exponent is positive and to the left if it is negative.

Thus, we can write

$$(4.71)(10^4) = 47,100 \qquad \text{Two zeros are needed for place value purposes.}$$
$$(1.78)(10^{-2}) = .0178 \qquad \text{One zero is needed for place value purposes.}$$

The use of scientific notation along with the properties of exponents can make some arithmetic problems much easier to evaluate. The following examples illustrate this point.

Example 1

Evaluate $(4,000)(.000012)$.

Solution

$$
\begin{aligned}
(4,000)(.000012) &= (4)(10^3)(1.2)(10^{-5}) \\
&= (4)(1.2)(10^3)(10^{-5}) \\
&= (4.8)(10^{-2}) \\
&= .048
\end{aligned}
$$

▲

Example 2

Evaluate $\dfrac{960,000}{.032}$.

Solution

$$
\begin{aligned}
\frac{960,000}{.032} &= \frac{(9.6)(10^5)}{(3.2)(10^{-2})} \\
&= (3)(10^7) \qquad \frac{10^5}{10^{-2}} = 10^{5-(-2)} = 10^7 \\
&= 30,000,000
\end{aligned}
$$

▲

Example 3   Evaluate $\dfrac{(6000)(.00008)}{(40,000)(.006)}$.

Solution

$$\dfrac{(6000)(.00008)}{(40,000)(.006)} = \dfrac{(6)(10^3)(8)(10^{-5})}{(4)(10^4)(6)(10^{-3})}$$

$$= \dfrac{(48)(10^{-2})}{(24)(10^1)}$$

$$= (2)(10^{-3}) \qquad \dfrac{10^{-2}}{10^1} = 10^{-2-1} = 10^{-3}$$

$$= .002$$

Problem Set 5.6

For Problems 1–30, evaluate each numerical expression.

1. 3^{-2}

2. 2^{-5}

3. 4^{-3}

4. 5^{-2}

5. $\left(\dfrac{3}{2}\right)^{-1}$

6. $\left(\dfrac{3}{4}\right)^{-2}$

7. $\dfrac{1}{2^{-4}}$

8. $\dfrac{1}{3^{-1}}$

9. $\left(-\dfrac{4}{3}\right)^0$

10. $\left(-\dfrac{1}{2}\right)^{-3}$

11. $\left(-\dfrac{2}{3}\right)^{-3}$

12. $(-16)^0$

13. $(-2)^{-2}$

14. $(-3)^{-2}$

15. $-(3^{-2})$

16. $-(2^{-2})$

17. $\dfrac{1}{\left(\dfrac{3}{4}\right)^{-3}}$

18. $\dfrac{1}{\left(\dfrac{3}{2}\right)^{-4}}$

19. $2^6 \cdot 2^{-9}$

20. $3^5 \cdot 3^{-2}$

21. $3^6 \cdot 3^{-3}$

22. $2^{-7} \cdot 2^2$

23. $\dfrac{10^2}{10^{-1}}$

24. $\dfrac{10^1}{10^{-3}}$

25. $\dfrac{10^{-1}}{10^2}$

26. $\dfrac{10^{-2}}{10^{-2}}$

27. $(2^{-1} \cdot 3^{-2})^{-1}$

28. $(3^{-1} \cdot 4^{-2})^{-1}$

29. $\left(\dfrac{4^{-1}}{3}\right)^{-2}$

30. $\left(\dfrac{3}{2^{-1}}\right)^{-3}$

For Problems 31–84, simplify each algebraic expression and express your answers using positive exponents only.

31. $x^6 x^{-1}$

32. $x^{-2} x^7$

33. $n^{-4} n^2$

34. $n^{-8} n^3$

35. $a^{-2} a^{-3}$

36. $a^{-4} a^{-6}$

37. $(2x^3)(4x^{-2})$

38. $(5x^{-4})(6x^7)$

39. $(3x^{-6})(9x^2)$

40. $(8x^{-8})(4x^2)$

41. $(5y^{-1})(-3y^{-2})$

42. $(-7y^{-3})(9y^{-4})$

43. $(8x^{-4})(12x^4)$

44. $(-3x^{-2})(-6x^2)$

45. $\dfrac{x^7}{x^{-3}}$

46. $\dfrac{x^2}{x^{-4}}$

47. $\dfrac{n^{-1}}{n^3}$

48. $\dfrac{n^{-2}}{n^5}$

49. $\dfrac{4n^{-1}}{2n^{-3}}$

50. $\dfrac{12n^{-2}}{3n^{-5}}$

51. $\dfrac{-24x^{-6}}{8x^{-2}}$

52. $\dfrac{56x^{-5}}{-7x^{-1}}$

53. $\dfrac{-52y^{-2}}{-13y^{-2}}$

54. $\dfrac{-91y^{-3}}{-7y^{-3}}$

55. $(x^{-3})^{-2}$

56. $(x^{-1})^{-5}$

57. $(x^2)^{-2}$

58. $(x^3)^{-1}$

59. $(x^3 y^4)^{-1}$

60. $(x^4 y^{-2})^{-2}$

61. $(x^{-2} y^{-1})^3$

62. $(x^{-3} y^{-4})^2$

63. $(2n^{-2})^3$

64. $(3n^{-1})^4$

65. $(4n^3)^{-2}$

66. $(2n^2)^{-3}$

67. $(3a^{-2})^4$

68. $(5a^{-1})^2$

69. $(5x^{-1})^{-2}$

70. $(4x^{-2})^{-2}$

71. $(2x^{-2}y^{-1})^{-1}$

72. $(3x^2y^{-3})^{-2}$

73. $\left(\dfrac{x^2}{y}\right)^{-1}$

74. $\left(\dfrac{y^2}{x^3}\right)^{-2}$

75. $\left(\dfrac{a^{-1}}{b^2}\right)^{-4}$

76. $\left(\dfrac{a^3}{b^{-2}}\right)^{-3}$

77. $\left(\dfrac{x^{-1}}{y^{-3}}\right)^{-2}$

78. $\left(\dfrac{x^{-3}}{y^{-4}}\right)^{-1}$

79. $\left(\dfrac{x^2}{x^3}\right)^{-1}$

80. $\left(\dfrac{x^4}{x}\right)^{-2}$

81. $\left(\dfrac{2x^{-1}}{x^{-2}}\right)^{-3}$

82. $\left(\dfrac{3x^{-2}}{x^{-5}}\right)^{-1}$

83. $\left(\dfrac{18x^{-1}}{9x}\right)^{-2}$

84. $\left(\dfrac{35x^2}{7x^{-1}}\right)^{-1}$

For Problems 85–94, write each number in scientific notation. For example, $786 = (7.86)(10^2)$.

85. 321

86. 74

87. 8000

88. 500

89. .00246

90. .017

91. .0000179

92. .00000049

93. 87,000,000

94. 623,000,000,000

For Problems 95–106, write each number in standard decimal form. For example, $(1.4)(10^3) = 1400$.

95. $(8)(10^3)$

96. $(6)(10^2)$

97. $(5.21)(10^4)$

98. $(7.2)(10^3)$

99. $(1.14)(10^7)$

100. $(5.64)(10^8)$

101. $(7)(10^{-2})$

102. $(8.14)(10^{-1})$

103. $(9.87)(10^{-4})$

104. $(4.37)(10^{-5})$

105. $(8.64)(10^{-6})$

106. $(3.14)(10^{-7})$

For Problems 107–118, use scientific notation and the properties of exponents to help evaluate each numerical expression.

107. $(.007)(120)$

108. $(.0004)(13)$

109. $(5,000,000)(.00009)$

110. $(800,000)(.0000006)$

111. $\dfrac{6000}{.0015}$

112. $\dfrac{480}{.012}$

113. $\dfrac{.00086}{4300}$

114. $\dfrac{.0057}{30,000}$

115. $\dfrac{.00039}{.0013}$

116. $\dfrac{.0000082}{.00041}$

117. $\dfrac{(.0008)(.07)}{(20,000)(.0004)}$

118. $\dfrac{(.006)(600)}{(.00004)(30)}$

THOUGHTS INTO WORDS

119. Is the following simplification process correct?

$$(2^{-2})^{-1} = \left(\frac{1}{2^2}\right)^{-1} = \left(\frac{1}{4}\right)^{-1} = \frac{1}{\left(\frac{1}{4}\right)^1} = 4$$

Could you suggest a better way to do the problem?

120. Explain the importance of scientific notation.

Further Investigations

121. Use your calculator to do Problems 1–16. Be sure that your answers are equivalent to the answers you obtained without the calculator.

122. Use your calculator to evaluate $(140,000)^2$. Your answer should be displayed in scientific notation; the format of the display depends on the particular calculator. For example, it may look like $\boxed{1.96 \quad 10}$

or 1.96E + 10 . Thus, in ordinary notation the answer is 19,600,000,000. Use your calculator to evaluate each of the following. Express final answers in ordinary notation.

(a) $(9000)^3$

(b) $(4000)^3$

(c) $(150,000)^2$

(d) $(170,000)^2$

(e) $(.012)^5$

(f) $(.0015)^4$

(g) $(.006)^3$

(h) $(.02)^6$

123. Use your calculator to check your answers to Problems 107–118.

SUMMARY

(5.1) Terms that contain variables with only whole numbers as exponents are called **monomials**. A **polynomial** is a monomial or a finite sum (or difference) of monomials. Polynomials of one term, two terms, and three terms are called **monomials**, **binomials**, and **trinomials**, respectively.

Addition and subtraction of polynomials are based on the distributive property and the process of combining similar terms.

(5.2) and (5.3) The following properties of exponents serve as a basis for multiplying polynomials.

1. $b^n \cdot b^m = b^{n+m}$

2. $(b^n)^m = b^{mn}$

3. $(ab)^n = a^n b^n$

(5.4) The following properties of exponents serve as a basis for dividing monomials.

1. $\dfrac{b^n}{b^m} = b^{n-m}$ when $n > m$

2. $\dfrac{b^n}{b^m} = 1$ when $n = m$

Dividing a polynomial by a monomial is based on the property $\dfrac{a+b}{c} = \dfrac{a}{c} + \dfrac{b}{c}$.

(5.5) To review the division of a polynomial by a binomial, turn to Section 5.5 and study the examples carefully.

(5.6) We use the following two definitions to extend our work with exponents to include zero and the negative integers.

Definition 5.2   If b is a nonzero real number, then $b^0 = 1$.

Definition 5.3   If n is a positive integer and b is a nonzero real number, then

$$b^{-n} = \frac{1}{b^n}.$$

The following properties of exponents are true for all integers.

1. $b^n \cdot b^m = b^{n+m}$
2. $(b^n)^m = b^{mn}$
3. $(ab)^n = a^n b^n$
4. $\left(\dfrac{a}{b}\right)^n = \dfrac{a^n}{b^n}$ $b \neq 0$ whenever it appears in a denominator
5. $\dfrac{b^n}{b^m} = b^{n-m}$

To represent a number in scientific notation, express it as the product of a number between 1 and 10 (including 1) and an integral power of 10.

Chapter 5 Review Problem Set

Perform the following additions and subtractions.

1. $(5x^2 - 6x + 4) + (3x^2 - 7x - 2)$
2. $(7y^2 + 9y - 3) - (4y^2 - 2y + 6)$
3. $(2x^2 + 3x - 4) + (4x^2 - 3x - 6) - (3x^2 - 2x - 1)$
4. $(-3x^2 - 2x + 4) - (x^2 - 5x - 6) - (4x^2 + 3x - 8)$

Remove parentheses and combine similar terms.

5. $5(2x - 1) + 7(x + 3) - 2(3x + 4)$
6. $3(2x^2 - 4x - 5) - 5(3x^2 - 4x + 1)$

7. $6(y^2 - 7y - 3) - 4(y^2 + 3y - 9)$

8. $3(a - 1) - 2(3a - 4) - 5(2a + 7)$
9. $-(a + 4) + 5(-a - 2) - 7(3a - 1)$

10. $-2(3n - 1) - 4(2n + 6) + 5(3n + 4)$
11. $3(n^2 - 2n - 4) - 4(2n^2 - n - 3)$
12. $-5(-n^2 + n - 1) + 3(4n^2 - 3n - 7)$

Find the indicated products.

13. $(5x^2)(7x^4)$ 14. $(-6x^3)(9x^5)$
15. $(-4xy^2)(-6x^2y^3)$
16. $(2a^3b^4)(-3ab^5)$
17. $(2a^2b^3)^3$ 18. $(-3xy^2)^2$
19. $5x(7x + 3)$
20. $(-3x^2)(8x - 1)$

Find the indicated products. Be sure to simplify answers.

21. $(x + 9)(x + 8)$
22. $(3x + 7)(x + 1)$

23. $(x - 5)(x + 2)$
24. $(y - 4)(y - 9)$
25. $(2x - 1)(7x + 3)$
26. $(4a - 7)(5a + 8)$
27. $(3a - 5)^2$
28. $(x + 6)(2x^2 + 5x - 4)$
29. $(5n - 1)(6n + 5)$
30. $(3n + 4)(4n - 1)$
31. $(2n + 1)(2n - 1)$
32. $(4n - 5)(4n + 5)$
33. $(2a + 7)^2$
34. $(3a + 5)^2$
35. $(x - 2)(x^2 - x + 6)$
36. $(2x - 1)(x^2 + 4x + 7)$
37. $(a + 5)^3$
38. $(a - 6)^3$
39. $(x^2 - x - 1)(x^2 + 2x + 5)$

40. $(n^2 + 2n + 4)(n^2 - 7n - 1)$

Perform the following divisions.

41. $\dfrac{36x^4y^5}{-3xy^2}$

42. $\dfrac{-56a^5b^7}{-8a^2b^3}$

43. $\dfrac{-18x^4y^3 - 54x^6y^2}{6x^2y^2}$

44. $\dfrac{-30a^5b^{10} + 39a^4b^8}{-3ab}$

45. $\dfrac{56x^4 - 40x^3 - 32x^2}{4x^2}$

46. $(x^2 + 9x - 1) \div (x + 5)$
47. $(21x^2 - 4x - 12) \div (3x + 2)$
48. $(2x^3 - 3x^2 + 2x - 4) \div (x - 2)$

Evaluate each of the following.

49. $3^2 + 2^2$
50. $(3 + 2)^2$
51. 2^{-4}
52. $(-5)^0$

53. -5^0
54. $\dfrac{1}{3^{-2}}$
55. $\left(\dfrac{3}{4}\right)^{-2}$
56. $\dfrac{1}{\left(\dfrac{1}{4}\right)^{-1}}$
57. $\dfrac{1}{(-2)^{-3}}$
58. $2^{-1} + 3^{-2}$
59. $3^0 + 2^{-2}$
60. $(2 + 3)^{-2}$

Simplify each of the following and express your answers using positive exponents only.

61. x^5x^{-8}
62. $(3x^5)(4x^{-2})$
63. $\dfrac{x^{-4}}{x^{-6}}$
64. $\dfrac{x^{-6}}{x^{-4}}$
65. $\dfrac{24a^5}{3a^{-1}}$
66. $\dfrac{48n^{-2}}{12n^{-1}}$
67. $(x^{-2}y)^{-1}$
68. $(a^2b^{-3})^{-2}$
69. $(2x)^{-1}$
70. $(3n^2)^{-2}$
71. $(2n^{-1})^{-3}$
72. $(4ab^{-1})(-3a^{-1}b^2)$

Write each of the following as a standard decimal fraction.

73. $(6.1)(10^2)$
74. $(5.6)(10^4)$
75. $(8)(10^{-2})$
76. $(9.2)(10^{-4})$

Write each of the following in scientific notation.

77. 9000
78. 47
79. .047
80. .00021

Use scientific notation and the properties of exponents to help evaluate each of the following.

81. $(.00004)(12,000)$
82. $(.0021)(2000)$
83. $\dfrac{.0056}{.0000028}$
84. $\dfrac{.00078}{39,000}$

CHAPTER 5 TEST

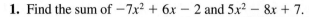

1. Find the sum of $-7x^2 + 6x - 2$ and $5x^2 - 8x + 7$.

2. Subtract $-x^2 + 9x - 14$ from $-4x^2 + 3x + 6$.

3. Remove parentheses and combine similar terms for the expression $3(2x - 1) - 6(3x - 2) - (x + 7)$.

4. Find the product $(-4xy^2)(7x^2y^3)$.

5. Find the product $(2x^2y)^2 \, (3xy^3)$.

For Problems 6–12, find the indicated products and express answers in simplest form.

6. $(x - 9)(x + 2)$

7. $(n + 14)(n - 7)$

8. $(5a + 3)(8a + 7)$

9. $(3x - 7y)^2$

10. $(x + 3)(2x^2 - 4x - 7)$

11. $(9x - 5y)(9x + 5y)$

12. $(3x - 7)(5x - 11)$

13. Find the indicated quotient: $\dfrac{-96x^4y^5}{-12x^2y}$.

14. Find the indicated quotient: $\dfrac{56x^2y - 72xy^2}{-8xy}$.

15. Find the indicated quotient: $(2x^3 - 5x^2 - 22x + 15) \div (2x - 3)$.

16. Find the indicated quotient: $(4x^3 + 23x^2 + 36) \div (x + 6)$.

17. Evaluate $\left(\dfrac{2}{3}\right)^{-3}$.

18. Evaluate $4^{-2} + 4^{-1} + 4^0$.

19. Evaluate $\dfrac{1}{2^{-4}}$.

20. Find the product $(-6x^{-4})(4x^2)$ and express the answer using a positive exponent.

21. Simplify $\left(\dfrac{8x^{-1}}{2x^2}\right)^{-1}$ and express the answer using a positive exponent.

22. Simplify $(x^{-3}y^5)^{-2}$ and express the answer using positive exponents.

23. Write .00027 in scientific notation.

24. Express $(9.2)(10^6)$ in standard decimal form.

25. Evaluate $(.000002)(3000)$.

C H A P T E R

Factoring and Solving Equations

A flower garden is in the shape of a right triangle with one leg 7 meters longer than the other leg and the hypotenuse is 1 meter longer than the longer leg. Find the lengths of all three sides of the right triangle. **A popular geometric formula, called the Pythagorean theorem, serves as a guideline for setting up an equation to solve this problem. We can use the equation $x^2 + (x + 7)^2 = (x + 8)^2$ to determine that the sides of the right triangle are 5 meters, 12 meters, and 13 meters long.**

The distributive property has played an important role in combining similar terms and multiplying polynomials. In this chapter we will see yet another use of the distributive property as we learn how to factor polynomials. The ability to factor polynomials will allow us to solve other kinds of equations, which will in turn help us to solve more kinds of word problems.

6.1 Factoring by Using the Distributive Property

In Chapter 1 we found the *greatest common factor* of two or more whole numbers by inspection or by using the prime factored form of the numbers. For example, by inspection we see that the greatest common factor of 8 and 12 is 4. This means that 4 is the largest whole number that is a factor of both 8 and 12. If it was difficult to determine the highest common factor by inspection then we used the prime factorization technique as follows.

$$42 = 2 \cdot 3 \cdot 7, \qquad 70 = 2 \cdot 5 \cdot 7$$

We see that $2 \cdot 7 = 14$ is the greatest common factor of 42 and 70.

It is meaningful to extend the concept of greatest common factor to monomials. Consider the following example.

Example 1

Find the greatest common factor of $8x^2$ and $12x^3$.

Solution

$$8x^2 = 2 \cdot 2 \cdot 2 \cdot x \cdot x$$
$$12x^3 = 2 \cdot 2 \cdot 3 \cdot x \cdot x \cdot x$$

Therefore, the greatest common factor is $2 \cdot 2 \cdot x \cdot x = 4x^2$.   ▲

By the greatest common factor of two or more monomials, we mean the monomial with the largest numerical coefficient and highest power of the variables, which is a factor of the given monomials.

Example 2

Find the greatest common factor of $16x^2y$, $24x^3y^2$, and $32xy$.

Solution

$$16x^2y = 2 \cdot 2 \cdot 2 \cdot 2 \cdot x \cdot x \cdot y$$
$$24x^3y^2 = 2 \cdot 2 \cdot 2 \cdot 3 \cdot x \cdot x \cdot x \cdot y \cdot y$$
$$32xy = 2 \cdot 2 \cdot 2 \cdot 2 \cdot 2 \cdot x \cdot y$$

Therefore, the greatest common factor is $2 \cdot 2 \cdot 2 \cdot x \cdot y = 8xy$.   ▲

You have used the distributive property to multiply a polynomial by a monomial. For example,

$$3x(x + 2) = 3x^2 + 6x.$$

Suppose that you start with $3x^2 + 6x$ and want to express it in factored form. Use the distributive property in the form $ab + ac = a(b + c)$.

$$3x^2 + 6x = 3x(x) + 3x(2) \qquad \text{\small 3x is the greatest common factor of } 3x^2 \text{ and } 6x.$$
$$= 3x(x + 2) \qquad\qquad \text{\small Use the distributive property.}$$

The next four examples further illustrate this process of **factoring out the greatest common monomial factor**.

Example 3

Solution

Factor $12x^3 - 8x^2$.

$$12x^3 - 8x^2 = 4x^2(3x) - 4x^2(2)$$
$$= 4x^2(3x - 2) \qquad ab - ac = a(b - c) \qquad \blacktriangle$$

Example 4

Solution

Factor $12x^2y + 18xy^2$.

$$12x^2y + 18xy^2 = 6xy(2x) + 6xy(3y)$$
$$= 6xy(2x + 3y) \qquad \blacktriangle$$

Example 5

Solution

Factor $24x^3 + 30x^4 - 42x^5$.

$$24x^3 + 30x^4 - 42x^5 = 6x^3(4) + 6x^3(5x) - 6x^3(7x^2)$$
$$= 6x^3(4 + 5x - 7x^2) \qquad \blacktriangle$$

Example 6

Solution

Factor $9x^2 + 9x$.

$$9x^2 + 9x = 9x(x) + 9x(1)$$
$$= 9x(x + 1) \qquad \blacktriangle$$

We want to emphasize the point made prior to Example 3. It is important to realize that we are factoring out the *greatest* common monomial factor. An expression such as $9x^2 + 9x$ in Example 6 could be factored as $9(x^2 + x)$, $3(3x^2 + 3x)$, $3x(3x + 3)$, or even as $\frac{1}{2}(18x^2 + 18x)$, but it is the form $9x(x + 1)$ that we want to obtain. We can accomplish this by factoring out the greatest common monomial factor; we sometimes refer to this process as **factoring completely**. A polynomial with integral coefficients is in completely factored form if:

1. It is expressed as a product of polynomials with integral coefficients;
2. No polynomial, other than a monomial, within the factored form can be further factored into polynomials with integral coefficients.

Thus, $9(x^2 + x)$, $3(3x^2 + 3x)$, and $3x(3x + 3)$ are not completely factored because they violate part 2. The form $\frac{1}{2}(18x^2 + 18x)$ violates both parts 1 and 2.

Sometimes there may be a **common binomial factor** rather than a common monomial factor. For example, each of the two terms of $x(y + 2) + z(y + 2)$ has a binomial factor of $(y + 2)$. Thus, we can factor $(y + 2)$ from each term and our result is as follows.

$$x(y + 2) + z(y + 2) = (y + 2)(x + z)$$

Consider a few more examples involving a common binomial factor.

$$a(b + c) - d(b + c) = (b + c)(a - d),$$
$$x(x + 2) + 3(x + 2) = (x + 2)(x + 3),$$
$$x(x + 5) - 4(x + 5) = (x + 5)(x - 4)$$

It may be that the original polynomial exhibits no apparent common monomial or binomial factor, which is the case with

$$ab + 3a + bc + 3c.$$

However, by factoring a from the first two terms and c from the last two terms, we see that

$$ab + 3a + bc + 3c = a(b + 3) + c(b + 3).$$

Now a common binomial factor of $(b + 3)$ is obvious and we can proceed as before.

$$a(b + 3) + c(b + 3) = (b + 3)(a + c)$$

This factoring process is called **factoring by grouping**. Let's consider two more examples of factoring by grouping.

$$x^2 - x + 5x - 5 = x(x - 1) + 5(x - 1)$$ Factor x from first two terms and 5 from last two terms

$$= (x - 1)(x + 5)$$ Factor common binomial factor of $(x - 1)$ from both terms

$$6x^2 - 4x - 3x + 2 = 2x(3x - 2) - 1(3x - 2)$$ Factor $2x$ from first two terms and -1 from last two terms

$$= (3x - 2)(2x - 1)$$ Factor common binomial factor of $(3x - 2)$ from both terms

Back to Solving Equations

Suppose we are told that the product of two numbers is zero. What do we know about the numbers? Do you agree we can conclude that at least one of the numbers must be zero? The following property formalizes this idea.

PROPERTY 6.1

For all real numbers a and b,

$$ab = 0 \quad \text{if and only if } a = 0 \text{ or } b = 0.$$

Property 6.1 provides us with another technique for solving equations.

Example 7

Solution

Solve $x^2 + 6x = 0$.

$$x^2 + 6x = 0$$
$$x(x + 6) = 0$$
$$x = 0 \quad \text{or} \quad x + 6 = 0 \qquad ab = 0 \text{ if and only if } a = 0 \text{ or } b = 0$$
$$x = 0 \quad \text{or} \quad x = -6$$

The solution set is $\{-6, 0\}$. (Be sure to check both of these in the original equation.) ▲

Example 8

Solution

Solve $x^2 = 12x$.

$$x^2 = 12x$$
$$x^2 - 12x = 0 \qquad \text{Added } -12x \text{ to both sides}$$
$$x(x - 12) = 0$$
$$x = 0 \quad \text{or} \quad x - 12 = 0 \qquad ab = 0 \text{ if and only if } a = 0 \text{ or } b = 0$$
$$x = 0 \quad \text{or} \quad x = 12$$

The solution set is $\{0, 12\}$. ▲

REMARK Notice in Example 8 we *did not* divide both sides of the original equation by x. This would cause us to lose the solution of 0. △

Example 9

Solution

Solve $4x^2 - 3x = 0$.

$$4x^2 - 3x = 0$$
$$x(4x - 3) = 0$$
$$x = 0 \quad \text{or} \quad 4x - 3 = 0 \qquad ab = 0 \text{ if and only if } a = 0 \text{ or } b = 0$$
$$x = 0 \quad \text{or} \quad 4x = 3$$
$$x = 0 \quad \text{or} \quad x = \frac{3}{4}$$

The solution set is $\left\{0, \frac{3}{4}\right\}$. ▲

Example 10

Solution

Solve $x(x + 2) + 3(x + 2) = 0$.

$$x(x + 2) + 3(x + 2) = 0$$
$$(x + 2)(x + 3) = 0$$
$$x + 2 = 0 \quad \text{or} \quad x + 3 = 0 \qquad ab = 0 \text{ if and only if } a = 0 \text{ or } b = 0$$
$$x = -2 \quad \text{or} \quad x = -3$$

The solution set is $\{-3, -2\}$. ▲

Each time that we extend our equation solving capabilities, we also gain more techniques for solving problems. Let's solve a geometric problem with the ideas we learned in this section.

Problem 1

The area of a square is numerically equal to twice its perimeter. Find the length of a side of the square.

Solution

Sketch a square and let s represent the length of each side (see Figure 6.1). Then the area is represented by s^2 and the perimeter by $4s$. Thus,

$$s^2 = 2(4s)$$
$$s^2 = 8s$$
$$s^2 - 8s = 0$$
$$s(s - 8) = 0$$
$$s = 0 \quad \text{or} \quad s - 8 = 0$$
$$s = 0 \quad \text{or} \quad s = 8.$$

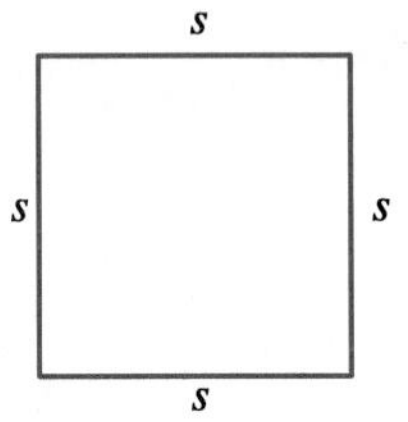

FIGURE 6.1

Since 0 is not a reasonable answer to the problem, the solution is 8. (Be sure to check this solution in the original statement of the problem!)   ▲

Problem Set 6.1

For Problems 1–10, find the greatest common factor of the given expressions.

1. $24y$ and $30xy$ **2.** $32x$ and $40xy$

3. $60x^2y$ and $84xy^2$

4. $72x^3$ and $63x^2$

5. $42ab^3$ and $70a^2b^2$

6. $48a^2b^2$ and $96ab^4$

7. $6x^3$, $8x$, and $24x^2$

8. $72xy$, $36x^2y$, and $84xy^2$

9. $16a^2b^2$, $40a^2b^3$, and $56a^3b^4$

10. $70a^3b^3$, $42a^2b^4$, and $49ab^5$

For Problems 11–46, factor each polynomial completely.

11. $8x + 12y$

12. $18x + 24y$

13. $14xy - 21y$

14. $24x - 40xy$

15. $18x^2 + 45x$

16. $12x + 28x^3$

17. $12xy^2 - 30x^2y$

18. $28x^2y^2 - 49x^2y$

19. $36a^2b - 60a^3b^4$

20. $65ab^3 - 45a^2b^2$

21. $16xy^3 + 25x^2y^2$

22. $12x^2y^2 + 29x^2y$

23. $64ab - 72cd$

24. $45xy - 72zw$

25. $9a^2b^4 - 27a^2b$

26. $7a^3b^5 - 42a^2b^6$

27. $52x^4y^2 + 60x^6y$

28. $70x^5y^3 - 42x^8y^2$

29. $40x^2y^2 + 8x^2y$

30. $84x^2y^3 + 12xy^3$

31. $12x + 15xy + 21x^2$

32. $30x^2y + 40xy + 55y$

33. $2x^3 - 3x^2 + 4x$

34. $x^4 + x^3 + x^2$

35. $44y^5 - 24y^3 - 20y^2$

36. $14a - 18a^3 - 26a^5$

37. $14a^2b^3 + 35ab^2 - 49a^3b$

38. $24a^3b^2 + 36a^2b^4 - 60a^4b^3$

39. $x(y + 1) + z(y + 1)$

40. $a(c + d) + 2(c + d)$

41. $a(b - 4) - c(b - 4)$

42. $x(y - 6) - 3(y - 6)$

43. $x(x + 3) + 6(x + 3)$

44. $x(x - 7) + 9(x - 7)$

45. $2x(x + 1) - 3(x + 1)$

46. $4x(x + 8) - 5(x + 8)$

For Problems 47–60, use the process of *factoring by grouping* to factor each of the following.

47. $5x + 5y + bx + by$

48. $7x + 7y + zx + zy$

49. $bx - by - cx + cy$

50. $2x - 2y - ax + ay$

51. $ac + bc + a + b$

52. $x + y + ax + ay$

53. $x^2 + 5x + 12x + 60$

54. $x^2 + 3x + 7x + 21$

55. $x^2 - 2x - 8x + 16$

56. $x^2 - 4x - 9x + 36$

57. $2x^2 + x - 10x - 5$

58. $3x^2 + 2x - 18x - 12$

59. $6n^2 - 3n - 8n + 4$

60. $20n^2 + 8n - 15n - 6$

For Problems 61–84, solve each equation.

61. $x^2 - 8x = 0$

62. $x^2 - 12x = 0$

63. $x^2 + x = 0$

64. $x^2 + 7x = 0$

65. $n^2 = 5n$

66. $n^2 = -2n$

67. $2y^2 - 3y = 0$

68. $4y^2 - 7y = 0$

69. $7x^2 = -3x$

70. $5x^2 = -2x$

71. $3n^2 + 15n = 0$

72. $6n^2 - 24n = 0$

73. $4x^2 = 6x$

74. $12x^2 = 8x$

75. $7x - x^2 = 0$

76. $9x - x^2 = 0$

77. $13x = x^2$

78. $15x = -x^2$

79. $5x = -2x^2$

80. $7x = -5x^2$

81. $x(x + 5) - 4(x + 5) = 0$

82. $x(3x - 2) - 7(3x - 2) = 0$

83. $4(x - 6) - x(x - 6) = 0$

84. $x(x + 9) = 2(x + 9)$

For Problems 85–91, set up an equation and solve each problem.

85. The square of a number equals nine times that number. Find the number.

86. Suppose that four times the square of a number equals 20 times that number. What is the number?

87. The area of a square is numerically equal to five times its perimeter. Find the length of a side of the square.

88. The area of a square is 14 times as large as the area of a triangle. One side of the triangle is 7 inches long and the altitude to that side is the same length as a side of the square. Find the length of a side of the square. Also find the areas of both figures and be sure that your answer checks.

89. Suppose that the area of a circle is numerically equal to the perimeter of a square and that the length of a radius of the circle is equal to the length of a side of the square. Find the length of a side of the square. Express your answer in terms of π.

90. One side of a parallelogram, an altitude to that side, and one side of a rectangle all have the same measure. If an adjacent side of the rectangle is 20 centimeters long and the area of the rectangle is twice the area of the parallelogram, find the area of both figures.

91. The area of a rectangle is twice the area of a square. If the rectangle is 6 inches long and the width of the rectangle is the same as the length of a side of the square, find the dimensions of both the rectangle and the square.

THOUGHTS INTO WORDS

92. Suppose that your friend factors $24x^2y + 36xy$ as follows:

$$24x^2y + 36xy = 4xy\,(6x + 9)$$
$$= (4xy)(3)(2x + 3)$$
$$= 12xy\,(2x + 3)$$

Is this correct? Would you make any suggestions?

93. The following solution is given for the equation $x(x - 10) = 0$:

$$x(x - 10) = 0$$
$$x^2 - 10x = 0$$
$$x(x - 10) = 0$$
$$x = 0 \text{ or } x - 10 = 0$$
$$x = 0 \text{ or } x = 10$$

The solution set is $\{0, 10\}$.

Is this a correct solution? Would you suggest any changes?

Further Investigations

94. The total surface area of a right circular cylinder is given by the formula $A = 2\pi r^2 + 2\pi rh$, where r represents the radius of a base and h represents the height of the cylinder. For computational purposes it may be more convenient to change the form of the right side of the formula by factoring it.

$$A = 2\pi r^2 + 2\pi rh$$
$$= 2\pi r(r + h)$$

Use $A = 2\pi r(r + h)$ to find the total surface area of each of the following cylinders. Also, use $\dfrac{22}{7}$ as an approximation for π.
(a) $r = 7$ centimeters and $h = 12$ centimeters;

(b) $r = 14$ meters and $h = 20$ meters;

(c) $r = 3$ feet and $h = 4$ feet;
(d) $r = 5$ yards and $h = 9$ yards.

95. The formula $A = P + Prt$ yields the total amount of money accumulated (A) when P dollars is invested at r percent simple interest for t years. For computational purposes it may be convenient to change the right side of the formula by factoring.

$$A = P + Prt$$
$$= P(1 + rt)$$

Use $A = P(1 + rt)$ to find the total amount of money accumulated for each of the following investments.
(a) $100 at 8% for 2 years;
(b) $200 at 9% for 3 years;
(c) $500 at 10% for 5 years;
(d) $1000 at 10% for 10 years.

For Problems 96–99, solve each equation for the indicated variable.

96. $ax + bx = c$ for x
97. $b^2x^2 - cx = 0$ for x
98. $5ay^2 = by$ for y
99. $y + ay - by - c = 0$ for y

6.2 Factoring the Difference of Two Squares

In Section 5.3 we noted some special multiplication patterns. One of these patterns was

$$(a - b)(a + b) = a^2 - b^2.$$

We can view this same pattern as the factoring pattern that follows.

Difference of Two Squares

$$a^2 - b^2 = (a - b)(a + b)$$

To apply the pattern is a fairly simple process, as these next examples illustrate. The steps inside the box are often performed mentally.

$$
\begin{aligned}
x^2 - 36 &= \boxed{(x)^2 - (6)^2} = (x - 6)(x + 6); \\
4x^2 - 25 &= \boxed{(2x)^2 - (5)^2} = (2x - 5)(2x + 5); \\
9x^2 - 16y^2 &= \boxed{(3x)^2 - (4y)^2} = (3x - 4y)(3x + 4y); \\
64 - y^2 &= \boxed{(8)^2 - (y)^2} = (8 - y)(8 + y).
\end{aligned}
$$

Since multiplication is commutative, the order of writing the factors is not important. For example, $(x - 6)(x + 6)$ can also be written as $(x + 6)(x - 6)$.

You must be careful not to assume an analogous factoring pattern for the *sum* of two squares; it does not exist. For example, $x^2 + 4 \neq (x + 2)(x + 2)$ since $(x + 2)(x + 2) = x^2 + 4x + 4$. We say that the **sum of two squares is not factorable using integers**. The phrase "using integers" is necessary because $x^2 + 4$ could be written as $\frac{1}{2}(2x^2 + 8)$, but such *factoring* is of no help. Furthermore, we do not consider $(1)(x^2 + 4)$ as factoring $x^2 + 4$.

It is possible that both the technique of *factoring out a common monomial factor* and the pattern *difference of two squares* can be applied to the same polynomial. In general, it is best to first look for a common monomial factor.

Example 1

Factor $2x^2 - 50$.

Solution

$$
\begin{aligned}
2x^2 - 50 &= 2(x^2 - 25) \quad &&\text{Common factor of 2} \\
&= 2(x - 5)(x + 5). \quad &&\text{Difference of squares}
\end{aligned}
$$

▲

By expressing $2x^2 - 50$ as $2(x - 5)(x + 5)$ we say that it has been **factored completely**. That means the factors 2, $x - 5$, and $x + 5$ cannot be factored any further using integers.

Example 2

Solution

Factor completely $18y^3 - 8y$.

$$18y^3 - 8y = 2y(9y^2 - 4) \qquad \text{Common factor of } 2y$$
$$= 2y(3y - 2)(3y + 2) \qquad \text{Difference of squares} \qquad \blacktriangle$$

Sometimes it is possible to apply the difference-of-squares pattern more than once.

Example 3

Solution

Factor completely $x^4 - 16$.

$$x^4 - 16 = (x^2 + 4)(x^2 - 4)$$
$$= (x^2 + 4)(x + 2)(x - 2) \qquad \blacktriangle$$

The following examples should help you to summarize our factoring ideas thus far.

$$5x^2 + 20 = 5(x^2 + 4)$$
$$25 - y^2 = (5 - y)(5 + y)$$
$$3 - 3x^2 = 3(1 - x^2) = 3(1 + x)(1 - x)$$
$$36x^2 - 49y^2 = (6x - 7y)(6x + 7y)$$
$a^2 + 9$ is not factorable using integers
$9x + 17y$ is not factorable using integers

Solving Equations

Each time that we pick up a new factoring technique we also develop more power for solving equations. Let's consider how we can use the difference-of-squares factoring pattern to help solve certain kinds of equations.

Example 4

Solution

Solve $x^2 = 25$.

$$x^2 = 25$$
$$x^2 - 25 = 0$$
$$(x + 5)(x - 5) = 0$$
$$x + 5 = 0 \quad \text{or} \quad x - 5 = 0 \qquad \text{Remember: } ab = 0 \text{ if and only if}$$
$$x = -5 \quad \text{or} \quad x = 5 \qquad a = 0 \text{ or } b = 0.$$

The solution set is $\{-5, 5\}$. Check these answers! $\blacktriangle$

Example 5

Solve $9x^2 = 25$.

Solution

$$9x^2 = 25$$
$$9x^2 - 25 = 0$$
$$(3x + 5)(3x - 5) = 0$$
$$3x + 5 = 0 \quad \text{or} \quad 3x - 5 = 0$$
$$3x = -5 \quad \text{or} \quad 3x = 5$$
$$x = -\frac{5}{3} \quad \text{or} \quad x = \frac{5}{3}$$

The solution set is $\left\{-\frac{5}{3}, \frac{5}{3}\right\}$.

▲

Example 6

Solve $5y^2 = 20$.

Solution

$$5y^2 = 20$$
$$\frac{5y^2}{5} = \frac{20}{5} \qquad \text{Divide both sides by 5.}$$
$$y^2 = 4$$
$$y^2 - 4 = 0$$
$$(y + 2)(y - 2) = 0$$
$$y + 2 = 0 \quad \text{or} \quad y - 2 = 0$$
$$y = -2 \quad \text{or} \quad y = 2$$

The solution set is $\{-2, 2\}$. Check it!

▲

Example 7

Solve $x^3 - 9x = 0$.

Solution

$$x^3 - 9x = 0$$
$$x(x^2 - 9) = 0$$
$$x(x - 3)(x + 3) = 0$$
$$x = 0 \quad \text{or} \quad x - 3 = 0 \quad \text{or} \quad x + 3 = 0$$
$$x = 0 \quad \text{or} \quad x = 3 \quad \text{or} \quad x = -3$$

The solution set is $\{-3, 0, 3\}$.

▲

The more we know about solving equations, the better off we are when we solve word problems.

Problem I

The combined area of two squares is 20 square centimeters. Each side of one square is twice as long as a side of the other square. Find the lengths of the sides of each square.

Solution

Let's sketch two squares and label the sides of the smaller square s (see Figure 6.2). Then the sides of the larger square are $2s$. Since the sum of the areas of the two squares is 20 square centimeters, we can set up and solve the following equation.

$$s^2 + (2s)^2 = 20$$
$$s^2 + 4s^2 = 20$$
$$5s^2 = 20$$
$$s^2 = 4$$
$$s^2 - 4 = 0$$
$$(s + 2)(s - 2) = 0$$
$$s + 2 = 0 \quad \text{or} \quad s - 2 = 0$$
$$s = -2 \quad \text{or} \quad s = 2$$

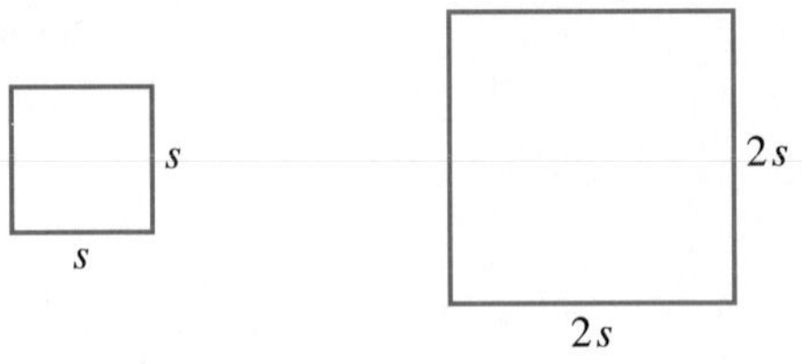

FIGURE 6.2

Since s represents the length of a side of a square, the solution -2 must be disregarded. Thus, one square has sides of length 2 centimeters and the other square has sides of length $2(2) = 4$ centimeters. ▲

Problem Set 6.2

For Problems 1–12, use the difference-of-squares pattern to factor each polynomial.

1. $x^2 - 1$

2. $x^2 - 25$

3. $x^2 - 100$

4. $x^2 - 121$

5. $x^2 - 4y^2$

6. $x^2 - 36y^2$

7. $9x^2 - y^2$

8. $49y^2 - 64x^2$

9. $36a^2 - 25b^2$

10. $4a^2 - 81b^2$

11. $1 - 4n^2$

12. $4 - 9n^2$

For Problems 13–40, factor each polynomial completely. Indicate any that are not factorable using integers. Don't forget to first look for a common monomial factor.

13. $5x^2 - 20$

14. $7x^2 - 7$

15. $8x^2 + 32$

16. $12x^2 + 60$

17. $2x^2 - 18y^2$

18. $8x^2 - 32y^2$

19. $x^3 - 25x$

20. $2x^3 - 2x$

21. $x^2 + 9y^2$

22. $18x - 42y$

23. $45x^2 - 36xy$

24. $16x^2 + 25y^2$

25. $36 - 4x^2$

26. $75 - 3x^2$

27. $4a^4 + 16a^2$

28. $9a^4 + 81a^2$

29. $x^4 - 81$

30. $16 - x^4$

31. $x^4 + x^2$

32. $x^5 + 2x^3$

33. $3x^3 + 48x$

34. $6x^3 + 24x$

35. $5x - 20x^3$

36. $4x - 36x^3$

37. $4x^2 - 64$

38. $9x^2 - 9$

39. $75x^3y - 12xy^3$

40. $32x^3y - 18xy^3$

For Problems 41–64, solve each equation.

41. $x^2 = 9$

42. $x^2 = 1$

43. $4 = n^2$

44. $144 = n^2$

45. $9x^2 = 16$

46. $4x^2 = 9$

47. $n^2 - 121 = 0$

48. $n^2 - 81 = 0$

49. $25x^2 = 4$

50. $49x^2 = 36$

51. $3x^2 = 75$

52. $7x^2 = 28$

53. $3x^3 - 48x = 0$

54. $x^3 - x = 0$

55. $n^3 = 16n$

56. $2n^3 = 8n$

57. $5 - 45x^2 = 0$

58. $3 - 12x^2 = 0$

59. $4x^3 - 400x = 0$

60. $2x^3 - 98x = 0$

61. $64x^2 = 81$

62. $81x^2 = 25$

63. $36x^3 = 9x$

64. $64x^3 = 4x$

For Problems 65–76, set up an equation and solve the problem.

65. Forty-nine less than the square of a number equals zero. Find the number.

66. The cube of a number equals nine times the number. Find the number.

67. Suppose that five times the cube of a number equals 80 times the number. Find the number.

68. Ten times the square of a number equals 40. Find the number.

69. The sum of the areas of two squares is 234 square inches. Each side of the larger square is five times the length of a side of the smaller square. Find the length of a side of each square.

70. The difference of the areas of two squares is 75 square feet. Each side of the larger square is twice the length of a side of the smaller square. Find the length of a side of each square.

71. Suppose that the length of a certain rectangle is $2\frac{1}{2}$ times its width and the area of that same rectangle is 160 square centimeters. Find the length and width of the rectangle.

72. Suppose that the width of a certain rectangle is three-fourths of its length and the area of that same rectangle is 108 square meters. Find the length and width of the rectangle.

73. The sum of the areas of two circles is 80π square meters. Find the length of a radius of each circle if one of them is twice as long as the other.

74. The area of a triangle is 98 square feet. If one side of the triangle and the altitude to that side are of equal length, find the length.

75. The total surface area of a right circular cylinder is 100π square centimeters. If a radius of the base and the altitude of the cylinder are of the same length, find the length of a radius.

76. The total surface area of a right circular cone is 192π square feet. If the slant height of the cone is equal in length to a diameter of the base, find the length of a radius.

77. How do we know that the equation $x^2 + 1 = 0$ has no solutions in the set of real numbers?

78. Why is the following factoring process incomplete?

$$16x^2 - 64 = (4x + 8)(4x - 8)$$

How should the factoring be done?

79. Consider the following solution.

$$4x^2 - 36 = 0$$
$$4(x^2 - 9) = 0$$
$$4(x + 3)(x - 3) = 0$$
$$4 = 0 \text{ or } x + 3 = 0 \quad \text{or} \quad x - 3 = 0$$
$$4 = 0 \text{ or } x = -3 \quad \text{or} \quad x = 3$$

The solution set is $\{-3, 3\}$.

Is this a correct solution? Would you have any suggestion to offer the person who did this problem?

Further Investigations

The following patterns can be used to factor the sum and difference of two cubes.

$$a^3 + b^3 = (a + b)(a^2 - ab + b^2)$$
$$a^3 - b^3 = (a - b)(a^2 + ab + b^2)$$

Consider the following examples.

$$x^3 + 8 = (x)^3 + (2)^3 = (x + 2)(x^2 - 2x + 4)$$
$$x^3 - 1 = (x)^3 - (1)^3 = (x - 1)(x^2 + x + 1)$$

Use the sum and difference-of-cubes patterns to factor each of the following.

80. $x^3 + 1$

81. $x^3 - 8$

82. $n^3 - 27$

83. $n^3 + 64$

84. $8x^3 + 27y^3$

85. $27a^3 - 64b^3$

86. $1 - 8x^3$

87. $1 + 27a^3$

88. $x^3 + 8y^3$

89. $8x^3 - y^3$

90. $a^3b^3 - 1$

91. $27x^3 - 8y^3$

92. $8 + n^3$

93. $125x^3 + 8y^3$

94. $27n^3 - 125$

95. $64 + x^3$

6.3 Factoring Trinomials of the Form $x^2 + bx + c$

One of the most common types of factoring used in algebra is the expression of a trinomial as the product of two binomials. In this section we want to consider trinomials where the coefficient of the squared term is 1; that is, trinomials of the form $x^2 + bx + c$.

Again, to develop a factoring technique we first look at some multiplication ideas. Consider the product $(x + r)(x + s)$ and use the distributive property to show how each term of the resulting trinomial is formed.

$$(x + r)(x + s) = x(x) + x(s) + r(x) + r(s)$$

$$x^2 + \quad (s + r)x \quad + \quad rs$$

Notice that the coefficient of the middle term is the *sum* of r and s and the last term is the *product* of r and s. These two relationships are used in the following examples.

Example 1

Solution

Factor $x^2 + 7x + 12$.

We need to complete the following with two numbers whose sum is 7 and whose product is 12.

$$x^2 + 7x + 12 = (x + \underline{\qquad})(x + \underline{\qquad})$$

This can be done by setting up a small table as follows.

Product	Sum
$1(12) = 12$	$1 + 12 = 13$
$2(6) = 12$	$2 + 6 = 8$
$3(4) = 12$	$3 + 4 = 7$

The bottom line contains the numbers that we need. Thus,

$$x^2 + 7x + 12 = (x + 3)(x + 4).$$

▲

Example 2

Solution

Factor $x^2 - 11x + 24$.

We need two numbers whose product is 24 and whose sum is -11.

Product	Sum
$(-1)(-24) = 24$	$-1 + (-24) = -25$
$(-2)(-12) = 24$	$-2 + (-12) = -14$
$(-3)(-8) = 24$	$-3 + (-8) = -11$
$(-4)(-6) = 24$	$-4 + (-6) = -10$

The third line contains the numbers that we want.

$$x^2 - 11x + 24 = (x - 3)(x - 8)$$

▲

Example 3

Factor $x^2 + 3x - 10$.

Solution

We need two numbers whose product is -10 and whose sum is 3.

Product	Sum
$1(-10) = -10$	$1 + (-10) = -9$
$-1(10) = -10$	$-1 + 10 = 9$
$2(-5) = -10$	$2 + (-5) = -3$
$-2(5) = -10$	$-2 + 5 = 3$

The bottom line is the key line. Thus,

$$x^2 + 3x - 10 = (x + 5)(x - 2).$$

▲

Example 4

Factor $x^2 - 2x - 8$.

Solution

We need two numbers whose product is -8 and whose sum is -2.

Product	Sum
$1(-8) = -8$	$1 + (-8) = -7$
$-1(8) = -8$	$-1 + 8 = 7$
$2(-4) = -8$	$2 + (-4) = -2$
$-2(4) = -8$	$-2 + 4 = 2$

The third line has the information we want.

$$x^2 - 2x - 8 = (x - 4)(x + 2)$$

▲

The tables in the four previous examples were used to illustrate one way of organizing your thoughts for such problems. We have shown complete tables. That is, for Example 4 the bottom line was included even though the desired numbers were obtained in the third line. If you use such tables, then keep in mind that as soon as the desired numbers are obtained, the table need not be completed any further. Furthermore, many times you may be able to find the numbers without using a table. The key ideas are the product and sum relationships.

Example 5

Factor $x^2 - 13x + 12$.

Solution

Product	Sum
$(-1)(-12) = 12$	$(-1) + (-12) = -13$

We need not complete the table.

$$x^2 - 13x + 12 = (x - 1)(x - 12)$$

▲

In the next example, we refer to the concept of absolute value. Recall that the absolute value refers to the number, with disregard for the sign. For example,

$$|4| = 4 \quad \text{and} \quad |-4| = 4.$$

Example 6

Factor $x^2 - x - 56$.

Solution

Notice that the coefficient of the middle term is -1. Therefore, we are looking for two numbers whose product is -56; since their sum is -1, the absolute value of the negative number must be one larger than the absolute value of the positive number. The numbers are -8 and 7 and we have

$$x^2 - x - 56 = (x - 8)(x + 7).$$

▲

Example 7

Factor $x^2 + 10x + 12$.

Solution

Product	Sum
$1(12) = 12$	$1 + 12 = 13$
$2(6) = 12$	$2 + 6 = 8$
$3(4) = 12$	$3 + 4 = 7$

Since the table is complete and no two factors of 12 produce a sum of 10, we conclude that

$$x^2 + 10x + 12$$

is not factorable using integers.

▲

In a problem such as Example 7 we need to be sure that all possibilities have been tried before we conclude that the trinomial is not factorable.

Back to Solving Equations

The property "$ab = 0$ if and only if $a = 0$ or $b = 0$" continues to play an important role as we solve equations that involve the factoring ideas of this section. Consider the following examples.

Example 8

Solve $x^2 + 8x + 15 = 0$.

Solution

$$x^2 + 8x + 15 = 0$$
$$(x + 3)(x + 5) = 0 \qquad \text{Factor the left side.}$$
$$x + 3 = 0 \quad \text{or} \quad x + 5 = 0 \qquad \text{Use } ab = 0 \text{ if and only if}$$
$$x = -3 \quad \text{or} \quad x = -5 \qquad a = 0 \text{ or } b = 0.$$

The solution set is $\{-5, -3\}$.   ▲

Example 9

Solve $x^2 + 5x - 6 = 0$.

Solution

$$x^2 + 5x - 6 = 0$$
$$(x + 6)(x - 1) = 0$$
$$x + 6 = 0 \quad \text{or} \quad x - 1 = 0$$
$$x = -6 \quad \text{or} \quad x = 1.$$

The solution set is $\{-6, 1\}$.   ▲

Example 10

Solve $y^2 - 4y = 45$.

Solution

$$y^2 - 4y = 45$$
$$y^2 - 4y - 45 = 0$$
$$(y - 9)(y + 5) = 0$$
$$y - 9 = 0 \quad \text{or} \quad y + 5 = 0$$
$$y = 9 \quad \text{or} \quad y = -5.$$

The solution set is $\{-5, 9\}$.   ▲

Don't forget that we can always check to be absolutely sure of our solutions. Let's check the solutions for Example 10.

If $y = 9$, then $y^2 - 4y = 45$ becomes

$$9^2 - 4(9) \stackrel{?}{=} 45$$
$$81 - 36 \stackrel{?}{=} 45$$
$$45 = 45.$$

If $y = -5$, then $y^2 - 4y = 45$ becomes

$$(-5)^2 - 4(-5) \stackrel{?}{=} 45$$
$$25 + 20 \stackrel{?}{=} 45$$
$$45 = 45.$$

Back to Problem Solving

The more we know about factoring and solving equations the more word problems we can solve.

Problem 1

Find two consecutive integers whose product is 72.

Solution

Let n represent one integer. Then $n + 1$ represents the next integer.

$$n(n + 1) = 72 \qquad \text{The product of the two integers is 72.}$$
$$n^2 + n = 72$$
$$n^2 + n - 72 = 0$$
$$(n + 9)(n - 8) = 0$$
$$n + 9 = 0 \qquad \text{or} \qquad n - 8 = 0$$
$$n = -9 \qquad \text{or} \qquad n = 8$$

If $n = -9$, then $n + 1 = -9 + 1 = -8$. If $n = 8$, then $n + 1 = 8 + 1 = 9$. Thus, the consecutive integers are -9 and -8 or 8 and 9. ▲

Problem 2

A rectangular plot is 6 meters longer than it is wide. The area of the plot is 16 square meters. Find the length and width of the plot.

Solution

Let w represent the width of the plot. Then $w + 6$ represents the length (see Figure 6.3).

FIGURE 6.3

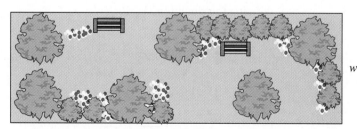

$w + 6$

Using the area formula ($A = lw$), we obtain

$$w(w + 6) = 16$$
$$w^2 + 6w = 16$$
$$w^2 + 6w - 16 = 0$$
$$(w + 8)(w - 2) = 0$$
$$w + 8 = 0 \quad \text{or} \quad w - 2 = 0$$
$$w = -8 \quad \text{or} \quad w = 2.$$

The solution of -8 is not possible for the width of a rectangle, so the plot is 2 meters wide and its length ($w + 6$) is 8 meters. ▲

The Pythagorean theorem, an important theorem pertaining to right triangles, can also serve as a guideline for solving certain types of problems. The Pythagorean theorem states that **in any right triangle** (see Figure 6.4), **the square of the longest side** (*called the hypotenuse*) **is equal to the sum of the squares of the other two sides** (*called legs*). Let's use this theorem to help solve a problem.

FIGURE 6.4

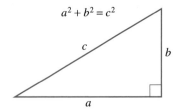

$a^2 + b^2 = c^2$

Problem 3

Suppose that the lengths of the three sides of a right triangle are consecutive whole numbers. Find the lengths of the three sides.

Solution

Let s represent the length of the shortest leg. Then $s + 1$ represents the length of the other leg and $s + 2$ represents the length of the hypotenuse. Using the Pythagorean theorem as a guideline, we obtain the following equation.

Sum of squares of two legs   Square of hypotenuse

$$s^2 + (s + 1)^2 \quad = \quad (s + 2)^2$$

Solving this equation yields

$$s^2 + s^2 + 2s + 1 = s^2 + 4s + 4$$
$$2s^2 + 2s + 1 = s^2 + 4s + 4$$
$$s^2 + 2s + 1 = 4s + 4$$
$$s^2 - 2s + 1 = 4$$
$$s^2 - 2s - 3 = 0$$
$$(s - 3)(s + 1) = 0$$

$$s - 3 = 0 \quad \text{or} \quad s + 1 = 0$$
$$s = 3 \quad \text{or} \quad s = -1.$$

The solution of -1 is not possible for the length of a side, so the shortest side (s) is of length 3. The other two sides ($s + 1$ and $s + 2$) have lengths of 4 and 5.

▲

Problem Set 6.3

For Problems 1–30, factor each trinomial completely. Indicate any that are not factorable using integers.

1. $x^2 + 10x + 24$

2. $x^2 + 9x + 14$

3. $x^2 + 13x + 40$

4. $x^2 + 11x + 24$

5. $x^2 - 11x + 18$

6. $x^2 - 5x + 4$

7. $n^2 - 11n + 28$

8. $n^2 - 7n + 10$

9. $n^2 + 6n - 27$

10. $n^2 + 3n - 18$

11. $n^2 - 6n - 40$

12. $n^2 - 4n - 45$

13. $t^2 + 12t + 24$

14. $t^2 + 20t + 96$

15. $x^2 - 18x + 72$

16. $x^2 - 14x + 32$

17. $x^2 + 5x - 66$

18. $x^2 + 11x - 42$

19. $y^2 - y - 72$

20. $y^2 - y - 30$

21. $x^2 + 21x + 80$

22. $x^2 + 21x + 90$

23. $x^2 + 6x - 72$

24. $x^2 - 8x - 36$

25. $x^2 - 10x - 48$

26. $x^2 - 12x - 64$

27. $x^2 + 3xy - 10y^2$

28. $x^2 - 4xy - 12y^2$

29. $a^2 - 4ab - 32b^2$

30. $a^2 + 3ab - 54b^2$

For Problems 31–50, solve each equation.

31. $x^2 + 10x + 21 = 0$

32. $x^2 + 9x + 20 = 0$

33. $x^2 - 9x + 18 = 0$

34. $x^2 - 9x + 8 = 0$

35. $x^2 - 3x - 10 = 0$

36. $x^2 - x - 12 = 0$

37. $n^2 + 5n - 36 = 0$

38. $n^2 + 3n - 18 = 0$

39. $n^2 - 6n - 40 = 0$

40. $n^2 - 8n - 48 = 0$

41. $t^2 + t - 56 = 0$

42. $t^2 + t - 72 = 0$

43. $x^2 - 16x + 28 = 0$

44. $x^2 - 18x + 45 = 0$

45. $x^2 + 11x = 12$

46. $x^2 + 8x = 20$

47. $x(x - 10) = -16$

48. $x(x - 12) = -35$

49. $-x^2 - 2x + 24 = 0$

50. $-x^2 + 6x + 16 = 0$

For Problems 51–68, set up an equation and solve each problem.

51. Find two consecutive integers whose product is 56.

52. Find two consecutive odd whole numbers whose product is 63.

53. Find two consecutive even whole numbers whose product is 168.

54. One number is two larger than another number. The sum of their squares is 100. Find the numbers.

55. Find four consecutive integers such that the product of the two larger integers is 22 less than twice the product of the two smaller integers.

56. Find three consecutive integers such that the product of the two smaller integers is two more than ten times the largest integer.

57. One number is three smaller than another number. The square of the larger number is nine larger than ten times the smaller number. Find the numbers.

58. The area of the floor of a rectangular room is 84 square feet. The length of the room is 5 feet more than its width. Find the length and width of the room.

59. Suppose that the width of a certain rectangle is 3 inches less than its length. The area is numerically six less than twice the perimeter. Find the length and width of the rectangle.

60. The sum of the areas of a square and a rectangle is 64 square centimeters. The length of the rectangle is 4 centimeters more than a side of the square and the width of the rectangle is 2 centimeters more than a side of the square. Find the dimensions of the square and the rectangle.

61. The perimeter of a rectangle is 30 centimeters and the area is 54 square centimeters. Find the length and width of the rectangle. [*Hint*: Let w represent the width; then $15 - w$ represents the length.]

62. The perimeter of a rectangle is 44 inches and its area is 120 square inches. Find the length and width of the rectangle.

63. An apple orchard contains 84 trees. The number of trees per row is five more than the number of rows. Find the number of rows.

64. A room contains 54 chairs. The number of rows is three less than the number of chairs per row. Find the number of rows.

65. Suppose that one leg of a right triangle is 7 feet shorter than the other leg. The hypotenuse is 2 feet longer than the longer leg. Find the lengths of all three sides of the right triangle.

66. Suppose that one leg of a right triangle is 7 meters longer than the other leg. The hypotenuse is 1 meter longer than the longer leg. Find the lengths of all three sides of the right triangle.

67. Suppose that the length of one leg of a right triangle is 2 inches less than the length of the other leg. If the length of the hypotenuse is 10 inches, find the length of each leg.

68. The length of one leg of a right triangle is 3 centimeters more than the length of the other leg. The length of the hypotenuse is 15 centimeters. Find the lengths of the two legs.

THOUGHTS INTO WORDS

69. What does the expression "not factorable using integers" mean to you?

70. Discuss the role that factoring plays in solving equations.

71. Explain how you would solve the equation $(x - 3)(x + 4) = 0$ and also how you would solve $(x - 3)(x + 4) = 8$.

Further Investigations

For Problems 72–75, factor each trinomial and assume that all variables appearing as exponents represent positive integers.

72. $x^{2a} + 10x^a + 24$

73. $x^{2a} + 13x^a + 40$

74. $x^{2a} - 2x^a - 8$

75. $x^{2a} + 6x^a - 27$

76. Suppose that we want to factor $n^2 + 26n + 168$ so that we can solve the equation $n^2 + 26n + 168 = 0$. We need to find two positive integers whose product is 168 and whose sum is 26. Since the constant term, 168, is rather large, let's look at it in prime factored form.

$$168 = 2 \cdot 2 \cdot 2 \cdot 3 \cdot 7$$

Now we can mentally form two numbers by using all of these factors in different combinations. Using two 2s and the 3 in one number, and the other 2 and the 7 in another number produces $2 \cdot 2 \cdot 3 = 12$ and $2 \cdot 7 = 14$. Therefore, we can solve the given equation as follows.

$$n^2 + 26n + 168 = 0$$
$$(n + 12)(n + 14) = 0$$
$$n + 12 = 0 \quad \text{or} \quad n + 14 = 0$$
$$n = -12 \quad \text{or} \quad n = -14$$

The solution set is $\{-14, -12\}$.

Solve each of the following equations.

(a) $n^2 + 30n + 216 = 0$
(b) $n^2 + 35n + 294 = 0$
(c) $n^2 - 40n + 384 = 0$
(d) $n^2 - 40n + 375 = 0$
(e) $n^2 + 6n - 432 = 0$
(f) $n^2 - 16n - 512 = 0$

6.4 Factoring Trinomials of the Form $ax^2 + bx + c$

Now let's consider factoring trinomials where the coefficient of the squared term is not 1. First, let's illustrate an informal trial and error technique that works quite well for certain types of trinomials. This technique simply relies on our knowledge of multiplication of binomials.

Example 1

Factor $2x^2 + 7x + 3$.

Solution

By looking at the first term, $2x^2$, and the positive signs of the other two terms, we know that the binomials are of the form

$$(2x + \underline{\quad})(x + \underline{\quad}).$$

Since the factors of the constant term, 3, are 1 and 3, we have only two possibilities to try.

$$(2x + 3)(x + 1) \quad \text{or} \quad (2x + 1)(x + 3)$$

By checking the middle term of both of these products we find the second one yields the correct middle term of $7x$. Therefore,

$$2x^2 + 7x + 3 = (2x + 1)(x + 3).$$

▲

Example 2

Factor $6x^2 - 17x + 5$.

Solution

First, note that $6x^2$ can be written as $2x \cdot 3x$ or $6x \cdot x$. Secondly, since the middle term of the trinomial is negative and the last term is positive, we know that the binomials are of the form

$$(2x - \underline{\hspace{1cm}})(3x - \underline{\hspace{1cm}}) \quad \text{or} \quad (6x - \underline{\hspace{1cm}})(x - \underline{\hspace{1cm}}).$$

Since the factors of the constant term, 5, are 1 and 5, the following possibilities exist.

$$(2x - 5)(3x - 1), \quad (2x - 1)(3x - 5),$$
$$(6x - 5)(x - 1), \quad (6x - 1)(x - 5)$$

By checking the middle term for each of these products we find that the product $(2x - 5)(3x - 1)$ produces the desired term of $-17x$. Therefore,

$$6x^2 - 17x + 5 = (2x - 5)(3x - 1). \qquad \blacktriangle$$

Example 3

Factor $4x^2 - 4x - 15$.

Solution

First, note that $4x^2$ can be written as $4x \cdot x$ or $2x \cdot 2x$. Secondly, the last term, -15, can be written as $(1)(-15)$, $(-1)(15)$, $(3)(-5)$, or $(-3)(5)$. Thus, we can generate the possibilities for the binomial factors as follows.

Using 1 and −15	Using −1 and 15
$(4x - 15)(x + 1)$	$(4x - 1)(x + 15)$
$(4x + 1)(x - 15)$	$(4x + 15)(x - 1)$
$(2x + 1)(2x - 15)$	$(2x - 1)(2x + 15)$

Using 3 and −5	Using −3 and 5
$(4x + 3)(x - 5)$	$(4x - 3)(x + 5)$
$(4x - 5)(x + 3)$	$(4x + 5)(x - 3)$
$\checkmark (2x - 5)(2x + 3)$	$(2x + 5)(2x - 3)$

By checking the middle term of each of these products we find that the product indicated with a check mark produces the desired middle term of $-4x$. Therefore,

$$4x^2 - 4x - 15 = (2x - 5)(2x + 3). \qquad \blacktriangle$$

Let's pause for a moment and look back over Examples 1, 2, and 3. Obviously, Example 3 created the most difficulty because we had to consider so many possibilities. We have suggested one possible format for considering the possibilities but as you practice such problems you may develop a format of your own that works better for you. Regardless of the format that you use, the key idea is to organize your work so that you consider all possibilities. Let's look at another example.

Example 4

Solution

Factor $4x^2 + 6x + 9$.

First, note that $4x^2$ can be written as $4x \cdot x$ or $2x \cdot 2x$. Secondly, since the middle term is positive and the last term is positive, we know that the binomials are of the form

$$(4x + \underline{\hspace{1cm}})(x + \underline{\hspace{1cm}}) \quad \text{or} \quad (2x + \underline{\hspace{1cm}})(2x + \underline{\hspace{1cm}}).$$

Since 9 can be written as $9 \cdot 1$ or $3 \cdot 3$, we have only the five following possibilities to try.

$$(4x + 9)(x + 1) \qquad (4x + 1)(x + 9)$$
$$(4x + 3)(x + 3) \qquad (2x + 1)(2x + 9)$$
$$(2x + 3)(2x + 3)$$

When we try all of these possibilities we find that none of them yields a middle term of $6x$. Therefore, $4x^2 + 6x + 9$ is *not factorable* using integers. ▲

REMARK   Example 4 illustrates the importance of organizing your work so that you try *all* possibilities before you conclude that a particular trinomial is not factorable.

△

Another Approach

There is another more systematic technique that you may wish to use with some trinomials. It is an extension of the method we used in the previous section. Recall that at the beginning of Section 6.3 we looked at the following product.

$$(x + r)(x + s) = x(x) + x(s) + r(x) + r(s)$$
$$= x^2 + (s + r)x + rs$$

Sum of r and s Product of r and s

Now let's look at the following product.

$$(px + r)(qx + s) = px(qx) + px(s) + r(qx) + r(s)$$
$$= (pq)x^2 + (ps + rq)x + rs$$

Notice that the product of the coefficient of the x^2 term, (pq), and the constant term, (rs), is $pqrs$. Likewise, the product of the two coefficients of x, $(ps$ and $rq)$, is also $pqrs$. Therefore, the two coefficients of x must have a sum of $ps + rq$ and a product of $pqrs$. This may seem a little confusing, but the next few examples illustrate how easy it is to apply.

Example 5

Solution

Factor $3x^2 + 14x + 8$.

$$3x^2 + 14x + 8 \qquad \text{Sum of 14}$$

Product of $3 \cdot 8 = 24$

We need to find two integers whose sum is 14 and whose product is 24. Obviously, 2 and 12 satisfy these conditions. Therefore, we will express the middle term of the trinomial, $14x$, as $2x + 12x$ and proceed as follows.

$$\begin{aligned} 3x^2 + 14x + 8 &= 3x^2 + 2x + 12x + 8 \\ &= x(3x + 2) + 4(3x + 2) \\ &= (3x + 2)(x + 4) \end{aligned}$$

▲

Example 6

Solution

Factor $16x^2 - 26x + 3$.

$$16x^2 - 26x + 3 \qquad \text{Sum of } -26$$

Product of $16(3) = 48$

We need two integers whose sum is -26 and whose product is 48. The integers -2 and -24 satisfy these conditions and allow us to express the middle term, $-26x$, as $-2x - 24x$. Then, we can factor as follows.

$$\begin{aligned} 16x^2 - 26x + 3 &= 16x^2 - 2x - 24x + 3 \\ &= 2x(8x - 1) - 3(8x - 1) \\ &= (8x - 1)(2x - 3) \end{aligned}$$

▲

Example 7

Solution

Factor $6x^2 - 5x - 6$.

$$6x^2 - 5x - 6 \qquad \text{Sum of } -5$$

Product of $6(-6) = -36$

We need two integers whose product is -36 and whose sum is -5. Furthermore, since the sum is negative, the absolute value of the negative number must be greater than the absolute value of the positive number. A little searching will determine that the numbers are -9 and 4. Thus, we can express the middle term of $-5x$ as $-9x + 4x$ and proceed as follows.

$$\begin{aligned} 6x^2 - 5x - 6 &= 6x^2 - 9x + 4x - 6 \\ &= 3x(2x - 3) + 2(2x - 3) \\ &= (2x - 3)(3x + 2) \end{aligned}$$

▲

Now that we have shown you two possible techniques for factoring trinomials of the form $ax^2 + bx + c$, the ball is in your court. Practice may not make you perfect at factoring, but it will surely help. We are not promoting one technique over the other; that is an individual choice. Many people find the trial and error technique we presented first very useful if the number of possibilities for the factors is fairly small. However, as the list of possibilities grows, the second technique does have the advantage of being systematic. So perhaps having both techniques at your fingertips is your best bet.

Now We Can Solve More Equations

The ability to factor certain trinomials of the form $ax^2 + bx + c$ provides us with greater equation solving capabilities. Consider the following examples.

Example 8

Solution

Solve $3x^2 + 17x + 10 = 0$.

$$3x^2 + 17x + 10 = 0$$
$$(x + 5)(3x + 2) = 0$$

Factoring $3x^2 + 17x + 10$ as $(x + 5)(3x + 2)$ may require some extra work on scratch paper.

$$x + 5 = 0 \quad \text{or} \quad 3x + 2 = 0 \qquad ab = 0 \text{ if and only if } a = 0 \text{ or } b = 0$$
$$x = -5 \quad \text{or} \quad 3x = -2$$
$$x = -5 \quad \text{or} \quad x = -\frac{2}{3}$$

The solution set is $\left\{-5, -\frac{2}{3}\right\}$. Check it!

Example 9

Solution

Solve $24x^2 + 2x - 15 = 0$.

$$24x^2 + 2x - 15 = 0$$
$$(4x - 3)(6x + 5) = 0$$
$$4x - 3 = 0 \quad \text{or} \quad 6x + 5 = 0$$
$$4x = 3 \quad \text{or} \quad 6x = -5$$
$$x = \frac{3}{4} \quad \text{or} \quad x = -\frac{5}{6}$$

The solution set is $\left\{-\frac{5}{6}, \frac{3}{4}\right\}$.

Problem Set 6.4

For Problems 1–50, factor each of the trinomials completely. Indicate any that are not factorable using integers.

1. $3x^2 + 7x + 2$
2. $2x^2 + 9x + 4$
3. $6x^2 + 19x + 10$
4. $12x^2 + 19x + 4$
5. $4x^2 - 25x + 6$
6. $5x^2 - 22x + 8$
7. $12x^2 - 31x + 20$
8. $8x^2 - 30x + 7$
9. $5y^2 - 33y - 14$
10. $3y^2 - 2y - 8$
11. $2n^2 + 13n - 24$
12. $4n^2 + 17n - 15$
13. $2x^2 + x + 7$
14. $7x^2 + 19x + 10$
15. $18x^2 + 45x + 7$
16. $10x^2 + x - 5$
17. $7x^2 - 30x + 8$
18. $6x^2 - 17x + 12$
19. $8x^2 + 2x - 21$
20. $9x^2 + 15x - 14$
21. $9t^2 - 15t - 14$
22. $12t^2 - 20t - 25$
23. $12y^2 + 79y - 35$
24. $9y^2 + 52y - 12$
25. $6n^2 + 2n - 5$
26. $20n^2 - 27n + 9$
27. $14x^2 + 55x + 21$
28. $15x^2 + 34x + 15$
29. $20x^2 - 31x + 12$
30. $8t^2 - 3t - 4$
31. $16n^2 - 8n - 15$
32. $25n^2 - 20n - 12$
33. $24x^2 - 50x + 25$
34. $24x^2 - 41x + 12$
35. $2x^2 + 25x + 72$
36. $2x^2 + 23x + 56$
37. $21a^2 + a - 2$
38. $14a^2 + 5a - 24$
39. $12a^2 - 31a - 15$
40. $10a^2 - 39a - 4$
41. $4x^2 + 12x + 9$
42. $9x^2 - 12x + 4$
43. $6x^2 - 5xy + y^2$
44. $12x^2 + 13xy + 3y^2$
45. $20x^2 + 7xy - 6y^2$
46. $8x^2 - 6xy - 35y^2$
47. $5x^2 - 32x + 12$
48. $3x^2 - 35x + 50$
49. $8x^2 - 55x - 7$
50. $12x^2 - 67x - 30$

For Problems 51–80, solve each equation.

51. $2x^2 + 13x + 6 = 0$
52. $3x^2 + 16x + 5 = 0$
53. $12x^2 + 11x + 2 = 0$
54. $15x^2 + 56x + 20 = 0$
55. $3x^2 - 25x + 8 = 0$
56. $4x^2 - 31x + 21 = 0$
57. $15n^2 - 41n + 14 = 0$
58. $6n^2 - 31n + 40 = 0$
59. $6t^2 + 37t - 35 = 0$
60. $2t^2 + 15t - 27 = 0$
61. $16y^2 - 18y - 9 = 0$
62. $9y^2 - 15y - 14 = 0$
63. $9x^2 - 6x - 8 = 0$
64. $12n^2 + 28n - 5 = 0$
65. $10x^2 - 29x + 10 = 0$
66. $4x^2 - 16x + 15 = 0$

67. $6x^2 + 19x = -10$

68. $12x^2 + 17x = -6$

69. $16x(x + 1) = 5$

70. $5x(5x + 2) = 8$

71. $35n^2 - 34n - 21 = 0$

72. $18n^2 - 3n - 28 = 0$

73. $4x^2 - 45x + 50 = 0$

74. $7x^2 - 65x + 18 = 0$

75. $7x^2 + 46x - 21 = 0$

76. $2x^2 + 7x - 30 = 0$

77. $12x^2 - 43x - 20 = 0$

78. $14x^2 - 13x - 12 = 0$

79. $18x^2 + 55x - 28 = 0$

80. $24x^2 + 17x - 20 = 0$

THOUGHTS INTO WORDS

81. Explain your thought process when factoring $24x^2 - 17x - 20$.

82. Your friend factors $8x^2 - 32x + 32$ as follows.

$$8x^2 - 32x + 32 = (4x - 8)(2x - 4)$$
$$= 4(x - 2)(2)(x - 2)$$
$$= 8(x - 2)(x - 2)$$

Is she correct? Do you have any suggestions for her?

83. Your friend solves the equation $8x^2 - 32x + 32 = 0$ as follows:

$$8x^2 - 32x + 32 = 0$$
$$(4x - 8)(2x - 4) = 0$$
$$4x - 8 = 0 \quad \text{or} \quad 2x - 4 = 0$$
$$4x = 8 \quad \text{or} \quad 2x = 4$$
$$x = 2 \quad \text{or} \quad x = 2$$

The solution set is $\{2\}$.

Is she correct? Would you have any suggestions for her?

Further Investigations

84. Consider the following approach to factoring $20x^2 + 39x + 18$.

$$20x^2 + 39x + 18 \qquad \text{Sum of 39}$$

Product of $20(18) = 360$

We need two integers whose sum is 39 and whose product is 360. To help find these integers, let's prime factor 360.

$$360 = 2 \cdot 2 \cdot 2 \cdot 3 \cdot 3 \cdot 5$$

Now by grouping these factors in various ways we find that $2 \cdot 2 \cdot 2 \cdot 3 = 24$, $3 \cdot 5 = 15$, and $24 + 15 = 39$. So the numbers are 15 and 24 and the middle term of the given trinomial, $39x$, can be expressed as $15x + 24x$. Therefore, we can complete the factoring as follows.

$$20x^2 + 39x + 18 = 20x^2 + 15x + 24x + 18$$
$$= 5x(4x + 3) + 6(4x + 3)$$
$$= (4x + 3)(5x + 6)$$

Factor each of the following trinomials.

(a) $20x^2 + 41x + 20$

(b) $24x^2 - 79x + 40$

(c) $30x^2 + 23x - 40$

(d) $36x^2 + 65x - 36$

6.5 Factoring, Solving Equations, and Problem Solving

Before we summarize our work with factoring techniques let's look at two more special factoring patterns. These patterns emerge when multiplying binomials. Consider the following examples.

$$(x + 5)^2 = (x + 5)(x + 5) = x^2 + 10x + 25;$$
$$(2x + 3)^2 = (2x + 3)(2x + 3) = 4x^2 + 12x + 9;$$
$$(4x + 7)^2 = (4x + 7)(4x + 7) = 16x^2 + 56x + 49.$$

In general, $(a + b)^2 = (a + b)(a + b) = a^2 + 2ab + b^2$.

$$(x - 6)^2 = (x - 6)(x - 6) = x^2 - 12x + 36;$$
$$(3x - 4)^2 = (3x - 4)(3x - 4) = 9x^2 - 24x + 16;$$
$$(5x - 2)^2 = (5x - 2)(5x - 2) = 25x^2 - 20x + 4.$$

In general, $(a - b)^2 = (a - b)(a - b) = a^2 - 2ab + b^2$. Thus, we have the following patterns.

Perfect Square Trinomials
$$a^2 + 2ab + b^2 = (a + b)^2,$$
$$a^2 - 2ab + b^2 = (a - b)^2$$

Trinomials of the form $a^2 + 2ab + b^2$ or $a^2 - 2ab + b^2$ are called **perfect square trinomials.** They are easy to recognize because of the nature of their terms. For example, $9x^2 + 30x + 25$ is a perfect square trinomial because:

1. The first term is a square: $(3x)^2$.

2. The last term is a square: $(5)^2$.

3. The middle term is twice the product of the quantities being squared in the first and last terms: $2(3x)(5)$.

Likewise, $25x^2 - 40xy + 16y^2$ is a perfect square trinomial because:

1. The first term is a square: $(5x)^2$.

2. The last term is a square: $(4y)^2$.

3. The middle term is twice the product of the quantities being squared in the first and last terms: $2(5x)(4y)$.

Once we know that we have a perfect square trinomial, then the factoring process follows immediately from the two basic patterns.

$$9x^2 + 30x + 25 = (3x + 5)^2$$

$$25x^2 - 40xy + 16y^2 = (5x - 4y)^2$$

Here are some additional examples of perfect square trinomials and their factored form.

$$
\begin{array}{lll}
x^2 - 16x + 64 = & (x)^2 - 2(x)(8) + (8)^2 & = (x - 8)^2, \\
16x^2 - 56x + 49 = & (4x)^2 - 2(4x)(7) + (7)^2 & = (4x - 7)^2, \\
25x^2 + 20xy + 4y^2 = & (5x)^2 + 2(5x)(2y) + (2y)^2 & = (5x + 2y)^2, \\
1 + 6y + 9y^2 = & (1)^2 + 2(1)(3y) + (3y)^2 & = (1 + 3y)^2, \\
4m^2 - 4mn + n^2 = & (2m)^2 - 2(2m)(n) + (n)^2 & = (2m - n)^2
\end{array}
$$

Perhaps you will want to do this step mentally after you feel comfortable with the process.

We have considered some basic factoring techniques in this chapter one at a time, but we must be able to apply them as needed in a variety of situations. So, let's first summarize the techniques and then consider some examples.

In this chapter we have discussed:

1. Factoring by using the distributive property to factor out the greatest common monomial or binomial factor;

2. Factoring by grouping;

3. Factoring by applying the difference-of-squares pattern;

4. Factoring by applying the perfect-square-trinomial pattern;

5. Factoring of trinomials of the form $x^2 + bx + c$ into the product of two binomials;

6. Factoring of trinomials of the form $ax^2 + bx + c$ into the product of two binomials.

As a general guideline, **always look for a greatest common monomial factor first**, and then proceed with the other factoring techniques.

In each of the following examples we have factored completely whenever possible. Study them carefully and notice the factoring techniques we used.

1. $2x^2 + 12x + 10 = 2(x^2 + 6x + 5) = 2(x + 1)(x + 5)$

2. $4x^2 + 36 = 4(x^2 + 9)$

Remember that the sum of two squares is not factorable using integers unless there is a common factor.

3. $4t^2 + 20t + 25 = (2t + 5)^2$ If you fail to recognize a perfect trinomial square, no harm is done. Simply proceed to factor into the product of two binomials and then you will recognize that the two binomials are the same.

4. $x^2 - 3x - 8$ is not factorable using integers. This becomes obvious in the following table.

Product	Sum
$1(-8) = -8$	$1 + (-8) = -7$
$-1(8) = -8$	$-1 + 8 = 7$
$2(-4) = -8$	$2 + (-4) = -2$
$-2(4) = -8$	$-2 + 4 = 2$

No two factors of -8 produce a sum of -3.

5. $6y^2 - 13y - 28 = (2y - 7)(3y + 4)$. The binomial factors can be found as follows.

$(y + \underline{\quad})(6y - \underline{\quad})$

$\qquad$ or $\qquad\qquad\qquad\qquad$ $1 \cdot 28$ $\quad$ or $\quad$ $28 \cdot 1$

$(y - \underline{\quad})(6y + \underline{\quad})$ $\qquad$ $2 \cdot 14$ $\quad$ or $\quad$ $14 \cdot 2$

$\qquad$ or $\qquad\qquad\qquad\qquad$ $4 \cdot 7$ $\quad$ or $\quad$ $\boxed{7 \cdot 4}$

$(2y - \underline{\quad})(3y + \underline{\quad})$ ←

$\qquad$ or

$(2y + \underline{\quad})(3y - \underline{\quad})$

6. $32x^2 - 50y^2 = 2(16x^2 - 25y^2) = 2(4x + 5y)(4x - 5y)$

Solving Equations by Factoring

Each time that we considered a new factoring technique in this chapter we used that technique to help solve some equations. It is important to be able to recognize which technique works for a particular type of equation.

Example 1

Solve $x^2 = 25x$.

Solution

$$x^2 = 25x$$
$$x^2 - 25x = 0$$
$$x(x - 25) = 0$$

$$x = 0 \quad \text{or} \quad x - 25 = 0$$
$$x = 0 \quad \text{or} \quad x = 25.$$

The solution set is $\{0, 25\}$. Check it! ▲

Example 2

Solution

Solve $x^3 - 36x = 0$.

$$x^3 - 36x = 0$$
$$x(x^2 - 36) = 0$$
$$x(x + 6)(x - 6) = 0$$
$$x = 0 \quad \text{or} \quad x + 6 = 0 \quad \text{or} \quad x - 6 = 0 \qquad \text{If } abc = 0$$
$$\text{then } a = 0 \text{ or}$$
$$b = 0 \text{ or } c = 0.$$

$$x = 0 \quad \text{or} \quad x = -6 \quad \text{or} \quad x = 6$$

The solution set is $\{-6, 0, 6\}$. Does it check? ▲

Example 3

Solution

Solve $10x^2 - 13x - 3 = 0$.

$$10x^2 - 13x - 3 = 0$$
$$(5x + 1)(2x - 3) = 0$$
$$5x + 1 = 0 \quad \text{or} \quad 2x - 3 = 0$$
$$5x = -1 \quad \text{or} \quad 2x = 3$$
$$x = -\frac{1}{5} \quad \text{or} \quad x = \frac{3}{2}$$

The solution set is $\left\{-\frac{1}{5}, \frac{3}{2}\right\}$. Does it check? ▲

Example 4

Solution

Solve $4x^2 - 28x + 49 = 0$.

$$4x^2 - 28x + 49 = 0$$
$$(2x - 7)^2 = 0$$
$$(2x - 7)(2x - 7) = 0$$
$$2x - 7 = 0 \quad \text{or} \quad 2x - 7 = 0$$
$$2x = 7 \quad \text{or} \quad 2x = 7$$
$$x = \frac{7}{2} \quad \text{or} \quad x = \frac{7}{2}$$

The solution set is $\left\{\frac{7}{2}\right\}$. ▲

Pay special attention to the next example. We need to change the form of the original equation before the property "$ab = 0$ if and only if $a = 0$ or $b = 0$" can be applied. The uniqueness of this property is that we have an indicated product set equal to zero.

Example 5

Solve $(x + 1)(x + 4) = 40$.

Solution

$$(x + 1)(x + 4) = 40$$
$$x^2 + 5x + 4 = 40$$
$$x^2 + 5x - 36 = 0$$
$$(x + 9)(x - 4) = 0$$
$$x + 9 = 0 \quad \text{or} \quad x - 4 = 0$$
$$x = -9 \quad \text{or} \quad x = 4$$

The solution set is $\{-9, 4\}$. Check it!   ▲

Example 6

Solve $2n^2 + 16n - 40 = 0$.

Solution

$$2n^2 + 16n - 40 = 0$$
$$2(n^2 + 8n - 20) = 0$$
$$n^2 + 8n - 20 = 0 \qquad \text{Multiplied both sides of equation by } \tfrac{1}{2}$$
$$(n + 10)(n - 2) = 0$$
$$n + 10 = 0 \quad \text{or} \quad n - 2 = 0$$
$$n = -10 \quad \text{or} \quad n = 2$$

The solution set is $\{-10, 2\}$. Does it check?   ▲

Problem Solving

The preface of this book states that a common thread throughout the book is *to learn a skill*, then *to use that skill to help solve equations*, and then *to use equations to help solve problems*. This thread should be very apparent in this chapter. Our new factoring skills have provided us with more ways of solving equations, which in turn gives us more power to solve word problems. Let's conclude the chapter by solving a few more problems.

Problem 1

Find two numbers whose product is 65 if one of the numbers is 3 more than twice the other number.

Solution

Let n represent one of the numbers; then $2n + 3$ represents the other number. Since their product is 65, we can set up and solve the following equation.

$$n(2n + 3) = 65$$
$$2n^2 + 3n - 65 = 0$$

$$(2n + 13)(n - 5) = 0$$

$$2n + 13 = 0 \quad \text{or} \quad n - 5 = 0$$

$$2n = -13 \quad \text{or} \quad n = 5$$

$$n = -\frac{13}{2} \quad \text{or} \quad n = 5$$

If $n = -\frac{13}{2}$, then $2n + 3 = 2\left(-\frac{13}{2}\right) + 3 = -10$. If $n = 5$, then $2n + 3 = 2(5)$ $+ 3 = 13$. Thus, the numbers are $-\frac{13}{2}$ and -10, or 5 and 13. ▲

Problem 2

The area of a triangular sheet of paper is 14 square inches. One side of the triangle is 3 inches longer than the altitude to that side. Find the length of the one side and the length of the altitude to that side.

Solution

Let h represent the altitude to the side. Then $h + 3$ represents the side of the triangle (see Figure 6.5).

FIGURE 6.5

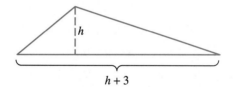

$$h + 3$$

Since the formula for finding the area of a triangle is $A = \frac{1}{2}bh$, we have

$$\frac{1}{2}h(h + 3) = 14$$

$$h(h + 3) = 28 \qquad \text{Multiplied both sides by 2}$$

$$h^2 + 3h = 28$$

$$h^2 + 3h - 28 = 0$$

$$(h + 7)(h - 4) = 0$$

$$h + 7 = 0 \quad \text{or} \quad h - 4 = 0$$

$$h = -7 \quad \text{or} \quad h = 4$$

The solution of -7 is not reasonable. Thus, the altitude is 4 inches and the length of the side to which that altitude is drawn is 7 inches. ▲

Problem 3

A strip of uniform width is shaded along both sides and both ends of a rectangular poster 12 inches by 16 inches. How wide is the strip if one-half of the poster is shaded?

Solution

Let x represent the width of the strip of the poster in Figure 6.6. The area of the strip is one-half of the area of the poster; therefore, it is $\frac{1}{2}(12)(16) = 96$ square

FIGURE 6.6

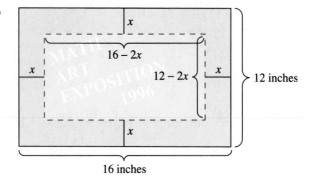

inches. Furthermore, we can represent the area of the strip around the poster by *the area of the poster minus the area of the unshaded portion.* Thus, we can set up and solve the following equation.

Area of poster − Area of unshaded = Area of strip
                          portion

$$16(12) - (16 - 2x)(12 - 2x) = 96$$
$$192 - (192 - 56x + 4x^2) = 96$$
$$192 - 192 + 56x - 4x^2 = 96$$
$$-4x^2 + 56x - 96 = 0$$
$$x^2 - 14x + 24 = 0$$
$$(x - 12)(x - 2) = 0$$
$$x - 12 = 0 \quad \text{or} \quad x - 2 = 0$$
$$x = 12 \quad \text{or} \quad x = 2$$

Obviously, the strip cannot be 12 inches wide since the total width of the poster is 12 inches. Thus, we must disregard the solution of 12 and conclude that the strip is 2 inches wide. ▲

Problem Set 6.5

For Problems 1–12, factor each of the perfect square trinomials.

1. $x^2 + 4x + 4$

2. $x^2 + 18x + 81$

3. $x^2 - 10x + 25$

4. $x^2 - 24x + 144$

5. $9n^2 + 12n + 4$

6. $25n^2 + 30n + 9$

7. $16a^2 - 8a + 1$

8. $36a^2 - 84a + 49$

9. $4 + 36x + 81x^2$

10. $1 - 4x + 4x^2$

11. $16x^2 - 24xy + 9y^2$

12. $64x^2 + 16xy + y^2$

For Problems 13–40, factor each polynomial completely. Indicate any that are not factorable using integers.

13. $2x^2 + 17x + 8$

14. $x^2 + 19x$

15. $2x^3 - 72x$

16. $30x^2 - x - 1$

17. $n^2 - 7n - 60$

18. $4n^3 - 100n$

19. $3a^2 - 7a - 4$

20. $a^2 + 7a - 30$

21. $8x^2 + 72$

22. $3y^3 - 36y^2 + 96y$

23. $9x^2 + 30x + 25$

24. $5x^2 - 5x - 6$

25. $15x^2 + 65x + 70$

26. $4x^2 - 20xy + 25y^2$

27. $24x^2 + 2x - 15$

28. $9x^2y - 27xy$

29. $xy + 5y - 8x - 40$

30. $xy - 3y + 9x - 27$

31. $20x^2 + 31xy - 7y^2$

32. $2x^2 - xy - 36y^2$

33. $24x^2 + 18x - 81$

34. $30x^2 + 55x - 50$

35. $12x^2 + 6x + 30$

36. $24x^2 - 8x + 32$

37. $5x^4 - 80$

38. $3x^5 - 3x$

39. $x^2 + 12xy + 36y^2$

40. $4x^2 - 28xy + 49y^2$

For Problems 41–70, solve each equation.

41. $4x^2 - 20x = 0$

42. $-3x^2 - 24x = 0$

43. $x^2 - 9x - 36 = 0$

44. $x^2 + 8x - 20 = 0$

45. $-2x^3 + 8x = 0$

46. $4x^3 - 36x = 0$

47. $6n^2 - 29n - 22 = 0$

48. $30n^2 - n - 1 = 0$

49. $(3n - 1)(4n - 3) = 0$

50. $(2n - 3)(7n + 1) = 0$

51. $(n - 2)(n + 6) = -15$

52. $(n + 3)(n - 7) = -25$

53. $2x^2 = 12x$

54. $-3x^2 = 15x$

55. $t^3 - 2t^2 - 24t = 0$

56. $2t^3 - 16t^2 - 18t = 0$

57. $12 - 40x + 25x^2 = 0$

58. $12 - 7x - 12x^2 = 0$

59. $n^2 - 28n + 192 = 0$

60. $n^2 + 33n + 270 = 0$

61. $(3n + 1)(n + 2) = 12$

62. $(2n + 5)(n + 4) = -1$

63. $x^3 = 6x^2$

64. $x^3 = -4x^2$

65. $9x^2 - 24x + 16 = 0$

66. $25x^2 + 60x + 36 = 0$

67. $x^3 + 10x^2 + 25x = 0$

68. $x^3 - 18x^2 + 81x = 0$

69. $24x^2 + 17x - 20 = 0$

70. $24x^2 + 74x - 35 = 0$

For Problems 71–88, set up an equation and solve each problem.

71. Find two numbers whose product is 15 such that one of the numbers is seven more than four times the other number.

72. Find two numbers whose product is 12 such that one of the numbers is four less than eight times the other number.

73. Find two numbers whose product is -1. One of the numbers is three more than twice the other number.

74. Suppose that the sum of the squares of three consecutive integers is 110. Find the integers.

75. One number is 1 more than twice another number. The sum of the squares of the two numbers is 97. Find the numbers.

76. One number is 1 less than three times another number. If the product of the two numbers is 102, find the numbers.

77. In an office building, a room contains 54 chairs. The number of chairs per row is three less than twice the number of rows. Find the number of rows and the number of chairs per row.

78. An apple orchard contains 85 trees. The number of trees in each row is three less than four times the number of rows. Find the number of rows and the number of trees per row.

79. Suppose that the combined area of two squares is 360 square feet. Each side of the larger square is three times as long as a side of the smaller square. How big is each square?

80. The area of a rectangular slab of sidewalk is 45 square feet. Its length is three more than four times its width. Find the length and width of the slab.

81. The length of a rectangular sheet of paper is 1 centimeter more than twice its width and the area of the rectangle is 55 square centimeters. Find the length and width of the rectangle.

82. Suppose that the length of a certain rectangle is three times its width. If the length is increased by 2 inches and the width increased by 1 inch, the newly formed rectangle has an area of 70 square inches. Find the length and width of the original rectangle.

83. The area of a triangle is 51 square inches. One side of the triangle is 1 inch less than three times the length of the altitude to that side. Find the length of that side and the length of the altitude to that side.

84. Suppose that a square and a rectangle have equal areas. Furthermore, suppose that the length of the rec-

tangle is twice the length of a side of the square and the width of the rectangle is 4 centimeters less than the length of a side of the square. Find the dimensions of both figures.

85. A strip of uniform width is to be cut off of both sides and both ends of a sheet of paper that is 8 inches by 11 inches in order to reduce the size of the paper to an area of 40 square inches. Find the width of the strip.

86. The sum of the areas of two circles is 100π square centimeters. The length of a radius of the larger circle is 2 centimeters more than the length of a radius of the smaller circle. Find the length of a radius of each circle.

87. The sum of the areas of two circles is 180π square inches. The length of a radius of the smaller circle is 6 inches less than the length of a radius of the larger circle. Find the length of a radius of each circle.

88. A strip of uniform width is shaded along both sides and both ends of a rectangular poster that is 18 inches by 14 inches. How wide is the strip if the unshaded portion of the poster has an area of 165 square inches?

THOUGHTS INTO WORDS

89. When factoring polynomials, why do you think that it is best to look for a greatest common monomial factor first?

90. Explain how you would solve $(4x - 3)(8x + 5) = 0$ and also how you would solve $(4x - 3)(8x + 5) = -9$.

91. Explain how you would solve $(x + 2)(x + 3) = (x + 2)(3x - 1)$. Do you see more than one approach to this problem?

SUMMARY

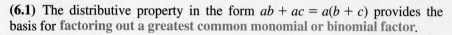

(6.1) The distributive property in the form $ab + ac = a(b + c)$ provides the basis for **factoring out a greatest common monomial or binomial factor**.

Rewriting an expression such as $ab + 3a + bc + 3c$ as $a(b + 3) + c(b + 3)$, and then factoring out the common binomial factor of $b + 3$ so that $a(b + 3) + c(b + 3)$ becomes $(b + 3)(a + c)$ is called **factoring by grouping**.

The property "$ab = 0$ if and only if $a = 0$ or $b = 0$" provides us with another technique for solving equations.

(6.2) The following factoring pattern is called the **difference of two squares**.

$$(a^2 - b^2) = (a - b)(a + b)$$

(6.3) The following multiplication pattern provides a technique for factoring trinomials of the form $x^2 + bx + c$.

$$(x + r)(x + s) = x^2 + rx + sx + rs$$
$$= x^2 + (r + s)x + rs$$

Sum of   Product of
r and s r and s

(6.4) We presented two different techniques for factoring trinomials of the form $ax^2 + bx + c$. To review these techniques, turn to Section 6.4 and study the examples.

(6.5) As a general guideline for **factoring completely**, always look for a greatest common monomial or binomial factor *first*, and then proceed with one or more of the following techniques.

1. Apply the difference-of-squares pattern.
2. Apply the perfect-square-trinomial pattern.
3. Factor a trinomial of the form $x^2 + bx + c$ into the product of two binomials.
4. Factor a trinomial of the form $ax^2 + bx + c$ into the product of two binomials.

Chapter 6 Review Problem Set

Factor completely. Indicate any that are not factorable using integers.

1. $x^2 - 9x + 14$
2. $3x^2 + 21x$
3. $9x^2 - 4$
4. $4x^2 + 8x - 5$
5. $25x^2 - 60x + 36$
6. $n^3 + 13n^2 + 40n$
7. $y^2 + 11y - 12$
8. $3xy^2 + 6x^2y$
9. $x^4 - 1$
10. $18n^2 + 9n - 5$
11. $x^2 + 7x + 24$
12. $4x^2 - 3x - 7$
13. $3n^2 + 3n - 90$
14. $x^3 - xy^2$
15. $2x^2 + 3xy - 2y^2$
16. $4n^2 - 6n - 40$
17. $5x + 5y + ax + ay$
18. $21t^2 - 5t - 4$
19. $2x^3 - 2x$
20. $3x^3 - 108x$
21. $16x^2 + 40x + 25$
22. $xy - 3x - 2y + 6$
23. $15x^2 - 7xy - 2y^2$
24. $6n^4 - 5n^3 + n^2$

Solve each of the following equations.

25. $x^2 + 4x - 12 = 0$
26. $x^2 = 11x$
27. $2x^2 + 3x - 20 = 0$
28. $9n^2 + 21n - 8 = 0$
29. $6n^2 = 24$
30. $16y^2 + 40y + 25 = 0$
31. $t^3 - t = 0$
32. $28x^2 + 71x + 18 = 0$

33. $x^2 + 3x - 28 = 0$
34. $(x - 2)(x + 2) = 21$
35. $5n^2 + 27n = 18$
36. $4n^2 + 10n = 14$
37. $2x^3 - 8x = 0$
38. $x^2 - 20x + 96 = 0$
39. $4t^2 + 17t - 15 = 0$
40. $3(x + 2) - x(x + 2) = 0$
41. $(2x - 5)(3x + 7) = 0$
42. $(x + 4)(x - 1) = 50$
43. $-7n - 2n^2 = -15$
44. $-23x + 6x^2 = -20$

Set up an equation and solve each of the following problems.

45. The larger of two numbers is one less than twice the smaller number. The difference of their squares is 33. Find the numbers.

46. The length of a rectangle is 2 centimeters less than five times the width of the rectangle. The area of the rectangle is 16 square centimeters. Find the length and width of the rectangle.

47. Suppose that the combined area of two squares is 104 square inches. Each side of the larger square is five times as long as a side of the smaller square. Find the size of each square.

48. The longer leg of a right triangle is one unit less than twice the length of the shorter leg. The hypotenuse is one unit more than twice the length of the shorter leg. Find the lengths of the three sides of the triangle.

49. The product of two numbers is 26 and one of the numbers is one larger than six times the other number. Find the numbers.

50. Find three consecutive positive odd whole numbers such that the sum of the squares of the two

smaller numbers is nine more than the square of the largest number.

51. The number of books per shelf in a bookcase is one less than nine times the number of shelves. If the bookcase contains 140 books, find the number of shelves.

52. The combined area of a square and a rectangle is 225 square yards. The length of the rectangle is 8 times the width of the rectangle and the length of a side of the square is the same as the width of the rectangle. Find the dimensions of the square and the rectangle.

53. Suppose that we want to find two consecutive integers such that the sum of their squares is 613. What are they?

54. If numerically the volume of a cube equals the total surface area of the cube, find the length of an edge of the cube.

55. The combined area of two circles is 53π square meters. The length of a radius of the larger circle is 1 meter more than three times the length of a radius of the smaller circle. Find the length of a radius of each circle.

56. The product of two consecutive odd whole numbers is one less than five times their sum. Find the integers.

57. Sandy has a photograph that is 14 centimeters long and 8 centimeters wide. She wants to reduce the length and width by the same amount so that the area is decreased by 40 square centimeters. By what amount should she reduce the length and width?

58. Suppose that a strip of uniform width is plowed along both sides and both ends of a garden that is 120 feet long and 90 feet wide (see Figure 6.7). How wide is the strip if the garden is one-half plowed?

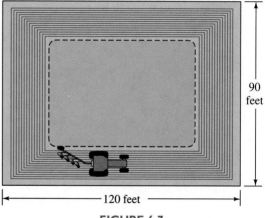

FIGURE 6.7

CHAPTER 6 TEST

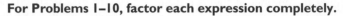

For Problems 1–10, factor each expression completely.

1. $x^2 + 3x - 10$ **2.** $x^2 - 5x - 24$

3. $2x^3 - 2x$ **4.** $x^2 + 21x + 108$

5. $18n^2 + 21n + 6$

6. $ax + ay + 2bx + 2by$

7. $4x^2 + 17x - 15$

8. $6x^2 + 24$

9. $30x^3 - 76x^2 + 48x$

10. $28 + 13x - 6x^2$

For Problems 11–21, solve each equation.

11. $7x^2 = 63$ **12.** $x^2 + 5x - 6 = 0$

13. $4n^2 = 32n$ **14.** $(3x - 2)(2x + 5) = 0$

15. $(x - 3)(x + 7) = -9$ **16.** $x^3 + 16x^2 + 48x = 0$

17. $9(x - 5) - x(x - 5) = 0$ **18.** $3t^2 + 35t = 12$

19. $8 - 10x - 3x^2 = 0$ **20.** $3x^3 = 75x$

21. $25n^2 - 70n + 49 = 0$

For Problems 22–25, set up an equation and solve each problem.

22. The length of a rectangle is 2 inches less than twice its width. If the area of the rectangle is 112 square inches, find the length of the rectangle.

23. The length of one leg of a right triangle is 4 centimeters more than the length of the other leg. The length of the hypotenuse is 8 centimeters more than the length of the shorter leg. Find the length of the shorter leg.

24. A room contains 112 chairs. The number of chairs per row is five less than three times the number of rows. Find the number of chairs per row.

25. If numerically the volume of a cube equals twice the total surface area, find the length of an edge of the cube.

Cumulative Review Problem Set

For Problems 1–6, evaluate each of the numerical expressions.

1. $(-3)^4$

2. -2^5

3. $\left(\dfrac{2}{3}\right)^{-1}$

4. $\dfrac{1}{4^{-2}}$

5. $\left(\dfrac{1}{2} - \dfrac{1}{3}\right)^{-2}$

6. $2^0 + 2^{-1} + 2^{-2}$

For Problems 7–12, evaluate each of the algebraic expressions for the given values of the variables.

7. $\dfrac{2x + 3y}{x - y}$ for $x = \dfrac{1}{2}$ and $y = -\dfrac{1}{3}$

8. $\dfrac{2}{5}n - \dfrac{1}{3}n - n + \dfrac{1}{2}n$ for $n = -\dfrac{3}{4}$

9. $\dfrac{3a - 2b - 4a + 7b}{-a - 3a + b - 2b}$ for $a = -1$ and $b = -\dfrac{1}{3}$

10. $-2(x - 4) + 3(2x - 1) - (3x - 2)$ for $x = -2$

11. $(x^2 + 2x - 4) - (x^2 - x - 2) + (2x^2 - 3x - 1)$ for $x = -1$

12. $2(n^2 - 3n - 1) - (n^2 + n + 4) - 3(2n - 1)$ for $n = 3$

For Problems 13–25, find the indicated products.

13. $(3x^2y^3)(-5xy^4)$

14. $(-6ab^4)(-2b^3)$

15. $(-2x^2y^5)^3$

16. $-3xy(2x - 5y)$

17. $(5x - 2)(3x - 1)$

18. $(7x - 1)(3x + 4)$

19. $(-x - 2)(2x + 3)$

20. $(7 - 2y)(7 + 2y)$

21. $(x - 2)(3x^2 - x - 4)$

22. $(2x - 5)(x^2 + x - 4)$

23. $(2n + 3)^3$

24. $(1 - 2n)^3$

25. $(x^2 - 2x + 6)(2x^2 + 5x - 6)$

For Problems 26–30, perform the indicated divisions.

26. $\dfrac{-52x^3y^4}{13xy^2}$

27. $\dfrac{-126a^3b^5}{-9a^2b^3}$

28. $\dfrac{56xy^2 - 64x^3y - 72x^4y^4}{8xy}$

29. $(2x^3 + 2x^2 - 19x - 21) \div (x + 3)$

30. $(3x^3 + 17x^2 + 6x - 4) \div (3x - 1)$

For Problems 31–34, simplify each expression and express your answers using positive exponents only.

31. $(-2x^3)(3x^{-4})$

32. $\dfrac{4x^{-2}}{2x^{-1}}$

33. $(3x^{-1}y^{-2})^{-1}$

34. $(xy^2z^{-1})^{-2}$

For Problems 35–37, use scientific notation and the properties of exponents to help evaluate each numerical expression.

35. $(.00003)(4000)$

36. $(.0002)(.003)^2$

37. $\dfrac{.00034}{.0000017}$

For Problems 38–43, factor each expression completely. Indicate any that are not factorable using integers.

38. $7x^2 - 28$

39. $2ac - ad + 2bc - bd$

40. $6x^2 + x - 35$

41. $3x^2 + 12x - 36$

42. $2x^2 - x - 4$

43. $16x^4 - 1$

For Problems 44–54, solve each of the equations.

44. $-3(x - 1) + 2(x + 3) = -4$

45. $\dfrac{3n + 1}{5} + \dfrac{n - 2}{3} = \dfrac{2}{15}$

46. $.06x + .08(1500 - x) = 110$

47. $2x^2 - 3x = 0$

48. $5n^2 = 125$

49. $x^2 + 2x - 48 = 0$

50. $14x^2 - 3x - 2 = 0$

51. $4x^3 - x^2 - 3x = 0$

52. $(n - 1)(n + 3) = 21$

53. $(5n - 3)(3n + 10) = 0$

54. $x(x + 2) - 4(x + 2) = 0$

For Problems 55–57, solve each of the inequalities.

55. $-5x + 3 > -4x + 5$

56. $\dfrac{3x}{4} - \dfrac{x}{2} \le \dfrac{5x}{6} - 1$

57. $.08(700 - x) + .11x \ge 65$

For Problems 58–67, set up an equation and solve each problem.

58. The sum of four and three times a certain number is the same as the sum of the number and ten. Find the number.

59. Fifteen percent of some number is six. Find the number.

60. Lou has 18 coins consisting of dimes and quarters. If the total value of the coins is $3.30, how many coins of each denomination does he have?

61. A sum of $1500 is invested, part of it at 8% interest and the remainder at 9%. If the total interest amounts to $128, find the amount invested at each rate.

62. How many gallons of water must be added to 15 gallons of a 12% salt solution to change it to a 10% salt solution?

63. Two airplanes leave Atlanta at the same time and fly in opposite directions. If one travels at 400 miles per hour and the other at 450 miles per hour, how long with it take them to be 2975 miles apart?

64. The length of a rectangle is one meter more than twice its width. If the perimeter of the rectangle is 44 meters, find the length and width.

65. The length of a rectangle is two feet less than twice the width. If the area of the rectangle is 144 square feet, find its length and width.

66. The combined area of two circles is 65π square inches. The length of a radius of the larger circle is 3 inches more than the length of a radius of the smaller circle. Find the length of a radius of each circle.

67. One leg of a right triangle is 4 centimeters longer than the other leg. The hypotenuse is 8 centimeters longer than the shorter leg. Find the lengths of the three sides of the right triangle.

Algebraic Fractions

One day Jeff rode his bicycle 40 miles out into the country. On the way back, he took a different route that was 2 miles longer and it took him an hour longer to return. If his rate on the way out into the country was 4 miles per hour faster than his rate back, find both rates. The fractional equation $\frac{40}{x} = \frac{42}{x-4} - 1$ can be used to determine that Jeff rode out into the country at 16 miles per hour and returned at 12 miles per hour.

In Chapter 2 our study of common fractions led naturally to some work with simple algebraic fractions. Then in Chapters 5 and 6 we discussed the basic operations that pertain to polynomials. Now we can use some ideas about polynomials—specifically the factoring techniques—to expand our study of algebraic fractions. This, in turn, gives us more techniques for solving equations, which increases our problem solving capabilities.

7.1 Simplifying Algebraic Fractions

If the numerator and denominator of a fraction are polynomials, then we call the fraction an **algebraic fraction** or a **rational expression**. The following are examples of algebraic fractions.

$$\frac{4}{x-2}, \qquad \frac{x^2 + 2x - 4}{x^2 - 9}, \qquad \frac{y + x^2}{xy - 3}, \qquad \frac{x^3 + 2x^2 - 3x - 4}{x^2 - 2x - 6}$$

Because we must avoid division by zero, no values can be assigned to variables that create a denominator of zero. Thus, the fraction $\frac{4}{x-2}$ is meaningful for all real number values of x except for $x = 2$. Rather than making a restriction for each individual fraction, we will simply assume that all denominators represent nonzero real numbers.

Recall that the **fundamental principle of fractions** $\left(\frac{ak}{bk} = \frac{a}{b}\right)$ provides the basis for expressing fractions in reduced (or simplified) form, as the next examples demonstrate.

$$\frac{18}{24} = \frac{3 \cdot \cancel{6}}{4 \cdot \cancel{6}} = \frac{3}{4}, \qquad\qquad \frac{-42xy}{77y} = -\frac{2 \cdot 3 \cdot \cancel{7} \cdot x \cdot \cancel{y}}{\cancel{7} \cdot 11 \cdot \cancel{y}} = -\frac{6x}{11},$$

$$\frac{15x}{25x} = \frac{3 \cdot \cancel{5} \cdot \cancel{x}}{5 \cdot \cancel{5} \cdot \cancel{x}} = \frac{3}{5}, \qquad\qquad \frac{28x^2y^2}{-63x^2y^3} = -\frac{4 \cdot \cancel{7} \cdot \cancel{x^2} \cdot \cancel{y^2}}{9 \cdot \cancel{7} \cdot \cancel{x^2} \cdot \cancel{y^3}} = -\frac{4}{9y}$$

The factoring techniques from Chapter 6 can be used to factor numerators and/or denominators so that the fundamental principle of fractions can be applied. Several examples should clarify this process.

Example 1

Simplify $\frac{x^2 + 6x}{x^2 - 36}$.

Solution

$$\frac{x^2 + 6x}{x^2 - 36} = \frac{x(\cancel{x + 6})}{(x - 6)(\cancel{x + 6})} = \frac{x}{x - 6}.$$ ▲

Example 2

Simplify $\frac{a + 2}{a^2 + 4a + 4}$.

Solution

$$\frac{a + 2}{a^2 + 4a + 4} = \frac{1(\cancel{a + 2})}{(a + 2)(\cancel{a + 2})} = \frac{1}{a + 2}.$$ ▲

Example 3

Simplify $\frac{x^2 + 4x - 21}{2x^2 + 15x + 7}$.

Solution

$$\frac{x^2 + 4x - 21}{2x^2 + 15x + 7} = \frac{(x - 3)(\cancel{x + 7})}{(2x + 1)(\cancel{x + 7})} = \frac{x - 3}{2x + 1}.$$ ▲

Example 4

Simplify $\dfrac{a^2b + ab^2}{ab + b^2}$.

Solution

$$\frac{a^2b + ab^2}{ab + b^2} = \frac{a\cancel{b}(a + b)}{\cancel{b}(a + b)} = a$$ ▲

Example 5

Simplify $\dfrac{4x^3y - 36xy}{2x^2 - 4x - 30}$.

Solution

$$\frac{4x^3y - 36xy}{2x^2 - 4x - 30} = \frac{4xy(x^2 - 9)}{2(x^2 - 2x - 15)}$$

$$= \frac{\overset{2}{\cancel{4}}xy(\cancel{x + 3})(x - 3)}{\cancel{2}(x - 5)(\cancel{x + 3})}$$

$$= \frac{2xy(x - 3)}{x - 5}$$ ▲

Notice in Example 5 that we left the numerator of the final fraction in factored form. We do this if polynomials other than monomials are involved. Either $\dfrac{2xy(x - 3)}{x - 5}$ or $\dfrac{2x^2y - 6xy}{x - 5}$ is an acceptable answer.

Remember that the quotient of any nonzero real number and its opposite is -1. For example, $\dfrac{7}{-7} = -1$ and $\dfrac{-9}{9} = -1$. Likewise, the indicated quotient of any polynomial and its opposite is equal to -1. For example,

$$\frac{x}{-x} = -1 \quad \text{because } x \text{ and } -x \text{ are opposites,}$$

$$\frac{x - y}{y - x} = -1 \quad \text{because } x - y \text{ and } y - x \text{ are opposites,}$$

$$\frac{a^2 - 9}{9 - a^2} = -1 \quad \text{because } a^2 - 9 \text{ and } 9 - a^2 \text{ are opposites.}$$

Use this idea to simplify algebraic fractions in the final examples of this section.

Example 6

Simplify $\dfrac{14 - 7n}{n - 2}$.

Solution

$$\frac{14 - 7n}{n - 2} = \frac{7(2 - n)}{n - 2}$$

$$= 7(-1) \qquad \frac{2 - n}{n - 2} = -1$$

$$= -7$$ ▲

Example 7   Simplify $\dfrac{x^2 + 4x - 21}{15 - 2x - x^2}$.

Solution
$$\frac{x^2 + 4x - 21}{15 - 2x - x^2} = \frac{(x + 7)(x - 3)}{(5 + x)(3 - x)}$$

$$= \left(\frac{x + 7}{x + 5}\right)(-1) \qquad \frac{x - 3}{3 - x} = -1$$

$$= -\frac{x + 7}{x + 5} \quad \text{or} \quad \frac{-x - 7}{x + 5} \qquad \blacktriangle$$

▼ Problem Set 7.1

For Problems 1–60, simplify each algebraic fraction.

1. $\dfrac{6x}{14y}$

2. $\dfrac{8y}{18x}$

3. $\dfrac{9xy}{24x}$

4. $\dfrac{12y}{20xy}$

5. $\dfrac{-15x^2y}{25x}$

6. $\dfrac{16x^3y^2}{-28x^2y}$

7. $\dfrac{-36x^4y^3}{-48x^6y^2}$

8. $\dfrac{-18x^3y}{-36xy^3}$

9. $\dfrac{12a^2b^5}{-54a^2b^3}$

10. $\dfrac{-24a^3b^3}{39a^5b^2}$

11. $\dfrac{32xy^2z^3}{72yz^4}$

12. $\dfrac{27x^2y^3z^4}{45x^3y^3z}$

13. $\dfrac{xy}{x^2 - 2x}$

14. $\dfrac{x^2 + 5x}{xy}$

15. $\dfrac{8x + 12y}{12}$

16. $\dfrac{8}{12x - 16y}$

17. $\dfrac{x^2 + 2x}{x^2 - 7x}$

18. $\dfrac{x^2 - 6x}{2x^2 + 6x}$

19. $\dfrac{7 - x}{x - 7}$

20. $\dfrac{x - 9}{9 - x}$

21. $\dfrac{15 - 3n}{n - 5}$

22. $\dfrac{2n^2 - 8n}{4 - n}$

23. $\dfrac{4x^3 - 4x}{1 - x^2}$

24. $\dfrac{9 - x^2}{3x^3 - 27x}$

25. $\dfrac{x^2 - 1}{3x^2 - 3x}$

26. $\dfrac{5x^2 + 25x}{x^2 - 25}$

27. $\dfrac{x^2 + xy}{x^2}$

28. $\dfrac{x^3}{x^3 - x^2y}$

29. $\dfrac{6x^3 - 15x^2y}{6x^2 + 24xy}$

30. $\dfrac{6x^2 + 42xy}{16x^3 - 8x^2y}$

31. $\dfrac{n^2 + 2n}{n^2 + 3n + 2}$

32. $\dfrac{n^2 + 9n + 18}{n^2 + 6n}$

33. $\dfrac{2n^2 + 5n - 3}{n^2 - 9}$

34. $\dfrac{3n^2 - 10n - 8}{n^2 - 16}$

35. $\dfrac{2x^2 + 17x + 35}{3x^2 + 19x + 20}$

36. $\dfrac{5x^2 - 32x + 12}{4x^2 - 27x + 18}$

37. $\dfrac{9(x - 1)^2}{12(x - 1)^3}$

38. $\dfrac{18(x + 2)^3}{16(x + 2)^2}$

39. $\dfrac{7x^2 + 61x - 18}{7x^2 + 19x - 6}$

40. $\dfrac{8x^2 - 51x + 18}{8x^2 + 29x - 12}$

41. $\dfrac{10a^2 + a - 3}{15a^2 + 4a - 3}$

42. $\dfrac{6a^2 - 11a - 10}{8a^2 - 22a + 5}$

43. $\dfrac{x^2 + 2xy - 3y^2}{2x^2 - xy - y^2}$

44. $\dfrac{x^2 - 3xy + 2y^2}{x^2 - 4y^2}$

45. $\dfrac{x^2 - 9}{-x^2 - 3x}$

46. $\dfrac{-x^2 - 2x}{x^2 - 4}$

47. $\dfrac{n^2 + 14n + 49}{8n + 56}$

48. $\dfrac{6n - 60}{n^2 - 20n + 100}$

49. $\dfrac{4n^2 - 12n + 9}{2n^2 - n - 3}$

50. $\dfrac{9n^2 + 30n + 25}{3n^2 - n - 10}$

51. $\dfrac{y^2 - 6y - 72}{y^2 - 8y - 84}$

52. $\dfrac{y^2 + 20y + 96}{y^2 + 23y + 120}$

53. $\dfrac{1 - x^2}{x - x^2}$

54. $\dfrac{2x + x^2}{4 - x^2}$

55. $\dfrac{6 - x - 2x^2}{12 + 7x - 10x^2}$

56. $\dfrac{15 + x - 2x^2}{21 - 10x + x^2}$

57. $\dfrac{x^2 + 7x - 18}{12 - 4x - x^2}$

58. $\dfrac{3x - 21}{28 - 4x}$

59. $\dfrac{5x - 40}{80 - 10x}$

60. $\dfrac{x^2 - x - 12}{8 + 2x - x^2}$

62. Which of the following simplification processes are correct? Explain your answers.

$$\frac{2x}{x} = 2 \qquad \frac{x + 2}{x} = 2 \qquad \frac{x(x + 2)}{x} = x + 2$$

Further Investigations

For Problems 63–66, simplify each fraction. You will need to use factoring by grouping.

63. $\dfrac{xy - 3x + 2y - 6}{xy + 5x + 2y + 10}$

64. $\dfrac{xy + 4x - y - 4}{xy + 4x - 4y - 16}$

65. $\dfrac{xy - 6x + y - 6}{xy - 6x + 5y - 30}$

66. $\dfrac{xy - 7x - 5y + 35}{xy - 9x - 5y + 45}$

The link between positive and negative exponents $\left(a^{-n} = \dfrac{1}{a^n}\right)$ along with the property, $\dfrac{a^n}{a^m} = a^{n-m}$, can also be used when reducing fractions. Consider the following example.

$$\frac{x^3}{x^7} = x^{3-7} = x^{-4} = \frac{1}{x^4}$$

For Problems 67–72, use this approach to help express each fraction in reduced form. Express all answers with positive exponents only.

67. $\dfrac{x^3}{x^9}$

68. $\dfrac{x^4}{x^8}$

69. $\dfrac{x^4 y^3}{x^7 y^5}$

70. $\dfrac{x^5 y^2}{x^6 y^3}$

71. $\dfrac{28a^2 b^3}{-7a^5 b^3}$

72. $\dfrac{-44a^3 b^4}{4a^3 b^6}$

THOUGHTS INTO WORDS

61. Explain the role factoring plays when simplifying algebraic fractions.

7.2 Multiplying and Dividing Algebraic Fractions

In Chapter 2 we defined the product of two rational numbers as $\dfrac{a}{b} \cdot \dfrac{c}{d} = \dfrac{ac}{bd}$. This definition extends to algebraic fractions in general.

DEFINITION 7.1

If $\dfrac{A}{B}$ and $\dfrac{C}{D}$ are rational expressions with $B \neq 0$ and $D \neq 0$, then

$$\frac{A}{B} \cdot \frac{C}{D} = \frac{AC}{BD}.$$

In other words, to multiply algebraic fractions we multiply the numerators, multiply the denominators, and **express the product in simplified form**. The following examples illustrate this concept.

1. $\dfrac{2x}{3y} \cdot \dfrac{5y}{4x} = \dfrac{\cancel{2} \cdot 5 \cdot \cancel{x} \cdot \cancel{y}}{3 \cdot \underset{2}{\cancel{4}} \cdot \cancel{x} \cdot \cancel{y}} = \dfrac{5}{6}.$

Notice that we used the commutative property of multiplication to rearrange factors in a more convenient form for recognizing common factors of the numerator and denominator.

2. $\dfrac{4a}{6b} \cdot \dfrac{8b}{12a^2} = \dfrac{\overset{4}{\cancel{4}} \cdot \cancel{8} \cdot \cancel{a} \cdot \cancel{b}}{\underset{3}{\cancel{6}} \cdot \underset{3}{\cancel{12}} \cdot \underset{a}{\cancel{a^2}} \cdot \cancel{b}} = \dfrac{4}{9a}.$

3. $\dfrac{-9x^2}{15xy} \cdot \dfrac{5y^2}{7x^2y^3} = -\dfrac{\overset{3}{\cancel{9}} \cdot \cancel{5} \cdot \cancel{x^2} \cdot \cancel{y^2}}{\underset{\underset{7}{\cancel{7}}}{\cancel{15}} \cdot 7 \cdot \underset{x}{\cancel{x^3}} \cdot \underset{y^2}{\cancel{y^4}}} = -\dfrac{3}{7xy^2}.$

When multiplying algebraic fractions, we sometimes need to factor the numerators and/or denominators so that we can recognize common factors. Consider the following examples.

Example 1

Multiply and simplify $\dfrac{x}{x^2 - 9} \cdot \dfrac{x + 3}{y}$.

Solution

$$\frac{x}{x^2 - 9} \cdot \frac{x + 3}{y} = \frac{x(\cancel{x + 3})}{(\cancel{x + 3})(x - 3)(y)}$$

$$= \frac{x}{y(x - 3)}$$

$\dfrac{x}{xy - 3y}$ is also an acceptable answer.

Example 2

Multiply and simplify $\dfrac{x}{x^2 + 2x} \cdot \dfrac{x^2 + 10x + 16}{5}$.

Solution

$$\frac{x}{x^2 + 2x} \cdot \frac{x^2 + 10x + 16}{5} = \frac{\cancel{x}(x+2)(x+8)}{\cancel{x}(x+2)(5)} = \frac{x + 8}{5}$$ ▲

Example 3

Multiply and simplify $\dfrac{a^2 - 3a}{a + 5} \cdot \dfrac{a^2 + 3a - 10}{a^2 - 5a + 6}$.

Solution

$$\frac{a^2 - 3a}{a + 5} \cdot \frac{a^2 + 3a - 10}{a^2 - 5a + 6} = \frac{a(a-3)(a+5)(a-2)}{(a+5)(a-2)(a-3)} = a$$ ▲

Example 4

Multiply and simplify $\dfrac{6n^2 + 7n - 3}{n + 1} \cdot \dfrac{n^2 - 1}{2n^2 + 3n}$.

Solution

$$\frac{6n^2 + 7n - 3}{n + 1} \cdot \frac{n^2 - 1}{2n^2 + 3n} = \frac{(2n+3)(3n-1)(n+1)(n-1)}{(n+1)(n)(2n+3)}$$

$$= \frac{(3n-1)(n-1)}{n}$$ ▲

Dividing Algebraic Fractions

Recall that to divide two rational numbers in $\dfrac{a}{b}$ form we *invert the divisor and multiply.* Symbolically, we express this as $\dfrac{a}{b} \div \dfrac{c}{d} = \dfrac{a}{b} \cdot \dfrac{d}{c}$. Furthermore, we call the numbers $\dfrac{c}{d}$ and $\dfrac{d}{c}$ reciprocals of each other because their product is one. Thus, we can also describe division as *to divide by a fraction, multiply by its reciprocal.* We define division of algebraic fractions in the same way using the same vocabulary.

DEFINITION 7.2

If $\dfrac{A}{B}$ and $\dfrac{C}{D}$ are rational expressions with $B \neq 0$, $D \neq 0$, and $C \neq 0$, then

$$\frac{A}{B} \div \frac{C}{D} = \frac{A}{B} \cdot \frac{D}{C} = \frac{AD}{BC}.$$

1. $\dfrac{4x}{7y} \div \dfrac{6x^2}{14y^2} = \dfrac{4x}{7y} \cdot \dfrac{14y^2}{6x^2} = \dfrac{\overset{2}{\cancel{4}} \cdot \overset{2}{\cancel{14}} \cdot \cancel{x} \cdot \overset{y}{\cancel{y^2}}}{\underset{3}{\cancel{7}} \cdot \cancel{6} \cdot \underset{x}{\cancel{x^2}} \cdot \cancel{y}} = \dfrac{4y}{3x}.$

2. $\dfrac{-8ab}{9b} \div \dfrac{18a^3}{15a^2b} = \dfrac{-8ab}{9b} \cdot \dfrac{15a^2b}{18a^3} = -\dfrac{\overset{4}{\cancel{8}} \cdot \overset{5}{\cancel{15}} \cdot \overset{}{\cancel{a^3}} \cdot \overset{b}{\cancel{b^2}}}{\underset{3}{\cancel{9}} \cdot \underset{9}{\cancel{18}} \cdot \cancel{a^3} \cdot \cancel{b}} = -\dfrac{20b}{27}.$

3. $\dfrac{x^2y^3}{4ab} \div \dfrac{5xy^2}{-9a^2b} = \dfrac{x^2y^3}{4ab} \cdot \dfrac{-9a^2b}{5xy^2} = -\dfrac{9 \cdot \overset{x}{\cancel{x^2}} \cdot \overset{y}{\cancel{y^3}} \cdot \overset{a}{\cancel{a^2}} \cdot \cancel{b}}{4 \cdot 5 \cdot \cancel{a} \cdot \cancel{b} \cdot \cancel{x} \cdot \cancel{y^2}} = -\dfrac{9axy}{20}.$

The key idea when dividing fractions is to *first* convert to an equivalent multiplication problem and then to proceed to factor numerator and denominator completely and look for common factors.

Example 5

Divide and simplify $\dfrac{x^2 - 4x}{xy} \div \dfrac{x^2 - 16}{y^3 + y^2}.$

Solution

$\dfrac{x^2 - 4x}{xy} \div \dfrac{x^2 - 16}{y^3 + y^2} = \dfrac{x^2 - 4x}{xy} \cdot \dfrac{y^3 + y^2}{x^2 - 16}$

$= \dfrac{\cancel{x}(x - 4)(\overset{y}{\cancel{y^2}})(y + 1)}{\cancel{x}\cancel{y}(x + 4)(x - 4)}$

$= \dfrac{y(y + 1)}{x + 4}$ ▲

Example 6

Divide and simplify $\dfrac{a^2 + 3a - 18}{a^2 + 4} \div \dfrac{1}{3a^2 + 12}.$

Solution

$\dfrac{a^2 + 3a - 18}{a^2 + 4} \div \dfrac{1}{3a^2 + 12} = \dfrac{a^2 + 3a - 18}{a^2 + 4} \cdot \dfrac{3a^2 + 12}{1}$

$= \dfrac{(a + 6)(a - 3)(3)(\cancel{a^2 + 4})}{\cancel{a^2 + 4}}$

$= 3(a + 6)(a - 3)$ ▲

Example 7

Divide and simplify $\dfrac{2n^2 - 7n - 4}{6n^2 + 7n + 2} \div (n - 4).$

Solution

$\dfrac{2n^2 - 7n - 4}{6n^2 + 7n + 2} \div (n - 4) = \dfrac{2n^2 - 7n - 4}{6n^2 + 7n + 2} \cdot \dfrac{1}{n - 4}$

$= \dfrac{(\cancel{2n + 1})(\cancel{n - 4})}{(\cancel{2n + 1})(3n + 2)(\cancel{n - 4})}$

$= \dfrac{1}{3n + 2}$ ▲

In a problem such as Example 7, it may be helpful to write the divisor with a denominator of 1. Thus, $n - 4$ can be written as $\dfrac{n - 4}{1}$; its reciprocal then is obviously $\dfrac{1}{n - 4}.$

Problem Set 7.2

For Problems 1–40, perform the indicated multiplications and divisions and express your answers in simplest form.

1. $\dfrac{5}{9} \cdot \dfrac{3}{10}$

2. $\dfrac{7}{8} \cdot \dfrac{12}{14}$

3. $\left(-\dfrac{3}{4}\right)\left(\dfrac{6}{7}\right)$

4. $\left(\dfrac{5}{6}\right)\left(-\dfrac{4}{15}\right)$

5. $\left(\dfrac{17}{9}\right) \div \left(-\dfrac{19}{9}\right)$

6. $\left(-\dfrac{15}{7}\right) \div \left(\dfrac{13}{14}\right)$

7. $\dfrac{8xy}{12y} \cdot \dfrac{6x}{14y}$

8. $\dfrac{9x}{15y} \cdot \dfrac{20xy}{18x}$

9. $\left(-\dfrac{5n^2}{18n}\right)\left(\dfrac{27n}{25}\right)$

10. $\left(\dfrac{4ab}{10}\right)\left(-\dfrac{30a}{22b}\right)$

11. $\dfrac{3a^2}{7} \div \dfrac{6a}{28}$

12. $\dfrac{4x}{11y} \div \dfrac{12x}{33}$

13. $\dfrac{18a^2b^2}{-27a} \div \dfrac{-9a}{5b}$

14. $\dfrac{24ab^2}{25b} \div \dfrac{-12ab}{15a^2}$

15. $24x^3 \div \dfrac{16x}{y}$

16. $14xy^2 \div \dfrac{7y}{9}$

17. $\dfrac{1}{15ab^3} \div \dfrac{-1}{12a}$

18. $\dfrac{-2}{7a^2b^3} \div \dfrac{1}{9ab^4}$

19. $\dfrac{18rs}{34} \div 9r$

20. $\dfrac{8rs}{3} \div 6s$

21. $\dfrac{y}{x+y} \cdot \dfrac{x^2-y^2}{xy}$

22. $\dfrac{x^2-9}{6} \cdot \dfrac{8}{x-3}$

23. $\dfrac{2x^2+xy}{xy} \cdot \dfrac{y}{10x+5y}$

24. $\dfrac{x^2+y^2}{x-y} \cdot \dfrac{x^2-xy}{3}$

25. $\dfrac{6ab}{4ab+4b^2} \div \dfrac{7a-7b}{a^2-b^2}$

26. $\dfrac{4ab}{2a^2-2ab} \div \dfrac{ab+b}{3a-3b}$

27. $\dfrac{x^2+11x+30}{x^2+4} \cdot \dfrac{5x^2+20}{x^2+14x+45}$

28. $\dfrac{x^2+15x+54}{x^2+2} \cdot \dfrac{3x^2+6}{x^2+10x+9}$

29. $\dfrac{2x^2-3xy+y^2}{4x^2y} \div \dfrac{x^2-y^2}{6x^2y^2}$

30. $\dfrac{2x^2+xy-y^2}{x^2y} \div \dfrac{5x^2+4xy-y^2}{y}$

31. $\dfrac{a+a^2}{15a^2+11a+2} \cdot \dfrac{1-a}{1-a^2}$

32. $\dfrac{2a^2-11a-21}{3a^2+a} \cdot \dfrac{3a^2-11a-4}{2a^2-5a-12}$

33. $\dfrac{2x^2-2xy}{x^2+4x-32} \cdot \dfrac{x^2-16}{5xy-5y^2}$

34. $\dfrac{x^3+3x^2}{x^2+4x+4} \cdot \dfrac{x^2-5x-14}{x^2+3x}$

35. $\dfrac{2x^2-xy-3y^2}{(x+y)^2} \div \dfrac{4x^2-12xy+9y^2}{10x-15y}$

36. $\dfrac{x^2+4xy+4y^2}{x^2} \div \dfrac{x^2-4y^2}{x^2-2xy}$

37. $\dfrac{(3t-1)^2}{45t-15} \div \dfrac{12t^2+5t-3}{20t+5}$

38. $\dfrac{5t^2-3t-2}{(t-1)^2} \div \dfrac{5t^2+32t+12}{4t^2-3t-1}$

39. $\dfrac{n^3-n}{n^2+7n+6} \cdot \dfrac{4n+24}{n^2-n}$

40. $\dfrac{2x^2-6x-36}{x^2+2x-48} \cdot \dfrac{x^2+5x-24}{2x^2-18}$

For Problems 41–46, perform the indicated operations and express the answers in simplest form.

Remember that multiplications and divisions are done in the order that they appear from left to right.

41. $\dfrac{6}{9y} \div \dfrac{30x}{12y^2} \cdot \dfrac{5xy}{4}$

42. $\dfrac{5xy^2}{12y} \cdot \dfrac{18x^2}{15y} \div \dfrac{3}{2xy}$

43. $\dfrac{8x^2}{xy - xy^2} \cdot \dfrac{x - 1}{8x^2 - 8y^2} \div \dfrac{xy}{x + y}$

44. $\dfrac{5x - 20}{x^2 - 9} \cdot \dfrac{x + 3}{x - 4} \div \dfrac{15}{x - 3}$

45. $\dfrac{x^2 + 9x + 18}{x^2 + 3x} \cdot \dfrac{x^2 + 5x}{x^2 - 25} \div \dfrac{x^2 + 8x}{x^2 + 3x - 40}$

46. $\dfrac{4x}{3x + 6y} \cdot \dfrac{5xy}{x^2 - 4} \div \dfrac{10}{x^2 + 4x + 4}$

THOUGHTS INTO WORDS

47. Give a step-by-step description of how to do the following multiplication problem.

$$\frac{x^2 - x}{x^2 - 1} \cdot \frac{x^2 + x - 6}{x^2 + 4x - 12}$$

48. Is $\left(\dfrac{x}{x + 1} \div \dfrac{x - 1}{x}\right) \div \dfrac{1}{x} = \dfrac{x}{x + 1} \div \left(\dfrac{x - 1}{x} \div \dfrac{1}{x}\right)$? Justify your answer.

49. Explain why the quotient $\dfrac{x - 2}{x + 1} \div \dfrac{x}{x - 1}$ is undefined for $x = -1$, $x = 1$, and $x = 0$ but is defined for $x = 2$.

7.3 Adding and Subtracting Algebraic Fractions

In Chapter 2 we defined addition and subtraction of rational numbers as $\dfrac{a}{b} + \dfrac{c}{b} = \dfrac{a + c}{b}$ and $\dfrac{a}{b} - \dfrac{c}{b} = \dfrac{a - c}{b}$, respectively. These definitions extend to algebraic fractions in general.

DEFINITION 7.3

If $\dfrac{A}{B}$ and $\dfrac{C}{B}$ are rational expressions with $B \neq 0$, then

$$\frac{A}{B} + \frac{C}{B} = \frac{A + C}{B} \qquad \text{and} \qquad \frac{A}{B} - \frac{C}{B} = \frac{A - C}{B}.$$

Thus, if the denominators of two algebraic fractions are the same, then we can add or subtract the fractions by adding or subtracting the numerators and placing the result over the common denominator.

$$\frac{5}{x} + \frac{7}{x} = \frac{5 + 7}{x} = \frac{12}{x},$$

$$\frac{8}{xy} - \frac{3}{xy} = \frac{8 - 3}{xy} = \frac{5}{xy},$$

$$\frac{14}{2x+1} + \frac{15}{2x+1} = \frac{14+15}{2x+1} = \frac{29}{2x+1},$$

$$\frac{3}{a-1} - \frac{4}{a-1} = \frac{3-4}{a-1} = \frac{-1}{a-1} \quad \text{or} \quad -\frac{1}{a-1}.$$

In the next examples notice how we put to use our previous work with simplifying polynomials.

$$\frac{x+3}{4} + \frac{2x-3}{4} = \frac{(x+3)+(2x-3)}{4} = \frac{3x}{4},$$

$$\frac{x+5}{7} - \frac{x+2}{7} = \frac{(x+5)-(x+2)}{7} = \frac{x+5-x-2}{7} = \frac{3}{7},$$

$$\frac{3x+1}{xy} + \frac{2x+3}{xy} = \frac{(3x+1)+(2x+3)}{xy} = \frac{5x+4}{xy},$$

$$\frac{2(3n+1)}{n} - \frac{3(n-1)}{n} = \frac{2(3n+1)-3(n-1)}{n} = \frac{6n+2-3n+3}{n} = \frac{3n+5}{n}$$

It may be necessary to simplify the fraction that results from adding or subtracting two fractions.

$$\frac{4x-3}{8} + \frac{2x+3}{8} = \frac{(4x-3)+(2x+3)}{8} = \frac{6x}{8} = \frac{3x}{4},$$

$$\frac{3n-1}{12} - \frac{n-5}{12} = \frac{(3n-1)-(n-5)}{12} = \frac{3n-1-n+5}{12}$$

$$= \frac{2n+4}{12} = \frac{2(n+2)}{12} = \frac{n+2}{6},$$

$$\frac{-2x+3}{x^2-4} + \frac{3x-1}{x^2-4} = \frac{(-2x+3)+(3x-1)}{x^2-4} = \frac{x+2}{x^2-4}$$

$$= \frac{\cancel{x+2}}{\cancel{(x+2)}(x-2)}$$

$$= \frac{1}{x-2}$$

Recall that to add or subtract rational numbers with different denominators we first change to equivalent fractions that have a common denominator. In fact, we found that by using the least common denominator (LCD) our work was easier. Let's carefully review the process because it will also work with algebraic fractions in general.

Example 1

Add $\frac{3}{5} + \frac{1}{4}$.

Solution

By inspection we see that the LCD is 20. Thus, we can change both fractions to equivalent fractions having a denominator of 20.

$$\frac{3}{5} + \frac{1}{4} = \frac{3}{5}\left(\frac{4}{4}\right) + \frac{1}{4}\left(\frac{5}{5}\right) = \frac{12}{20} + \frac{5}{20} = \frac{17}{20}$$

Form   Form
of I   of I

Example 2

Subtract $\dfrac{5}{18} - \dfrac{7}{24}$.

Solution

If the LCD cannot be found by inspection, then the prime factorization forms can be used.

$$\left.\begin{array}{l} 18 = 2 \cdot 3 \cdot 3 \\ 24 = 2 \cdot 2 \cdot 2 \cdot 3 \end{array}\right\} \longrightarrow \text{LCD} = 2 \cdot 2 \cdot 2 \cdot 3 \cdot 3 = 72.$$

$$\frac{5}{18} - \frac{7}{24} = \frac{5}{18}\left(\frac{4}{4}\right) - \frac{7}{24}\left(\frac{3}{3}\right) = \frac{20}{72} - \frac{21}{72} = -\frac{1}{72}$$

Now let's consider adding and subtracting algebraic fractions with different denominators.

Example 3

Add $\dfrac{x-2}{4} + \dfrac{3x+1}{3}$.

Solution

By inspection we see that the LCD is 12.

$$\frac{x-2}{4} + \frac{3x+1}{3} = \left(\frac{x-2}{4}\right)\left(\frac{3}{3}\right) + \left(\frac{3x+1}{3}\right)\left(\frac{4}{4}\right)$$

$$= \frac{3(x-2)}{12} + \frac{4(3x+1)}{12}$$

$$= \frac{3(x-2) + 4(3x+1)}{12}$$

$$= \frac{3x - 6 + 12x + 4}{12}$$

$$= \frac{15x - 2}{12}$$

Example 4

Subtract $\dfrac{n-2}{2} - \dfrac{n-6}{6}$.

Solution

By inspection, the LCD is 6.

$$\frac{n-2}{2} - \frac{n-6}{6} = \left(\frac{n-2}{2}\right)\left(\frac{3}{3}\right) - \frac{n-6}{6}$$

$$= \frac{3(n-2)}{6} - \frac{(n-6)}{6}$$

$$= \frac{3(n-2) - (n-6)}{6}$$

$$= \frac{3n - 6 - n + 6}{6}$$

$$= \frac{2n}{6}$$

$$= \frac{n}{3} \qquad \text{Don't forget to simplify!} \qquad \blacktriangle$$

It does not create any serious difficulties when the denominators contain variables; our approach remains basically the same.

Example 5

Add $\dfrac{3}{4x} + \dfrac{7}{3x}$.

Solution

By inspection the LCD is $12x$.

$$\frac{3}{4x} + \frac{7}{3x} = \frac{3}{4x}\left(\frac{3}{3}\right) + \frac{7}{3x}\left(\frac{4}{4}\right) = \frac{9}{12x} + \frac{28}{12x} = \frac{9 + 28}{12x} = \frac{37}{12x} \qquad \blacktriangle$$

Example 6

Subtract $\dfrac{11}{12x} - \dfrac{5}{14x}$.

Solution

$$\left.\begin{array}{l} 12x = 2 \cdot 2 \cdot 3 \cdot x \\ 14x = 2 \cdot 7 \cdot x \end{array}\right\} \longrightarrow \text{LCD} = 2 \cdot 2 \cdot 3 \cdot 7 \cdot x = 84x.$$

$$\frac{11}{12x} - \frac{5}{14x} = \frac{11}{12x}\left(\frac{7}{7}\right) - \frac{5}{14x}\left(\frac{6}{6}\right)$$

$$= \frac{77}{84x} - \frac{30}{84x} = \frac{77 - 30}{84x} = \frac{47}{84x} \qquad \blacktriangle$$

Example 7

Add $\dfrac{2}{y} + \dfrac{4}{y-2}$.

Solution

By inspection the LCD is $y(y-2)$.

$$\frac{2}{y} + \frac{4}{y-2} = \frac{2}{y}\left(\frac{y-2}{y-2}\right) + \frac{4}{y-2}\left(\frac{y}{y}\right)$$

$$\begin{array}{cc} \uparrow & \uparrow \\ \text{Form} & \text{Form} \\ \text{of I} & \text{of I} \end{array}$$

$$= \frac{2(y-2)}{y(y-2)} + \frac{4y}{y(y-2)}$$

$$= \frac{2(y-2) + 4y}{y(y-2)}$$

$$= \frac{2y - 4 + 4y}{y(y-2)} = \frac{6y - 4}{y(y-2)}$$ ▲

Notice the final result in Example 7. The numerator, $6y - 4$, can be factored into $2(3y - 2)$. However, because this produces no common factors with the denominator, the fraction cannot be simplified. Thus, the final answer can be left as $\frac{6y - 4}{y(y-2)}$, or it would also be acceptable to express it as $\frac{2(3y-2)}{y(y-2)}$.

Example 8　　　Subtract $\dfrac{4}{x+2} - \dfrac{7}{x+3}$.

Solution　　　By inspection the LCD is $(x+2)(x+3)$.

$$\frac{4}{x+2} - \frac{7}{x+3} = \left(\frac{4}{x+2}\right)\left(\frac{x+3}{x+3}\right) - \left(\frac{7}{x+3}\right)\left(\frac{x+2}{x+2}\right)$$

$$= \frac{4(x+3)}{(x+2)(x+3)} - \frac{7(x+2)}{(x+3)(x+2)}$$

$$= \frac{4(x+3) - 7(x+2)}{(x+2)(x+3)}$$

$$= \frac{4x + 12 - 7x - 14}{(x+2)(x+3)}$$

$$= \frac{-3x - 2}{(x+2)(x+3)}$$ ▲

Problem Set 7.3

For Problems 1–34, add or subtract as indicated. Be sure to express your answers in simplest form.

1. $\dfrac{5}{x} + \dfrac{12}{x}$

2. $\dfrac{17}{x} - \dfrac{13}{x}$

3. $\dfrac{7}{3x} - \dfrac{5}{3x}$

4. $\dfrac{4}{5x} + \dfrac{3}{5x}$

5. $\dfrac{7}{2n} + \dfrac{1}{2n}$

6. $\dfrac{5}{3n} + \dfrac{4}{3n}$

7. $\dfrac{9}{4x^2} - \dfrac{13}{4x^2}$

8. $\dfrac{12}{5x^2} - \dfrac{22}{5x^2}$

9. $\dfrac{x+1}{x} + \dfrac{3}{x}$

10. $\dfrac{x-2}{x} + \dfrac{4}{x}$

11. $\dfrac{3}{x-1} - \dfrac{6}{x-1}$

12. $\dfrac{8}{x+4} - \dfrac{10}{x+4}$

13. $\dfrac{x+1}{x} - \dfrac{1}{x}$

14. $\dfrac{2x+3}{x} - \dfrac{3}{x}$

15. $\dfrac{3t-1}{4} + \dfrac{2t+3}{4}$

16. $\dfrac{4t-1}{7} + \dfrac{8t-5}{7}$

17. $\dfrac{7a+2}{3} - \dfrac{4a-6}{3}$

18. $\dfrac{9a - 1}{6} - \dfrac{4a - 2}{6}$

19. $\dfrac{4n + 3}{8} + \dfrac{6n + 5}{8}$

20. $\dfrac{2n - 5}{10} + \dfrac{6n - 1}{10}$

21. $\dfrac{3n - 7}{6} - \dfrac{9n - 1}{6}$

22. $\dfrac{2n - 6}{5} - \dfrac{7n - 1}{5}$

23. $\dfrac{5x - 2}{7x} - \dfrac{8x + 3}{7x}$

24. $\dfrac{4x + 1}{3x} - \dfrac{2x + 5}{3x}$

25. $\dfrac{3(x + 2)}{4x} + \dfrac{6(x - 1)}{4x}$

26. $\dfrac{4(x - 3)}{5x} + \dfrac{2(x + 6)}{5x}$

27. $\dfrac{6(n - 1)}{3n} + \dfrac{3(n + 2)}{3n}$

28. $\dfrac{2(n - 4)}{3n} + \dfrac{4(n + 2)}{3n}$

29. $\dfrac{2(3x - 4)}{7x^2} - \dfrac{7x - 8}{7x^2}$

30. $\dfrac{3(4x - 3)}{8x^2} - \dfrac{11x - 9}{8x^2}$

31. $\dfrac{a^2}{a + 2} - \dfrac{4}{a + 2}$

32. $\dfrac{n^2}{n - 4} - \dfrac{16}{n - 4}$

33. $\dfrac{3x}{(x - 6)^2} - \dfrac{18}{(x - 6)^2}$

34. $\dfrac{x^2 + 5x}{(x + 1)^2} + \dfrac{4}{(x + 1)^2}$

For Problems 35–80, add or subtract as indicated and express your answers in simplest form.

35. $\dfrac{3x}{8} + \dfrac{5x}{4}$

36. $\dfrac{5x}{3} + \dfrac{2x}{9}$

37. $\dfrac{7n}{12} - \dfrac{4n}{3}$

38. $\dfrac{n}{6} - \dfrac{7n}{12}$

39. $\dfrac{y}{6} + \dfrac{3y}{4}$

40. $\dfrac{3y}{4} + \dfrac{7y}{5}$

41. $\dfrac{8x}{3} - \dfrac{3x}{7}$

42. $\dfrac{5y}{6} - \dfrac{3y}{8}$

43. $\dfrac{2x}{6} + \dfrac{3x}{5}$

44. $\dfrac{6x}{9} + \dfrac{7x}{12}$

45. $\dfrac{7n}{8} - \dfrac{3n}{9}$

46. $\dfrac{8n}{10} - \dfrac{7n}{15}$

47. $\dfrac{x + 3}{5} + \dfrac{x - 4}{2}$

48. $\dfrac{x - 2}{5} + \dfrac{x + 1}{6}$

49. $\dfrac{x - 6}{9} + \dfrac{x + 2}{3}$

50. $\dfrac{x - 2}{4} + \dfrac{x + 4}{8}$

51. $\dfrac{3n - 1}{3} + \dfrac{2n + 5}{4}$

52. $\dfrac{2n + 3}{4} + \dfrac{4n - 1}{7}$

53. $\dfrac{4n - 3}{6} - \dfrac{3n + 5}{18}$

54. $\dfrac{5n - 2}{12} - \dfrac{4n + 7}{6}$

55. $\dfrac{3x}{4} + \dfrac{x}{6} - \dfrac{5x}{8}$

56. $\dfrac{5x}{2} - \dfrac{3x}{4} - \dfrac{7x}{6}$

57. $\dfrac{x}{5} - \dfrac{3}{10} - \dfrac{7x}{12}$

58. $\dfrac{4x}{3} + \dfrac{5}{9} - \dfrac{11x}{6}$

59. $\dfrac{5}{8x} + \dfrac{1}{6x}$

60. $\dfrac{7}{8x} + \dfrac{5}{12x}$

61. $\dfrac{5}{6y} - \dfrac{7}{9y}$

62. $\dfrac{11}{9y} - \dfrac{8}{15y}$

63. $\dfrac{5}{12x} - \dfrac{11}{16x^2}$

64. $\dfrac{4}{9x} - \dfrac{7}{6x^2}$

65. $\dfrac{3}{2x} - \dfrac{2}{3x} + \dfrac{5}{4x}$

66. $\dfrac{3}{4x} - \dfrac{5}{6x} + \dfrac{10}{9x}$

67. $\dfrac{3}{x - 5} + \dfrac{7}{x}$

68. $\dfrac{4}{x - 8} + \dfrac{9}{x}$

69. $\dfrac{2}{n - 1} - \dfrac{3}{n}$

70. $\dfrac{5}{n + 3} - \dfrac{7}{n}$

71. $\dfrac{4}{n} - \dfrac{6}{n + 4}$

72. $\dfrac{8}{n} - \dfrac{3}{n - 9}$

73. $\dfrac{6}{x} - \dfrac{12}{2x + 1}$

74. $\dfrac{2}{x} - \dfrac{6}{3x - 2}$

75. $\dfrac{4}{x + 4} + \dfrac{6}{x - 3}$

76. $\dfrac{7}{x - 2} + \dfrac{8}{x + 1}$

77. $\dfrac{3}{x - 2} - \dfrac{9}{x + 1}$

78. $\dfrac{5}{x - 1} - \dfrac{4}{x + 6}$

79. $\dfrac{3}{2x - 1} - \dfrac{4}{3x + 1}$

80. $\dfrac{6}{3x - 4} - \dfrac{4}{2x + 3}$

THOUGHTS INTO WORDS

81. Give a step-by-step description of how to do the following addition problem.

$\dfrac{3x - 1}{6} + \dfrac{2x + 3}{9}$

82. Why are $\dfrac{3}{x - 2}$ and $\dfrac{3}{2 - x}$ opposites of each other? What should be the result of adding $\dfrac{3}{x - 2}$ and $\dfrac{3}{2 - x}$?

83. Suppose that your friend does an addition problem as follows.

$$\dfrac{5}{8} + \dfrac{7}{12} = \dfrac{5(12) + 8(7)}{8(12)} = \dfrac{60 + 56}{96} = \dfrac{116}{96} = \dfrac{29}{24}$$

Is this answer correct? What advice would you offer your friend?

Further Investigations

Consider the addition problem $\dfrac{9}{x - 2} + \dfrac{4}{2 - x}$. Notice that the denominators, $x - 2$ and $2 - x$, are opposites of each other; that is $-1(2 - x) = (x - 2)$. In such cases, add the fractions as follows.

$$\dfrac{9}{x - 2} + \dfrac{4}{2 - x} = \dfrac{9}{x - 2} + \dfrac{4}{2 - x}\left(\dfrac{-1}{-1}\right)$$

$$\underset{\text{Form of I}}{\uparrow}$$

$$= \dfrac{9}{x - 2} + \dfrac{-4}{x - 2} = \dfrac{9 + (-4)}{x - 2} = \dfrac{5}{x - 2}$$

For Problems 84–89, use this approach to help with the additions and subtractions.

84. $\dfrac{7}{x - 1} - \dfrac{2}{1 - x}$

85. $\dfrac{5}{x - 3} + \dfrac{1}{3 - x}$

86. $\dfrac{x}{x - 4} + \dfrac{4}{4 - x}$

87. $\dfrac{-4}{a - 1} + \dfrac{2}{1 - a}$

88. $\dfrac{1}{x^2 - 9} + \dfrac{2}{x + 3} - \dfrac{3}{3 - x}$

89. $\dfrac{n}{2n - 1} - \dfrac{3}{1 - 2n}$

7.4 More on Addition and Subtraction of Algebraic Fractions

To add and subtract algebraic fractions you need both a good understanding of the process and the ability to perform the various computations. Regarding the computational work, you should adopt a carefully organized format showing as many steps as you need in order to minimize the chances of making careless errors. Don't be eager to find a shortcut for doing these kinds of problems.

Study the following examples very carefully. Notice the same basic procedure for each problem: (1) Find the LCD; (2) Change each fraction to an equivalent fraction having the LCD as its denominator; (3) Add or subtract numerators and place this result over the LCD; and (4) Look for possibilities to simplify the resulting fraction.

Example 1

Add $\dfrac{3}{x^2 + 2x} + \dfrac{5}{x}$.

Solution

1st denominator: $x^2 + 2x = x(x + 2)$
2nd denominator: x $\rightarrow$ LCD is $x(x + 2)$

$$\dfrac{3}{x^2 + 2x} + \dfrac{5}{x} = \dfrac{3}{x(x + 2)} + \dfrac{5}{x}\left(\dfrac{x + 2}{x + 2}\right)$$

This fraction has the LCD as its denominator.   Form of 1

$$= \dfrac{3}{x(x + 2)} + \dfrac{5(x + 2)}{x(x + 2)} = \dfrac{3 + 5(x + 2)}{x(x + 2)}$$

$$= \dfrac{3 + 5x + 10}{x(x + 2)} = \dfrac{5x + 13}{x(x + 2)}$$

▲

Example 2

Subtract $\dfrac{4}{x^2 - 4} - \dfrac{1}{x - 2}$.

Solution

$x^2 - 4 = (x + 2)(x - 2)$
$x - 2 = x - 2$ $\rightarrow$ LCD is $(x + 2)(x - 2)$

$$\dfrac{4}{x^2 - 4} - \dfrac{x}{x - 2} = \dfrac{4}{(x + 2)(x - 2)} - \left(\dfrac{1}{x - 2}\right)\left(\dfrac{x + 2}{x + 2}\right)$$

$$= \dfrac{4}{(x + 2)(x - 2)} - \dfrac{1(x + 2)}{(x + 2)(x - 2)}$$

$$= \dfrac{4 - 1(x + 2)}{(x + 2)(x - 2)} = \dfrac{4 - x - 2}{(x + 2)(x - 2)}$$

$$= \frac{-x + 2}{(x + 2)(x - 2)}$$

$$= \frac{-1(x - 2)}{(x + 2)(x - 2)} \quad \rightarrow \quad \begin{array}{l} \text{Note the changing of} \\ -x + 2 \text{ to } -1(x - 2). \end{array}$$

$$= -\frac{1}{x + 2}$$

▲

Example 3

Add $\dfrac{2}{a^2 - 9} + \dfrac{3}{a^2 + 5a + 6}$.

Solution

$$\left. \begin{array}{l} a^2 - 9 = (a + 3)(a - 3) \\ a^2 + 5a + 6 = (a + 3)(a + 2) \end{array} \right\} \rightarrow \text{LCD is } (a + 3)(a - 3)(a + 2)$$

$$\frac{2}{a^2 - 9} + \frac{3}{a^2 + 5a + 6}$$

$$= \left(\frac{2}{(a + 3)(a - 3)} \right)\left(\frac{a + 2}{a + 2} \right) + \left(\frac{3}{(a + 3)(a + 2)} \right)\left(\frac{a - 3}{a - 3} \right)$$

$$\qquad\qquad\qquad \underset{\substack{\uparrow \\ \text{Form} \\ \text{of I}}}{} \qquad\qquad\qquad\qquad\qquad \underset{\substack{\uparrow \\ \text{Form} \\ \text{of I}}}{}$$

$$= \frac{2(a + 2)}{(a + 3)(a - 3)(a + 2)} + \frac{3(a - 3)}{(a + 3)(a - 3)(a + 2)}$$

$$= \frac{2(a + 2) + 3(a - 3)}{(a + 3)(a - 3)(a + 2)} = \frac{2a + 4 + 3a - 9}{(a + 3)(a - 3)(a + 2)}$$

$$= \frac{5a - 5}{(a + 3)(a - 3)(a + 2)} \quad \text{or} \quad \frac{5(a - 1)}{(a + 3)(a - 3)(a + 2)}$$

▲

Example 4

Perform the indicated operations.

$$\frac{2x}{x^2 - y^2} + \frac{3}{x + y} - \frac{2}{x - y}$$

Solution

$$\left. \begin{array}{l} x^2 - y^2 = (x + y)(x - y) \\ x + y = x + y \\ x - y = x - y \end{array} \right\} \rightarrow \text{LCD is } (x + y)(x - y)$$

$$\frac{2x}{x^2 - y^2} + \frac{3}{x + y} - \frac{2}{x - y}$$

$$= \frac{2x}{(x + y)(x - y)} + \left(\frac{3}{x + y} \right)\left(\frac{x - y}{x - y} \right) - \left(\frac{2}{x - y} \right)\left(\frac{x + y}{x + y} \right)$$

$$= \frac{2x}{(x + y)(x - y)} + \frac{3(x - y)}{(x + y)(x - y)} - \frac{2(x + y)}{(x + y)(x - y)}$$

$$= \frac{2x + 3(x - y) - 2(x + y)}{(x + y)(x - y)}$$

$$= \frac{2x + 3x - 3y - 2x - 2y}{(x + y)(x - y)}$$

$$= \frac{3x - 5y}{(x + y)(x - y)}$$

Complex Fractions

Fractional forms that contain fractions in the numerators and/or denominators are called **complex fractions**. The following are examples of complex fractions.

$$\frac{\dfrac{2}{3}}{\dfrac{4}{5}}, \quad \frac{\dfrac{1}{x}}{\dfrac{3}{y}}, \quad \frac{\dfrac{1}{2} + \dfrac{1}{3}}{\dfrac{5}{6} - \dfrac{1}{4}}, \quad \frac{\dfrac{2}{x} + \dfrac{2}{y}}{\dfrac{5}{x} - \dfrac{1}{y^2}}$$

It is often necessary to *simplify* a complex fraction, that is, to express it as a simple fraction. Let's illustrate this process with the next four examples.

Example 5

Simplify $\dfrac{\dfrac{2}{3}}{\dfrac{4}{5}}$.

Solution

This type of problem creates no difficulty since it is merely a division problem. Thus,

$$\frac{\dfrac{2}{3}}{\dfrac{4}{5}} = \frac{2}{3} \div \frac{4}{5} = \frac{\overset{1}{\cancel{2}}}{3} \cdot \frac{5}{\underset{2}{\cancel{4}}} = \frac{5}{6}.$$

Example 6

Simplify $\dfrac{\dfrac{1}{x}}{\dfrac{3}{y}}$.

Solution

$$\frac{\dfrac{1}{x}}{\dfrac{3}{y}} = \frac{1}{x} \div \frac{3}{y} = \frac{1}{x} \cdot \frac{y}{3} = \frac{y}{3x}$$

Example 7

Simplify $\dfrac{\dfrac{1}{2} + \dfrac{1}{3}}{\dfrac{5}{6} - \dfrac{1}{4}}$.

Solution A

$$\dfrac{\dfrac{1}{2} + \dfrac{1}{3}}{\dfrac{5}{6} - \dfrac{1}{4}} = \dfrac{\dfrac{3}{6} + \dfrac{2}{6}}{\dfrac{10}{12} - \dfrac{3}{12}} = \dfrac{\dfrac{5}{6}}{\dfrac{7}{12}} = \dfrac{5}{\cancel{6}_{1}} \cdot \dfrac{\cancel{12}^{2}}{7} = \dfrac{10}{7}$$

Invert divisor
and multiply

Solution B

The least common multiple of all four denominators (2, 3, 6, and 4) is 12. We shall multiply the entire complex fraction by a form of 1, namely, $\dfrac{12}{12}$.

$$\dfrac{\dfrac{1}{2} + \dfrac{1}{3}}{\dfrac{5}{6} - \dfrac{1}{4}} = \left(\dfrac{12}{12}\right)\left(\dfrac{\dfrac{1}{2} + \dfrac{1}{3}}{\dfrac{5}{6} - \dfrac{1}{4}}\right)$$

$$= \dfrac{12\left(\dfrac{1}{2} + \dfrac{1}{3}\right)}{12\left(\dfrac{5}{6} - \dfrac{1}{4}\right)} = \dfrac{6 + 4}{10 - 3} = \dfrac{10}{7}$$

▲

Example 8

Simplify $\dfrac{\dfrac{2}{x} + \dfrac{3}{y}}{\dfrac{5}{x} - \dfrac{1}{y^2}}$.

Solution A

$$\dfrac{\dfrac{2}{x} + \dfrac{3}{y}}{\dfrac{5}{x} - \dfrac{1}{y^2}} = \dfrac{\dfrac{2}{x}\left(\dfrac{y}{y}\right) + \dfrac{3}{y}\left(\dfrac{x}{x}\right)}{\dfrac{5}{x}\left(\dfrac{y^2}{y^2}\right) - \dfrac{1}{y^2}\left(\dfrac{x}{x}\right)} = \dfrac{\dfrac{2y}{xy} + \dfrac{3x}{xy}}{\dfrac{5y^2}{xy^2} - \dfrac{x}{xy^2}}$$

$$= \dfrac{\dfrac{2y + 3x}{xy}}{\dfrac{5y^2 - x}{xy^2}}$$

$$= \dfrac{2y + 3x}{\cancel{xy}} \cdot \dfrac{\cancel{xy^2}^{\,y}}{5y^2 - x}$$

$$= \dfrac{y(2y + 3x)}{5y^2 - x}$$

Solution B

The least common multiple of all four denominators $(x, y, x, \text{ and } y^2)$ is xy^2. We shall multiply the entire complex fraction by a form of 1, namely $\dfrac{xy^2}{xy^2}$.

$$\frac{\dfrac{2}{x} + \dfrac{3}{y}}{\dfrac{5}{x} - \dfrac{1}{y^2}} = \left(\frac{xy^2}{xy^2}\right)\left(\frac{\dfrac{2}{x} + \dfrac{3}{y}}{\dfrac{5}{x} - \dfrac{1}{y^2}}\right)$$

$$= \frac{xy^2\left(\dfrac{2}{x} + \dfrac{3}{y}\right)}{xy^2\left(\dfrac{5}{x} - \dfrac{1}{y^2}\right)}$$

$$= \frac{2y^2 + 3xy}{5y^2 - x} \qquad \text{or} \qquad \frac{y(2y + 3x)}{5y^2 - x} \qquad \blacktriangle$$

Certainly either approach (Solution A or Solution B) will work for problems such as Examples 7 and 8. You should carefully examine Solution B of each example. This approach works very effectively with algebraic complex fractions where the LCD of all the denominators of the simple fractions is easy to find. The two methods for simplifying a complex fraction can be summarized as follows.

1. Simplify the numerator and denominator of the fraction separately. Then divide the simplified numerator by the simplified denominator; or

2. Multiply the numerator and denominator of the complex fraction by the least common multiple of all of the denominators that appear in the complex fraction.

Example 9

Simplify $\dfrac{\dfrac{2}{x} - 3}{4 + \dfrac{5}{y}}$.

Solution

$$\frac{\dfrac{2}{x} - 3}{4 + \dfrac{5}{y}} = \left(\frac{xy}{xy}\right)\left(\frac{\dfrac{2}{x} - 3}{4 + \dfrac{5}{y}}\right)$$

$$= \frac{(xy)\left(\dfrac{2}{x}\right) - (xy)(3)}{(xy)(4) + (xy)\left(\dfrac{5}{y}\right)}$$

$$= \frac{2y - 3xy}{4xy + 5x} \qquad \text{or} \qquad \frac{y(2 - 3x)}{x(4y + 5)} \qquad \blacktriangle$$

▼ Problem Set 7.4

For Problems 1–40, perform the indicated operations and express answers in simplest form.

1. $\dfrac{4}{x^2 - 4x} + \dfrac{3}{x}$

2. $\dfrac{3}{x^2 + 2x} + \dfrac{7}{x}$

3. $\dfrac{7}{x^2 + 2x} - \dfrac{5}{x}$

4. $\dfrac{9}{x^2 - 5x} - \dfrac{1}{x}$

5. $\dfrac{8}{n} - \dfrac{2}{n^2 - 6n}$

6. $\dfrac{6}{n} - \dfrac{4}{n^2 + 6n}$

7. $\dfrac{4}{n^2 + n} - \dfrac{4}{n}$

8. $\dfrac{8}{n^2 - 2n} + \dfrac{4}{n}$

9. $\dfrac{7}{2x} - \dfrac{x}{x^2 - x}$

10. $\dfrac{3x}{x^2 + 2x} + \dfrac{4}{5x}$

11. $\dfrac{3}{x^2 - 16} + \dfrac{5}{x + 4}$

12. $\dfrac{6}{x^2 - 9} + \dfrac{9}{x - 3}$

13. $\dfrac{8x}{x^2 - 1} - \dfrac{4}{x - 1}$

14. $\dfrac{6x}{x^2 - 4} - \dfrac{3}{x + 2}$

15. $\dfrac{4}{a^2 - 2a} + \dfrac{7}{a^2 + 2a}$

16. $\dfrac{3}{a^2 + 4a} + \dfrac{5}{a^2 - 4a}$

17. $\dfrac{1}{x^2 - 6x} - \dfrac{1}{x^2 + 6x}$

18. $\dfrac{3}{x^2 + 5x} - \dfrac{4}{x^2 - 5x}$

19. $\dfrac{n}{n^2 - 16} - \dfrac{2}{3n + 12}$

20. $\dfrac{n}{n^2 - 25} - \dfrac{2}{3n - 15}$

21. $\dfrac{5x}{6x + 4} + \dfrac{2x}{9x + 6}$

22. $\dfrac{7x}{3x - 12} + \dfrac{3x}{4x - 16}$

23. $\dfrac{x - 1}{5x + 5} - \dfrac{x - 4}{3x + 3}$

24. $\dfrac{2x + 1}{6x + 12} - \dfrac{3x - 4}{8x + 16}$

25. $\dfrac{2}{x^2 + 7x + 12} + \dfrac{3}{x^2 - 9}$

26. $\dfrac{x}{x^2 - 1} + \dfrac{3}{x^2 + 5x + 4}$

27. $\dfrac{x}{x^2 + 6x + 8} - \dfrac{5}{x^2 - 3x - 10}$

28. $\dfrac{x}{x^2 - x - 30} - \dfrac{7}{x^2 - 7x + 6}$

29. $\dfrac{a}{ab + b^2} - \dfrac{b}{a^2 + ab}$

30. $\dfrac{2x}{xy + y^2} - \dfrac{2y}{x^2 + xy}$

31. $\dfrac{3}{x - 5} - \dfrac{4}{x^2 - 25} + \dfrac{5}{x + 5}$

32. $\dfrac{2}{x + 1} + \dfrac{3}{x^2 - 1} - \dfrac{5}{x - 1}$

33. $\dfrac{10}{x^2 - 2x} + \dfrac{8}{x^2 + 2x} - \dfrac{3}{x^2 - 4}$

34. $\dfrac{1}{x^2 + 7x} - \dfrac{2}{x^2 - 7x} - \dfrac{5}{x^2 - 49}$

35. $\dfrac{3x}{x^2 + 7x + 10} - \dfrac{2}{x + 2} + \dfrac{3}{x + 5}$

36. $\dfrac{4}{x-3} - \dfrac{3}{x-6} + \dfrac{x}{x^2-9x+18}$

37. $\dfrac{5x}{3x^2+7x-20} - \dfrac{1}{3x-5} - \dfrac{2}{x+4}$

38. $\dfrac{4x}{6x^2+7x+2} - \dfrac{2}{2x+1} - \dfrac{4}{3x+2}$

39. $\dfrac{2}{x+4} - \dfrac{1}{x-3} + \dfrac{2x+1}{x^2+x-12}$

40. $\dfrac{3}{x-5} - \dfrac{4}{x+7} + \dfrac{3x-27}{x^2+2x-35}$

For Problems 41–60, simplify each of the complex fractions.

41. $\dfrac{\dfrac{1}{2} - \dfrac{3}{4}}{\dfrac{1}{6} + \dfrac{1}{3}}$

42. $\dfrac{\dfrac{3}{8} + \dfrac{1}{4}}{\dfrac{1}{2} + \dfrac{3}{16}}$

43. $\dfrac{\dfrac{2}{9} + \dfrac{1}{3}}{\dfrac{5}{6} - \dfrac{2}{3}}$

44. $\dfrac{\dfrac{7}{8} - \dfrac{1}{3}}{\dfrac{1}{6} + \dfrac{3}{4}}$

45. $\dfrac{3 - \dfrac{2}{3}}{2 + \dfrac{1}{4}}$

46. $\dfrac{4 + \dfrac{3}{5}}{\dfrac{1}{3} - 2}$

47. $\dfrac{\dfrac{3}{x}}{\dfrac{9}{y}}$

48. $\dfrac{\dfrac{-6}{a}}{\dfrac{8}{b}}$

49. $\dfrac{\dfrac{2}{x} + \dfrac{3}{y}}{\dfrac{5}{x} - \dfrac{1}{y}}$

50. $\dfrac{\dfrac{3}{x^2} - \dfrac{2}{x}}{\dfrac{4}{x} - \dfrac{7}{x^2}}$

51. $\dfrac{\dfrac{1}{y} - \dfrac{4}{x^2}}{\dfrac{7}{x} - \dfrac{3}{y}}$

52. $\dfrac{\dfrac{4}{ab} + \dfrac{2}{b}}{\dfrac{8}{a} + \dfrac{1}{b}}$

53. $\dfrac{\dfrac{6}{x} + 2}{\dfrac{3}{x} + 4}$

54. $\dfrac{1 - \dfrac{6}{y}}{3 - \dfrac{2}{y}}$

55. $\dfrac{\dfrac{3}{2x^2} - \dfrac{4}{x}}{\dfrac{5}{3x} + \dfrac{7}{x^2}}$

56. $\dfrac{\dfrac{4}{3x} + \dfrac{5}{x^2}}{\dfrac{7}{4x} - \dfrac{9}{x}}$

57. $\dfrac{\dfrac{x+2}{4}}{\dfrac{1}{x} + \dfrac{3}{2}}$

58. $\dfrac{\dfrac{3}{x+1} + 2}{-4 + \dfrac{2}{x+1}}$

59. $\dfrac{\dfrac{1}{x-1} - 2}{\dfrac{3}{x-1} + 4}$

60. $\dfrac{\dfrac{3}{x-2} + \dfrac{2}{x+2}}{\dfrac{4}{x+2} - \dfrac{5}{x-2}}$

For Problems 61–71, answer each question with an algebraic fraction.

61. If by jogging at a constant rate Joan can complete a race in 40 minutes, how much of the course has she completed at the end of *m* minutes?

62. If Kent can mow the entire lawn in *m* minutes, what fractional part of the lawn has he mowed at the end of 20 minutes?

63. If Sandy drove k kilometers at a rate of r kilometers per hour, how long did it take her to make the trip?

64. If Roy traveled m miles in h hours, what was his rate in miles per hour?

65. If l liters of gasoline cost d dollars, what is the price per liter?

66. If p pounds of candy cost c cents, what is the price per pound?

67. Suppose that the product of two numbers is 34 and one of the numbers is n. What is the other number?

68. If a cold water faucet, when opened, can fill a tank in 3 hours, how much of the tank is filled at the end of h hours (see Figure 7.1)?

FIGURE 7.1

69. If the area of a rectangle is 47 square inches and the length is l inches, what is the width of the rectangle?

70. If the area of a rectangle is 56 square centimeters and the width is w centimeters, what is the length of the rectangle?

71. If the area of a triangle is 48 square feet and the length of one side is b feet, what is the length of the altitude to that side?

THOUGHTS INTO WORDS

72. Which of the two techniques presented in the text would you use to simplify $\dfrac{\dfrac{1}{4} + \dfrac{1}{3}}{\dfrac{3}{4} - \dfrac{1}{6}}$? Which technique would you use to simplify $\dfrac{\dfrac{3}{8} - \dfrac{5}{7}}{\dfrac{7}{9} + \dfrac{6}{25}}$?

Explain your choice for each problem.

Further Investigations

For Problems 73–76, simplify each complex fraction.

73. $1 - \dfrac{n}{1 - \dfrac{1}{n}}$

74. $2 - \dfrac{3n}{1 + \dfrac{4}{n}}$

75. $\dfrac{\dfrac{3x}{4 - \dfrac{2}{x}} - 1}{}$

76. $\dfrac{5x}{3 + \dfrac{1}{x}} + 2$

Fractional Equations and Problem Solving

We shall consider fractional equations of two basic types in this text. One type has only constants as denominators, and the other type has some variables in the denominators. In Chapter 3, we considered fractional equations involving only constants in the denominators. Let's review our approach to such equations since we will be using that same basic technique to solve any fractional equation.

Example 1

Solve $\dfrac{x-2}{3} + \dfrac{x+1}{4} = \dfrac{1}{6}$.

Solution

$$\dfrac{x-2}{3} + \dfrac{x+1}{4} = \dfrac{1}{6}$$

$$12\left(\dfrac{x-2}{3} + \dfrac{x+1}{4}\right) = 12\left(\dfrac{1}{6}\right) \qquad \text{Mulitiply both sides by 12, which is the LCD of all three denominators.}$$

$$12\left(\dfrac{x-2}{3}\right) + 12\left(\dfrac{x+1}{4}\right) = 12\left(\dfrac{1}{6}\right)$$

$$4(x-2) + 3(x+1) = 2$$

$$4x - 8 + 3x + 3 = 2$$

$$7x - 5 = 2$$

$$7x = 7$$

$$x = 1$$

The solution set is $\{1\}$. (Check it!) ▲

If an equation contains a variable in one or more denominators, then we proceed in essentially the same way *except we must avoid any value of the variable that makes a denominator zero*. Consider the following example.

Example 2

Solve $\dfrac{3}{x} + \dfrac{1}{2} = \dfrac{5}{x}$.

Solution

First, we need to realize that *x cannot equal zero*. Then, we can proceed in the usual way.

$$\dfrac{3}{x} + \dfrac{1}{2} = \dfrac{5}{x}$$

$$2x\left(\dfrac{3}{x} + \dfrac{1}{2}\right) = 2x\left(\dfrac{5}{x}\right) \qquad \text{Multiply both sides by 2x, which is the LCD of all denominators.}$$

$$6 + x = 10$$

$$x = 4$$

✔ **Check**

$\dfrac{3}{x} + \dfrac{1}{2} = \dfrac{5}{x}$ becomes $\dfrac{3}{4} + \dfrac{1}{2} \stackrel{?}{=} \dfrac{5}{4}$, when $x = 4$

$$\dfrac{3}{4} + \dfrac{2}{4} \stackrel{?}{=} \dfrac{5}{4}$$

$$\dfrac{5}{4} = \dfrac{5}{4}.$$

The solution set is $\{4\}$. ▲

Example 3

Solve $\dfrac{5}{x+2} = \dfrac{2}{x-1}$.

Solution

Since neither denominator can be zero, we know that $x \neq -2$ and $x \neq 1$.

$$\frac{5}{x+2} = \frac{2}{x-1}$$

$$(x+2)(x-1)\left(\frac{5}{x+2}\right) = (x+2)(x-1)\left(\frac{2}{x-1}\right)$$ Multiply both sides by $(x+2)(x-1)$, which is the LCD.

$$5(x-1) = 2(x+2)$$

$$5x - 5 = 2x + 4$$

$$3x = 9$$

$$x = 3$$

Since the only restrictions are $x \neq -2$ and $x \neq 1$, the solution set is $\{3\}$. (Check it!)   ▲

Example 4

Solve $\dfrac{2}{x-2} + 2 = \dfrac{x}{x-2}$.

Solution

Since no denominator can be zero, $x \neq 2$.

$$\frac{2}{x-2} + 2 = \frac{x}{x-2}$$

$$(x-2)\left(\frac{2}{x-2} + 2\right) = (x-2)\left(\frac{x}{x-2}\right)$$ Multiply by $x-2$, which is the LCD.

$$2 + 2(x-2) = x$$

$$2 + 2x - 4 = x$$

$$2x - 2 = x$$

$$x = 2$$

Since 2 cannot be used because it will produce a denominator of zero, there is no solution to the given equation. The solution set is $\varnothing$.   ▲

Example 4 illustrates the importance of recognizing the restrictions that must be placed on possible values of a variable. We will always indicate such restrictions at the beginning of our solution.

Example 5

Solve $\dfrac{125 - n}{n} = 4 + \dfrac{10}{n}$.

Solution

$$\frac{125 - n}{n} = 4 + \frac{10}{n}, \qquad n \neq 0$$ Note the necessary restriction.

$$n\left(\frac{125 - n}{n}\right) = n\left(4 + \frac{10}{n}\right)$$ Multiply both sides by n.

$$125 - n = 4n + 10$$
$$115 = 5n$$
$$23 = n$$

Since the only restriction is $n \neq 0$, the solution set is $\{23\}$. ▲

Back to Problem Solving

We are now ready to solve more problems, specifically those that translate into fractional equations.

Problem 1

One number is 10 larger than another number. The indicated quotient of the smaller number divided by the larger number reduces to $\frac{3}{5}$. Find the numbers.

Solution

Let n represent the smaller number. Then $n + 10$ represents the larger number. The second sentence in the statement of the problem translates into the following equation.

$$\frac{n}{n + 10} = \frac{3}{5}, \qquad n \neq -10$$

$$5n = 3(n + 10) \qquad \text{Cross products are equal.}$$
$$5n = 3n + 30$$
$$2n = 30$$
$$n = 15$$

If n is 15, then $n + 10$ is 25. Thus, the numbers are 15 and 25. To check, consider the quotient of the smaller number divided by the larger number.

$$\frac{15}{25} = \frac{3 \cdot \cancel{5}}{5 \cdot \cancel{5}} = \frac{3}{5}$$

▲

Before we tackle the next word problem, let's review a basic relationship pertaining to division. We can always check a division problem by *multiplying the divisor times the quotient and adding the remainder*. This can be expressed as

$$\text{Dividend} = (\text{Divisor})(\text{Quotient}) + \text{Remainder}.$$

Sometimes the remainder is expressed as a fractional part of the divisor. The above relationship then becomes

$$\frac{\text{Dividend}}{\text{Divisor}} = \text{Quotient} + \frac{\text{Remainder}}{\text{Divisor}}.$$

Problem 2

The sum of two numbers is 57. If the larger number is divided by the smaller number, the quotient is 10 and the remainder is 2. Find the numbers.

Solution A

Let n represent the smaller number. Then $57 - n$ represents the larger number. The relationship

$$\frac{\text{Dividend}}{\text{Divisor}} = \text{Quotient} + \frac{\text{Remainder}}{\text{Divisor}}$$

can be used as a guideline.

$$\frac{57 - n}{n} = 10 + \frac{2}{n}, \qquad n \neq 0$$

$$n\left(\frac{57 - n}{n}\right) = n\left(10 + \frac{2}{n}\right)$$

$$57 - n = 10n + 2$$

$$55 = 11n$$

$$5 = n$$

If n is 5, then $57 - n$ is 52. Thus, the numbers are 5 and 52. (Do they check?)

Solution B

Let n represent the larger number. Then $57 - n$ represents the smaller number.

$$\frac{n}{57 - n} = 10 + \frac{2}{57 - n}, \qquad n \neq 57$$

$$(57 - n)\left(\frac{n}{57 - n}\right) = (57 - n)\left(10 + \frac{2}{57 - n}\right)$$

$$n = 10(57 - n) + 2$$

$$n = 570 - 10n + 2$$

$$11n = 572$$

$$n = 52$$

If n is 52 then $57 - n$ is 5. Thus, the numbers are 5 and 52.   ▲

Solutions A and B are both acceptable ways of solving the problem. However, notice that the equation generated in Solution A is a little easier to solve than the one in Solution B. Sometimes it pays to look ahead as you declare what the variable is to represent. You may be able to make the computational work a little easier by a wise choice of what the variable represents.

In Chapter 4 we solved some uniform motion problems in which the formula $d = rt$ played an important role. Let's consider another one of those problems; keep in mind that the formula $d = rt$ can also be written as $\frac{d}{r} = t$ or $\frac{d}{t} = r$.

Problem 3

Wendy rides her bicycle 30 miles in the same time that it takes Kim to ride her bicycle 20 miles. If Wendy rides 5 miles per hour faster than Kim, find the rate of each.

Solution

Let r represent Kim's rate; then $r + 5$ represents Wendy's rate. The fact that their times are equal can be used as a guideline.

$$\underset{\text{Time of Kim}}{\frac{\text{Distance Kim rides}}{\text{Rate Kim rides}}} \underset{\text{Equals}}{=} \underset{\text{Time of Wendy}}{\frac{\text{Distance Wendy rides}}{\text{Rate Wendy rides}}}$$

$$\frac{20}{r} = \frac{30}{r + 5}, \qquad r \neq 0 \text{ and } r \neq -5$$

$$20(r + 5) = 30r$$

$$20r + 100 = 30r$$

$$100 = 10r$$

$$10 = r$$

Therefore, Kim rides at 10 miles per hour and Wendy rides at $10 + 5 = 15$ miles per hour. ▲

Problem Set 7.5

For Problems 1–40, solve each of the equations.

1. $\dfrac{x}{2} + \dfrac{x}{3} = 10$

2. $\dfrac{x}{8} - \dfrac{x}{6} = -1$

3. $\dfrac{x}{6} - \dfrac{4x}{3} = \dfrac{1}{9}$

4. $\dfrac{3x}{4} + \dfrac{x}{5} = \dfrac{3}{10}$

5. $\dfrac{n}{2} + \dfrac{n-1}{6} = \dfrac{5}{2}$

6. $\dfrac{n+2}{7} + \dfrac{n}{3} = \dfrac{12}{7}$

7. $\dfrac{t-3}{4} + \dfrac{t+1}{9} = -1$

8. $\dfrac{t-2}{4} - \dfrac{t+3}{7} = 1$

9. $\dfrac{2x+3}{3} + \dfrac{3x-4}{4} = \dfrac{17}{4}$

10. $\dfrac{3x-1}{4} + \dfrac{2x-3}{5} = -2$

11. $\dfrac{x-4}{8} - \dfrac{x+5}{4} = 3$

12. $\dfrac{x+6}{9} - \dfrac{x-2}{5} = \dfrac{7}{15}$

13. $\dfrac{3x+2}{5} - \dfrac{2x-1}{6} = \dfrac{2}{15}$

14. $\dfrac{4x-1}{3} - \dfrac{2x+5}{8} = \dfrac{1}{6}$

15. $\dfrac{1}{x} + \dfrac{2}{3} = \dfrac{7}{6}$

16. $\dfrac{2}{x} + \dfrac{1}{4} = \dfrac{13}{20}$

17. $\dfrac{5}{3n} - \dfrac{1}{9} = \dfrac{1}{n}$

18. $\dfrac{9}{n} - \dfrac{1}{4} = \dfrac{7}{n}$

19. $\dfrac{1}{2x} + 3 = \dfrac{4}{3x}$

20. $\dfrac{2}{3x} + 1 = \dfrac{5}{4x}$

21. $\dfrac{4}{5t} - 1 = \dfrac{3}{2t}$

22. $\dfrac{1}{6t} - 2 = \dfrac{7}{8t}$

23. $\dfrac{-5}{4h} + \dfrac{7}{6h} = \dfrac{1}{4}$

24. $\dfrac{3}{h} + \dfrac{5}{2h} = 1$

25. $\dfrac{90 - n}{n} = 10 + \dfrac{2}{n}$

26. $\dfrac{51 - n}{n} = 7 + \dfrac{3}{n}$

27. $\dfrac{n}{49 - n} = 3 + \dfrac{1}{49 - n}$

28. $\dfrac{n}{57 - n} = 10 + \dfrac{2}{57 - n}$

29. $\dfrac{x}{x + 3} - 2 = \dfrac{-3}{x + 3}$

30. $\dfrac{4}{x - 2} = \dfrac{5}{x + 6}$

31. $\dfrac{7}{x + 3} = \dfrac{5}{x - 9}$

32. $\dfrac{x}{x - 4} - 2 = \dfrac{4}{x - 4}$

33. $\dfrac{x}{x + 2} + 3 = \dfrac{1}{x + 2}$

34. $\dfrac{x}{x - 5} - 4 = \dfrac{3}{x - 5}$

35. $-1 - \dfrac{5}{x - 2} = \dfrac{3}{x - 2}$

36. $\dfrac{x - 1}{x} - 2 = \dfrac{3}{2}$

37. $1 + \dfrac{n + 1}{2n} = \dfrac{3}{4}$

38. $\dfrac{3}{n - 1} + 4 = \dfrac{2}{n - 1}$

39. $\dfrac{h}{2} - \dfrac{h}{4} + \dfrac{h}{3} = 1$

40. $\dfrac{h}{4} + \dfrac{h}{5} - \dfrac{h}{6} = 1$

For Problems 41–52, set up an equation and solve each problem.

41. The numerator of a fraction is 8 less than the denominator. The fraction in its simplest form is $\dfrac{5}{6}$. Find the fraction.

42. One number is 12 larger than another number. The indicated quotient of the smaller number divided by the larger reduces to $\dfrac{2}{3}$. Find the numbers.

43. What number must be added to the numerator and denominator of $\dfrac{2}{5}$ to produce a fraction equivalent to $\dfrac{4}{5}$?

44. What number must be subtracted from the numerator and denominator of $\dfrac{29}{31}$ to produce a fraction equivalent to $\dfrac{11}{12}$?

45. The sum of two numbers is 65. If the larger number is divided by the smaller, the quotient is 8 and the remainder is 2. Find the numbers.

46. One number is 57 larger than another number. If the larger number is divided by the smaller, the quotient is 5 and the remainder is 5. Find the numbers.

47. The denominator of a fraction is 4 less than the numerator. If the numerator is increased by 6 and the denominator is doubled, the resulting fraction is equivalent to 1. Find the original fraction.

48. One number is 74 smaller than another number. If the larger number is divided by the smaller number, the quotient is 6 and the remainder is 4. Find the numbers.

49. It took Heidi 3 hours and 20 minutes more to ride her bicycle 125 miles than it took Abby to ride 75 miles. If they both rode at the same rate, find this rate.

50. Two trains left a depot traveling in opposite directions at the same rate. One train traveled 338 miles in 2 hours more time than it took the other train to travel 234 miles. Find the rate of the trains.

51. Kent drives his Mazda 270 miles in the same time that Dave drives his Datsun 250 miles. If Kent averages 4 miles per hour faster than Dave, find their rates.

52. An airplane travels 2050 miles in the same time that a car travels 260 miles. If the rate of the plane is 358 miles per hour greater than the rate of the car, find the rate of each.

THOUGHTS INTO WORDS

53. **(a)** Explain how to do the addition problem $\dfrac{3}{x+2} + \dfrac{5}{x-1}$.

    (b) Explain how to solve the equation $\dfrac{3}{x+2} + \dfrac{5}{x-1} = 0$.

54. How can you tell by inspection that $\dfrac{x}{x-4} = \dfrac{4}{x-4}$ has no solution?

55. How would you help someone solve the equation $\dfrac{1}{x} + \dfrac{2}{x} = \dfrac{3}{x}$?

 **Further Investigations**

For Problems 56–59, solve each of the equations.

56. $\dfrac{3}{2n} + \dfrac{1}{n} = \dfrac{5}{3n}$

57. $\dfrac{1}{2n} + \dfrac{4}{n} = \dfrac{9}{2n}$

58. $\dfrac{n+1}{2} + \dfrac{n}{3} = \dfrac{1}{2}$

59. $\dfrac{1}{n+2} + \dfrac{2}{n+3} = \dfrac{3n+7}{(n+2)(n+3)}$

 ## 7.6 More Fractional Equations and Problem Solving

Let's begin this section by considering a few more fractional equations. We will continue to solve them using the same basic techniques as in the previous section.

Example 1

Solve $\dfrac{10}{8x-2} - \dfrac{6}{4x-1} = \dfrac{1}{9}$.

Solution

$$\frac{10}{8x-2} - \frac{6}{4x-1} = \frac{1}{9}, \qquad x \neq \frac{1}{4} \qquad \text{Do you see why } x \text{ cannot equal } \tfrac{1}{4}\text{?}$$

$$\frac{10}{2(4x-1)} - \frac{6}{4x-1} = \frac{1}{9} \qquad \text{Factor the first denominator.}$$

$$18(4x-1)\left(\frac{10}{2(4x-1)} - \frac{6}{4x-1}\right) = 18(4x-1)\left(\frac{1}{9}\right) \qquad \text{Multiply both sides by } 18(4x-1), \text{ which is the LCD.}$$

$$9(10) - 18(6) = 2(4x-1)$$

$$90 - 108 = 8x - 2$$

$$-18 = 8x - 2$$

$$-16 = 8x$$

$$-2 = x \qquad \text{Be sure that the solution } -2 \text{ checks!}$$

The solution set is $\{-2\}$.　▲

REMARK

In the second step of the solution for Example 1 you may choose to reduce $\dfrac{10}{2(4x - 1)}$ to $\dfrac{5}{4x - 1}$. Then the left side, $\dfrac{5}{4x - 1} - \dfrac{6}{4x - 1}$, simplifies to $\dfrac{-1}{4x - 1}$. This forms the proportion $\dfrac{-1}{4x - 1} = \dfrac{1}{9}$, which can be easily solved using the *cross multiplication method.* △

Example 2

Solve $\dfrac{2n}{n + 3} + \dfrac{5n}{n^2 - 9} = 2.$

Solution

$$\frac{2n}{n + 3} + \frac{5n}{n^2 - 9} = 2, \qquad n \neq -3 \text{ and } n \neq 3$$

$$\frac{2n}{n + 3} + \frac{5n}{(n + 3)(n - 3)} = 2$$

$$(n + 3)(n - 3)\left(\frac{2n}{n + 3} + \frac{5n}{(n + 3)(n - 3)}\right) = (n + 3)(n - 3)(2)$$

$$2n(n - 3) + 5n = 2(n^2 - 9)$$

$$2n^2 - 6n + 5n = 2n^2 - 18$$

$$-6n + 5n = -18 \qquad \text{Add } -2n^2 \text{ to both sides.}$$

$$-n = -18$$

$$n = 18$$

The solution set is $\{18\}$. ▲

Example 3

Solve $n + \dfrac{1}{n} = \dfrac{10}{3}.$

Solution

$$n + \frac{1}{n} = \frac{10}{3}, \qquad n \neq 0$$

$$3n\left(n + \frac{1}{n}\right) = 3n\left(\frac{10}{3}\right)$$

$$3n^2 + 3 = 10n$$

$$3n^2 - 10n + 3 = 0$$

$$(3n - 1)(n - 3) = 0$$

$$3n - 1 = 0 \qquad \text{or} \qquad n - 3 = 0$$

$$3n = 1 \qquad \text{or} \qquad n = 3$$

$$n = \frac{1}{3} \qquad \text{or} \qquad n = 3$$

Remember when we were using the factoring techniques to help solve equations of this type in Chapter 5?

The solution set is $\left\{\dfrac{1}{3}, 3\right\}$. ▲

Problem Solving

Recall that $\frac{2}{3}$ and $\frac{3}{2}$ are called multiplicative inverses, or *reciprocals*, of each other because their product is one. In general, the reciprocal of any nonzero real number n is the number $\frac{1}{n}$. Let's use this idea to solve a problem.

Problem 1

The sum of a number and its reciprocal is $\frac{26}{5}$. Find the number.

Solution

Let n represent the number. Then $\frac{1}{n}$ represents its reciprocal.

$$\text{Number} + \text{Its reciprocal} = \frac{26}{5}$$

$$n + \frac{1}{n} = \frac{26}{5}, \qquad n \neq 0$$

$$5n\left(n + \frac{1}{n}\right) = 5n\left(\frac{26}{5}\right)$$

$$5n^2 + 5 = 26n$$

$$5n^2 - 26n + 5 = 0$$

$$(5n - 1)(n - 5) = 0$$

$$5n - 1 = 0 \qquad \text{or} \qquad n - 5 = 0$$

$$5n = 1 \qquad \text{or} \qquad n = 5$$

$$n = \frac{1}{5} \qquad \text{or} \qquad n = 5$$

If the number is $\frac{1}{5}$, its reciprocal is $\frac{1}{\frac{1}{5}} = 5$. If the number is 5, its reciprocal is $\frac{1}{5}$. ▲

Now let's consider another uniform motion problem, which is a slight variation of those we studied in the previous section. Again keep in mind that the distance-rate-time relationships are constantly being used in these problems.

Problem 2

To travel 60 miles, it takes Sue, riding a moped, 2 hours less than it takes LeAnn, riding a bicycle, to travel 50 miles (see Figure 7.2). Sue travels 10 miles per hour faster than LeAnn. Find the times and rates of both girls.

FIGURE 7.2

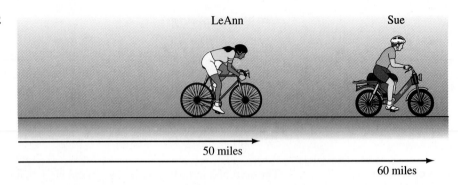

LeAnn   Sue

50 miles

60 miles

Solution

Let t represent LeAnn's time; then $t - 2$ represents Sue's time. We can record the information from the problem in the following table.

	Distance	Time	Rate $\left(r = \dfrac{d}{t}\right)$
LeAnn	50	t	$\dfrac{50}{t}$
Sue	60	$t - 2$	$\dfrac{60}{t - 2}$

The fact that Sue travels 10 miles per hour faster than LeAnn can be used as a guideline to set up an equation.

Rate of Sue   Equals   Rate of LeAnn plus 10

$$\frac{60}{t - 2} \quad = \quad \frac{50}{t} + 10, \qquad t \neq 2 \text{ and } t \neq 0$$

Solving this equation yields

$$\frac{60}{t - 2} = \frac{50}{t} + 10$$

$$t(t - 2)\left(\frac{60}{t - 2}\right) = t(t - 2)\left(\frac{50}{t} + 10\right)$$

$$60t = 50(t - 2) + 10t(t - 2)$$

$$60t = 50t - 100 + 10t^2 - 20t$$

$$0 = 10t^2 - 30t - 100$$

$$0 = t^2 - 3t - 10$$

$$0 = (t - 5)(t + 2)$$

$$t - 5 = 0 \quad \text{or} \quad t + 2 = 0$$
$$t = 5 \quad \text{or} \quad t = -2$$

The negative solution must be disregarded, so LeAnn's time is 5 hours and Sue's time is $5 - 2 = 3$ hours. LeAnn's rate is $\frac{50}{5} = 10$ miles per hour, and Sue's rate is $\frac{60}{3} = 20$ miles per hour. (Be sure that all of these results check back into the original problem!) ▲

There is another class of problems we commonly refer to as *work* problems, or sometimes as *rate-time* problems. For example, if a certain machine produces 120 items in 10 minutes, then we say that it is producing at a rate of $\frac{120}{10} = 12$ items per minute. Likewise, if a person can do a certain job in 5 hours, then he (she) is working at a rate of $\frac{1}{5}$ of the job per hour. In general, if Q is the quantity of something done in t units of time, then the rate, r, is given by $r = \frac{Q}{t}$. The rate is stated in terms of *so much quantity per unit of time*. The uniform motion problems we discussed earlier are a special kind of rate-time problems where the *quantity* is distance. The use of tables to organize information, as we illustrated with the uniform motion problems, is a convenient aid for some rate-time problems. Let's consider some problems.

Problem 3

Printing press A can produce 35 fliers per minute and press B can produce 50 fliers per minute. Suppose that 2225 fliers are produced by first using A by itself for 15 minutes, and then using both A and B until the job is done. How long would the printing press B need to be used?

Solution

Let m represent the number of minutes that printing press B is used. Then $m + 15$ represents the number of minutes that press A is used. The information in the problem can be organized as follows.

	Quantity	Time	Rate
A	$35(m + 15)$	$m + 15$	35
B	$50m$	m	50

Since the total quantity (total number of fliers) is 2225 fliers, we can set up and solve the following equation.

$$35(m + 15) + 50m = 2225$$
$$35m + 525 + 50m = 2225$$

$$85m = 1700$$
$$m = 20$$

Therefore, printing press B must be used for 20 minutes.   ▲

Problem 4

Bill can mow a lawn in 45 minutes and Jennifer can mow the same lawn in 30 minutes. How long would it take the two of them working together to mow the lawn (see Figure 7.3)?

FIGURE 7.3

REMARK   Before you look at the solution of this problem, *estimate* the answer. Remember that Jennifer can mow the lawn by herself in 30 minutes.   △

Solution   Bill's rate is $\dfrac{1}{45}$ of the lawn per minute and Jennifer's rate is $\dfrac{1}{30}$ of the lawn per minute. If we let m represent the number of minutes that they work together, then $\dfrac{1}{m}$ represents the rate when working together. Therefore, since the sum of the individual rates must equal the rate working together, we can set up and solve the following equation.

$$\frac{1}{30} + \frac{1}{45} = \frac{1}{m}, \qquad m \neq 0$$
$$90m\left(\frac{1}{30} + \frac{1}{45}\right) = 90m\left(\frac{1}{m}\right)$$
$$3m + 2m = 90$$
$$5m = 90$$
$$m = 18$$

It should take them 18 minutes to mow the lawn when working together. (How close was your estimate?)   ▲

Problem 5

It takes Amy twice as long to deliver papers as it does Nancy. How long would it take each girl by herself if they can deliver the papers together in 40 minutes?

Solution

Let m represent the number of minutes that it takes Nancy by herself; then $2m$ represents Amy's time by herself. Therefore, Nancy's rate is $\dfrac{1}{m}$ and Amy's rate is $\dfrac{1}{2m}$. Since the combined rate is $\dfrac{1}{40}$, we can set up and solve the following equation.

$$\frac{1}{m} + \frac{1}{2m} = \frac{1}{40}, \qquad m \neq 0$$

$$40m\left(\frac{1}{m} + \frac{1}{2m}\right) = 40m\left(\frac{1}{40}\right)$$

$$40 + 20 = m$$

$$60 = m.$$

Therefore, Nancy can deliver the papers by herself in 60 minutes and Amy can deliver them by herself in $2(60) = 120$ minutes. ▲

One final example of this section outlines another approach that some people find meaningful for work problems. This approach represents the fractional parts of a job. For example, if a person can do a certain job in 7 hours, then at the end of 3 hours he (she) has finished $\dfrac{3}{7}$ of the job. (Again, a constant rate of work is being assumed.) At the end of 5 hours, $\dfrac{5}{7}$ of the job has been done—in general, at the end of h hours, $\dfrac{h}{7}$ of the job has been completed. Let's use this idea to solve a work problem.

Problem 6

Together Al and Larry can paint a house in $3\dfrac{3}{5}$ days. Al can paint the house alone in 6 days. How long would it take Larry to paint the house by himself?

Solution

Let n represent the number of days that it would take Larry by himself. Now we can use the following guideline to set up an equation. (Notice that "1" represents the one job to be done, which is painting the house.)

| Fractional part of job done by Al when working together | $+$ | Fractional part of job done by Larry when working together | $=$ Total job |

$$\frac{3\frac{3}{5}}{6} + \frac{3\frac{3}{5}}{n} = 1, \qquad n \neq 0$$

$$\frac{\frac{18}{5}}{6} + \frac{\frac{18}{5}}{n} = 1 \qquad \text{Remember our work with complex fractions!}$$

$$\frac{18}{30} + \frac{18}{5n} = 1$$

$$\frac{3}{5} + \frac{18}{5n} = 1$$

$$5n\left(\frac{3}{5} + \frac{18}{5n}\right) = 5n(1) \qquad \text{Multiply both sides by } 5n.$$

$$3n + 18 = 5n$$

$$18 = 2n$$

$$9 = n$$

It would take Larry 9 days to paint the house by himself.   ▲

Let's emphasize a point we made earlier. Don't become discouraged if solving word problems is still giving you trouble. The development of problem skills is a long-term objective. If you continue to work hard and give it your best shot, you will gradually become more and more confident in your approach to solving problems. Don't be afraid to try some different approaches on your own. Our problem solving suggestions simply provide a framework for you to build on.

▼ Problem Set 7.6

For Problems 1–32, solve each equation.

1. $\frac{4}{x} + \frac{7}{6} = \frac{1}{x} + \frac{2}{3x}$

2. $\frac{2}{3x} - \frac{9}{x} = -\frac{25}{9}$

3. $\frac{3}{2x + 2} + \frac{4}{x + 1} = \frac{11}{12}$

4. $\frac{5}{2x - 6} + \frac{1}{x - 3} = \frac{7}{2}$

5. $\frac{5}{2n - 10} - \frac{3}{n - 5} = 1$

6. $\frac{7}{3x + 6} - \frac{2}{x + 2} = 2$

7. $\frac{3}{2t} - \frac{5}{t} = \frac{7}{5t} + 1$

8. $\frac{2}{3t} + \frac{3}{4t} = 1 - \frac{5}{2t}$

9. $\frac{x}{x - 2} + \frac{4}{x + 2} = 1$

10. $\frac{2x}{x + 1} - \frac{3}{x - 1} = 2$

11. $\frac{x}{x - 4} - \frac{2x}{x + 4} = -1$

12. $\frac{2x}{x + 2} + \frac{x}{x - 2} = 3$

13. $\frac{3n}{n + 3} - \frac{n}{n - 3} = 2$

14. $\frac{4n}{n - 5} - \frac{2n}{n + 5} = 2$

15. $\frac{3}{t^2 - 4} + \frac{5}{t + 2} = \frac{2}{t - 2}$

16. $\frac{t}{2t - 8} + \frac{16}{t^2 - 16} = \frac{1}{2}$

17. $\frac{4}{x - 1} - \frac{2x - 3}{x^2 - 1} = \frac{6}{x + 1}$

18. $\frac{3x - 1}{x^2 - 9} + \frac{4}{x + 3} = \frac{5}{x - 3}$

19. $8 + \dfrac{5}{y^2 + 2y} = \dfrac{3}{y + 2}$

20. $2 + \dfrac{4}{y - 1} = \dfrac{4}{y^2 - y}$

21. $n + \dfrac{1}{n} = \dfrac{17}{4}$

22. $n + \dfrac{3}{n} = 4$

23. $\dfrac{15}{4n} + \dfrac{15}{4(n + 4)} = 1$

24. $\dfrac{10}{7x} + \dfrac{10}{7(x + 3)} = 1$

25. $x - \dfrac{5x}{x - 2} = \dfrac{-10}{x - 2}$

26. $\dfrac{x + 1}{x - 3} - \dfrac{3}{x} = \dfrac{12}{x^2 - 3x}$

27. $\dfrac{t}{4t - 4} + \dfrac{5}{t^2 - 1} = \dfrac{1}{4}$

28. $\dfrac{x}{3x - 6} + \dfrac{4}{x^2 - 4} = \dfrac{1}{3}$

29. $\dfrac{3}{n - 5} + \dfrac{4}{n + 7} = \dfrac{2n + 11}{n^2 + 2n - 35}$

30. $\dfrac{2}{n + 3} + \dfrac{3}{n - 4} = \dfrac{2n - 1}{n^2 - n - 12}$

31. $\dfrac{a}{a + 2} + \dfrac{3}{a + 4} = \dfrac{14}{a^2 + 6a + 8}$

32. $3 + \dfrac{6}{t - 3} = \dfrac{6}{t^2 - 3t}$

For Problems 33–50, set up an equation and solve the problem.

33. The sum of a number and twice its reciprocal is $\dfrac{9}{2}$. Find the number.

34. The sum of a number and three times its reciprocal is 4. Find the number.

35. A number is $\dfrac{21}{10}$ larger than its reciprocal. Find the number.

36. Suppose that the reciprocal of a number subtracted from the number yields $\dfrac{5}{6}$. Find the number.

37. Suppose that Celia rides her bicycle 60 miles in 2 hours less than it takes Tom to ride his bicycle 85 miles. If Celia rides 3 miles per hour faster than Tom, find their respective rates.

38. To travel 300 miles, it takes a freight train 2 hours more than it does an express train to travel 280 miles. The rate of the express train is 20 miles per hour greater than the rate of the freight train. Find the rates of both trains.

39. One day Jeff rides his bicycle out into the country 40 miles (see Figure 7.4). On the way back, he takes a different route that is 2 miles longer and it takes him an hour longer to return. If his rate on the way out into the country is 4 miles per hour faster than his rate back, find both rates.

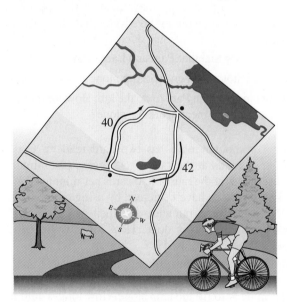

FIGURE 7.4

40. Rita jogs for 8 miles and then walks an additional 12 miles. She jogs at a rate twice her walking rate and she covered the entire distance of 20 miles in 4 hours. Find the rate she jogs and the rate that she walks.

41. A water tank can be filled by an inlet pipe in 5 minutes. A drain pipe will empty the tank in 6 minutes. If by mistake the drain is left open as the tank is being filled, how long will it take before the tank overflows?

42. Mark can do a job in 10 minutes. Phil can do the same job in 15 minutes. If they work together, how long will it take them to complete the job?

43. It takes Barry twice as long to deliver papers as it does Mike. How long would it take each if they can deliver the papers together in 40 minutes?

44. Working together Cindy and Sharon can address envelopes in 12 minutes. Cindy could do the addressing by herself in 20 minutes. How long would it take Sharon to address the envelopes by herself?

45. It takes Pat 12 hours to complete a task. After he had been working for 3 hours, his brother Mike joined him and together they finished the job in 5 hours. How long would it take Mike to do the job by himself?

46. Working together Pam and Laura can complete a job in $1\frac{1}{2}$ hours. When working alone it takes Laura 4 hours longer than Pam to do the job. How long does it take each of them working alone?

47. A computer room has two card readers. Card reader A reads at the rate of 600 cards per minute and reader B at the rate of 850 cards per minute. Suppose 9400 cards have been read by using card reader A by itself for 6 minutes, and then using both of them until the job is done. How long would card reader B need to be used?

48. It takes two pipes 3 hours to fill a water tank. Pipe B can fill the tank alone in 8 hours more than it takes pipe A to fill the tank alone. How long would it take each pipe to fill the tank by itself?

49. Suppose that Amelia can type 600 words in 5 minutes less than it takes Paul to type 600 words. If Amelia types at a rate 20 words per minute faster than Paul, find the rate of each.

50. A computer company markets two card readers. Card reader B can read 8000 cards in 2 minutes less time than it takes card reader A to read 7500 cards. If the rate of card reader A is 250 cards per minute less than the rate of reader B, find each rate.

THOUGHTS INTO WORDS

51. Write a paragraph or two summarizing the new ideas about problem solving that you have acquired thus far in this course.

Further Investigations

For Problems 52–54, solve each equation.

52. $\dfrac{3x-1}{x^2-9} + \dfrac{4}{x+3} = \dfrac{7}{x-3}$

53. $\dfrac{x-2}{x^2-1} + \dfrac{3}{x+1} = \dfrac{-5}{x-1}$

54. $\dfrac{7x-12}{x^2-16} - \dfrac{5}{x+4} = \dfrac{2}{x-4}$

SUMMARY

(7.1) The **fundamental principle of fractions** $\left(\dfrac{ak}{bk} = \dfrac{a}{b}\right)$ provides the basis for simplifying algebraic fractions.

(7.2) To **multiply** algebraic fractions, *multiply the numerators, multiply the denominators, and express the product in simplified form.*

To **divide** algebraic fractions, *invert the divisor and multiply.*

(7.3) **Addition** and **subtraction** of algebraic fractions are based on the following.

$$\frac{a}{b} + \frac{c}{b} = \frac{a+c}{b} \qquad \text{Addition}$$

$$\frac{a}{b} - \frac{c}{b} = \frac{a-c}{b} \qquad \text{Subtraction}$$

(7.4) Use the following procedure when adding and subtracting fractions.

1. Find the least common denominator.
2. Change each fraction to an equivalent fraction that has the LCD as its denominator.
3. Add or subtract the numerators and place this result over the LCD.
4. Look for possibilities to simplify the final fraction.

Fractional forms that contain fractions in the numerators and/or denominators are called **complex fractions**.

(7.5) and **(7.6)** To solve a fractional equation it is often convenient to begin by multiplying both sides of the equation by the LCD, which *clears the equation of all fractions.*

Chapter 7 Review Problem Set

For Problems 1–4, simplify each algebraic fraction.

1. $\dfrac{56x^3y}{72xy^3}$

2. $\dfrac{x^2 - 9x}{x^2 - 6x - 27}$

3. $\dfrac{3n^2 - n - 10}{n^2 - 4}$

4. $\dfrac{16a^2 + 24a + 9}{20a^2 + 7a - 6}$

For Problems 5–15, perform the indicated operations and express your answers in simplest form.

5. $\dfrac{7x^2y^2}{12y^3} \cdot \dfrac{18y}{28x}$

6. $\dfrac{x^2y}{x^2 + 2x} \cdot \dfrac{x^2 - x - 6}{y}$

7. $\dfrac{n^2 - 2n - 24}{n^2 + 11n + 28} \div \dfrac{n^3 - 6n^2}{n^2 - 49}$

8. $\dfrac{4a^2 + 4a + 1}{(a + 6)^2} \div \dfrac{6a^2 - 5a - 4}{3a^2 + 14a - 24}$

9. $\dfrac{3x + 4}{5} + \dfrac{2x - 7}{4}$

10. $\dfrac{7}{3x} + \dfrac{5}{4x} - \dfrac{2}{8x^2}$

11. $\dfrac{7}{n} + \dfrac{3}{n - 1}$

12. $\dfrac{2}{a - 4} - \dfrac{3}{a - 2}$

13. $\dfrac{2x}{x^2 - 3x} - \dfrac{3}{4x}$

14. $\dfrac{2}{x^2 + 7x + 10} + \dfrac{3}{x^2 - 25}$

15. $\dfrac{5x}{x^2 - 4x - 21} - \dfrac{3}{x - 7} + \dfrac{4}{x + 3}$

For Problems 16 and 17, simplify each complex fraction.

16. $\dfrac{\dfrac{3}{x} - \dfrac{4}{y^2}}{\dfrac{4}{y} + \dfrac{5}{x}}$

17. $\dfrac{\dfrac{2}{x} - 1}{3 + \dfrac{5}{y}}$

For Problems 18–29, solve each equation.

18. $\dfrac{2x - 1}{3} + \dfrac{3x - 2}{4} = \dfrac{5}{6}$

19. $\dfrac{5}{3x} - 2 = \dfrac{7}{2x} + \dfrac{1}{5x}$

20. $\dfrac{67 - x}{x} = 6 + \dfrac{4}{x}$

21. $\dfrac{5}{2n + 3} = \dfrac{6}{3n - 2}$

22. $\dfrac{x}{x - 3} + \dfrac{5}{x + 3} = 1$

23. $n + \dfrac{1}{n} = 2$

24. $\dfrac{n - 1}{n^2 + 8n - 9} - \dfrac{n}{n + 9} = 4$

25. $\dfrac{6}{7x} - \dfrac{1}{6} = \dfrac{5}{6x}$

26. $n + \dfrac{1}{n} = \dfrac{5}{2}$

27. $\dfrac{n}{5} = \dfrac{10}{n - 5}$

28. $\dfrac{-1}{2x - 5} + \dfrac{2x - 4}{4x^2 - 25} = \dfrac{5}{6x + 15}$

29. $1 + \dfrac{1}{n - 1} = \dfrac{1}{n^2 - n}$

For Problems 30–37, set up an equation and solve each problem.

30. The sum of two numbers is 75. If the larger number is divided by the smaller, the quotient is 9 and the remainder is 5. Find the numbers.

31. It takes Nancy three times as long to complete a task as it does Becky. How long would it take each of them to complete the task if working together they can do it in 2 hours?

32. The sum of a number and twice its reciprocal is 3. Find the number.

33. The denominator of a fraction is twice the numerator. If 4 is added to the numerator and 18 to the denominator, a fraction that is equivalent to $\dfrac{4}{9}$ is produced. Find the original fraction.

34. Lanette can ride her moped 44 miles in the same time that Todd rides his bicycle 30 miles. If Lanette rides 7 miles per hour faster than Todd, find their rates.

35. Jim rode his bicycle 36 miles in 4 hours. For the first 20 miles he rode at a constant rate and then for the last 16 miles he reduced his rate by 2 miles per hour. Find his rate for the last 16 miles.

36. An inlet pipe can fill a tank in 10 minutes. A drain can empty the tank in 12 minutes. If the tank is empty and both the inlet pipe and drain are open, how long before the tank overflows?

37. Sue can type 1000 words in 5 minutes more than it takes Corinne to type 840 words. If Sue's rate is 6 words per minute less than Corinne's rate, find their typing speeds.

CHAPTER 7 TEST

For Problems 1–4, simplify each algebraic fraction.

1. $\dfrac{72x^4y^5}{81x^2y^4}$

2. $\dfrac{x^2 + 6x}{x^2 - 36}$

3. $\dfrac{2n^2 - 7n - 4}{3n^2 - 8n - 16}$

4. $\dfrac{2x^3 + 7x^2 - 15x}{x^3 - 25x}$

For Problems 5–14, perform the indicated operations and express answers in simplest form.

5. $\left(\dfrac{8x^2y}{7x}\right)\left(\dfrac{21xy^3}{12y^2}\right)$

6. $\dfrac{x^2 - 49}{x^2 + 7x} \div \dfrac{x^2 - 4x - 21}{x^2 - 2x}$

7. $\dfrac{x^2 - 5x - 36}{x^2 - 15x + 54} \cdot \dfrac{x^2 - 2x - 24}{x^2 + 7x}$

8. $\dfrac{3x - 1}{6} - \dfrac{2x - 3}{8}$

9. $\dfrac{n + 2}{3} - \dfrac{n - 1}{5} + \dfrac{n - 6}{6}$

10. $\dfrac{3}{2x} - \dfrac{5}{6} + \dfrac{7}{9x}$

11. $\dfrac{6}{n} - \dfrac{4}{n - 1}$

12. $\dfrac{2x}{x^2 + 6x} - \dfrac{3}{4x}$

13. $\dfrac{9}{x^2 + 4x - 32} + \dfrac{5}{x + 8}$

14. $\dfrac{-3}{6x^2 - 7x - 20} - \dfrac{5}{3x^2 - 14x - 24}$

For Problems 15–22, solve each of the equations.

15. $\dfrac{x + 3}{5} - \dfrac{x - 2}{6} = \dfrac{23}{30}$

16. $\dfrac{5}{8x} - 2 = \dfrac{3}{x}$

17. $n + \dfrac{4}{n} = \dfrac{13}{3}$

18. $\dfrac{x}{8} = \dfrac{6}{x - 2}$

19. $\dfrac{x}{x - 1} + \dfrac{2}{x + 1} = \dfrac{8}{3}$

20. $\dfrac{3}{2x + 1} = \dfrac{5}{3x - 6}$

21. $\dfrac{4}{n^2 - n} - \dfrac{3}{n - 1} = -1$

22. $\dfrac{3n - 1}{3} + \dfrac{2n + 5}{4} = \dfrac{4n - 6}{9}$

(continued on next page)

CHAPTER 7 TEST *(continued)*

For Problems 23–25, set up an equation and solve the problem.

23. The sum of a number and twice its reciprocal is $3\frac{2}{3}$. Find the number.

24. Wendy can ride her bicycle 42 miles in the same time that it takes Betty to ride her bicycle 36 miles. Wendy rides 2 miles per hour faster than Betty. Find Wendy's rate.

25. Garth can mow a lawn in 20 minutes and Alex can mow the same lawn in 30 minutes. How long would it take the two of them working together to mow the lawn?

▽ Cumulative Review Problem Set

For Problems 1–8, evaluate each of the algebraic expressions for the given values of the variables. You may want to first simplify the expression or change its form by factoring.

1. $3x - 2xy - 7x + 5xy$ for $x = \frac{1}{2}$ and $y = 3$

2. $7(a - b) - 3(a - b) - (a - b)$ for $a = -3$ and $b = -5$

3. $\dfrac{xy + yz}{y}$ for $x = \frac{2}{3}$, $y = \frac{5}{6}$, and $z = \frac{3}{4}$

4. $ab + b^2$ for $a = .4$ and $b = .6$

5. $x^2 - y^2$ for $x = -6$ and $y = 4$

6. $x^2 + 5x - 36$ for $x = -9$

7. $\dfrac{x^2 + 2x}{x^2 + 5x + 6}$ for $x = -6$

8. $\dfrac{x^2 + 3x - 10}{x^2 - 9x + 14}$ for $x = 4$

For Problems 9–16, evaluate each of the expressions.

9. 3^{-3}

10. $\left(\dfrac{2}{3}\right)^{-1}$

11. $\left(\dfrac{1}{2} + \dfrac{1}{3}\right)^0$

12. $\left(\dfrac{1}{3} + \dfrac{1}{4}\right)^{-1}$

13. -4^{-2}

14. $\left(\dfrac{2}{3}\right)^{-2}$

15. $\dfrac{1}{\left(\dfrac{2}{5}\right)^{-2}}$

16. $(-3)^{-3}$

For Problems 17–32, perform the indicated operations and express your answers in simplest form.

17. $\dfrac{7}{5x} + \dfrac{2}{x} - \dfrac{3}{2x}$

18. $\dfrac{4x}{5y} \div \dfrac{12x^2}{10y^2}$

19. $\dfrac{4}{x - 6} + \dfrac{3}{x + 4}$

20. $\dfrac{2}{x^2 - 4x} - \dfrac{3}{x^2}$

21. $\dfrac{x^2 - 8x}{x^2 - x - 56} \cdot \dfrac{x^2 - 49}{3xy}$

22. $\dfrac{5}{x^2 - x - 12} - \dfrac{3}{x - 4}$

23. $(-5x^2y)(7x^3y^4)$

24. $(9ab^3)^2$

25. $(-3n^2)(5n^2 + 6n - 2)$

26. $(5x - 1)(3x + 4)$

27. $(2x + 5)^2$

28. $(x + 2)(2x^2 - 3x - 1)$

29. $(x^2 - x - 1)(x^2 + 2x - 3)$

30. $(-2x - 1)(3x - 7)$

31. $\dfrac{24x^2y^3 - 48x^4y^5}{8xy^2}$

32. $(28x^2 - 19x - 20) \div (4x - 5)$

For Problems 33–42, factor completely.

33. $3x^3 + 15x^2 + 27x$

34. $x^2 - 100$

35. $5x^2 - 22x + 8$

36. $8x^2 - 22x - 63$

37. $n^2 + 25n + 144$

38. $nx + ny - 2x - 2y$

39. $3x^3 - 3x$

40. $2x^3 - 6x^2 - 108x$

41. $36x^2 - 60x + 25$

42. $3x^2 - 5xy - 2y^2$

For Problems 43–57, solve each of the equations.

43. $3(x - 2) - 2(x + 6) = -2(x + 1)$

44. $x^2 = -11x$

45. $.2x - 3(x - .4) = 1$

46. $\dfrac{3n - 1}{4} = \dfrac{5n + 2}{7}$

47. $5n^2 - 5 = 0$

48. $x^2 + 5x - 6 = 0$

49. $n + \dfrac{4}{n} = 4$

50. $\dfrac{2x + 1}{2} + \dfrac{3x - 4}{3} = 1$

51. $2(x - 1) - x(x - 1) = 0$

52. $\dfrac{3}{2x} - 1 = \dfrac{5}{3x} + 2$

53. $6t^2 + 19t - 7 = 0$

54. $(2x - 1)(x - 8) = 0$

55. $(x + 1)(x + 6) = 24$

56. $\dfrac{x}{x - 2} - \dfrac{7}{x + 1} = 1$

57. $\dfrac{1}{n} - \dfrac{2}{n - 1} = \dfrac{3}{n}$

C H A P T E R

Coordinate Geometry and Linear Systems

Punch that contains 10% grapefruit juice is mixed with punch that contains 5% grapefruit juice to produce 5 quarts of punch that is 8% grapefruit juice. How many quarts of 10% and 5% grapefruit juice punch must be used? **Use the two equations $x + y = 5$ and $.10x + .05y = .08(5)$ to determine that 3 quarts of the punch that is 10% grapefruit juice and 2 quarts of the punch that is 5% grapefruit juice should be used.**

On a real number line, a one-to-one correspondence is established between the set of real numbers and the points on a line. This means that to each point on a line there corresponds a unique real number and to each real number there corresponds a point on the line. We used the real number line in Section 3.4 to graph the solutions for inequalities in one variable. In this chapter we will extend the idea of associating points with numbers so that we can establish a correspondence between pairs of real numbers and points in a plane. This correspondence will allow us to graph the solutions of equations and inequalities that contain two variables. Furthermore, we will solve systems of two linear equations in two variables, which will give us more problem solving power.

8.1   Cartesian Coordinate System

Consider two number lines (one vertical and one horizontal) perpendicular to each other at the point we associate with zero on both lines (Figure 8.1). We refer to these number lines as the horizontal and vertical **axes** or together as the **coordinate axes**. They partition the plane into four parts called **quadrants**. The quadrants are numbered counterclockwise from I to IV as indicated in Figure 8.1. The point of intersection of the two axes is called the **origin**.

FIGURE 8.1

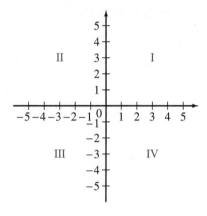

It is now possible to set up a one-to-one correspondence between ordered pairs of real numbers and the points in a plane. To each ordered pair of real numbers there corresponds a unique point in the plane and to each point there corresponds a unique ordered pair of real numbers. We have indicated a part of this correspondence in Figure 8.2. The ordered pair (3, 1) corresponds to a point A

FIGURE 8.2

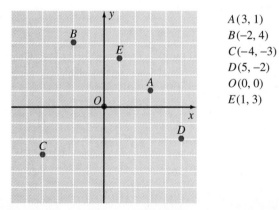

$A(3, 1)$
$B(-2, 4)$
$C(-4, -3)$
$D(5, -2)$
$O(0, 0)$
$E(1, 3)$

and means that the point A is located 3 units to the right of and one unit up from the origin. (The ordered pair (0, 0) corresponds to the origin.) The ordered pair $(-2, 4)$ corresponds to point B and that means that point B is located 2 units to the left and 4 units up from the origin. Make sure that you agree with all of the other locations in Figure 8.2.

In general, we refer to the real numbers a and b in an ordered pair (a, b), associated with a point as the **coordinates of the point**. The first number, a, called the **abscissa**, is the directed distance of the point from the vertical axis measured parallel to the horizontal axis. The second number, b, called the **ordinate**, is the directed distance of the point from the horizontal axis measured parallel to the vertical axis (Figure 8.3(a)). Thus, in the first quadrant all points have a positive abscissa and a positive ordinate. In the second quadrant all points have a negative abscissa and a positive ordinate. We have indicated the sign situations for all four quadrants in Figure 8.3(b). This system of associating points with ordered pairs of real numbers is called the **Cartesian coordinate system** or the **rectangular coordinate system**.

FIGURE 8.3

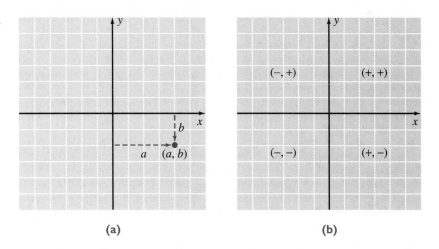

(a)   (b)

Historically, the branch of mathematics we call **analytic geometry** or **coordinate geometry** is based on the Cartesian coordinate system. With this system, René Descartes, a 17th century French mathematician, was able to transform geometric problems into an algebraic setting and then use the tools of algebra to solve the problems.

Basically, there are two kinds of problems in analytic (coordinate) geometry.

1. Given an algebraic equation, find its geometric graph.
2. Given a set of conditions pertaining to a geometric figure, find its algebraic equation.

We will consider a few problems of type 1 at this time.

Let's begin by considering the solutions for the equation $y = x + 3$. A *solution* of an equation in two variables is an ordered pair of real numbers that satisfies the equation. When using the variables x and y, we agree that the first number of an ordered pair is a value for x and the second number is a value for y. We see that $(1, 4)$ is a solution for $y = x + 3$ because when x is replaced by 1 and y by 4, the true numerical statement $4 = 1 + 3$ is obtained. Likewise, $(-1, 2)$ is a solution for $y = x + 3$ because $2 = -1 + 3$ is a true statement. Infinitely many pairs of real numbers that satisfy $y = x + 3$ can be found by arbitrarily choosing values for x and for each value of x chosen, determining a corresponding value for y. Let's use a table to record some of the solutions for $y = x + 3$.

Choose x	Determine y from y = x + 3	Solution for y = x + 3
0	3	$(0, 3)$
1	4	$(1, 4)$
3	6	$(3, 6)$
5	8	$(5, 8)$
−1	2	$(-1, 2)$
−3	0	$(-3, 0)$
−5	−2	$(-5, -2)$

Now we can locate the point associated with each ordered pair on a rectangular coordinate system; we label the horizontal axis as the *x*-axis and the vertical axis as the *y*-axis (Figure 8.4(a)). The straight line in Figure 8.4(b) that contains the points is called the *graph of the equation $y = x + 3$.*

FIGURE 8.4

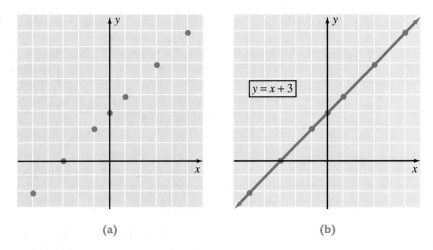

(a)   (b)

The following examples further illustrate the process of graphing equations.

Example 1

Solution

Graph $y = x^2$.

First we set up a table of some of the solutions.

x	y	Solutions (x, y)
0	0	(0, 0)
1	1	(1, 1)
2	4	(2, 4)
3	9	(3, 9)
−1	1	(−1, 1)
−2	4	(−2, 4)
−3	9	(−3, 9)

Then we plot the points associated with the solutions (Figure 8.5 (a)). Finally, we connect the points with a smooth curve (Figure 8.5 (b)). The curve in Figure 8.5 (b) is called a **parabola**, which is not important to us in this course, but does become a major topic of study in other mathematics courses.

FIGURE 8.5

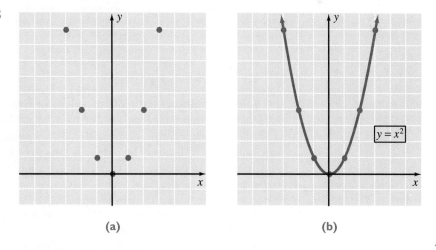

(a)   (b)

How many solutions do we need to have in a table of values? There is no definite answer to this question other than "a sufficient number so that the graph of the equation can be determined." In other words, we need to plot points until we can tell the nature of the curve.

Example 2

Graph $2x + 3y = 6$.

Solution

First, let's change the form of the equation to make it easier to find solutions. We can either solve for x in terms of y or for y in terms of x. With the latter, we obtain

$$2x + 3y = 6$$
$$3y = 6 - 2x$$
$$y = \frac{6 - 2x}{3}.$$

Now we can set up a table of values. Plotting these points and connecting them produces Figure 8.6.

FIGURE 8.6

x	y
0	2
3	0
6	-2
-3	4
-6	6

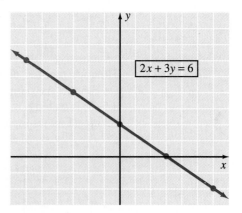

Look carefully at the table of values in Example 2. Notice that we chose values of x so that integers were obtained for y. This is not necessary but it does make things easier from a computational viewpoint. Also, plotting points associated with pairs of integers is more exact than getting involved with fractions.

To graph an equation in two variables x and y, keep in mind the following steps.

1. Solve the equation for y in terms of x or for x in terms of y, if it is not already in such a form.
2. Set up a table of ordered pairs that satisfy the equation.
3. Plot the points associated with the ordered pairs.
4. Connect the points with a smooth curve.

Let's conclude this section with two more examples that illustrate step 1.

Example 3

Solve $4x + 9y = 12$ for y.

Solution

$$4x + 9y = 12$$
$$9y = 12 - 4x \qquad \text{Subtracted } 4x \text{ from both sides}$$
$$y = \frac{12 - 4x}{9} \qquad \text{Divided both sides by 9}$$

▲

Example 4

Solve $4x - 5y = 6$ for y.

Solution

$$4x - 5y = 6$$
$$-5y = 6 - 4x \qquad \text{Subtracted } 4x \text{ from both sides}$$
$$y = \frac{6 - 4x}{-5} \qquad \text{Divided both sides by } -5$$
$$y = \frac{4x - 6}{5} \qquad \frac{6 - 4x}{-5} \text{ can be changed to } \frac{-6 + 4x}{5} \text{ by multiplying}$$
$$\text{numerator and denominator by } -1$$

▲

Problem Set 8.1

For Problems 1–10, solve the given equation for the variable indicated.

1. $3x + 7y = 13$ for y
2. $5x + 9y = 17$ for y
3. $x - 3y = 9$ for x
4. $2x - 7y = 5$ for x
5. $-x + 5y = 14$ for y
6. $-2x - y = 9$ for y
7. $-3x + y = 7$ for x
8. $-x - y = 9$ for x
9. $-2x + 3y = -5$ for y
10. $3x - 4y = -7$ for y

For Problems 11–34, graph each of the equations.

11. $y = x + 1$
12. $y = x + 4$
13. $y = x - 2$
14. $y = -x - 1$
15. $y = (x - 2)^2$
16. $y = (x + 1)^2$

17. $y = x^2 - 2$
18. $y = x^2 + 1$
19. $y = \frac{1}{2}x + 3$
20. $y = \frac{1}{2}x - 2$
21. $x + 2y = 4$
22. $x + 3y = 6$
23. $2x - 5y = 10$
24. $5x - 2y = 10$
25. $y = x^3$
26. $y = x^4$
27. $y = -x^2$
28. $y = -x^3$
29. $y = x$
30. $y = -x$
31. $y = -3x + 2$
32. $3x - y = 4$
33. $y = 2x^2$
34. $y = -3x^2$

35. How would you convince someone that there are infinitely many ordered pairs of real numbers that satisfy the equation $x + y = 9$?

36. Explain why no points of the graph of the equation $y = x^2 + 1$ will lie below the x-axis.

Further Investigations

37. (a) Graph the equations $y = x^2 + 2$, $y = x^2 + 4$, and $y = x^2 - 3$ on the same set of axes.

(b) Based on your graphs in part (a), sketch a graph of $y = x^2 - 1$ without plotting any points.

38. (a) Graph the equations $y = (x - 2)^2$, $y = (x - 4)^2$, and $y = (x + 3)^2$ on the same set of axes.

(b) Based on your graphs in part (a), sketch a graph of $y = (x + 5)^2$ without plotting any points.

39. (a) Graph the equations $y = (x - 1)^2 + 2$, $y = (x - 3)^2 - 2$, and $y = (x + 2)^2 + 3$ on the same set of axes.

(b) Based on your graphs in part (a), sketch a graph of $y = (x + 1)^2 - 4$ without plotting any points.

8.2 Linear Equations in Two Variables

The following table summarizes some of our results from graphing equations in the previous section and accompanying problem set.

Equation	Type of graph produced
$y = x + 3$	Straight line
$y = x^2$	Parabola
$2x + 3y = 6$	Straight line
$y = -3x + 2$	Straight line
$y = x^2 - 2$	Parabola
$y = (x - 2)^2$	Parabola
$5x - 2y = 10$	Straight line
$3x - y = 4$	Straight line
$y = x^3$	No name will be given at this time, but not a straight line
$y = x$	Straight line
$y = \frac{1}{2}x + 3$	Straight line

In this table pay special attention to the equations that produced a straight line graph. They are called **linear equations in two variables**. In general, any equation of the form $Ax + By = C$, where A, B, and C are constants (A and B not both zero) and x and y are variables, is a linear equation in two variables and its graph is a straight line.

We should clarify two points about the previous description of a linear equation in two variables. First, the choice of x and y for variables is arbitrary. Any two letters could be used to represent the variables. An equation such as $3m + 2n = 7$ can be considered a linear equation in two variables. So that we are not constantly changing the labeling of the coordinate axes when graphing equations, however, it is much easier to use the same two variables in all equations. Thus, we will go along with convention and use x and y as our variables. Secondly, the statement "any equation of the form $Ax + By = C$" technically means "any equation of the form $Ax + By = C$ or *equivalent* to the form." For example, the equation $y = x + 3$, which has a straight line graph, is equivalent to $-x + y = 3$.

The knowledge that any equation of the form $Ax + By = C$ produces a straight line graph, along with the fact that two points determine a straight line, makes graphing linear equations in two variables a simple process. We merely find two solutions, plot the corresponding points, and connect the points with a straight line. It is probably wise to find a third point as a check point. Let's consider an example.

Example 1

Solution

Graph $2x - 3y = 6$.

Let $x = 0$, then $2(0) - 3y = 6$

$$-3y = 6$$
$$y = -2.$$

Thus, $(0, -2)$ is a solution. Let $y = 0$, then $2x - 3(0) = 6$

$$2x = 6$$
$$x = 3.$$

Thus, $(3, 0)$ is a solution. Let $x = -3$, then $2(-3) - 3y = 6$

$$-6 - 3y = 6$$
$$-3y = 12$$
$$y = -4.$$

Thus, $(-3, -4)$ is a solution. Plot the points associated with these three solutions and connect them with a straight line to produce the graph of $2x - 3y = 6$ in Figure 8.7.

FIGURE 8.7

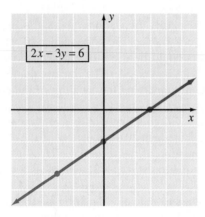

$2x - 3y = 6$

Let us briefly review our approach to Example 1. Notice that we did not begin our solution by either solving for y in terms of x or for x in terms of y. The reason for this is that we know the graph is a straight line and therefore there is no need for an extensive table of values. Thus, there is no real benefit to changing the form of the original equation. The first two solutions indicate where the line intersects the coordinate axes. The ordinate of the point $(0, -2)$ is called the *y-intercept* and the abscissa of the point $(3, 0)$ the *x-intercept* of this graph. That is, the graph of the equation $2x - 3y = 6$ has a y-intercept of -2 and an x-intercept of 3. In general, the intercepts are often easy to find. You can let $x = 0$ and solve for y to find the y-intercept and let $y = 0$ and solve for x to find the x-intercept. The third solution, $(-3, -4)$, serves as a check point. If $(-3, -4)$ had not been on the line determined by the two intercepts, then we would have known that an error had been committed.

Example 2

Graph $x + 2y = 4$.

Solution

Without showing all of our work, the following table indicates the intercepts and a check point.

x	y	
0	2	Intercepts
4	0	
2	1	Check point

Plot the points (0, 2), (4, 0), and (2, 1), and connect them with a straight line to produce the graph in Figure 8.8.

FIGURE 8.8

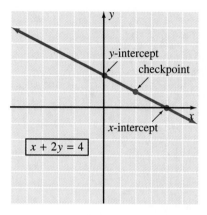

$x + 2y = 4$

Example 3

Graph $2x + 3y = 7$.

Solution

The intercepts and a check point are given in the following table. Finding intercepts may involve fractions, but the computation is usually easy. The points from the table are plotted and the graph of $2x + 3y = 7$ is shown in Figure 8.9.

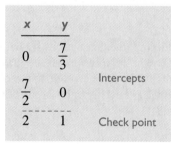

x	y	
0	$\dfrac{7}{3}$	
$\dfrac{7}{2}$	0	Intercepts
2	1	Check point

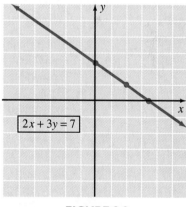

$2x + 3y = 7$

FIGURE 8.9

Example 4

Graph $y = 2x$.

Solution

Notice that (0, 0) is a solution; thus, this line intersects both axes at the origin. Since both the x-intercept and the y-intercept are determined by the origin, (0, 0), another point is necessary to graph the line. Then a third point should be found as a check point. These results are summarized in the following table. The graph of $y = 2x$ is shown in Figure 8.10.

x	y	
0	0	Intercept
2	4	Additional point
−1	−2	Check point

FIGURE 8.10

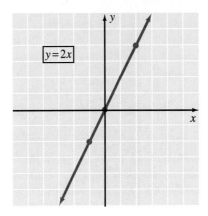

Example 5   Graph $x = 3$.

Solution   Since we are considering linear equations in *two variables*, the equation $x = 3$ is equivalent to $x + 0(y) = 3$. Now we can see that any value of y can be used, but the x-value must always be 3. Therefore, some of the solutions are $(3, 0)$, $(3, 1)$, $(3, 2)$, $(3, -1)$, and $(3, -2)$. The graph of all of the solutions is the vertical line indicated in Figure 8.11.

FIGURE 8.11

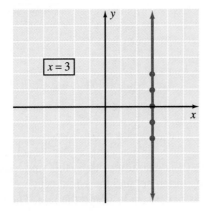

Problem Set 8.2

Use the techniques of this section to sketch a graph for each of the following equations.

1. $x + y = 2$
2. $x + y = 4$
3. $x - y = 3$
4. $x - y = 1$
5. $x - y = -4$
6. $-x + y = 5$
7. $x + 2y = 2$
8. $x + 3y = 5$
9. $3x - y = 6$
10. $2x - y = -4$
11. $3x - 2y = 6$
12. $2x - 3y = 4$
13. $x - y = 0$
14. $x + y = 0$
15. $y = 3x$
16. $y = -2x$
17. $x = -2$
18. $y = 3$
19. $y = 0$
20. $x = 0$
21. $y = -2x - 1$
22. $y = 3x - 4$
23. $y = \frac{1}{2}x + 1$
24. $y = \frac{2}{3}x - 2$
25. $y = -\frac{1}{3}x - 2$
26. $y = -\frac{3}{4}x - 1$
27. $4x + 5y = -10$
28. $3x + 5y = -9$
29. $-2x + y = -4$
30. $-3x + y = -5$

31. $3x - 4y = 7$
32. $4x - 3y = 10$
33. $y + 4x = 0$
34. $y - 5x = 0$
35. $x = 2y$
36. $x = -3y$

THOUGHTS INTO WORDS

37. Your friend is having trouble understanding why the graph of the equation $y = 3$ is a horizontal line that contains the point $(0, 3)$. What might you do to help him?

38. How do we know that the graph of $y = -4x$ is a straight line that contains the origin?

39. Do all graphs of linear equations have x-intercepts? Explain your answer.

40. How do we know that the graphs of $x - y = 4$ and $-x + y = -4$ are the same line?

Further Investigations

From our previous work with absolute value we know that $|x + y| = 2$ is equivalent to $x + y = 2$ or $x + y = -2$. Therefore, the graph of $|x + y| = 2$ consists of the two lines $x + y = 2$ and $x + y = -2$. Graph each of the following.

41. $|x + y| = 1$
42. $|x - y| = 2$
43. $|2x + y| = 4$
44. $|3x - y| = 6$

45. Graph $y = \frac{2}{3}x + 2$, $y = \frac{2}{3}x - 4$, and $y = \frac{2}{3}x + 5$ on the same set of axes. What observation do you make about these lines?

46. Graph $y = -2x + 5$, $y = -2x - 4$, $y = -2x + 1$, and $y = -2x - 2$ on the same set of axes. What observation do you make about these lines?

8.3   Slope of a Line

In Figure 8.12 note that the line associated with $4x - y = 4$ is *steeper* than the line associated with $2x - 3y = 6$. Mathematically, the concept of **slope** is used to discuss the *steepness* of lines. The slope of a line is the ratio of the vertical

FIGURE 8.12

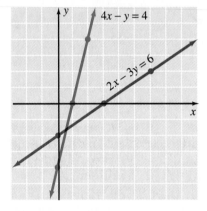

change compared to the horizontal change as we move from one point on a line to another point. We indicate this in Figure 8.13 using the points P_1 and P_2.

FIGURE 8.13

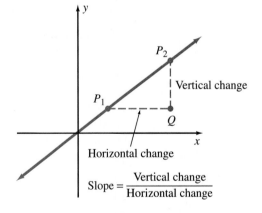

A precise definition for slope can be given by considering the coordinates of the points P_1, P_2, and Q in Figure 8.14. Since P_1 and P_2 represent any two points on the line, we will assign the coordinates (x_1, y_1) to P_1 and (x_2, y_2) to P_2. The point Q is the same distance from the y-axis as P_2 and the same distance from the x-axis as P_1. Thus, we assign the coordinates (x_2, y_1) to Q (Figure 8.14). It should now be apparent that the vertical change is $y_2 - y_1$ and the horizontal change is $x_2 - x_1$. Thus, the following definition for slope is given.

FIGURE 8.14

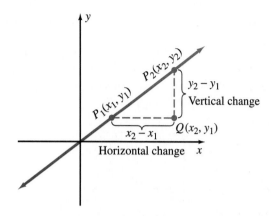

DEFINITION 8.1

If points P_1 and P_2 with coordinates (x_1, y_1) and (x_2, y_2) respectively, are any two different points on a line, then the slope of the line (denoted by m) is

$$m = \frac{y_2 - y_1}{x_2 - x_1}, \qquad x_1 \neq x_2.$$

Using Definition 8.1, you can easily determine the slope of a line if the coordinates of two points on the line are known.

Example 1

Find the slope of the line determined by each of the following pairs of points.

(**a**) $(2, 1)$ and $(4, 6)$ (**b**) $(3, 2)$ and $(-4, 5)$
(**c**) $(-4, -3)$ and $(-1, -3)$

Solution

(**a**) Let $(2, 1)$ be P_1 and $(4, 6)$ be P_2 (Figure 8.15).

$$m = \frac{y_2 - y_1}{x_2 - x_1} = \frac{6 - 1}{4 - 2} = \frac{5}{2}$$

FIGURE 8.15

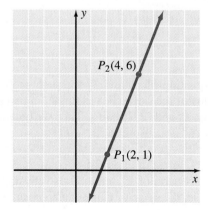

(b) Let $(3, 2)$ be P_1 and $(-4, 5)$ be P_2 (Figure 8.16).

$$m = \frac{y_2 - y_1}{x_2 - x_1} = \frac{5 - 2}{-4 - 3} = \frac{3}{-7} = -\frac{3}{7}$$

FIGURE 8.16

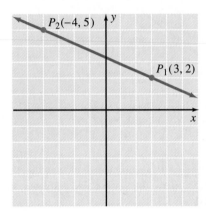

(c) Let $(-4, -3)$ be P_1 and $(-1, -3)$ be P_2 (Figure 8.17).

$$m = \frac{y_2 - y_1}{x_2 - x_1} = \frac{-3 - (-3)}{-1 - (-4)} = \frac{0}{3} = 0$$

FIGURE 8.17

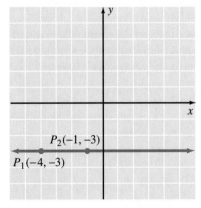

The designation of P_1 and P_2 for such problems is arbitrary and will not affect the value of the slope. For example, in part (a) of Example 1 suppose that we let $(4, 6)$ be P_1 and $(2, 1)$ be P_2. Then we obtain

$$m = \frac{y_2 - y_1}{x_2 - x_1} = \frac{1 - 6}{2 - 4} = \frac{-5}{-2} = \frac{5}{2}.$$

The various parts of Example 1 also illustrate the three basic possibilities for slope; that is, the slope of a line can be *positive, negative,* or *zero*. A line that has a positive slope rises as we move from left to right as in part (a). A line that has a negative slope falls as we move from left to right, as in part (b). A horizontal line, as in part (c), has a slope of 0. Finally, we need to realize that **the concept of slope is undefined for vertical lines.** This is due to the fact that for any vertical line the change in x as we move from one point to another is zero. Thus, the ratio $\dfrac{y_2 - y_1}{x_2 - x_1}$ will have a denominator of zero and be undefined. So in Definition 8.1 the restriction $x_1 \neq x_2$ is made.

Example 2

Find the slope of the line determined by the equation $3x + 4y = 12$.

Solution

Since any two points on the line can be used to determine the slope of the line, let's find the intercepts.

$$\text{If } x = 0, \text{ then } 3(0) + 4y = 12$$
$$4y = 12$$
$$y = 3.$$

$$\text{If } y = 0, \text{ then } 3x + 4(0) = 12$$
$$3x = 12$$
$$x = 4.$$

Using $(0, 3)$ as P_1 and $(4, 0)$ as P_2, we obtain

$$m = \frac{y_2 - y_1}{x_2 - x_1} = \frac{0 - 3}{4 - 0} = \frac{-3}{4} = -\frac{3}{4}.$$

We need to emphasize one final idea pertaining to the concept of slope. The slope of a line is a **ratio** of vertical change compared to horizontal change. A slope of $\dfrac{3}{4}$ means that for every 3 units of vertical change there must be a corresponding 4 units of horizontal change. So starting at some point on the line, we could move to other points on the line as follows.

$\dfrac{3}{4} = \dfrac{6}{8}$:   by moving 6 units *up* and 8 units to the *right*;

$\dfrac{3}{4} = \dfrac{15}{20}$:   by moving 15 units *up* and 20 units to the *right*;

$\dfrac{3}{4} = \dfrac{\frac{3}{2}}{2}$:   by moving $1\dfrac{1}{2}$ units *up* and 2 units to the *right*;

$\dfrac{3}{4} = \dfrac{-3}{-4}$:   by moving 3 units *down* and 4 units to the *left*.

Likewise, a slope of $-\dfrac{5}{6}$ indicates that starting at some point on the line we could move to other points on the line as follows.

$$-\frac{5}{6} = \frac{-5}{6}: \qquad \text{by moving 5 units } down \text{ and 6 units to the } right;$$

$$-\frac{5}{6} = \frac{5}{-6}: \qquad \text{by moving 5 units } up \text{ and 6 units to the } left;$$

$$-\frac{5}{6} = \frac{-10}{12}: \qquad \text{by moving 10 units } down \text{ and 12 units to the } right;$$

$$-\frac{5}{6} = \frac{15}{-18}: \qquad \text{by moving 15 units } up \text{ and 18 units to the } left.$$

Applications of Slope

The concept of slope has many real-world applications even though the word "slope" is often not used. For example, in Figure 8.18 the highway is said to have a *grade* of 17%. This means that for every horizontal distance of 100 feet, the highway rises or drops 17 feet. In other words, the slope of the highway is $\frac{17}{100}$.

FIGURE 8.18

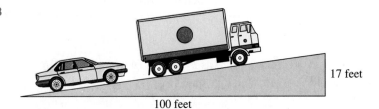

17 feet

100 feet

Example 3

A certain highway has a 3% grade. How many feet does it rise in a horizontal distance of one mile?

Solution

A 3% grade means a slope of $\frac{3}{100}$. Therefore, if we let y represent the unknown vertical distance and use the fact that 1 mile = 5280 feet, we can set up and solve the following proportion.

$$\frac{3}{100} = \frac{y}{5280}$$
$$100y = 3(5280) = 15{,}840$$
$$y = 158.4$$

The highway rises 158.4 feet in a horizontal distance of one mile. ▲

A roofer, when making an estimate to replace a roof, is not only concerned about the total area to be covered but also the *pitch* of the roof. In Figure 8.19, the two roofs might require the same amount of shingles, but the roof on the left will take longer to complete because the pitch is so great that scaffolding will be required.

FIGURE 8.19

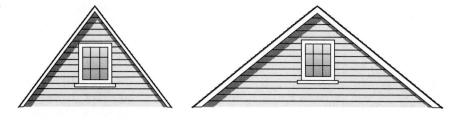

The concept of slope is also used in the construction of flights of stairs (Figure 8.20). The terms *rise* and *run* are commonly used and the steepness (slope) of the stairs can be expressed as the ratio of *rise to run*. In Figure 8.20, the stairs on the left with the ratio of $\frac{10}{11}$ are steeper than the stairs on the right that have a ratio of $\frac{7}{11}$.

FIGURE 8.20

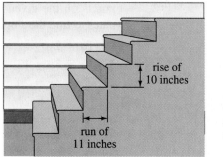
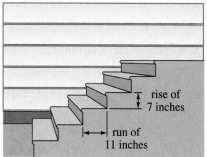

Technically, the concept of slope is involved in most situations where the idea of an incline is used. Hospital beds are constructed so that both the head-end and foot-end can be raised or lowered; that is, the slope of either end of the bed can be changed. Likewise, treadmills are designed so that the incline (slope) of the platform can be raised or lowered as desired. Perhaps you can think of several other applications of the concept of slope.

Problem Set 8.3

For Problems 1–20, find the slope of the line determined by each pair of points.

1. $(7, 5), (3, 2)$

2. $(9, 10), (6, 2)$

3. $(-1, 3), (-6, -4)$

4. $(-2, 5), (-7, -1)$

5. $(2, 8), (7, 2)$

6. $(3, 9), (8, 4)$

7. $(-2, 5), (1, -5)$

8. $(-3, 4), (2, -6)$

9. $(4, -1), (-4, -7)$

10. $(5, -3), (-5 \ -9)$

11. $(3, -4), (2, -4)$

12. $(-3, -6), (5, -6)$

13. $(-6, -1), (-2, -7)$

14. $(-8, -3), (-2, -11)$

15. $(-2, 4), (-2, -6)$

16. $(-4, -5), (-4, 9)$

17. $(-1, 10), (-9, 2)$

18. $(-2, 12), (-10, 2)$

19. $(a, b), (c, d)$

20. $(a, 0), (0, b)$

21. Find y if the line through $(7, 8)$ and $(2, y)$ has a slope of $\frac{4}{5}$.

22. Find y if the line through $(12, 14)$ and $(3, y)$ has a slope of $\frac{4}{3}$.

23. Find x if the line through $(-2, -4)$ and $(x, 2)$ has a slope of $-\frac{3}{2}$.

24. Find x if the line through $(6, -4)$ and $(x, 6)$ has a slope of $-\frac{5}{4}$.

For Problems 25–32, you are given one point on a line and the slope of the line. Find the coordinates of three other points on the line.

25. $(3, 2), m = \frac{2}{3}$

26. $(4, 1), m = \frac{5}{6}$

27. $(-2, -4), m = \frac{1}{2}$

28. $(-6, -2), m = \frac{2}{5}$

29. $(-3, 4), m = -\frac{3}{4}$

30. $(-2, 6), m = -\frac{3}{7}$

31. $(4, -5), m = -2$

32. $(6, -2), m = 4$

For Problems 33–40, sketch the line determined by each pair of points and decide whether the slope of the line is *positive*, *negative*, or *zero*.

33. $(2, 8), (7, 1)$

34. $(1, -2), (7, -8)$

35. $(-1, 3), (-6, -2)$

36. $(7, 3), (4, -6)$

37. $(-2, 4), (6, 4)$

38. $(-3, -4), (5, -4)$

39. $(-3, 5), (2, -7)$

40. $(-1, -1), (1, -9)$

For Problems 41–60, find the coordinates of two points on the given line and then use those coordinates to find the slope of the line.

41. $3x + 2y = 6$

42. $4x + 3y = 12$

43. $5x - 4y = 20$

44. $7x - 3y = 21$

45. $x + 5y = 6$

46. $2x + y = 4$

47. $2x - y = -7$

48. $x - 4y = -6$

49. $y = 3$

50. $x = 6$

51. $-2x + 5y = 9$

52. $-3x - 7y = 10$

53. $6x - 5y = -30$

54. $7x - 6y = -42$

55. $y = -3x - 1$

56. $y = -2x + 5$

57. $y = 4x$

58. $y = 6x$

59. $y = \frac{2}{3}x - \frac{1}{2} =$

60. $y = -\frac{3}{4}x + \frac{1}{5}$

61. Suppose that a highway rises a distance of 135 feet in a horizontal distance of 2640 feet. Express the grade of the highway to the nearest tenth of a percent.

62. The grade of a highway up a hill is 27%. How much change in horizontal distance is there if the vertical height of the hill is 550 feet? Express the answer to the nearest foot.

63. If the ratio of rise to run is to be $\frac{3}{5}$ for some stairs and the measure of the rise is 19 centimeters, find the measure of the run to the nearest centimeter.

64. If the ratio of rise to run is to be $\frac{2}{3}$ for some stairs and the measure of the run is 28 centimeters, find the measure of the rise to the nearest centimeter.

65. A county ordinance requires a $2\frac{1}{4}\%$ "fall" for a sewage pipe from the house to the main pipe at the street. How much vertical drop must there be for a horizontal distance of 45 feet? Express the answer to the nearest tenth of a foot.

THOUGHTS INTO WORDS

66. How would you explain the concept of slope to someone who was absent from class the day it was discussed?

67. If one line has a slope of $\frac{2}{3}$ and another line has a slope of 2, which line is steeper? Explain your answer.

68. Why do we say that the slope of a vertical line is undefined?

69. Suppose that a line has a slope of $\frac{3}{4}$ and contains the point (5, 2). Are the points $(-3, -4)$ and (14, 9) also on the line? Explain your answer.

8.4 Writing Equations of Lines

As we stated in Section 8.1, basically there are two types of problems in analytic or coordinate geometry.

1. Given an algebraic equation, find its geometric graph.

2. Given a set of conditions pertaining to a geometric figure, determine its algebraic equation.

Problems of type 1 were a part of the first two sections of this chapter. At this time we want to consider a few problems of type 2 that deal with straight lines. In other words, given certain facts about a line, we need to be able to write its algebraic equation.

Example 1

Find the equation of the line that has a slope of $\frac{3}{4}$ and contains the point (1, 2).

Solution

First, let's draw the line as indicated in Figure 8.21. Since the slope is $\frac{3}{4}$, a second point can be determined by moving 3 units up and 4 units to the right of the given point (1, 2). (The point (5, 5) merely helps to draw the line; it will not be used in analyzing the problem.) Now let's choose a point (x, y) that represents all points on the line other than the given point $(1, 2)$. The slope determined by $(1, 2)$ and (x, y) is $\dfrac{3}{4}$. Thus,

$$\frac{y - 2}{x - 1} = \frac{3}{4}$$
$$3(x - 1) = 4(y - 2)$$
$$3x - 3 = 4y - 8$$
$$3x - 4y = -5.$$

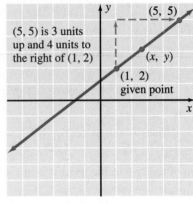

$(5, 5)$ is 3 units up and 4 units to the right of $(1, 2)$

(x, y)

$(1, 2)$ given point

FIGURE 8.21   ▲

Example 2

Find the equation of the line that contains $(3, 4)$ and $(-2, 5)$.

Solution

First, let's draw the line determined by the two given points (Figure 8.22).

FIGURE 8.22

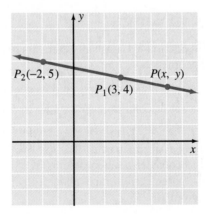

$P_2(-2, 5)$

$P(x, y)$

$P_1(3, 4)$

Since we know two points, we can find the slope.

$$m = \frac{y_2 - y_1}{x_2 - x_1} = \frac{5 - 4}{-2 - 3} = \frac{1}{-5} = -\frac{1}{5}$$

Now we can use the same approach as in Example 1. Form an equation using a variable point (x, y), one of the two given points—we choose P_1, and the slope of $-\dfrac{1}{5}$.

$$\frac{y - 4}{x - 3} = \frac{1}{-5} \qquad -\frac{1}{5} = \frac{1}{-5}$$
$$x - 3 = -5y + 20$$
$$x + 5y = 23$$

▲

Example 3

Find the equation of the line that has a slope of $\frac{1}{4}$ and a y-intercept of 2.

Solution

A y-intercept of 2 means that the point $(0, 2)$ is on the line. Since the slope is $\frac{1}{4}$, we can find another point by moving 1 unit up and 4 units to the right of $(0, 2)$. The line is drawn in Figure 8.23. We choose variable point (x, y) and proceed as in the previous examples.

$$\frac{y - 2}{x - 0} = \frac{1}{4}$$
$$x = 4y - 8$$
$$x - 4y = -8$$

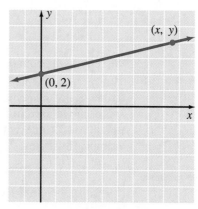

FIGURE 8.23

▲

Perhaps it would be helpful for you to pause for a moment and look back over Examples 1, 2, and 3. Notice that we used the same basic approach in all three examples; that is, we chose a variable point (x, y) and used it, along with another known point, to determine the equation of the line. You should also recognize that the approach we take in these examples can be generalized to produce some special forms of equations for straight lines.

Point-Slope Form

Example 4

Find the equation of the line that has a slope of m and contains the point (x_1, y_1).

Solution

Choose (x, y) to represent any other point on the line (Figure 8.24) and the slope of the line given by

$$m = \frac{y - y_1}{x - x_1}$$

enables us to obtain

$$y - y_1 = m(x - x_1).$$

FIGURE 8.24

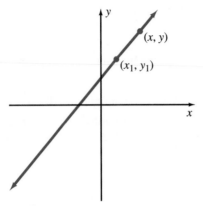

We refer to the equation $\boxed{y - y_1 = m(x - x_1)}$

as the **point-slope form** of the equation of a straight line. Instead of the approach we used in Example 1, we could use the point-slope form to write the equation of a line with a given slope that contains a given point, as the next example illustrates.

Example 5

Write the equation of the line that has a slope of $\dfrac{3}{5}$ and contains the point $(2, -4)$.

Solution

Substituting $\dfrac{3}{5}$ for m, 2 for x_1, and -4 for y_1 in the point-slope form, we obtain

$$y - y_1 = m(x - x_1)$$

$$y - (-4) = \frac{3}{5}(x - 2)$$

$$y + 4 = \frac{3}{5}(x - 2)$$

$$5(y + 4) = 3(x - 2)$$

$$5y + 20 = 3x - 6$$

$$26 = 3x - 5y.$$

Slope-Intercept form

Now consider the equation of a line that has a slope of m and a y-intercept of b (Figure 8.25).

FIGURE 8.25

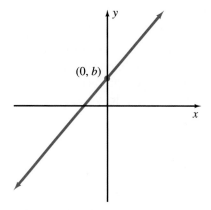

A y-intercept of b means that the line contains the point $(0, b)$; therefore, we can use the point-slope form as follows.

$$y - y_1 = m(x - x_1)$$
$$y - b = m(x - 0) \qquad y_1 = b \text{ and } x_1 = 0$$
$$y - b = mx$$
$$y = mx + b$$

The equation $\boxed{y = mx + b}$

is called the **slope-intercept form** of the equation of a straight line. We use it for two primary purposes, as the next two examples illustrate.

Example 6

Find the equation of the line that has a slope of $\dfrac{1}{4}$ and a y-intercept of 2.

Solution

This is a restatement of Example 3, but this time we will use the slope-intercept form of a line $(y = mx + b)$ to write its equation. From the statement of the problem we know that $m = \dfrac{1}{4}$ and $b = 2$. Thus, substituting these values for m and b in $y = mx + b$, we obtain

$$y = mx + b$$
$$y = \frac{1}{4}x + 2$$
$$4y = x + 8$$
$$x - 4y = -8. \qquad \text{Same result as in Example 3} \qquad \blacktriangle$$

REMARK It is acceptable to occasionally leave answers in slope-intercept form. We did not do that in Example 6 because we wanted to show that it was the same result as in Example 3. $\triangle$

Example 7

Find the slope of the line whose equation is $2x + 3y = 4$.

Solution

In Section 8.3 we solved this type of problem when we determined the coordinates of two points on the line and then used the slope formula. Now we have another approach for such a problem. Let's change the given equation to slope-intercept form; that is to say, let's solve the equation for y in terms of x.

$$2x + 3y = 4$$
$$3y = -2x + 4$$
$$y = -\frac{2}{3} + \frac{4}{3}$$

Compare this result to $y = mx + b$, and you see that $m = -\frac{2}{3}$ and $b = \frac{4}{3}$. ▲

In general, **if the equation of a nonvertical line is written in slope-intercept form, the coefficient of x is the slope of the line and the constant term is the y-intercept.** (Remember that the concept of slope is not defined for a vertical line.) Let's use that concept in a few more examples.

Example 8

Find the slope and y-intercept of each of the following lines.
(*a*) $5x - 4y = 12$ (*b*) $-y = 3x - 4$ (*c*) $y = 2$

Solution

(*a*) Change $5x - 4y = 12$ to slope-intercept form to obtain

$$5x - 4y = 12$$
$$-4y = -5x + 12$$
$$4y = 5x - 12$$
$$y = \frac{5}{4}x - 3.$$

The slope of the line is $\frac{5}{4}$ (the coefficient of x) and the y-intercept is -3 (the constant term).
(*b*) Multiply both sides of the given equation by -1 to change it to slope-intercept form.

$$-y = 3x - 4$$
$$y = -3x + 4.$$

The slope of the line is -3 and the y-intercept is 4.
(*c*) The equation $y = 2$ can be written as

$$y = 0(x) + 2.$$

The slope of the line is 0 and the y-intercept is 2. (A line with a slope of 0 is horizontal.) ▲

Problem Set 8.4

For Problems 1–12, find the equation of the line that contains the given point and has the given slope. Express equations in the form $Ax + By = C$, where A, B, and C are integers.

1. $(2, 3)$, $m = \dfrac{2}{3}$

2. $(5, 2)$, $m = \dfrac{3}{7}$

3. $(-3, -5)$, $m = \dfrac{1}{2}$

4. $(5, -6)$, $m = \dfrac{3}{5}$

5. $(-4, 8)$, $m = -\dfrac{1}{3}$

6. $(-2, -4)$, $m = -\dfrac{5}{6}$

7. $(3, -7)$, $m = 0$

8. $(-3, 9)$, $m = 0$

9. $(0, 0)$, $m = -\dfrac{4}{9}$

10. $(0, 0)$, $m = \dfrac{5}{11}$

11. $(-6, -2)$, $m = 3$

12. $(2, -10)$, $m = -2$

For Problems 13–22, find the equation of the line that contains the two given points. Express equations in the form $Ax + By = C$, where A, B, and C are integers.

13. $(2, 3)$ and $(7, 10)$

14. $(1, 4)$ and $(9, 10)$

15. $(3, -2)$ and $(-1, 4)$

16. $(-2, 8)$ and $(4, -2)$

17. $(-1, -2)$ and $(-6, -7)$

18. $(-8, -7)$ and $(-3, -1)$

19. $(0, 0)$ and $(-3, -5)$

20. $(5, -8)$ and $(0, 0)$

21. $(0, 4)$ and $(7, 0)$

22. $(-2, 0)$ and $(0, -9)$

For Problems 23–32, find the equation of the line with the given slope and y-intercept. Leave your answers in slope-intercept form.

23. $m = \dfrac{3}{5}$ and $b = 2$

24. $m = \dfrac{5}{9}$ and $b = 4$

25. $m = 2$ and $b = -1$

26. $m = 4$ and $b = -3$

27. $m = -\dfrac{1}{6}$ and $b = -4$

28. $m = -\dfrac{5}{7}$ and $b = -1$

29. $m = -1$ and $b = \dfrac{5}{2}$

30. $m = -2$ and $b = \dfrac{7}{3}$

31. $m = -\dfrac{5}{9}$ and $b = -\dfrac{1}{2}$

32. $m = -\dfrac{7}{12}$ and $b = -\dfrac{2}{3}$

For Problems 33–44, determine the slope and y-intercept of the line represented by the given equation.

33. $y = -2x - 5$

34. $y = \dfrac{2}{3}x + 4$

35. $3x - 5y = 15$

36. $7x + 5y = 35$

37. $-4x + 9y = 18$

38. $-6x + 7y = -14$

39. $-y = -\dfrac{3}{4}x + 4$

40. $5x - 2y = 0$

41. $-2x - 11y = 11$

42. $-y = \dfrac{2}{3}x + \dfrac{11}{2}$

43. $9x + 7y = 0$

44. $-5x - 13y = 26$

THOUGHTS INTO WORDS

45. Explain the importance of the slope-intercept form ($y = mx + b$) of the equation of a line.

46. What does it mean to say that two points *determine* a line?

47. How would you describe coordinate geometry to a group of elementary algebra students?

48. How can you tell by inspection that $y = 2x - 4$ and $y = -3x - 1$ are not parallel lines?

Further Investigations

The following two properties pertain to parallel and perpendicular lines.

(a) Two nonvertical lines are parallel if and only if their slopes are equal.

(b) Two nonvertical lines are perpendicular if and only if the product of their slopes is -1.

For Problems 49–54, determine whether each pair of lines is (a) parallel, (b) perpendicular, or (c) intersecting lines that are not perpendicular.

49. $\begin{pmatrix} 5x - 2y = 6 \\ 2x + 5y = 9 \end{pmatrix}$

50. $\begin{pmatrix} 2x - y = 9 \\ 6x - 3y = 5 \end{pmatrix}$

51. $\begin{pmatrix} 4x - 3y = 12 \\ 3x - 4y = 12 \end{pmatrix}$

52. $\begin{pmatrix} 9x + 2y = 18 \\ 2x - 9y = 13 \end{pmatrix}$

53. $\begin{pmatrix} x - 3y = 7 \\ 5x - 15y = 9 \end{pmatrix}$

54. $\begin{pmatrix} 7x - 2y = 14 \\ 7x + 2y = 5 \end{pmatrix}$

55. Write the equation of the line that contains the point (4, 3) and is parallel to the line $2x - 3y = 6$.

56. Write the equation of the line that contains the point (−1, 3) and is perpendicular to the line $3x - y = 4$.

8.5   Solving Linear Systems by Graphing

Suppose that we graph $x - 2y = 4$ and $x + 2y = 8$ on the same set of axes, as indicated in Figure 8.26. The ordered pair, (6, 1), which is associated with the point of intersection of the two lines, satisfies both equations. That is to say, (6, 1) is the solution for $x - 2y = 4$ and $x + 2y = 8$. To check this, we can substitute 6 for x and 1 for y in both equations.

$$x - 2y = 4 \quad \text{becomes } 6 - 2(1) = 4,$$
$$x + 2y = 8 \quad \text{becomes } 6 + 2(1) = 8$$

FIGURE 8.26

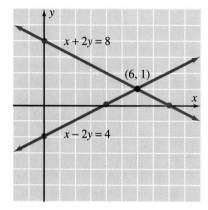

Thus, we say that $\{(6, 1)\}$ is the solution set of the system

$$\begin{pmatrix} x - 2y = 4 \\ x + 2y = 8 \end{pmatrix}.$$

Two or more linear equations in two variables considered together are called a **system of linear equations**. The following are systems of linear equations.

$$\begin{pmatrix} x - 2y = 4 \\ x + 2y = 8 \end{pmatrix}, \quad \begin{pmatrix} 5x - 3y = 9 \\ 3x + 7y = 12 \end{pmatrix}, \quad \begin{pmatrix} 4x - y = 5 \\ 2x + y = 9 \\ 7x - 2y = 13 \end{pmatrix}$$

To **solve a system of linear equations** means to find all of the ordered pairs that are solutions of all of the equations in the system. In this book we will consider only systems of *two* linear equations in two variables. There are several techniques for solving systems of linear equations. We will use three of them in this chapter—a graphing method in this section, and two other methods in the following sections.

To solve a system of linear equations by **graphing**, you proceed as in the opening discussion of this section. We graph the equations on the same set of axes and then the ordered pairs associated with any points of intersection are the solutions to the system. Let's consider another example.

Example I

Solve the system $\begin{pmatrix} x + y = 5 \\ x - 2y = -4 \end{pmatrix}.$

Solution

Let's find the intercepts and a check point for each of the lines.

$x + y = 5$		
x	y	
0	5	Intercepts
5	0	
2	3	Check point

$x - 2y = -4$		
x	y	
0	2	Intercepts
-4	0	
-2	1	Check point

Figure 8.27 shows the graphs of the two equations.

FIGURE 8.27

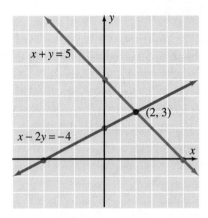

It appears that (2, 3) is the solution of the system. To check it we can substitute 2 for x and 3 for y in both equations.

$$x + y \quad \text{becomes } 2 + 3 = 5, \qquad \text{A true statement}$$
$$x - 2y = -4 \quad \text{becomes } 2 - 2(3) = -4 \qquad \text{A true statement}$$

Therefore, $\{(2, 3)\}$ is the solution set.   ▲

It should be evident that solving systems of equations by graphing requires accurate graphs. In fact, unless the solutions are integers it is really quite difficult to obtain exact solutions from a graph. For this reason the systems in this section have integral solutions. Furthermore, checking a solution takes on additional significance when using the graphing approach. By checking you can be absolutely sure that you are *reading* the correct solution from the graph.

Figure 8.28 shows the three possible cases for the graph of a system of two linear equations in two variables.

CASE I   The graphs of the two equations are two lines intersecting in one point. There is *one solution* and we call the system a **consistent system.**

CASE II   The graphs of the two equations are parallel lines. There is *no solution* and we call the system an **inconsistent system**.

CASE III   The graphs of the two equations are the same line. There *are infinitely many* solutions to the system. Any pair of real numbers that satisfies one of the equations will also satisfy the other equation, and we say the equations are **dependent**.

Case I
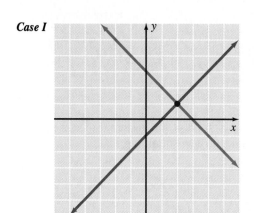

Case II
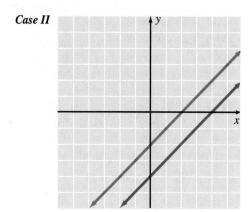

Case III
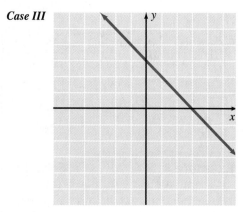

FIGURE 8.28

Thus, as we solve a system of two linear equations in two variables we know what to expect. The system will have no solutions, one ordered pair as a solution, or infinitely many ordered pairs as solutions. Most of the systems that we will be working with in this text will have one solution.

An example of Case I was given in Example 1 (Figure 8.27). The next two examples illustrate the other cases.

Example 2   Solve the system $\left(\begin{array}{l} 2x + 3y = 6 \\ 2x + 3y = 12 \end{array} \right)$.

Solution

2x + 3y = 6	
x	**y**
0	2
3	0
−3	4

2x + 3y = 12	
x	**y**
0	4
6	0
3	2

Figure 8.29 shows the graph of the system. Since the lines are parallel, there is *no solution* to the system. The solution set is ∅.

FIGURE 8.29

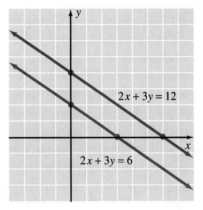

Example 3   Solve the system $\left(\begin{array}{l} x + y = 3 \\ 2x + 2y = 6 \end{array} \right)$.

Solution

x + y = 3	
x	**y**
0	3
3	0
1	2

2x + 2y = 6	
x	**y**
0	3
3	0
1	2

Figure 8.30 shows the graph of this system. Since the graph of both equations is the same line, there are *infinitely many solutions* to the system. Any ordered pair of real numbers that satisfies one equation will also satisfy the other equation.

FIGURE 8.30

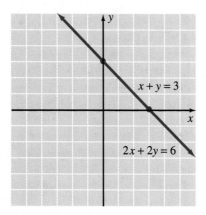

$x + y = 3$

$2x + 2y = 6$

Problem Set 8.5

For Problems 1–10, decide whether the given ordered pair is a solution of the given system of equations.

1. $\begin{pmatrix} 5x + y = 9 \\ 3x - 2y = 4 \end{pmatrix} (1, 4)$

2. $\begin{pmatrix} 3x - 2y = 8 \\ 2x + y = 3 \end{pmatrix} (2, -1)$

3. $\begin{pmatrix} x - 3y = 17 \\ 2x + 5y = -21 \end{pmatrix} (2, -5)$

4. $\begin{pmatrix} 2x + 7y = -5 \\ 4x - y = 6 \end{pmatrix} (1, -1)$

5. $\begin{pmatrix} y = 2x \\ 3x - 4y = 5 \end{pmatrix} (-1, -2)$

6. $\begin{pmatrix} -x + 2y = 8 \\ x - 2y = 8 \end{pmatrix} (-2, 3)$

7. $\begin{pmatrix} 6x - 5y = 5 \\ 3x + 4y = -4 \end{pmatrix} (0, -1)$

8. $\begin{pmatrix} y = 3x - 1 \\ 5x - 2y = -1 \end{pmatrix} (3, 8)$

9. $\begin{pmatrix} -3x - y = 4 \\ -2x + 3y = -23 \end{pmatrix} (4, -5)$

10. $\begin{pmatrix} 5x + 4y = -15 \\ 2x - 7y = -6 \end{pmatrix} (-3, 0)$

For Problems 11–30, use the graphing method to solve each system.

11. $\begin{pmatrix} x + y = 1 \\ x - y = 3 \end{pmatrix}$

12. $\begin{pmatrix} x - y = 2 \\ x + y = -4 \end{pmatrix}$

13. $\begin{pmatrix} x + 2y = 4 \\ 2x - y = 3 \end{pmatrix}$

14. $\begin{pmatrix} 2x - y = -8 \\ x + y = 2 \end{pmatrix}$

15. $\begin{pmatrix} x + 3y = 6 \\ x + 3y = 3 \end{pmatrix}$

16. $\begin{pmatrix} y = -2x \\ y - 3x = 0 \end{pmatrix}$

17. $\begin{pmatrix} x + y = 0 \\ x - y = 0 \end{pmatrix}$

18. $\begin{pmatrix} 3x - y = 3 \\ 3x - y = -3 \end{pmatrix}$

19. $\begin{pmatrix} 3x - 2y = 5 \\ 2x + 5y = -3 \end{pmatrix}$

20. $\begin{pmatrix} 2x + 3y = 1 \\ 4x - 3y = -7 \end{pmatrix}$

21. $\begin{pmatrix} y = -2x + 3 \\ 6x + 3y = 9 \end{pmatrix}$

22. $\begin{pmatrix} y = 2x + 5 \\ x + 3y = -6 \end{pmatrix}$

23. $\begin{pmatrix} y = 5x - 2 \\ 4x + 3y = 13 \end{pmatrix}$

24. $\begin{pmatrix} y = x - 2 \\ 2x - 2y = 4 \end{pmatrix}$

25. $\begin{pmatrix} y = 4 - 2x \\ y = 7 - 3x \end{pmatrix}$

26. $\begin{pmatrix} y = 3x + 4 \\ y = 5x + 8 \end{pmatrix}$

27. $\begin{pmatrix} y = 2x \\ 3x - 2y = -2 \end{pmatrix}$

28. $\begin{pmatrix} y = 3x \\ 4x - 3y = 5 \end{pmatrix}$

29. $\begin{pmatrix} 7x - 2y = -8 \\ x = -2 \end{pmatrix}$

30. $\begin{pmatrix} 3x + 8y = -1 \\ y = -2 \end{pmatrix}$

THOUGHTS INTO WORDS

31. Discuss the strengths and weaknesses of solving a system of linear equations by graphing.

32. Determine a system of two linear equations for which the solution is (5, 7). Are there other systems with the same solution set? If so, find at least one more system.

33. Is it possible for a system of two linear equations to have exactly two solutions? Defend your answer.

8.6 Elimination-by-Addition Method

We have used the addition property of equality (if $a = b$, then $a + c = b + c$) to help solve equations containing one variable. An extension of the addition property forms the basis for another method of solving systems of linear equations. Property 8.1 states that two equations can be added and the resulting equation will be equivalent to the original ones. Let's use this property to help solve a system of equations.

PROPERTY 8.1

For all real numbers a, b, c, and d, if $a = b$, and $c = d$, then

$$a + c = b + d.$$

Example 1

Solve $\begin{pmatrix} x + y = 12 \\ x - y = 2 \end{pmatrix}$.

Solution

$$\begin{array}{r} x + y = 12 \\ x - y = 2 \\ \hline 2x = 14 \end{array}$$ This is the result of adding the two equations.

Solving this new equation in one variable we obtain

$$2x = 14$$
$$\boxed{x = 7.}$$

Now we can substitute the value of 7 for x in one of the original equations. Thus,

$$x + y = 12$$
$$7 + y = 12$$
$$\boxed{y = 5.}$$

To check, we see that

$$\begin{array}{ccc} x + y = 12 & \text{and} & x - y = 2, \\ 7 + 5 = 12 & \text{and} & 7 - 5 = 2. \end{array}$$

Thus, the solution set of the system is $\{(7, 5)\}$. ▲

Note in Example 1 that by adding the two original equations, a simple equation that contains only one variable was obtained. Adding equations to eliminate a variable is the key idea behind the **elimination-by-addition method** for solving systems of linear equations. The next example further illustrates this point.

Example 2

Solve $\begin{pmatrix} 2x + 3y = -26 \\ 4x - 3y = 2 \end{pmatrix}$.

Solution

$$\begin{array}{r} 2x + 3y = -26 \\ 4x - 3y = 2 \\ \hline 6x = -24 \end{array}$$ Added the two equations

$$\boxed{x = -4}$$

Substitute -4 for x in one of the two original equations.

$$2x + 3y = -26$$
$$2(-4) + 3y = -26$$
$$-8 + 3y = -26$$
$$3y = -18$$
$$\boxed{y = -6}$$

The solution set is $\{(-4, -6)\}$. ▲

It may be necessary to change the form of one, or perhaps both, of the original equations before adding them. The next example demonstrates this idea.

Example 3

Solve $\begin{pmatrix} y = x - 21 \\ x + y = -3 \end{pmatrix}$.

Solution

$$y = x - 21 \quad \underrightarrow{\text{Subtract } x \text{ from both sides}} \quad -x + y = -21$$
$$x + y = -3 \quad \underrightarrow{\text{Leave alone}} \quad \underline{\quad x + y = \ -3}$$
$$2y = -24$$
$$\boxed{y = -12}$$

Substitute -12 for y in one of the original equations.

$$x + y = -3$$
$$x + (-12) = -3$$
$$\boxed{x = 9}$$

The solution set is $\{(9, -12)\}$.   ▲

Frequently, the multiplication property of equality needs to be applied first so that adding the equations will eliminate a variable, as the next example illustrates.

Example 4

Solve $\begin{pmatrix} 2x + 5y = 29 \\ 3x - y = 1 \end{pmatrix}$.

Solution

Notice that adding the equations as they are would not eliminate a variable. However, we should observe that multiplying the bottom equation by 5 and then adding this newly formed, but equivalent, equation to the top equation will eliminate the y's.

$$2x + 5y = 29 \quad \underrightarrow{\text{Leave alone}} \quad 2x + 5y = 29$$
$$3x - y = 1 \quad \underrightarrow{\text{Multiply both sides by 5}} \quad \underline{15x - 5y = \ 5}$$
$$17x \qquad = 34$$
$$\boxed{x = 2}$$

Substitute 2 for x in one of the original equations.

$$3x - y = 1$$
$$3(2) - y = 1$$
$$6 - y = 1$$
$$-y = -5$$
$$\boxed{y = 5}$$

The solution set is $\{(2, 5)\}$.   ▲

Notice in these problems that after finding the value of one of the variables, we substitute this number into one of the *original* equations to find the value of the other variable. It doesn't matter which of the two original equations you use, so pick the easiest one to solve.

Sometimes the multiplication property needs to be applied to both equations. Let's look at an example of this type.

Example 5

Solve $\begin{pmatrix} 2x + 3y = & 4 \\ 9x - 2y = & -13 \end{pmatrix}$.

Solution A

$$2x + 3y = 4 \quad \xrightarrow{\text{Multiply both sides by 9}} \quad 18x + 27y = 36$$
$$9x - 2y = -13 \quad \xrightarrow{\text{Multiply both sides by } -2} \quad -18x + 4y = 26$$
$$31y = 62$$
$$\boxed{y = 2}$$

Substitute 2 for y in one of the original equations.

$$2x + 3y = 4$$
$$2x + 3(2) = 4$$
$$2x + 6 = 4$$
$$2x = -2$$
$$\boxed{x = -1}$$

The solution set is $\{(-1, 2)\}$.

Solution B

$$2x + 3y = 4 \quad \xrightarrow{\text{Multiply both sides by 2}} \quad 4x + 6y = 8$$
$$9x - 2y = -13 \quad \xrightarrow{\text{Multiply both sides by 3}} \quad 27x - 6y = -39$$
$$31x = -31$$
$$\boxed{x = -1}$$

Substitute -1 for x in one of the original equations.

$$2x + 3y = 4$$
$$2(-1) + 3y = 4$$
$$-2 + 3y = 4$$
$$3y = 6$$
$$\boxed{y = 2}$$

The solution set is $\{(-1, 2)\}$. ▲

Look back over Solutions A and B for Example 5 carefully. Especially notice the first steps where the multiplication property of equality is applied to the two equations. In Solution A we multiplied by numbers so that adding the resulting equations eliminated the x variable. In Solution B we multiplied so that the y variable was eliminated when we added the resulting equations. Either approach will work; pick the one that involves the easiest computation.

Example 6

Solve $\begin{pmatrix} 3x - 4y = 7 \\ 5x + 3y = 9 \end{pmatrix}$.

Solution

$3x - 4y = 7$ Multiply both sides by 3 $9x - 12y = 21$

$5x + 3y = 9$ Multiply both sides by 4 $20x + 12y = 36$

$$29x \qquad = 57$$

$$x = \frac{57}{29}$$

Since substituting $\frac{57}{29}$ for x into one of the original equations will produce some messy calculations, let's solve for y by eliminating the x's.

$3x - 4y = 7$ Multiply both sides by -5 $-15x + 20y = -35$

$5x + 3y = 9$ Multiply both sides by 3 $15x + 9y = 27$

$$29y = -8$$

$$y = -\frac{8}{29}$$

The solution set is $\left\{ \left(\frac{57}{29}, -\frac{8}{29} \right) \right\}$. ▲

Problem Solving

Many word problems that we solved earlier in this text using one equation in one variable can also be solved using a system of two linear equations in two variables. In fact, many times you may find that it seems quite natural to use two variables. It may also seem more meaningful at times to use variables other than x and y. Let's consider some examples.

Problem 1

The cost of 3 tennis balls and 2 golf balls is $7. Furthermore, the cost of 6 tennis balls and 3 golf balls is $12. Find the cost of 1 tennis ball and the cost of 1 golf ball.

Solution

Let's use t to represent the cost of one tennis ball and g the cost of one golf ball. The problem translates into the following system of equations.

$3t + 2g = 7$ The cost of 3 tennis balls and 2 golf balls is $7.

$6t + 3g = 12$ The cost of 6 tennis balls and 3 golf balls is $12.

Solving this system by the elimination-by-addition method, we obtain

$3t + 2g = 7$ Multiply both sides by -2 $-6t - 4g = -14$

$6t + 3g = 12$ Leave alone $6t + 3g = 12$

$$-g = -2$$

$$g = 2.$$

Substitute 2 for g in one of the original equations.

$$3t + 2g = 7$$
$$3t + 2(2) = 7$$
$$3t + 4 = 7$$
$$3t = 3$$
$$\boxed{t = 1}$$

The cost of a tennis ball is \$1 and the cost of a golf ball is \$2.

Problem 2

Niki invested \$500, part of it at 9% interest and the rest at 10%. Her total interest earned for a year was \$48. How much did she invest at each rate?

Solution

Let x represent the amount invested at 9%. Let y represent the amount invested at 10%. The problem translates into the following system.

$$x + y = 500 \qquad \text{Niki invested \$500.}$$
$$.09x + .10y = 48 \qquad \text{Her total interest earned for a year was \$48.}$$

Multiplying the second equation by 100 produces $9x + 10y = 4800$. Then, we have the following equivalent system to solve.

$$\begin{pmatrix} x + \quad y = 500 \\ 9x + 10y = 4800 \end{pmatrix}$$

Using the elimination-by-addition method, we can proceed as follows.

$x + \quad y = 500$	Multiply both sides by -9	$-9x - \quad 9y = -4500$
$9x + 10y = 4800$	Leave alone	$9x + 10y = \quad 4800$

$$\boxed{y = 300}$$

Substituting 300 for y in $x + y = 500$ yields

$$x + y = 500$$
$$x + 300 = 500$$
$$\boxed{x = 200.}$$

Therefore, we know that \$200 was invested at 9% interest and \$300 at 10%.

Before you tackle the word problems in this next problem set, it might be helpful to review the problem solving suggestions we offered in Section 4.2. Those suggestions continue to apply here except that now we have the flexibility of using two equations and two unknowns. Don't forget that to check a word problem you need to see if your answers satisfy the conditions stated in the problem.

Problem Set 8.6

For Problems 1–24, solve each system using the elimination-by-addition method.

1. $\begin{pmatrix} x + y = & 14 \\ x - y = & -2 \end{pmatrix}$

2. $\begin{pmatrix} x - y = -14 \\ x + y = & 6 \end{pmatrix}$

3. $\begin{pmatrix} x + 4y = -21 \\ 3x - 4y = & 1 \end{pmatrix}$

4. $\begin{pmatrix} -3x + 2y = -21 \\ 3x - 7y = & 36 \end{pmatrix}$

5. $\begin{pmatrix} y = 6 - x \\ x - y = -18 \end{pmatrix}$

6. $\begin{pmatrix} x + y = -10 \\ x = y + 6 \end{pmatrix}$

7. $\begin{pmatrix} 5x + & y = 23 \\ 3x - 2y = 19 \end{pmatrix}$

8. $\begin{pmatrix} 4x - 5y = -36 \\ x + 2y = & 30 \end{pmatrix}$

9. $\begin{pmatrix} x + 2y = 5 \\ 3x - 2y = 6 \end{pmatrix}$

10. $\begin{pmatrix} 2x - & y = 3 \\ -2x + 5y = 7 \end{pmatrix}$

11. $\begin{pmatrix} y = -x \\ 2x - y = -2 \end{pmatrix}$

12. $\begin{pmatrix} 3x = 2y \\ 8x + 20y = 19 \end{pmatrix}$

13. $\begin{pmatrix} 4x + 5y = & 9 \\ 5x - 6y = -50 \end{pmatrix}$

14. $\begin{pmatrix} 2x + 7y = & 46 \\ 3x - 4y = -18 \end{pmatrix}$

15. $\begin{pmatrix} 9x - 7y = 29 \\ 5x - 3y = 17 \end{pmatrix}$

16. $\begin{pmatrix} 7x + 5y = -6 \\ 4x + 3y = -4 \end{pmatrix}$

17. $\begin{pmatrix} 6x + 5y = -6 \\ 8x - 3y = & 21 \end{pmatrix}$

18. $\begin{pmatrix} 3x - 8y = -23 \\ 7x + 4y = -48 \end{pmatrix}$

19. $\begin{pmatrix} 2x - 7y = -1 \\ 9x + 4y = -2 \end{pmatrix}$

20. $\begin{pmatrix} 6x - 2y = -3 \\ 5x - 9y = & 1 \end{pmatrix}$

21. $\begin{pmatrix} x + & y = 750 \\ .07x + .08y = & 57.5 \end{pmatrix}$

22. $\begin{pmatrix} x + & y = 700 \\ .06x + .09y = & 54 \end{pmatrix}$

23. $\begin{pmatrix} .09x + .11y = 31 \\ y = x + 100 \end{pmatrix}$

24. $\begin{pmatrix} .08x + .1y = 56 \\ y = 2x \end{pmatrix}$

For Problems 25–40, solve each problem by setting up and solving a system of two linear equations in two variables.

25. The sum of two numbers is 30 and their difference is 12. Find the numbers.

26. The sum of two numbers is 20. If twice the smaller is subtracted from the larger, the result is 2. Find the numbers.

27. The difference of two numbers is 7. If three times the smaller is subtracted from twice the larger, the result is 6. Find the numbers.

28. The difference of two numbers is 17. If the larger is increased by three times the smaller, the result is 37. Find the numbers.

29. One number is twice another number. The sum of three times the smaller and five times the larger is 78. Find the numbers.

30. One number is three times another number. The sum of four times the smaller and seven times the larger is 175. Find the numbers.

31. Three lemons and 2 apples cost $1.05. Two lemons and 3 apples cost $1.20. Find the cost of 1 lemon and also the cost of 1 apple.

32. Betty bought 30 stamps for $6. Some of them were 17-cent stamps and the rest were 22-cent stamps. How many of each kind did she buy?

33. Larry has $1.45 in change in his pocket, consisting of dimes and quarters. If he has a total of 10 coins, how many of each kind does he have?

34. Suppose that Cindy has 90 cents in change in her purse, consisting of nickels and dimes. If she has a total of 13 coins, how many of each kind does she have in her purse?

35. Suppose that a library buys a total of 35 books, which cost $462. Some of the books cost $12 per book and the rest cost $14 per book. How many books of each price did they buy?

36. Some punch that contains 10% grapefruit juice is mixed with some punch that contains 5% grapefruit juice to produce 5 quarts of punch that is 8% grapefruit juice. How many quarts of 10% and 5% grapefruit juice must be used?

37. A 10% salt solution is to be mixed with a 15% salt solution to produce 10 gallons of a 13% salt solution. How many gallons of the 10% solution and how many gallons of the 15% solution will be needed?

38. The income from a student production was $21,000. The price of a student ticket was $3 and nonstudent tickets were $5 each. Five thousand people attended the production. How many tickets of each kind were sold?

39. Sidney invested $1300, part of it at 10% and the rest at 12%. If his total yearly interest was $146, how much did he invest at each rate?

40. Heather invested $1100, part of it at 8% and the rest at 9%. Her total interest earned for a year was $95. How much did she invest at each rate?

41. Explain how you would solve the system $\begin{pmatrix} 2x - 3y = 5 \\ 4x + 7y = 9 \end{pmatrix}$ using the elimination-by-addition method.

42. Give a general description of how to apply the elimination-by-addition method.

Further Investigations

For Problems 43–46, solve each system using the elimination-by-addition method.

43. $\begin{pmatrix} \dfrac{1}{2}x - \dfrac{1}{3}y = -2 \\ \dfrac{3}{2}x + \dfrac{2}{3}y = 34 \end{pmatrix}$

44. $\begin{pmatrix} x - 2y = 0 \\ 3x + 5y = 0 \end{pmatrix}$

45. $\begin{pmatrix} x - 4y = 6 \\ 2x - 8y = 3 \end{pmatrix}$

46. $\begin{pmatrix} y = 1 - 2x \\ 6x + 3y = 3 \end{pmatrix}$

8.7   Substitution Method

There is a third method of solving systems of equations that is called the **substitution method**. Like the addition method, it produces exact solutions and can be used on any system of linear equations; however, we will find that some systems lend themselves more to the substitution method than others. Let's consider a few examples to demonstrate the use of the substitution method.

Example 1

Solve $\begin{pmatrix} y = x + 10 \\ x + y = 14 \end{pmatrix}$.

Solution

Since the first equation states that y equals $x + 10$, we can substitute $x + 10$ for y in the second equation.

$$x + y = 14 \qquad \text{Substitute } x + 10 \text{ for } y \qquad x + (x + 10) = 14$$

Now we have an equation with one variable that can be solved in the usual way.

$$x + (x + 10) = 14$$
$$2x + 10 = 14$$
$$2x = 4$$
$$\boxed{x = 2}$$

Substituting 2 for x in one of the original equations we can find the value of y.

$$y = x + 10$$
$$y = 2 + 10$$
$$\boxed{y = 12}$$

The solution set is $\{(2, 12)\}$.   ▲

Example 2

Solve $\begin{pmatrix} 3x + 5y = -7 \\ x = 2y + 5 \end{pmatrix}$.

Solution

Since the second equation states that x equals $2y + 5$ we can substitute $2y + 5$ for x in the first equation.

$$3x + 5y = -7 \qquad \text{Substitute } 2y + 5 \text{ for } x \qquad 3(2y + 5) + 5y = -7$$

Solving this equation we obtain

$$3(2y + 5) + 5y = -7$$
$$6y + 15 + 5y = -7$$
$$11y + 15 = -7$$
$$11y = -22$$
$$\boxed{y = -2.}$$

Substituting -2 for y in one of the two original equations produces

$$x = 2y + 5$$
$$x = 2(-2) + 5$$
$$x = -4 + 5$$
$$\boxed{x = 1.}$$

The solution set is $\{(1, -2)\}$.

Note that the key idea behind the substitution method is the elimination of a variable, but the elimination is done by a substitution rather than by addition of the equations. The substitution method is especially convenient to use if at least one of the equations is of the form *y equals* or *x equals*. In Example 1, the first equation is of the form *y equals* and in Example 2 the second equation is of the form *x equals*. Let's consider another example using the substitution method.

Example 3

Solve $\begin{pmatrix} 2x + 3y = -30 \\ y = \dfrac{2}{3}x - 6 \end{pmatrix}$.

Solution

The second equation allows us to substitute $\dfrac{2}{3}x - 6$ for y in the first equation.

$$2x + 3y = -30 \qquad \xrightarrow{\text{Substitute } \frac{2}{3}x - 6 \text{ for } y} \qquad 2x + 3\left(\frac{2}{3}x - 6\right) = -30$$

Solving this equation produces

$$2x + 3\left(\frac{2}{3}x - 6\right) = -30$$
$$2x + 2x - 18 = -30$$
$$4x - 18 = -30$$
$$4x = -12$$
$$\boxed{x = -3.}$$

Now we can substitute -3 for x in one of the original equations.

$$y = \frac{2}{3}x - 6$$
$$y = \frac{2}{3}(-3) - 6$$
$$y = -2 - 6$$
$$\boxed{y = -8}$$

The solution set is $\{(-3, -8)\}$.

It may be necessary to change the form of one of the equations before a substitution can be made. The following examples clarify this point.

Example 4

Solve $\begin{pmatrix} 4x - 5y = 55 \\ x + y = -2 \end{pmatrix}$.

Solution

The form of the second equation can easily be changed to make it ready for the substitution method.

$$x + y = -2$$
$$y = -2 - x \qquad \text{Added } -x \text{ to both sides}$$

Now, $-2 - x$ can be substituted for y in the first equation.

$$4x - 5y = 55 \qquad \xrightarrow{\text{Substitute } -2 - x \text{ for } y} \qquad 4x - 5(-2 - x) = 55$$

Solving this equation, we obtain

$$4x - 5(-2 - x) = 55$$
$$4x + 10 + 5x = 55$$
$$9x + 10 = 55$$
$$9x = 45$$
$$x = 5.$$

Substituting 5 for x in one of the *original* equations produces

$$x + y = -2$$
$$5 + y = -2$$
$$y = -7.$$

The solution set is $\{(5, -7)\}$.   ▲

In Example 4, we could have started by changing the form of the first equation to make it ready for substitution. However, you should be able to *look ahead* and see that this would produce a fractional form to substitute. We were able to avoid any messy calculations with fractions by changing the form of the second equation instead of the first. Sometimes when using the substitution method, you cannot avoid the use of fractional forms. The next example is a case in point.

Example 5

Solve $\begin{pmatrix} 3x + 2y = 8 \\ 2x - 3y = -38 \end{pmatrix}$.

Solution

Looking ahead we see that changing the form of either equation will produce a fractional form. Therefore, let's merely pick the first equation and solve for y.

$$3x + 2y = 8$$
$$2y = 8 - 3x \qquad \text{Added } -3x \text{ to both sides}$$
$$y = \frac{8 - 3x}{2} \qquad \text{Multiplied both sides by } \frac{1}{2}$$

Now we can substitute $\dfrac{8 - 3x}{2}$ for y in the second equation and determine the value of x.

$$2x - 3y = -38$$

$$2x - 3\left(\dfrac{8 - 3x}{2}\right) = -38$$

$$2x - \dfrac{24 - 9x}{2} = -38$$

$$4x - 24 + 9x = -76 \qquad \text{Multiplied both sides by 2}$$

$$13x - 24 = -76$$

$$13x = -52$$

$$\boxed{x = -4}$$

Substituting -4 for x in one of the original equations, we obtain

$$3x + 2y = 8$$

$$3(-4) + 2y = 8$$

$$-12 + 2y = 8$$

$$2y = 20$$

$$\boxed{y = 10.}$$

The solution set is $\{(-4, 10)\}$.

Which Method to Use

We have now studied three methods of solving systems of linear equations—the *graphing method*, the *elimination-by-addition method*, and the *substitution method*. As we indicated earlier, the graphing method is quite restrictive and works well only if the solutions are integers or if only approximate answers are needed. Both the elimination-by-addition method and the substitution method can be used to obtain exact solutions for any system of linear equations in two variables. The method you choose may depend upon the original form of the equations. Let's consider two examples to illustrate this point.

Example 6

Solve $\begin{pmatrix} 7x - 5y = -52 \\ y = 3x - 4 \end{pmatrix}$.

Solution

Since the second equation indicates that $3x - 4$ can be substituted for y, this system lends itself to the substitution method.

$$7x - 5y = -52 \qquad \xrightarrow{\text{Substitute } 3x - 4 \text{ for } y} \qquad 7x - 5(3x - 4) = -52$$

Solving this equation, we obtain

$$7x - 5(3x - 4) = -52$$

$$7x - 15x + 20 = -52$$

$$-8x + 20 = -52$$
$$-8x = -72$$
$$\boxed{x = 9.}$$

Substituting 9 for x in one of the original equations produces

$$y = 3x - 4$$
$$y = 3(9) - 4$$
$$y = 27 - 4$$
$$\boxed{y = 23.}$$

The solution set is $\{(9, 23)\}$. ▲

Example 7

Solve $\begin{pmatrix} 10x + 7y = 19 \\ 2x - 6y = -11 \end{pmatrix}$.

Solution

Because changing the form of either of the two equations in preparation for the substitution method would produce a fractional form, we are probably better off to use the elimination-by-addition method. Furthermore, we should notice that the coefficients of x lend themselves to this method.

$$
\begin{array}{lll}
10x + 7y = 19 & \xrightarrow{\text{Leave alone}} & 10x + 7y = 19 \\
2x - 6y = -11 & \xrightarrow{\text{Multiply by } -5} & -10x + 30y = 55 \\
& & \rule{3cm}{0.4pt} \\
& & 37y = 74 \\
& & \boxed{y = 2}
\end{array}
$$

Substituting 2 for y in the first equation of the given system produces

$$10x + 7y = 19$$
$$10x + 7(2) = 19$$
$$10x + 14 = 19$$
$$10x = 5$$
$$x = \frac{5}{10} = \frac{1}{2}.$$

The solution set is $\left\{\left(\frac{1}{2}, 2\right)\right\}$. ▲

In Section 8.5 we explained that you can tell by graphing the equations whether the system has no solutions, one solution, or infinitely many solutions. That is, the two lines may be parallel (no solutions), or they may intersect in one point (one solution), or they may coincide (infinitely many solutions). From a practical viewpoint, the systems that have one solution deserve most of our attention. However, we do need to be able to deal with the other situations as they arise. Let's use two examples to demonstrate what occurs when we hit a "no solution" or "infinitely many solutions" situation when we are using either the elimination-by-addition or substitution method.

Example 8

Solve the system $\begin{pmatrix} y = 2x - 1 \\ 6x - 3y = 7 \end{pmatrix}$.

Solution

Since the first equation indicates that $2x - 1$ can be substituted for y, this system lends itself to the substitution method.

$6x - 3y = 7$ $\xrightarrow{\text{Substitute } 2x - 1 \text{ for } y}$ $6x - 3(2x - 1) = 7$

Now let's solve this equation.

$$6x - 6x + 3 = 7$$
$$0 + 3 = 7$$
$$\boxed{3 = 7}$$

The false numerical statement, $3 = 7$, implies that the system has *no solutions*. Thus, the solution set is $\varnothing$. (You may want to graph the two lines to verify this conclusion!)   ▲

Example 9

Solve the system $\begin{pmatrix} 2x - 3y = 4 \\ 10x - 15y = 20 \end{pmatrix}$.

Solution

Let's use the elimination-by-addition method.

$2x - 3y = 4$ $\xrightarrow{\text{Multiply both sides by } -5}$ $-10x + 15y = -20$
$10x - 15y = 20$ $\xrightarrow{\text{Leave alone}}$ $\underline{10x - 15y = 20}$
$ 0 + 0 = 0$

The true numerical statement, $0 + 0 = 0$, implies that the system has *infinitely many solutions*. Any ordered pair that satisfies one of the equations will also satisfy the other equation.   ▲

Problem Solving

Let's conclude this section with three more word problems.

Problem 1

The length of a rectangle is 1 centimeter less than three times the width. The perimeter of the rectangle is 94 centimeters. Find the length and width of the rectangle.

Solution
Let w represent the width of the rectangle. Let l represent the length of the rectangle (Figure 8.31).

FIGURE 8.31

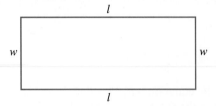

The problem translates into the following system of equations.

$l = 3w - 1,$ The length of a rectangle is 1 centimeter less than 3 times the width.

$2l + 2w = 94.$ The perimeter of the rectangle is 94 centimeters.

Multiplying both sides of the second equation by one-half produces the equivalent equation $l + w = 47$; so we have the following system to solve.

$$\begin{pmatrix} l = 3w - 1 \\ l + w = 47 \end{pmatrix}$$

The first equation indicates we can substitute $3w - 1$ for l in the second equation.

$l + w = 47$ — Substitute $3w - 1$ for l → $3w - 1 + w = 47$

Solving this equation yields

$$3w - 1 + w = 47$$
$$4w - 1 = 47$$
$$4w = 48$$
$$w = 12.$$

Substituting 12 for w in one of the original equations produces

$$l = 3w - 1$$
$$l = 3(12) - 1$$
$$l = 36 - 1$$
$$l = 35.$$

The rectangle is 12 centimeters wide and 35 centimeters long.   ▲

Problem 2
The sum of the digits of a two-digit number is 7. The tens digit is 1 more than twice the units digit. Find the number.

Solution
Let t represent the tens digit. Let u represent the units digit.

$$t + u = 7, \qquad \text{The sum of the digits is 7.}$$
$$t = 2u + 1 \qquad \text{The tens digit is one more than twice the units digit.}$$

The second equation indicates that $2u + 1$ can be substituted for t in the first equation.

$$t + u = 7$$
$$(2u + 1) + u = 7$$
$$3u + 1 = 7$$
$$3u = 6$$
$$\boxed{u = 2}$$

Substituting 2 for u in one of the original equations yields

$$t = 2u + 1$$
$$t = 2(2) + 1$$
$$\boxed{t = 5.}$$

The tens digit is 5 and the units digit is 2; the number is 52. ▲

The two-variable expression $10t + u$ can be used to represent any two-digit number. The t represents the tens digit and u the units digit. For example, if $t = 4$ and $u = 7$, then $10t + u$ becomes $10(4) + 7$ or 47. Now let's consider the following problem.

Problem 3

A two-digit number is 7 times its units digit. The sum of the digits is 8. Find the number.

Solution

Let t represent the tens digit. Let u represent the units digit.

$$10t + u = 7u, \qquad \text{A two-digit number is 7 times its units digit.}$$
$$t + u = 8 \qquad \text{The sum of the digits is 8.}$$

The first equation can be simplifed to $5t - 3u = 0$; thus we have the following system to solve.

$$\begin{pmatrix} 5t - 3u = 0 \\ t + u = 8 \end{pmatrix}$$

Using the elimination-by-addition method, we proceed as follows.

$$5t - 3u = 0 \qquad \xrightarrow{\text{Leave alone}} \qquad 5t - 3u = 0$$
$$t + u = 8 \qquad \xrightarrow{\text{Multiply both sides by 3}} \qquad 3t + 3u = 24$$
$$\overline{\ 8t = 24}$$
$$\boxed{t = 3}$$

Substituting 3 for t in $t + u = 8$ yields

$$t + u = 8$$
$$3 + u = 8$$
$$\boxed{u = 5.}$$

The tens digit is 3 and the units digit is 5; the number is 35. ▲

▼ Problem Set 8.7

For Problems 1–26, solve each system using the substitution method.

1. $\begin{pmatrix} y = 2x - 1 \\ x + y = 14 \end{pmatrix}$

2. $\begin{pmatrix} y = 3x + 4 \\ x + y = 52 \end{pmatrix}$

3. $\begin{pmatrix} x - y = -14 \\ y = -3x - 2 \end{pmatrix}$

4. $\begin{pmatrix} x - y = -23 \\ y = -2x + 5 \end{pmatrix}$

5. $\begin{pmatrix} 4x - 3y = -6 \\ y = -2x + 7 \end{pmatrix}$

6. $\begin{pmatrix} 8x - y = -8 \\ y = 4x + 5 \end{pmatrix}$

7. $\begin{pmatrix} x + y = 1 \\ 3x + 6y = 7 \end{pmatrix}$

8. $\begin{pmatrix} 2x - 4y = -9 \\ x + y = 3 \end{pmatrix}$

9. $\begin{pmatrix} 2x - y = 12 \\ x = \dfrac{3}{4}y \end{pmatrix}$

10. $\begin{pmatrix} 4x - 5y = 6 \\ y = \dfrac{2}{3}x \end{pmatrix}$

11. $\begin{pmatrix} y = \dfrac{3}{2}x \\ 6x - 5y = 15 \end{pmatrix}$

12. $\begin{pmatrix} x = \dfrac{3}{5}y \\ 4x - 3y = 12 \end{pmatrix}$

13. $\begin{pmatrix} 4y - 1 = x \\ 2x - 8y = 3 \end{pmatrix}$

14. $\begin{pmatrix} y = 5x - 2 \\ y = 2x + 7 \end{pmatrix}$

15. $\begin{pmatrix} 7x + 2y = -2 \\ 6x + 5y = 18 \end{pmatrix}$

16. $\begin{pmatrix} 2x + 3y = 31 \\ 3x + 5y = -20 \end{pmatrix}$

17. $\begin{pmatrix} 8x - 3y = -9 \\ x + 5y = -71 \end{pmatrix}$

18. $\begin{pmatrix} 9x - 2y = -18 \\ 4x - y = -7 \end{pmatrix}$

19. $\begin{pmatrix} 4x - 6y = 1 \\ 2x + 3y = 4 \end{pmatrix}$

20. $\begin{pmatrix} 2x + 3y = 1 \\ 4x + 6y = 2 \end{pmatrix}$

21. $\begin{pmatrix} 5x + 7y = 3 \\ 3x - 2y = 0 \end{pmatrix}$

22. $\begin{pmatrix} 7x - 3y = 4 \\ 2x + 5y = 0 \end{pmatrix}$

23. $\begin{pmatrix} .05x + .07y = 33 \\ y = x + 300 \end{pmatrix}$

24. $\begin{pmatrix} .06x + .08y = 15 \\ y = x + 100 \end{pmatrix}$

25. $\begin{pmatrix} x + y = 13 \\ .05x + .1y = 1.15 \end{pmatrix}$

26. $\begin{pmatrix} x + y = 17 \\ .1x + .25y = 3.2 \end{pmatrix}$

For Problems 27–46, solve each system using either the elimination-by-addition or the substitution method, whichever seems more appropriate to you.

27. $\begin{pmatrix} 5x - 4y = 14 \\ 7x + 3y = -32 \end{pmatrix}$

28. $\begin{pmatrix} 2x + 3y = 13 \\ 3x - 5y = -28 \end{pmatrix}$

29. $\begin{pmatrix} 2x + 9y = 6 \\ y = -x \end{pmatrix}$

30. $\begin{pmatrix} y = 3x + 2 \\ 4x - 3y = -21 \end{pmatrix}$

31. $\begin{pmatrix} x + y = 22 \\ .6x + .5y = 12 \end{pmatrix}$

32. $\begin{pmatrix} x + y = 13 \\ .4x + .5y = 6 \end{pmatrix}$

33. $\begin{pmatrix} 4x - y = 0 \\ 7x + 2y = 9 \end{pmatrix}$

34. $\begin{pmatrix} x - 2y = 0 \\ 5x = 8y + 12 \end{pmatrix}$

35. $\begin{pmatrix} 2x + y = 1 \\ 6x - 7y = -57 \end{pmatrix}$

36. $\begin{pmatrix} 7x - 9y = 11 \\ x - 3y = 3 \end{pmatrix}$

37. $\begin{pmatrix} 6x - y = -1 \\ 10x + 2y = 13 \end{pmatrix}$

38. $\begin{pmatrix} x + 4y = -5 \\ 6x - 5y = -1 \end{pmatrix}$

39. $\begin{pmatrix} 4x + 8y = 20 \\ x + 2y = 5 \end{pmatrix}$

40. $\begin{pmatrix} x = 5y - 5 \\ 2x - 10y = 2 \end{pmatrix}$

41. $\begin{pmatrix} 3x - 8y = -5 \\ x = 2y \end{pmatrix}$

42. $\begin{pmatrix} x + y = -3 \\ 5x + 6y = -22 \end{pmatrix}$

43. $\begin{pmatrix} 5y - 2x = -4 \\ 10y = 3x + 4 \end{pmatrix}$

44. $\begin{pmatrix} 3x - 7 = 2y + 15 \\ 3x = 6y + 18 \end{pmatrix}$

45. $\begin{pmatrix} x = -y - 1 \\ 6x - 5y = 4 \end{pmatrix}$

46. $\begin{pmatrix} 4x + 7y = 1 \\ y = 2x + 3 \end{pmatrix}$

For Problems 47–60, solve each problem by setting up and solving a system of two linear equations in two variables. Use either the elimination-by-addition or the substitution method to solve the systems.

47. Suppose that the sum of two numbers is 46 and the difference of the numbers is 22. Find the numbers.

48. The sum of two numbers is 52. The larger number is two more than four times the smaller number. Find the numbers.

49. Suppose that a certain motel rents double rooms at $28 per day and single rooms at $19 per day. If a total of 50 rooms was rented one day for $1265, how many of each kind were rented?

50. The total receipts from a concert amounted to $2600. Student tickets were sold at $4 each and nonstudent tickets at $6 each. The number of student tickets sold was five times the number of nonstudent tickets sold. How many student tickets and how many nonstudent tickets were sold?

51. The sum of the digits of a two-digit number is 9. The two-digit number is nine times its units digit. Find the number.

52. The sum of the digits of a two-digit number is 13. If the digits are reversed, the new number is nine larger than the original number. Find the number.

53. Suppose that Larry has a number of dimes and quarters totaling $12.05. The number of quarters is five more than twice the number of dimes. How many coins of each kind does he have?

54. Suppose that Sue has three times as many nickels as pennies in her collection. Together her pennies and nickels have a value of $4.80. How many pennies and how many nickels does she have?

55. The sum of the digits of a two-digit number is 12. The tens digit is three times as large as the units digit. Find the number.

56. The units digit of a two-digit nunber is one more than four times the tens digit. The sum of the digits is 11. Find the number.

57. Tina invested some money at 8% interest and some money at 9%. She invested $250 more at 9% than she invested at 8%. Her total yearly interest from the two investments was $48. How much did Tina invest at each rate?

58. One solution contains 40% alcohol and a second solution contains 70% alcohol. How many liters of each solution should be mixed to make 30 liters that contain 50% alcohol?

59. One solution contains 30% alcohol and a second solution contains 70% alcohol. How many liters of each solution should be mixed to make 10 liters containing 40% alcohol?

60. The length of a rectangle is 3 meters less than four times the width. The perimeter of the rectangle is 74 meters. Find the length and width of the rectangle.

THOUGHTS INTO WORDS

61. Explain how you would solve the system $\begin{pmatrix} 5x - 4y = 10 \\ 3x - y = 6 \end{pmatrix}$ using the substitution method.

62. How do you decide whether to solve a system of linear equations by using the elimination-by-addition method or the substitution method?

63. What do you see as the strengths and weaknesses of the elimination-by-addition method and the substitution method?

Further Investigations

For Problems 64–67, use the substitution method to solve each system.

64. $\begin{pmatrix} y = \dfrac{2}{3}x - \dfrac{1}{2} \\ \dfrac{3}{2}x - y = -\dfrac{1}{3} \end{pmatrix}$

65. $\begin{pmatrix} 5x - 3y = 0 \\ 4x + 7y = 0 \end{pmatrix}$

66. $\begin{pmatrix} y = 2x - 1 \\ 6x - 3y = 3 \end{pmatrix}$

67. $\begin{pmatrix} x = -4y + 5 \\ 2x + 8y = -1 \end{pmatrix}$

SUMMARY

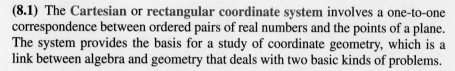

(8.1) The **Cartesian** or **rectangular coordinate system** involves a one-to-one correspondence between ordered pairs of real numbers and the points of a plane. The system provides the basis for a study of coordinate geometry, which is a link between algebra and geometry that deals with two basic kinds of problems.

1. Given an algebraic equation find its geometric graph.
2. Given a set of conditions pertaining to a geometric figure, find its algebraic equation.

One graphing technique is to plot a sufficient number of points until the graph of the equation is determined.

(8.2) Any equation of the form $Ax + By = C$, where A, B, and C are constants (A and B not both zero) and x and y are variables, is a **linear equation in two variables** and its graph is a **straight line**.

To **graph** a linear equation we can find two solutions (the intercepts are usually easy to determine), plot the corresponding points, and connect the points with a straight line.

(8.3) If points P_1 and P_2 with coordinates (x_1, y_1) and (x_2, y_2), respectively, are any two points on a line, then the **slope** of the line (denoted by m) is given by

$$m = \frac{y_2 - y_1}{x_2 - x_1}, \qquad x_1 \neq x_2.$$

The slope of a line is a **ratio** of vertical change relative to horizontal change. The slope of a line can be negative, positive, or zero. The concept of slope is not defined for vertical lines.

(8.4) You should review Examples 1, 2, and 3 of this section to pull together a general approach to writing equations of lines given certain conditions. The equation

$$y - y_1 = m(x - x_1)$$

is called the **point-slope form** of the equation of a straight line. The equation

$$y = mx + b$$

is called the **slope-intercept form** of the equation of a straight line. If the equation of a nonvertical line is written in slope-intercept form, then the coefficient of x is the slope of the line and the constant term is the y-intercept.

(8.5) Solving a system of two linear equations by graphing produces one of the following three possibilities.

1. The graphs of the two equations are two intersecting lines, which indicates one solution for the system, called a **consistent system**.

2. The graphs of the two equations are two parallel lines, which indicates **no solution** for the system, called an **inconsistent system**.

3. The graphs of the two equations are the same line, which indicates **infinitely many solutions** for the system. We refer to the equations as a set of **dependent** equations.

(8.6) The **elimination-by-addition method** for solving a system of two linear equations relies on the property "if $a = b$ and $c = d$, then $a + c = b + d$." Equations are added to eliminate a variable.

(8.7) The **substitution method** for solving a system of two linear equations involves solving one of the two equations for one variable and **substituting** this expression into the other equation.

To review the problem of choosing a method for solving a particular system of two linear equations, return to Section 8.7 and study Examples 6 and 7 one more time.

▼ Chapter 8 Review Problem Set

For Problems 1–6, graph each of the equations.

1. $2x - 5y = 10$
2. $y = -\frac{1}{3}x + 1$
3. $y = 2x^2 + 1$
4. $y = -x^2 - 2$
5. $2x - 3y = 0$
6. $y = -x^3$

7. Find the slope of the line determined by $(3, -4)$ and $(-2, 5)$.

8. Find the slope of the line $5x - 6y = 30$.

9. Write the equation of the line that has a slope of $-\frac{5}{7}$ and contains the point $(2, -3)$.

10. Write the equation of the line that contains $(2, 5)$ and $(-1, -3)$.

11. Write the equation of the line that has a slope of $\frac{2}{9}$ and a y-intercept of -1.

12. Write the equation of the line that contains $(2, 4)$ and which is perpendicular to the x-axis.

13. Solve the system $\begin{pmatrix} 2x + y = 4 \\ x - y = 5 \end{pmatrix}$ by using the graphing method.

For Problems 14–25, solve each system by using either the elimination-by-addition method or the substitution method.

14. $\begin{pmatrix} 2x - y = 1 \\ 3x - 2y = -5 \end{pmatrix}$

15. $\begin{pmatrix} 2x + 5y = 7 \\ x = -3y + 1 \end{pmatrix}$

16. $\begin{pmatrix} 3x + 2y = 7 \\ 4x - 5y = 3 \end{pmatrix}$

17. $\begin{pmatrix} 9x + 2y = 140 \\ x + 5y = 135 \end{pmatrix}$

18. $\begin{pmatrix} \dfrac{1}{2}x + \dfrac{1}{4}y = -5 \\ \dfrac{2}{3}x - \dfrac{1}{2}y = 0 \end{pmatrix}$

19. $\begin{pmatrix} x + y = 1000 \\ .07x + .09y = 82 \end{pmatrix}$

20. $\begin{pmatrix} y = 5x + 2 \\ 10x - 2y = 1 \end{pmatrix}$

21. $\begin{pmatrix} 5x - 7y = 9 \\ y = 3x - 2 \end{pmatrix}$

22. $\begin{pmatrix} 10t + u = 6u \\ t + u = 12 \end{pmatrix}$

23. $\begin{pmatrix} t = 2u \\ 10t + u - 36 = 10u + t \end{pmatrix}$

24. $\begin{pmatrix} u = 2t + 1 \\ 10t + u + 10u + t = 110 \end{pmatrix}$

25. $\begin{pmatrix} y = -\dfrac{2}{3}x \\ \dfrac{1}{3}x - y = -9 \end{pmatrix}$

Solve each of the following problems by setting up and solving a system of two linear equations in two variables.

26. The sum of two numbers is 113. The larger number is one less than twice the smaller number. Find the numbers.

27. Last year Mark invested a certain amount of money at 9% annual interest and $50 more than that amount at 11%. He received $55.50 in interest. How much did he invest at each rate?

28. Cindy has 43 coins consisting of nickels and dimes. The total value of the coins is $3.40. How many coins of each kind does she have?

29. The length of a rectangle is 1 inch more than three times the width. If the perimeter of the rectangle is 50 inches, find the length and width.

30. The sum of the digits of a two-digit number is 13 and the units digit is one more than twice the tens digit. Find the number.

31. The sum of the digits of a two-digit number is 9. The number formed by reversing the digits is 27 larger than the original number. Find the original number.

32. Two angles are complementary and one of them is 6° less than twice the other one. Find the measure of each angle.

33. Two angles are supplementary and the larger angle is 20° less than three times the smaller angle. Find the measure of each angle.

34. Four cheeseburgers and five milkshakes cost a total of $8.35. Two milkshakes cost $.35 more than one cheeseburger. Find the cost of a cheeseburger and also find the cost of a milkshake.

35. Three cans of prune juice and two cans of tomato juice cost $3.85. On the other hand, two cans of prune juice and three cans of tomato juice cost $3.55. Find the cost per can of each.

CHAPTER 8 TEST

For Problems 1–4, graph each of the equations.

1. $5x + 3y = 15$

2. $-2x + y = -4$

3. $y = -\dfrac{1}{2}x - 2$

4. $y = 2x^2 - 3$

5. Is $(-2, -3)$ a solution of $7x - 2y = -8$?

6. Is $(-1, -5)$ a solution of $y = -x^2 - 4$?

7. Find the y-intercept of the graph of $-3x + 4y = -12$.

8. Find the x-intercepts of the graph of $y = -x^2 + 16$.

9. Find the slope of the line determined by the points $(2, -4)$ and $(-5, -1)$.

10. Find the slope of the line determined by the equation $4x - 5y = 20$.

For Problems 11–13, express each equation in $Ax + By = C$ form, where A, B, and C are integers.

11. Determine the equation of the line that has a slope of $-\dfrac{3}{5}$ and a y-intercept of 4.

12. Determine the equation of the line containing $(4, -2)$ and having a slope of $\dfrac{4}{9}$.

13. Determine the equation of the line that contains the points $(4, 6)$ and $(-2, -3)$.

14. Solve the system $\begin{pmatrix} 3x - 2y = -4 \\ 2x + 3y = 19 \end{pmatrix}$ by graphing.

15. Solve the system $\begin{pmatrix} x - 3y = -9 \\ 4x + 7y = 40 \end{pmatrix}$ using the elimination-by-addition method.

16. Solve the system $\begin{pmatrix} 5x + y = -14 \\ 6x - 7y = -66 \end{pmatrix}$ using the substitution method.

17. Solve the system $\begin{pmatrix} 2x - 7y = 26 \\ 3x + 2y = -11 \end{pmatrix}$.

18. Solve the system $\begin{pmatrix} 8x + 5y = -6 \\ 4x - y = 18 \end{pmatrix}$.

(continued on next page)

CHAPTER 8 TEST (continued)

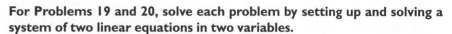

For Problems 19 and 20, solve each problem by setting up and solving a system of two linear equations in two variables.

19. The sum of the digits of a two digit number is 11 and the units digit is one more than four times the tens digit. Find the number.

20. The length of a rectangle is 1 inch less than twice the width of the rectangle. If the perimeter of the rectangle is 40 inches, find the length of the rectangle.

Square Roots and Radicals

Suppose that a car is traveling on a highway during a rainstorm at 65 miles per hour. Suddenly something darts across the highway and the driver hits the brake pedal. How far will the car skid on the wet pavement? The formula $S = \sqrt{30Df}$, where S represents the speed of the car, D the length of skid marks, and f a coefficient of friction can be used to determine that the car will skid approximately 400 feet.

In Section 2.3, we used $\sqrt{2}$ and $\sqrt{3}$ as examples of irrational numbers. Irrational numbers in decimal form are nonrepeating decimals. For example, $\sqrt{2} = 1.414213562373\ldots$, where the three dots indicate that the decimal expansion continues indefinitely. In Chapter 2 we stated that we would return to the irrationals in Chapter 9. The time has come for us to extend our skills relative to the set of irrational numbers.

▼ 9.1   Square Roots and Radicals

To **square a number** means to raise it to the second power—that is, to use the number as a factor twice.

$$3^2 = 3 \cdot 3 = 9, \qquad \text{"Three squared equals nine"}$$
$$7^2 = 7 \cdot 7 = 49,$$
$$(-3)^2 = (-3)(-3) = 9,$$
$$(-7)^2 = (-7)(-7) = 49$$

A **square root of a number** is one of its two equal factors. Thus, 3 is a square root of 9 because $3 \cdot 3 = 9$. Likewise, -3 is also a square root of 9 because $(-3)(-3) = 9$. We define the concept of square root as follows.

DEFINITION 9.1 a is a square root of b if $a^2 = b$.

The following generalizations are a direct consequence of Definition 9.1.

1. Every positive real number has two square roots; one is positive and the other is negative. They are opposites of each other.

2. Negative real numbers have no real number square roots. (This follows from Definition 9.1 since any nonzero real number is positive when squared.)

3. The square root of 0 is 0.

The symbol $\sqrt{}$, called a **radical sign**, is used to indicate the nonnegative square root. The number under the radical sign is called the **radicand**. The entire expression, such as $\sqrt{49}$, is called a **radical**.

$\sqrt{49} = 7$ $\sqrt{49}$ indicates the **nonnegative** or **principal** square root of 49.

$-\sqrt{49} = -7$ $-\sqrt{49}$ indicates the negative square root of 49.

$\sqrt{0} = 0$ Zero has only one square root. Technically, we could write $-\sqrt{0} = -0 = 0$.

$\sqrt{-4}$ $\sqrt{-4}$ is not a real number.

$-\sqrt{-4}$ $-\sqrt{-4}$ is not a real number.

If a is a nonnegative integer that is the square of an integer, then $\sqrt{a}$ and $-\sqrt{a}$ are rational numbers. For example, $\sqrt{1}$, $\sqrt{4}$, and $\sqrt{25}$ are the rational numbers 1, 2, and 5, respectively. The numbers 1, 4, and 25 are called perfect squares because each one represents the square of some integer. The following chart contains the squares of the whole numbers from 1 through 20, inclusive. You should know these values so that you can immediately recognize such square roots as $\sqrt{81} = 9$, $\sqrt{144} = 12$, $\sqrt{289} = 17$, and so on from the list. Furthermore, perfect squares of multiples of 10 are easy to recognize. For example, since $30^2 = 900$, we know that $\sqrt{900} = 30$.

$1^2 = 1$,	$8^2 = 64$,	$15^2 = 225$,
$2^2 = 4$,	$9^2 = 81$,	$16^2 = 256$,
$3^2 = 9$,	$10^2 = 100$,	$17^2 = 289$,
$4^2 = 16$,	$11^2 = 121$,	$18^2 = 324$,
$5^2 = 25$,	$12^2 = 144$,	$19^2 = 361$,
$6^2 = 36$,	$13^2 = 169$,	$20^2 = 400$
$7^2 = 49$,	$14^2 = 196$,	

Knowledge of the previous listing of perfect squares also helps with square roots of some fractions. Consider the following examples.

$$\sqrt{\frac{16}{25}} = \frac{4}{5} \quad \text{because} \left(\frac{4}{5}\right)^2 = \frac{16}{25},$$

$$\sqrt{\frac{36}{49}} = \frac{6}{7} \quad \text{because} \left(\frac{6}{7}\right)^2 = \frac{36}{49},$$

$$\sqrt{.09} = .3 \quad \text{because } (.3)^2 = .09$$

If a is a positive integer that is *not* the square of an integer, then $\sqrt{a}$ and $-\sqrt{a}$ are irrational numbers. For example, $\sqrt{2}$, $-\sqrt{2}$, $\sqrt{23}$, $\sqrt{31}$, $\sqrt{52}$, and $-\sqrt{75}$ are irrational numbers. Remember that irrational numbers have nonrepeating decimal representations. For example, $\sqrt{2} = 1.414213562373 \ldots$, where the decimal never repeats a block of digits.

For practical purposes, we often need to use a rational approximation of an irrational number. The calculator becomes a very useful tool for finding such approximations. Be sure that you can use your calculator to find the following approximate square roots. Each approximate square root has been rounded to the nearest thousandth.

$$\sqrt{19} = 4.359, \qquad \sqrt{38} = 6.164, \qquad \sqrt{72} = 8.485, \qquad \sqrt{93} = 9.644$$

REMARK If you don't have your calculator available, turn to the appendix and use the table of square roots. △

Adding and Subtracting Square Roots

Recall our use of the distributive property as the basis for combining similar terms. For example,

$$3x + 2x = (3 + 2)x = 5x, \qquad 7y - 4y = (7 - 4)y = 3y,$$
$$9a^2 + 5a^2 = (9 + 5)a^2 = 14a^2.$$

In a like manner, we can often simplify expressions that contain radicals by using the distributive property as follows.

$$5\sqrt{2} + 7\sqrt{2} = (5 + 7)\sqrt{2} = 12\sqrt{2},$$
$$8\sqrt{5} - 2\sqrt{5} = (8 - 2)\sqrt{5} = 6\sqrt{5},$$
$$4\sqrt{7} + 6\sqrt{7} + 3\sqrt{11} - \sqrt{11} = (4 + 6)\sqrt{7} + (3 - 1)\sqrt{11} = 10\sqrt{7} + 2\sqrt{11}$$

Note that **to add or subtract square roots they must have the same radicand.** Also note the form we use to indicate multiplication when a radical is involved. For example, $5 \cdot \sqrt{2}$ is written as $5\sqrt{2}$.

Now suppose that we need to evaluate $5\sqrt{2} - \sqrt{2} + 4\sqrt{2} - 2\sqrt{2}$, to the nearest tenth. We can either evaluate the expression as it stands or first simplify it by combining radicals and then evaluate that result. Let's use the latter approach. (It would probably be a good idea for you to do it both ways for checking purposes.)

$$5\sqrt{2} - \sqrt{2} + 4\sqrt{2} - 2\sqrt{2} = (5 - 1 + 4 - 2)\sqrt{2}$$
$$= 6\sqrt{2}$$
$$= 8.5, \quad \text{to the nearest tenth}$$

Example 1

Find a rational approximation, to the nearest tenth, for

$$7\sqrt{3} + 9\sqrt{5} + 2\sqrt{3} - 3\sqrt{5} + 13\sqrt{3}.$$

Solution

First, let's simplify the given expression and then evaluate that result.

$$7\sqrt{3} + 9\sqrt{5} + 2\sqrt{3} - 3\sqrt{5} + 13\sqrt{3} = (7 + 2 + 13)\sqrt{3} + (9 - 3)\sqrt{5}$$
$$= 22\sqrt{3} + 6\sqrt{5}$$
$$= 51.5, \quad \text{to the nearest tenth} \quad \blacktriangle$$

Applications of Radicals

Many real-world applications involve radical expressions. For example, the *period* of a pendulum is the time it takes to swing from one side to the other and back (see Figure 9.1). A formula for the period is

$$T = 2\pi\sqrt{\frac{L}{32}},$$

where T is expressed in seconds and L is in feet.

FIGURE 9.1

Example 2

Solution

Find the period, to the nearest tenth of a second, of a pendulum of length 2.5 feet.

Let's use 3.14 as an approximation for π and substitute 2.5 for L in the formula.

$$T = 2\pi\sqrt{\frac{L}{32}}$$

$$= 2(3.14)\sqrt{\frac{2.5}{32}}$$

$$= 1.8 \text{ to the nearest tenth}$$

The period is approximately 1.8 seconds. ▲

Police use the formula $S = \sqrt{30Df}$ to estimate a car's speed based on the length of skid marks (see Figure 9.2). In this formula, S represents the car's speed in miles per hour, D the length of skid marks measured in feet, and f represents a coefficient of friction. For a particular situation, the coefficient of friction is a constant depending on the type and condition of the road surface.

FIGURE 9.2

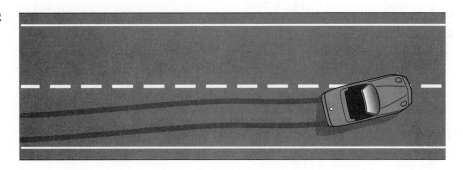

Example 3

Using .40 as a coefficient of friction, find how fast a car was moving if it skidded 225 feet. Express the answer to the nearest mile per hour.

Solution

Substitute .40 for f and 225 for D in the formula $S = \sqrt{30Df}$.

$$S = \sqrt{30(225)(.40)} = 52 \text{ to the nearest whole number}$$

The car was traveling at approximately 52 miles per hour. ▲

▼ Problem Set 9.1

For Problems 1–18, evaluate each radical without using a calculator or a table.

1. $\sqrt{49}$ 2. $\sqrt{100}$ 3. $-\sqrt{64}$

4. $-\sqrt{36}$ 5. $\sqrt{121}$

6. $\sqrt{144}$

7. $\sqrt{3600}$

8. $\sqrt{2500}$

9. $-\sqrt{1600}$

10. $-\sqrt{900}$

11. $\sqrt{6400}$

12. $\sqrt{400}$

13. $\sqrt{324}$

14. $-\sqrt{361}$

15. $\sqrt{\dfrac{25}{9}}$

16. $\sqrt{\dfrac{1}{225}}$

17. $\sqrt{.16}$

18. $\sqrt{.0121}$

For Problems 19–28, use your calculator or the table in the appendix to evaluate each radical.

19. $\sqrt{576}$ **20.** $\sqrt{7569}$

21. $\sqrt{2304}$ **22.** $\sqrt{9801}$

23. $\sqrt{784}$ **24.** $\sqrt{1849}$

25. $\sqrt{4225}$ **26.** $\sqrt{2704}$

27. $\sqrt{3364}$ **28.** $\sqrt{1444}$

For Problems 29–36, use your calculator or the table in the appendix to find a rational approximation of each square root. Express your answers to the nearest hundredth.

29. $\sqrt{19}$ **30.** $\sqrt{34}$

31. $\sqrt{50}$ **32.** $\sqrt{66}$

33. $\sqrt{75}$ **34.** $\sqrt{90}$

35. $\sqrt{95}$ **36.** $\sqrt{98}$

For Problems 37–46, use your calculator or the table in the appendix to find a whole number approximation for each of the following.

37. $\sqrt{4325}$ **38.** $\sqrt{7500}$

39. $\sqrt{1175}$ **40.** $\sqrt{1700}$

41. $\sqrt{9501}$ **42.** $\sqrt{8050}$

43. $\sqrt{6614}$ **44.** $\sqrt{5825}$

45. $\sqrt{3400}$ **46.** $\sqrt{2250}$

For Problems 47–58, simplify each expression by using the distributive property.

47. $7\sqrt{2} + 14\sqrt{2}$

48. $9\sqrt{3} + 4\sqrt{3}$

49. $17\sqrt{7} - 9\sqrt{7}$

50. $19\sqrt{5} - 8\sqrt{5}$

51. $6\sqrt{3} - 15\sqrt{3}$

52. $7\sqrt{6} - 21\sqrt{6}$

53. $9\sqrt{5} + 3\sqrt{5} - 6\sqrt{5}$

54. $8\sqrt{7} + 13\sqrt{7} - 9\sqrt{7}$

55. $8\sqrt{2} - 4\sqrt{3} - 9\sqrt{2} + 6\sqrt{3}$

56. $7\sqrt{5} - 9\sqrt{6} + 14\sqrt{5} - 2\sqrt{6}$

57. $6\sqrt{7} + 5\sqrt{10} - 8\sqrt{10} - 4\sqrt{7} - 11\sqrt{7} + \sqrt{10}$

58. $\sqrt{3} - \sqrt{5} + 4\sqrt{5} - 3\sqrt{3} - 9\sqrt{5} - 16\sqrt{3}$

For Problems 59–72, find a rational approximation, to the nearest one tenth, for each of the radical expressions.

59. $9\sqrt{3} + \sqrt{3}$

60. $6\sqrt{2} + 14\sqrt{2}$

61. $9\sqrt{5} - 3\sqrt{5}$

62. $18\sqrt{6} - 12\sqrt{6}$

63. $14\sqrt{2} - 15\sqrt{2}$

64. $7\sqrt{3} - 12\sqrt{3}$

65. $8\sqrt{7} - 4\sqrt{7} + 6\sqrt{7}$

66. $9\sqrt{5} - 2\sqrt{5} + \sqrt{5}$

67. $4\sqrt{3} - 2\sqrt{2}$

68. $3\sqrt{2} + \sqrt{3} - \sqrt{5}$

69. $9\sqrt{6} - 3\sqrt{5} + 2\sqrt{6} - 7\sqrt{5} - \sqrt{6}$

70. $8\sqrt{7} - 2\sqrt{10} + 4\sqrt{7} - 3\sqrt{10} - 7\sqrt{7} + 4\sqrt{10}$

71. $4\sqrt{11} - 5\sqrt{11} - 7\sqrt{11} + 2\sqrt{11} - 3\sqrt{11}$

72. $14\sqrt{13} - 17\sqrt{13} + 3\sqrt{13} - 4\sqrt{13} - 5\sqrt{13}$

73. Use the formula from Example 2 to find the period of pendulums that have lengths of 2 feet, 3.5

feet, and 4 feet. Express the answers to the nearest tenth of a second.

74. Use the formula from Example 3 with a coefficient of friction of .35 and find the speed of cars that leave skid marks of lengths 150 feet, 200 feet, and 275 feet. Express answers to the nearest mile per hour.

75. The time T, measured in seconds, that it takes for an object to fall d feet (neglecting air resistance) is given by the formula $T = \sqrt{\dfrac{d}{16}}$. Find the times that it takes objects to fall 75 feet, 125 feet, and 5280 feet. Express answers to the nearest tenth of a second.

THOUGHTS INTO WORDS

76. Why is $\sqrt{-4}$ not a real number?

77. How could you find a whole number approximation for $\sqrt{1450}$ if you did not have a calculator or table available?

Further Investigations

78. Since $2^3 = 8$, we say that 2 is the cube root of 8, which is written as $\sqrt[3]{8} = 2$. (The small 3 in the expression $\sqrt[3]{8}$ is called the index of the radical.) Likewise, since $(-2)^3 = -8$, we have $\sqrt[3]{-8} = -2$. Evaluate each of the following.

(a) $\sqrt[3]{1}$ **(b)** $\sqrt[3]{27}$

(c) $\sqrt[3]{64}$ **(d)** $\sqrt[3]{\dfrac{8}{27}}$

(e) $\sqrt[3]{\dfrac{1}{64}}$ **(f)** $\sqrt[3]{\dfrac{8}{125}}$

(g) $\sqrt[3]{-27}$ **(h)** $\sqrt[3]{-125}$

9.2 Simplifying Radicals

Note the following facts that pertain to square roots.

$$\sqrt{4 \cdot 9} = \sqrt{36} = 6$$

and

$$\sqrt{4}\,\sqrt{9} = 2 \cdot 3 = 6.$$

Thus, we observe $\sqrt{4 \cdot 9} = \sqrt{4}\sqrt{9}$. Such facts illustrate a general property.

PROPERTY 9.1

For any nonnegative real numbers a and b,

$$\sqrt{ab} = \sqrt{a}\,\sqrt{b}.$$

In other words, we say that **the square root of a product is equal to the product of the square roots**.

    Property 9.1 and the definition of square root provide the basis for expressing radical expressions in simplest radical form. For now, *simplest radical form means that the radicand contains no factors other than 1 that are perfect squares.* Let's use a few examples to illustrate this meaning of simplest radical form.

1. $\sqrt{8} = \sqrt{4 \cdot 2} = \sqrt{4}\sqrt{2} = 2\sqrt{2}.$

    ↑   ↑

    4 is a $\sqrt{4} = 2$

    perfect

    square

2. $\sqrt{45} = \sqrt{9 \cdot 5} = \sqrt{9}\sqrt{5} = 3\sqrt{5}.$

    ↑   ↑

    9 is a $\sqrt{9} = 3$

    perfect

    square

3. $\sqrt{48} = \sqrt{16 \cdot 3} = \sqrt{16}\sqrt{3} = 4\sqrt{3}.$

    ↑   ↑

    16 is a $\sqrt{16} = 4$

    perfect

    square

The first step in each example is to express the radicand of the given radical as the product of two factors, at least one of which is a perfect square other than 1. Also observe the radicands of the final radicals. In each case, the radicand *cannot* be expressed as the product of two factors, at least one of which is a perfect square other than 1. We say that the final radicals, $2\sqrt{2}$, $3\sqrt{5}$, and $4\sqrt{3}$, are in simplest radical form.

You may vary the steps somewhat in changing to simplest radical form, but the final result should be the same. Consider another sequence of steps to change $\sqrt{48}$ to simplest form.

$$\sqrt{48} = \sqrt{4 \cdot 12} = \sqrt{4}\sqrt{12} = 2\sqrt{12} = 2\sqrt{4 \cdot 3} = 2\sqrt{4}\sqrt{3} = 2 \cdot 2 \cdot \sqrt{3} = 4\sqrt{3}$$

    ↑   ↑   ↑   ↑

    4 is a   This a not   4 is a   Same result

    perfect   in simplest   perfect   as in

    square   form   square   example (3)

Another variation of the technique for changing radicals to simplest form is to prime factor the radicand and then to look for perfect squares in exponential form. Let's try this and redo the previous examples.

4. $\sqrt{8} = \sqrt{2 \cdot 2 \cdot 2} = \sqrt{2^2 \cdot 2} = \sqrt{2^2}\sqrt{2} = 2\sqrt{2}.$

    ↑   ↑

    Prime 2^2 is a

    factors   perfect

    of 8   square

5. $\sqrt{45} = \sqrt{3 \cdot 3 \cdot 5} = \sqrt{3^2 \cdot 5} = \sqrt{3^2}\sqrt{5} = 3\sqrt{5}.$

    ↑   ↑

  Prime 3^2 is a
  factors   perfect
  of 45   square

6. $\sqrt{48} = \sqrt{2 \cdot 2 \cdot 2 \cdot 2 \cdot 3} = \sqrt{2^4 \cdot 3} = \sqrt{2^4}\sqrt{3} = 2^2\sqrt{3} = 4\sqrt{3}.$

    ↑   ↑   ↑

  Prime 2^4 is a $\sqrt{2^4} = 2^2$
  factors   perfect   since
  of 48   square $2^2 \cdot 2^2 = 2^4$

The following examples further illustrate the process of changing to simplest radical form. Only the major steps are shown, so be sure that you can fill in the details.

7. $\sqrt{56} = \sqrt{4}\sqrt{14} = 2\sqrt{14}.$

8. $\sqrt{75} = \sqrt{25}\sqrt{3} = 5\sqrt{3}.$

9. $\sqrt{108} = \sqrt{2 \cdot 2 \cdot 3 \cdot 3 \cdot 3} = \sqrt{2^2 \cdot 3^2}\sqrt{3} = 6\sqrt{3}.$

10. $5\sqrt{12} = 5\sqrt{4}\sqrt{3} = 5 \cdot 2 \cdot \sqrt{3} = 10\sqrt{3}.$

We use the following property when changing radicals with variables in the radicand to simplest radical form.

PROPERTY 9.2

If a is any nonnegative real number, then

$$\sqrt{a^2} = a.$$

The next few examples illustrate the use of Property 9.2 along with our previous knowledge of radicals. The variables represent positive real numbers.

11. $\sqrt{x^2 y} = \sqrt{x^2}\sqrt{y} = x\sqrt{y}.$

12. $\sqrt{4x^3} = \sqrt{4x^2}\sqrt{x} = 2x\sqrt{x}.$

          ↑

   $4x^2$ is a perfect
   square since
   $(2x)(2x) = 4x^2$

13. $\sqrt{8xy^3} = \sqrt{4y^2}\sqrt{2xy} = 2y\sqrt{2xy}.$

14. $\sqrt{27x^5 y^3} = \sqrt{9x^4 y^2}\sqrt{3xy} = 3x^2 y\sqrt{3xy},$

                    ↑

           $\sqrt{9x^4 y^2} = 3x^2 y$

If the numerical coefficient of the radicand is quite large, you may want to look at it in prime factored form. The next example demonstrates this idea.

15. $\sqrt{180a^6b^3} = \sqrt{2 \cdot 2 \cdot 3 \cdot 3 \cdot 5 \cdot a^6 \cdot b^2 \cdot b}$

$\qquad\qquad\quad = \sqrt{36a^6b^2}\sqrt{5b}$

$\qquad\qquad\quad = 6a^3b\sqrt{5b}.$

When simplifying expressions that contain radicals, it is often necessary to first change the radicals to simplest form and then apply the distributive property.

Example 1

Simplify $5\sqrt{8} + 3\sqrt{2}$.

Solution

$5\sqrt{8} + 3\sqrt{2} = 5\sqrt{4}\,\sqrt{2} + 3\sqrt{2}$

$\qquad\qquad\quad = 5 \cdot 2 \cdot \sqrt{2} + 3\sqrt{2}$

$\qquad\qquad\quad = 10\sqrt{2} + 3\sqrt{2}$

$\qquad\qquad\quad = (10 + 3)\sqrt{2}$

$\qquad\qquad\quad = 13\sqrt{2}$ ▲

Example 2

Simplify $2\sqrt{27} - 5\sqrt{48} + 4\sqrt{3}$.

Solution

$2\sqrt{27} - 5\sqrt{48} + 4\sqrt{3} = 2\sqrt{9}\sqrt{3} - 5\sqrt{16}\sqrt{3} + 4\sqrt{3}$

$\qquad\qquad\qquad\qquad\quad = 2 \cdot 3 \cdot \sqrt{3} - 5 \cdot 4 \cdot \sqrt{3} + 4\sqrt{3}$

$\qquad\qquad\qquad\qquad\quad = 6\sqrt{3} - 20\sqrt{3} + 4\sqrt{3}$

$\qquad\qquad\qquad\qquad\quad = (6 - 20 + 4)\sqrt{3}$

$\qquad\qquad\qquad\qquad\quad = -10\sqrt{3}$ ▲

Example 3

Simplify $\frac{1}{4}\sqrt{45} + \frac{1}{3}\sqrt{20}$.

Solution

$\frac{1}{4}\sqrt{45} + \frac{1}{3}\sqrt{20} = \frac{1}{4}\sqrt{9}\sqrt{5} + \frac{1}{3}\sqrt{4}\sqrt{5}$

$\qquad\qquad\qquad\quad = \frac{1}{4} \cdot 3 \cdot \sqrt{5} + \frac{1}{3} \cdot 2 \cdot \sqrt{5}$

$\qquad\qquad\qquad\quad = \frac{3}{4}\sqrt{5} + \frac{2}{3}\sqrt{5}$

$\qquad\qquad\qquad\quad = \left(\frac{3}{4} + \frac{2}{3}\right)\sqrt{5} = \left(\frac{9}{12} + \frac{8}{12}\right)\sqrt{5}$

$\qquad\qquad\qquad\quad = \frac{17}{12}\sqrt{5}$ ▲

Problem Set 9.2

For Problems 1−30, change each radical to simplest radical form.

1. $\sqrt{24}$

2. $\sqrt{54}$

3. $\sqrt{18}$

4. $\sqrt{50}$

5. $\sqrt{27}$

6. $\sqrt{12}$

7. $\sqrt{40}$

8. $\sqrt{90}$

9. $\sqrt{28}$

10. $\sqrt{63}$

11. $\sqrt{80}$

12. $\sqrt{125}$

13. $\sqrt{117}$

14. $\sqrt{126}$

15. $4\sqrt{72}$

16. $8\sqrt{98}$

17. $3\sqrt{75}$

18. $5\sqrt{108}$

19. $-5\sqrt{20}$

20. $-6\sqrt{45}$

21. $-8\sqrt{96}$

22. $-4\sqrt{54}$

23. $\frac{3}{2}\sqrt{8}$

24. $\frac{5}{2}\sqrt{32}$

25. $\frac{3}{4}\sqrt{12}$

26. $\frac{4}{5}\sqrt{27}$

27. $-\frac{2}{3}\sqrt{45}$

28. $-\frac{3}{5}\sqrt{125}$

29. $-\sqrt{150}$

30. $-\sqrt{112}$

For Problems 31−52, express each radical in simplest radical form. All variables represent nonnegative real numbers.

31. $\sqrt{x^2 y^3}$

32. $\sqrt{xy^4}$

33. $\sqrt{2x^2 y}$

34. $\sqrt{3x^2 y^2}$

35. $\sqrt{8x^2}$

36. $\sqrt{24x^3}$

37. $\sqrt{27a^3 b}$

38. $\sqrt{45a^2 b^4}$

39. $\sqrt{144ab^5}$

40. $\sqrt{121a^3 b^6}$

41. $\sqrt{63x^4 y^2}$

42. $\sqrt{28x^3 y}$

43. $3\sqrt{48x^2}$

44. $5\sqrt{12x^2 y^2}$

45. $-6\sqrt{72x^7}$

46. $-8\sqrt{80y^9}$

47. $\frac{2}{9}\sqrt{54xy}$

48. $\frac{4}{3}\sqrt{20xy}$

49. $\frac{3}{4}\sqrt{52x^5 y^6}$

50. $\frac{5}{4}\sqrt{56x^5 y^7}$

51. $-\frac{2}{3}\sqrt{169a^8}$

52. $-\frac{2}{7}\sqrt{196a^{10}}$

For Problems 53−66, simplify each expression.

53. $7\sqrt{32} + 5\sqrt{2}$

54. $6\sqrt{48} + 5\sqrt{3}$

55. $4\sqrt{45} - 9\sqrt{5}$

56. $7\sqrt{24} - 12\sqrt{6}$

57. $3\sqrt{50} + 4\sqrt{18}$

58. $6\sqrt{98} + 3\sqrt{32}$

59. $4\sqrt{63} - 7\sqrt{28}$

60. $2\sqrt{40} - 7\sqrt{90}$

61. $5\sqrt{12} + 3\sqrt{27} - 2\sqrt{75}$

62. $4\sqrt{18} - 6\sqrt{50} - 3\sqrt{72}$

63. $\frac{1}{2}\sqrt{20} + \frac{2}{3}\sqrt{45} - \frac{1}{4}\sqrt{80}$

64. $\frac{1}{3}\sqrt{12} - \frac{3}{2}\sqrt{48} + \frac{3}{4}\sqrt{108}$

65. $3\sqrt{8} - 5\sqrt{20} - 7\sqrt{18} - 9\sqrt{125}$

66. $5\sqrt{27} - 3\sqrt{24} + 8\sqrt{54} - 7\sqrt{75}$

67. Explain how you would help someone express $5\sqrt{72}$ in simplest radical form.

68. Explain your thought process when expressing $\sqrt{153}$ in simplest radical form.

Further Investigations

69. Express each of the following in simplest radical form. The divisibility rules given in Problem Set 1.2 should be of some help.

(a) $\sqrt{162}$ (b) $\sqrt{279}$

(c) $\sqrt{275}$ (d) $\sqrt{212}$

70. Use your calculator and evaluate each expression in Problems 53–66. Then evaluate the simplified expression that you obtained when doing those problems. Your two results for each problem should be the same.

71. Sometimes a fairly good estimate can be made of a radical expression by using whole number approximations. For example, $5\sqrt{35} + 7\sqrt{50}$ is approximately $5(6) + 7(7) = 79$. Using a calculator we find that $5\sqrt{35} + 7\sqrt{50} = 79.1$ to the nearest tenth. In this case our whole number estimate is very good.

For each of the following, first make a whole number estimate, and then use your calculator to see how well you estimated.

(a) $3\sqrt{10} - 4\sqrt{24} + 6\sqrt{65}$

(b) $9\sqrt{27} + 5\sqrt{37} - 3\sqrt{80}$

(c) $12\sqrt{5} + 13\sqrt{18} + 9\sqrt{47}$

(d) $3\sqrt{98} - 4\sqrt{83} - 7\sqrt{120}$

(e) $4\sqrt{170} + 2\sqrt{198} + 5\sqrt{227}$

(f) $-3\sqrt{256} - 6\sqrt{287} + 11\sqrt{321}$

72. Evaluate $\sqrt{x^2}$ for $x = 5$, $x = 4$, $x = -3$, $x = 9$, $x = -8$, and $x = -11$. For which values of x does $\sqrt{x^2} = x$? For which values of x does $\sqrt{x^2} = -x$? For which values of x does $\sqrt{x^2} = |x|$?

9.3   More on Simplifying Radicals

Another property of square roots is motivated by the following examples.

$$\sqrt{\frac{36}{9}} = \sqrt{4} = 2$$

and

$$\frac{\sqrt{36}}{\sqrt{9}} = \frac{6}{3} = 2$$

Thus, we see that $\sqrt{\dfrac{36}{9}} = \dfrac{\sqrt{36}}{\sqrt{9}}$. We can state the following general property.

PROPERTY 9.3

For any nonnegative real numbers a and b, b not zero,

$$\sqrt{\frac{a}{b}} = \frac{\sqrt{a}}{\sqrt{b}}.$$

Property 9.3 states that **the square root of the quotient is equal to the quotient of the square roots**.

To evaluate a radical such as $\sqrt{\dfrac{25}{4}}$, for which the numerator and denominator of the fractional radicand are perfect squares, you may use Property 9.3 or merely rely on the definition of square root.

$$\sqrt{\frac{25}{4}} = \frac{\sqrt{25}}{\sqrt{4}} = \frac{5}{2}$$

or

$$\sqrt{\frac{25}{4}} = \frac{5}{2} \quad \text{because } \frac{5}{2} \cdot \frac{5}{2} = \frac{25}{4}$$

Sometimes it is easier to do the indicated division first and then to find the square root, as we see in the next example.

$$\sqrt{\frac{324}{9}} = \sqrt{36} = 6$$

At this time let's extend our concept of simplest radical form. An algebraic expression that contains only square root radicals is said to be in **simplest radical form** if the following conditions are satisfied.

1. No fraction appears within a radical sign. $\left(\sqrt{\dfrac{2}{3}} \text{ violates this condition.} \right)$

2. No radical appears in the denominator. $\left(\dfrac{5}{\sqrt{8}} \text{ violates this condition.} \right)$

3. No radicand when expressed in prime factored form contains a factor raised to a power greater than 1. ($\sqrt{8} = \sqrt{2^3}$ violates this condition.)

Condition 3 can also be stated as "no radicand contains a factor that is a perfect square other than 1." Again $\sqrt{8}$ violates this condition since $\sqrt{8} = \sqrt{4 \cdot 2}$ and 4 is a perfect square.

The next three examples present situations where only the denominator of the radicand is a perfect square.

Example 1

Simplify $\sqrt{\dfrac{13}{4}}$.

Solution

$$\sqrt{\frac{13}{4}} = \frac{\sqrt{13}}{\sqrt{4}} = \frac{\sqrt{13}}{2}$$

▲

Example 2

Simplify $\sqrt{\dfrac{28}{9}}$.

Solution

$$\sqrt{\dfrac{28}{9}} = \dfrac{\sqrt{28}}{\sqrt{9}} = \dfrac{\sqrt{28}}{3} = \dfrac{\sqrt{4}\sqrt{7}}{3} = \dfrac{2\sqrt{7}}{3}$$

↑
Don't stop here.
The radical in the
numerator can be
simplified.

▲

Example 3

Simplify $\dfrac{\sqrt{12}}{\sqrt{16}}$.

Solution A

$$\dfrac{\sqrt{12}}{\sqrt{16}} = \dfrac{\sqrt{12}}{4} = \dfrac{\sqrt{4}\sqrt{3}}{4} = \dfrac{2\sqrt{3}}{4} = \dfrac{\sqrt{3}}{2}$$

Solution B

$$\dfrac{\sqrt{12}}{\sqrt{16}} = \sqrt{\dfrac{12}{16}} = \sqrt{\dfrac{3}{4}} = \dfrac{\sqrt{3}}{\sqrt{4}} = \dfrac{\sqrt{3}}{2}$$

Reduce the
fraction.

▲

The two approaches to Example 3 illustrate the need to *think first* and then *push the pencil*. You may find one approach easier than another.

Now let's consider an example where neither the numerator nor the denominator of the radicand is a perfect square.

Example 4

Simplify $\sqrt{\dfrac{2}{3}}$.

— Form of I

Solution A

$$\sqrt{\dfrac{2}{3}} = \dfrac{\sqrt{2}}{\sqrt{3}} = \dfrac{\sqrt{2}}{\sqrt{3}} \cdot \dfrac{\sqrt{3}}{\sqrt{3}} = \dfrac{\sqrt{6}}{3}$$

Solution B

$$\sqrt{\dfrac{2}{3}} = \sqrt{\dfrac{2}{3} \cdot \dfrac{3}{3}} = \sqrt{\dfrac{6}{9}} = \dfrac{\sqrt{6}}{\sqrt{9}} = \dfrac{\sqrt{6}}{3}$$

▲

We refer to the process we used to simplify the radical in Example 4 as **rationalizing the denominator**. Notice that the denominator becomes a rational number. There is more than one way to rationalize the denominator, as the next example shows.

Example 5

Simplify $\dfrac{\sqrt{5}}{\sqrt{8}}$.

Solution A

$$\dfrac{\sqrt{5}}{\sqrt{8}} = \dfrac{\sqrt{5}}{\sqrt{8}} \cdot \dfrac{\sqrt{8}}{\sqrt{8}} = \dfrac{\sqrt{40}}{8} = \dfrac{\sqrt{4}\sqrt{10}}{8} = \dfrac{2\sqrt{10}}{8} = \dfrac{\sqrt{10}}{4}$$

Solution B

$$\frac{\sqrt{5}}{\sqrt{8}} = \frac{\sqrt{5}}{\sqrt{8}} \cdot \frac{\sqrt{2}}{\sqrt{2}} = \frac{\sqrt{10}}{\sqrt{16}} = \frac{\sqrt{10}}{4}$$

Solution C

$$\frac{\sqrt{5}}{\sqrt{8}} = \frac{\sqrt{5}}{\sqrt{4}\sqrt{2}} = \frac{\sqrt{5}}{2\sqrt{2}} = \frac{\sqrt{5}}{2\sqrt{2}} \cdot \frac{\sqrt{2}}{\sqrt{2}} = \frac{\sqrt{10}}{4}$$ ▲

Study the following examples and check our final radicals according to the three conditions we listed on page 387.

Example 6

Simplify each of the following.

(a) $\dfrac{3}{\sqrt{x}}$ (b) $\sqrt{\dfrac{2x}{3y}}$ (c) $\dfrac{3\sqrt{5}}{\sqrt{6}}$ (d) $\sqrt{\dfrac{4x^2}{9y}}$

Solution

(a) $\dfrac{3}{\sqrt{x}} = \dfrac{3}{\sqrt{x}} \cdot \dfrac{\sqrt{x}}{\sqrt{x}} = \dfrac{3\sqrt{x}}{x}.$

(b) $\sqrt{\dfrac{2x}{3y}} = \dfrac{\sqrt{2x}}{\sqrt{3y}} = \dfrac{\sqrt{2x}}{\sqrt{3y}} \cdot \dfrac{\sqrt{3y}}{\sqrt{3y}} = \dfrac{\sqrt{6xy}}{3y}.$

(c) $\dfrac{3\sqrt{5}}{\sqrt{6}} = \dfrac{3\sqrt{5}}{\sqrt{6}} \cdot \dfrac{\sqrt{6}}{\sqrt{6}} = \dfrac{3\sqrt{30}}{6} = \dfrac{\sqrt{30}}{2}.$

(d) $\sqrt{\dfrac{4x^2}{9y}} = \dfrac{\sqrt{4x^2}}{\sqrt{9y}} = \dfrac{2x}{\sqrt{9}\sqrt{y}} = \dfrac{2x}{3\sqrt{y}} = \dfrac{2x}{3\sqrt{y}} \cdot \dfrac{\sqrt{y}}{\sqrt{y}} = \dfrac{2x\sqrt{y}}{3y}.$ ▲

Let's return again to the idea of simplifying expressions that contain radicals. Sometimes it may appear as if no simplifying can be done; however, after the individual radicals have been changed to simplest form then the distributive property may apply. Consider the following examples.

Example 7

Simplify $5\sqrt{2} + \dfrac{3}{\sqrt{2}}.$

Solution

$$5\sqrt{2} + \frac{3}{\sqrt{2}} = 5\sqrt{2} + \frac{3}{\sqrt{2}} \cdot \frac{\sqrt{2}}{\sqrt{2}} = 5\sqrt{2} + \frac{3\sqrt{2}}{2}$$

$$= \left(5 + \frac{3}{2}\right)\sqrt{2} = \left(\frac{10}{2} + \frac{3}{2}\right)\sqrt{2}$$

$$= \frac{13}{2}\sqrt{2} \quad \text{or} \quad \frac{13\sqrt{2}}{2}$$ ▲

Example 8

Simplify $\sqrt{\dfrac{3}{2}} + \sqrt{24}.$

Solution

$$\sqrt{\frac{3}{2}} + \sqrt{24} = \frac{\sqrt{3}}{\sqrt{2}} + \sqrt{24} = \frac{\sqrt{3}}{\sqrt{2}} \cdot \frac{\sqrt{2}}{\sqrt{2}} + \sqrt{4}\sqrt{6}$$

$$= \frac{\sqrt{6}}{2} + 2\sqrt{6}$$

$$= \left(\frac{1}{2} + 2\right)\sqrt{6}$$

$$= \left(\frac{1}{2} + \frac{4}{2}\right)\sqrt{6}$$

$$= \frac{5}{2}\sqrt{6}$$

▲

Example 9   Simplify $3\sqrt{\frac{7}{4}} - 14\sqrt{\frac{1}{7}} + 5\sqrt{28}$.

Solution

$$3\sqrt{\frac{7}{4}} - 14\sqrt{\frac{1}{7}} + 5\sqrt{28} = \frac{3\sqrt{7}}{\sqrt{4}} - \frac{14\sqrt{1}}{\sqrt{7}} + 5\sqrt{4}\sqrt{7}$$

$$= \frac{3\sqrt{7}}{2} - \frac{14}{\sqrt{7}} \cdot \frac{\sqrt{7}}{\sqrt{7}} + 10\sqrt{7}$$

$$= \frac{3\sqrt{7}}{2} - \frac{14\sqrt{7}}{7} + 10\sqrt{7}$$

$$= \frac{3\sqrt{7}}{2} - 2\sqrt{7} + 10\sqrt{7}$$

$$= \left(\frac{3}{2} - 2 + 10\right)\sqrt{7}$$

$$= \left(\frac{3}{2} - \frac{4}{2} + \frac{20}{2}\right)\sqrt{7} = \frac{19}{2}\sqrt{7}$$

▲

Problem Set 9.3

For Problems 1–10, evaluate each radical.

1. $\sqrt{\frac{16}{25}}$

2. $\sqrt{\frac{4}{49}}$

3. $-\sqrt{\frac{81}{9}}$

4. $-\sqrt{\frac{64}{16}}$

5. $\sqrt{\frac{1}{64}}$

6. $\sqrt{\frac{100}{121}}$

7. $\sqrt{\frac{169}{144}}$

8. $-\sqrt{\frac{196}{144}}$

9. $-\sqrt{\frac{25}{256}}$

10. $-\sqrt{\frac{289}{225}}$

For Problems 11–40, change each radical to simplest radical form.

11. $\sqrt{\frac{19}{25}}$

12. $\sqrt{\frac{17}{4}}$

13. $\sqrt{\frac{8}{49}}$

14. $\sqrt{\frac{24}{25}}$

15. $\frac{\sqrt{28}}{\sqrt{9}}$

16. $\frac{\sqrt{45}}{\sqrt{16}}$

17. $\frac{\sqrt{12}}{\sqrt{36}}$

18. $\frac{\sqrt{20}}{\sqrt{64}}$

19. $\sqrt{\dfrac{3}{2}}$

20. $\sqrt{\dfrac{2}{5}}$

21. $\sqrt{\dfrac{5}{8}}$

22. $\sqrt{\dfrac{7}{12}}$

23. $\dfrac{\sqrt{56}}{\sqrt{8}}$

24. $\dfrac{\sqrt{55}}{\sqrt{11}}$

25. $\dfrac{\sqrt{63}}{\sqrt{7}}$

26. $\dfrac{\sqrt{96}}{\sqrt{6}}$

27. $\dfrac{\sqrt{5}}{\sqrt{18}}$

28. $\dfrac{\sqrt{3}}{\sqrt{32}}$

29. $\dfrac{\sqrt{4}}{\sqrt{27}}$

30. $\dfrac{\sqrt{9}}{\sqrt{48}}$

31. $\sqrt{\dfrac{1}{24}}$

32. $\sqrt{\dfrac{1}{12}}$

33. $\dfrac{2\sqrt{3}}{\sqrt{5}}$

34. $\dfrac{3\sqrt{2}}{\sqrt{6}}$

35. $\dfrac{4\sqrt{2}}{3\sqrt{3}}$

36. $\dfrac{2\sqrt{5}}{7\sqrt{8}}$

37. $\dfrac{3\sqrt{7}}{4\sqrt{12}}$

38. $\dfrac{6\sqrt{12}}{5\sqrt{24}}$

39. $\sqrt{4\dfrac{1}{9}}$

40. $\sqrt{3\dfrac{1}{4}}$

For Problems 41–60, change each radical to simplest radical form. All variables represent positive real numbers.

41. $\dfrac{3}{\sqrt{x}}$

42. $\dfrac{2}{\sqrt{xy}}$

43. $\dfrac{5}{\sqrt{2x}}$

44. $\dfrac{7}{\sqrt{3y}}$

45. $\sqrt{\dfrac{3}{x}}$

46. $\sqrt{\dfrac{8}{x}}$

47. $\sqrt{\dfrac{12}{x^2}}$

48. $\sqrt{\dfrac{27}{4y^2}}$

49. $\dfrac{\sqrt{2x}}{\sqrt{5y}}$

50. $\dfrac{\sqrt{3y}}{\sqrt{32x}}$

51. $\dfrac{\sqrt{5x}}{\sqrt{27y}}$

52. $\dfrac{\sqrt{3x^2}}{\sqrt{5y^2}}$

53. $\dfrac{\sqrt{2x^3}}{\sqrt{8y}}$

54. $\dfrac{\sqrt{5x^2}}{\sqrt{45y^3}}$

55. $\sqrt{\dfrac{9}{x^3}}$

56. $\sqrt{\dfrac{25}{y^5}}$

57. $\dfrac{4}{\sqrt{x^7}}$

58. $\dfrac{14}{\sqrt{x^5}}$

59. $\dfrac{3\sqrt{x}}{2\sqrt{y^3}}$

60. $\dfrac{5\sqrt{x}}{7\sqrt{xy}}$

For Problems 61–72, simplify each expression.

61. $7\sqrt{3} + \sqrt{\dfrac{1}{3}}$

62. $-3\sqrt{2} + \sqrt{\dfrac{1}{2}}$

63. $4\sqrt{10} - \sqrt{\dfrac{2}{5}}$

64. $8\sqrt{5} - 3\sqrt{\dfrac{1}{5}}$

65. $-2\sqrt{5} - 5\sqrt{\dfrac{1}{5}}$

66. $6\sqrt{7} + 4\sqrt{\dfrac{1}{7}}$

67. $-3\sqrt{6} - \dfrac{5\sqrt{2}}{\sqrt{3}}$

68. $4\sqrt{8} - \dfrac{6}{\sqrt{2}}$

69. $4\sqrt{12} + \dfrac{3}{\sqrt{3}} - 5\sqrt{27}$

70. $3\sqrt{8} - 4\sqrt{18} - \dfrac{6}{\sqrt{2}}$

71. $\dfrac{9\sqrt{5}}{\sqrt{3}} - 6\sqrt{60} + \dfrac{10\sqrt{3}}{\sqrt{5}}$

72. $-2\sqrt{3} - 3\sqrt{48} + \dfrac{3}{\sqrt{3}}$

73. Your friend simplifies $\sqrt{\dfrac{6}{8}}$ as follows:

$$\sqrt{\frac{6}{8}} = \frac{\sqrt{6}}{\sqrt{8}} = \frac{\sqrt{6}}{\sqrt{8}} \cdot \frac{\sqrt{8}}{\sqrt{8}} = \frac{\sqrt{48}}{8} = \frac{\sqrt{16}\sqrt{3}}{8}$$

$$= \frac{4\sqrt{3}}{8} = \frac{\sqrt{3}}{2}$$

Could you show him a much shorter way to simplify this expression?

74. Is the expression $3\sqrt{2} + \sqrt{50}$ in simplest radical form? Why or why not?

Further Investigations

75. Use your calculator and evaluate each expression in Problems 61–72. Then evaluate the simplified expression that you obtained when doing those problems. Your two results for each problem should be the same.

9.4 Products and Quotients Involving Radicals

We use Property 9.1 ($\sqrt{ab} = \sqrt{a}\sqrt{b}$) to multiply square roots and in some cases to simplify the resulting radical. The following examples represent several types of multiplication problems that involve radicals.

Example I

Multiply and simplify where possible.

    (a) $\sqrt{3}\sqrt{12}$ *(b)* $\sqrt{3}\sqrt{15}$ *(c)* $\sqrt{7}\sqrt{8}$

    (d) $\sqrt{5}\sqrt{11}$ *(e)* $(3\sqrt{2})(4\sqrt{3})$

Solution

    (a) $\sqrt{3}\sqrt{12} = \sqrt{36} = 6.$

    (b) $\sqrt{3}\sqrt{15} = \sqrt{45} = \sqrt{9}\sqrt{5} = 3\sqrt{5}.$

    (c) $\sqrt{7}\sqrt{8} = \sqrt{56} = \sqrt{4}\,\sqrt{14} = 2\sqrt{14}.$

    (d) $\sqrt{5}\sqrt{11} = \sqrt{55}.$

    (e) $(3\sqrt{2})(4\sqrt{3}) = 3 \cdot 4 \cdot \sqrt{2} \cdot \sqrt{3} = 12\sqrt{6}.$ ▲

Recall our use of the distributive property when we need to find the product of a monomial and a polynomial. For example, $2x(3x + 4) = 2x(3x) + 2x(4) = 6x^2 + 8x$. Likewise, the distributive property and Property 9.1 provide the basis for finding certain special products involving radicals. The following examples demonstrate this idea.

Example 2

Multiply and simplify where possible.

    (a) $\sqrt{2}(\sqrt{3} + \sqrt{5})$ *(b)* $\sqrt{3}(\sqrt{12} - \sqrt{6})$

    (c) $\sqrt{8}(\sqrt{2} - 3)$ *(d)* $\sqrt{x}(\sqrt{x} + \sqrt{y})$

Solution

(a) $\sqrt{2}(\sqrt{3} + \sqrt{5}) = \sqrt{2}\sqrt{3} + \sqrt{2}\sqrt{5} = \sqrt{6} + \sqrt{10}.$

(b) $\sqrt{3}(\sqrt{12} - \sqrt{6}) = \sqrt{3}\sqrt{12} - \sqrt{3}\sqrt{6}$

$\qquad\qquad\qquad = \sqrt{36} - \sqrt{18}$

$\qquad\qquad\qquad = 6 - \sqrt{9}\sqrt{2}$

$\qquad\qquad\qquad = 6 - 3\sqrt{2}.$

(c) $\sqrt{8}(\sqrt{2} - 3) = \sqrt{8}\sqrt{2} - \sqrt{8}(3)$

$\qquad\qquad\qquad = \sqrt{16} - 3\sqrt{8}$

$\qquad\qquad\qquad = 4 - 3\sqrt{4}\sqrt{2}$

$\qquad\qquad\qquad = 4 - 6\sqrt{2}.$

(d) $\sqrt{x}(\sqrt{x} + \sqrt{y}) = \sqrt{x}\sqrt{x} + \sqrt{x}\sqrt{y}$

$\qquad\qquad\qquad = x + \sqrt{xy}.$ ▲

The distributive property plays a central role in finding the product of two binomials. For example, $(x + 2)(x + 3) = x(x + 3) + 2(x + 3) = x^2 + 3x + 2x + 6 = x^2 + 5x + 6$. We can find the product of two binomial expressions involving radicals in a similar fashion.

Example 3

Multiply and simplify.

(a) $(\sqrt{3} + \sqrt{5})(\sqrt{2} + \sqrt{6})$ $\qquad$ (b) $(\sqrt{7} - 3)(\sqrt{7} + 6)$

Solution

(a) $(\sqrt{3} + \sqrt{5})(\sqrt{2} + \sqrt{6}) = \sqrt{3}(\sqrt{2} + \sqrt{6}) + \sqrt{5}(\sqrt{2} + \sqrt{6})$

$\qquad\qquad\qquad\qquad\qquad = \sqrt{3}\sqrt{2} + \sqrt{3}\sqrt{6} + \sqrt{5}\sqrt{2} + \sqrt{5}\sqrt{6}$

$\qquad\qquad\qquad\qquad\qquad = \sqrt{6} + \sqrt{18} + \sqrt{10} + \sqrt{30}$

$\qquad\qquad\qquad\qquad\qquad = \sqrt{6} + 3\sqrt{2} + \sqrt{10} + \sqrt{30}.$

(b) $(\sqrt{7} - 3)(\sqrt{7} + 6) = \sqrt{7}(\sqrt{7} + 6) - 3(\sqrt{7} + 6)$

$\qquad\qquad\qquad\qquad\quad = \sqrt{7}\sqrt{7} + 6\sqrt{7} - 3\sqrt{7} - 18$

$\qquad\qquad\qquad\qquad\quad = 7 + 6\sqrt{7} - 3\sqrt{7} - 18$

$\qquad\qquad\qquad\qquad\quad = -11 + 3\sqrt{7}.$ ▲

If the binomials are of the form $(a + b)(a - b)$, then we can use the multiplication pattern $(a + b)(a - b) = a^2 - b^2$.

Example 4

Multiply and simplify.

(a) $(\sqrt{6} + 2)(\sqrt{6} - 2)$ $\qquad$ (b) $(3 - \sqrt{5})(3 + \sqrt{5})$

(c) $(\sqrt{8} + \sqrt{5})(\sqrt{8} - \sqrt{5})$

Solution

(a) $(\sqrt{6} + 2)(\sqrt{6} - 2) = (\sqrt{6})^2 - 2^2 = 6 - 4 = 2$

(b) $(3 - \sqrt{5})(3 + \sqrt{5}) = 3^2 - (\sqrt{5})^2 = 9 - 5 = 4$

(c) $(\sqrt{8} + \sqrt{5})(\sqrt{8} - \sqrt{5}) = (\sqrt{8})^2 - (\sqrt{5})^2 = 8 - 5 = 3$ ▲

Note that in each part of Example 4 the final product contains no radicals. This will happen whenever we multiply expressions such as $\sqrt{x} + \sqrt{y}$ and $\sqrt{x} - \sqrt{y}$, where x and y are rational numbers.

$$(\sqrt{x} + \sqrt{y})(\sqrt{x} - \sqrt{y}) = (\sqrt{x})^2 - (\sqrt{y})^2 = x - y$$

Expressions such as $\sqrt{8} + \sqrt{5}$ and $\sqrt{8} - \sqrt{5}$ are called **conjugates** of each other. Likewise, $\sqrt{6} + 2$ and $\sqrt{6} - 2$ are conjugates as are $3 - \sqrt{5}$ and $3 + \sqrt{5}$. Now let's see how conjugates can be used to rationalize denominators.

Example 5

Simplify $\dfrac{4}{\sqrt{5} + \sqrt{2}}$.

Solution

$$\frac{4}{\sqrt{5} + \sqrt{2}} = \frac{4}{\sqrt{5} + \sqrt{2}} \cdot \frac{\sqrt{5} - \sqrt{2}}{\sqrt{5} - \sqrt{2}} \qquad \frac{\sqrt{5} - \sqrt{2}}{\sqrt{5} - \sqrt{2}} \text{ is merely a form of I}$$

$$= \frac{4(\sqrt{5} - \sqrt{2})}{(\sqrt{5} + \sqrt{2})(\sqrt{5} - \sqrt{2})}$$

$$= \frac{4(\sqrt{5} - \sqrt{2})}{5 - 2}$$

$$= \frac{4(\sqrt{5} - \sqrt{2})}{3} \qquad \text{or} \qquad \frac{4\sqrt{5} - 4\sqrt{2}}{3}$$

Either answer
is acceptable.   ▲

The next four examples further illustrate the process of rationalizing and simplifying expressions that contain binomial denominators.

Example 6

Rationalize the denominator and simplify $\dfrac{\sqrt{3}}{\sqrt{6} - 2}$.

Solution

$$\frac{\sqrt{3}}{\sqrt{6} - 2} = \frac{\sqrt{3}}{\sqrt{6} - 2} \cdot \frac{\sqrt{6} + 2}{\sqrt{6} + 2}$$

$$= \frac{\sqrt{3}(\sqrt{6} + 2)}{(\sqrt{6} - 2)(\sqrt{6} + 2)}$$

$$= \frac{\sqrt{18} + 2\sqrt{3}}{6 - 4}$$

$$= \frac{3\sqrt{2} + 2\sqrt{3}}{2} \qquad \sqrt{18} = \sqrt{9}\sqrt{2} = 3\sqrt{2}$$

▲

Example 7

Rationalize the denominator and simplify $\dfrac{14}{2\sqrt{3} + \sqrt{5}}$.

Solution

$$\frac{14}{2\sqrt{3} + \sqrt{5}} = \frac{14}{2\sqrt{3} + \sqrt{5}} \cdot \frac{2\sqrt{3} - \sqrt{5}}{2\sqrt{3} - \sqrt{5}}$$

$$= \frac{14(2\sqrt{3} - \sqrt{5})}{(2\sqrt{3} + \sqrt{5})(2\sqrt{3} - \sqrt{5})}$$

$$= \frac{14(2\sqrt{3} - \sqrt{5})}{12 - 5} = \frac{14(2\sqrt{3} - \sqrt{5})}{7}$$

$$= 2(2\sqrt{3} - \sqrt{5}) \quad \text{or} \quad 4\sqrt{3} - 2\sqrt{5}$$

▲

Example 8

Rationalize the denominator and simplify $\dfrac{\sqrt{x} + 2}{\sqrt{x} - 3}$.

Solution

$$\frac{\sqrt{x} + 2}{\sqrt{x} - 3} = \frac{\sqrt{x} + 2}{\sqrt{x} - 3} \cdot \frac{\sqrt{x} + 3}{\sqrt{x} + 3}$$

$$= \frac{(\sqrt{x} + 2)(\sqrt{x} + 3)}{(\sqrt{x} - 3)(\sqrt{x} + 3)}$$

$$= \frac{x + 2\sqrt{x} + 3\sqrt{x} + 6}{x - 9}$$

$$= \frac{x + 5\sqrt{x} + 6}{x - 9}$$

▲

Example 9

Rationalize the denominator and simplify $\dfrac{3 + \sqrt{2}}{\sqrt{2} - 6}$.

Solution

$$\frac{3 + \sqrt{2}}{\sqrt{2} - 6} = \frac{3 + \sqrt{2}}{\sqrt{2} - 6} \cdot \frac{\sqrt{2} + 6}{\sqrt{2} + 6}$$

$$= \frac{(3 + \sqrt{2})(\sqrt{2} + 6)}{(\sqrt{2} - 6)(\sqrt{2} + 6)}$$

$$= \frac{3\sqrt{2} + 18 + 2 + 6\sqrt{2}}{2 - 36}$$

$$= \frac{9\sqrt{2} + 20}{-34}$$

$$= -\frac{9\sqrt{2} + 20}{34} \qquad \frac{a}{-b} = -\frac{a}{b}$$

▲

Problem Set 9.4

For Problems 1–20, multiply and simplify where possible.

1. $\sqrt{7}\sqrt{5}$

2. $\sqrt{3}\sqrt{10}$

3. $\sqrt{6}\sqrt{8}$

4. $\sqrt{6}\sqrt{12}$

5. $\sqrt{5}\sqrt{10}$

6. $\sqrt{2}\sqrt{12}$

7. $\sqrt{3}\sqrt{27}$

8. $\sqrt{2}\sqrt{32}$

9. $\sqrt{8}\sqrt{12}$

10. $\sqrt{12}\sqrt{20}$

11. $(3\sqrt{3})(5\sqrt{7})$

12. $(4\sqrt{2})(7\sqrt{5})$

13. $(-2\sqrt{2})(3\sqrt{7})$

14. $(5\sqrt{3})(-6\sqrt{10})$

15. $(3\sqrt{6})(4\sqrt{6})$

16. $(2\sqrt{7})(5\sqrt{7})$

17. $(5\sqrt{2})(4\sqrt{12})$

18. $(3\sqrt{2})(2\sqrt{27})$

19. $(4\sqrt{3})(2\sqrt{15})$

20. $(7\sqrt{7})(2\sqrt{14})$

For Problems 21–42, find the products by applying the distributive property. Express your answers in simplest radical form.

21. $\sqrt{2}(\sqrt{3} + \sqrt{5})$

22. $\sqrt{3}(\sqrt{5} + \sqrt{7})$

23. $\sqrt{6}(\sqrt{2} - 5)$

24. $\sqrt{8}(\sqrt{3} - 4)$

25. $\sqrt{3}(\sqrt{6} + \sqrt{7})$

26. $\sqrt{2}(\sqrt{8} + \sqrt{10})$

27. $\sqrt{12}(\sqrt{6} - \sqrt{8})$

28. $\sqrt{15}(\sqrt{3} - \sqrt{5})$

29. $4\sqrt{3}(\sqrt{2} - 2\sqrt{5})$

30. $3\sqrt{2}(3\sqrt{5} - 2\sqrt{7})$

31. $(\sqrt{2} + 6)(\sqrt{2} + 9)$

32. $(\sqrt{3} + 4)(\sqrt{3} + 7)$

33. $(\sqrt{6} - 5)(\sqrt{6} + 3)$

34. $(\sqrt{7} - 6)(\sqrt{7} + 1)$

35. $(\sqrt{3} + \sqrt{6})(\sqrt{6} + \sqrt{8})$

36. $(\sqrt{2} + \sqrt{6})(\sqrt{8} + \sqrt{5})$

37. $(5 + \sqrt{10})(5 - \sqrt{10})$

38. $(\sqrt{11} - 2)(\sqrt{11} + 2)$

39. $(3\sqrt{2} - \sqrt{3})(3\sqrt{2} + \sqrt{3})$

40. $(5\sqrt{6} - \sqrt{2})(5\sqrt{6} + \sqrt{2})$

41. $(5\sqrt{3} + 2\sqrt{6})(5\sqrt{3} - 2\sqrt{6})$

42. $(4\sqrt{5} + 5\sqrt{7})(4\sqrt{5} - 5\sqrt{7})$

For Problems 43–54, find each product and express your answers in simplest radical form. All variables represent nonnegative real numbers.

43. $\sqrt{xy}\sqrt{x}$

44. $\sqrt{x^2y}\sqrt{y}$

45. $\sqrt{3x}\sqrt{6y}$

46. $\sqrt{2a}\sqrt{6b}$

47. $(4\sqrt{a})(3\sqrt{ab})$

48. $(5\sqrt{a})(6\sqrt{a})$

49. $\sqrt{2x}(\sqrt{3x} - \sqrt{6y})$

50. $\sqrt{6x}(\sqrt{2x} - \sqrt{4y})$

51. $(\sqrt{x} + 5)(\sqrt{x} - 3)$

52. $(3 + \sqrt{x})(7 - \sqrt{x})$

53. $(\sqrt{x} + 7)(\sqrt{x} - 7)$

54. $(\sqrt{x} + 9)(\sqrt{x} - 9)$

For Problems 55–70, rationalize the denominators and simplify. All variables represent positive real numbers.

55. $\dfrac{3}{\sqrt{2} + 4}$

56. $\dfrac{5}{\sqrt{3}+7}$

57. $\dfrac{8}{\sqrt{6}-2}$

58. $\dfrac{10}{3-\sqrt{7}}$

59. $\dfrac{2}{\sqrt{5}+\sqrt{3}}$

60. $\dfrac{3}{\sqrt{6}+\sqrt{5}}$

61. $\dfrac{10}{2-3\sqrt{3}}$

62. $\dfrac{5}{3\sqrt{2}-4}$

63. $\dfrac{4}{\sqrt{x}-2}$

64. $\dfrac{7}{\sqrt{x}+4}$

65. $\dfrac{\sqrt{x}}{\sqrt{x}+3}$

66. $\dfrac{\sqrt{y}}{\sqrt{y}-6}$

67. $\dfrac{\sqrt{a}+2}{\sqrt{a}-5}$

68. $\dfrac{\sqrt{a}-3}{\sqrt{a}+1}$

69. $\dfrac{2+\sqrt{3}}{3-\sqrt{2}}$

70. $\dfrac{3-\sqrt{5}}{4+\sqrt{8}}$

THOUGHTS INTO WORDS

71. Explain how the distributive property has been used in this chapter.

72. How would you help someone rationalize the denominator and simplify the expression $\dfrac{\sqrt{4}}{\sqrt{12}+\sqrt{8}}$?

Further Investigations

73. Return to Problems 55–62 and use your calculator to evaluate each given expression. Then evaluate the result that you obtained when you rationalized the denominators.

9.5 **Solving Radical Equations**

Equations that contain radicals with variables in the radicand are called **radical equations**. The following are examples of radical equations that involve one variable.

$$\sqrt{x}=3, \qquad \sqrt{2x+1}=5, \qquad \sqrt{3x+4}=-4,$$
$$\sqrt{5s-2}=\sqrt{2s+19}, \qquad \sqrt{2y-4}=y-2, \qquad \sqrt{x+6}=x$$

In order to solve such equations we need the following property of equality.

PROPERTY 9.4   For real numbers a and b, if $a = b$, then

$$a^2 = b^2.$$

Property 9.4 states that we can **square both sides of an equation**. However, squaring both sides of an equation sometimes produces results that do not satisfy the original equation. Let's consider two examples to illustrate the point.

Example 1

Solve $\sqrt{x} = 3$.

Solution

$$\sqrt{x} = 3$$
$$(\sqrt{x})^2 = 3^2 \qquad \text{Square both sides.}$$
$$x = 9$$

Since $\sqrt{9} = 3$, the solution set is $\{9\}$.   ▲

Example 2

Solve $\sqrt{x} = -3$.

Solution

$$\sqrt{x} = -3$$
$$(\sqrt{x})^2 = (-3)^2 \qquad \text{Square both sides.}$$
$$x = 9$$

Since $\sqrt{9} \neq -3$, 9 is not a solution and the solution set is $\varnothing$.   ▲

In general, squaring both sides of an equation will produce an equation that has all of the solutions of the original equation, but it may also have some extra solutions that will not satisfy the original equation. (Such extra solutions are called **extraneous solutions** or **roots**.) Therefore, when using the "squaring" property (Property 9.4) you *must* check each potential solution in the original equation.

Let's consider some examples to demonstrate different situations that arise when solving radical equations.

Example 3

Solve $\sqrt{2x + 1} = 5$.

Solution

$$\sqrt{2x + 1} = 5$$
$$(\sqrt{2x + 1})^2 = 5^2 \qquad \text{Square both sides.}$$
$$2x + 1 = 25$$
$$2x = 24$$
$$x = 12$$

✔ **Check**

$$\sqrt{2x + 1} = 5$$
$$\sqrt{2(12) + 1} \stackrel{?}{=} 5$$
$$\sqrt{24 + 1} \stackrel{?}{=} 5$$
$$\sqrt{25} \stackrel{?}{=} 5$$
$$5 = 5$$

The solution set is $\{12\}$.   ▲

Example 4

Solution

Solve $\sqrt{3x + 4} = -4$.

$$\sqrt{3x + 4} = -4$$
$$(\sqrt{3x + 4})^2 = (-4)^2 \qquad \text{Square both sides.}$$
$$3x + 4 = 16$$
$$3x = 12$$
$$x = 4$$

 Check

$$\sqrt{3x + 4} = -4$$
$$\sqrt{3(4) + 4} \stackrel{?}{=} -4$$
$$\sqrt{16} \stackrel{?}{=} -4$$
$$4 \neq -4$$

Since 4 does not check (4 is an extraneous root), the equation $\sqrt{3x + 4} = -4$ has no real number solutions. The solution set is $\varnothing$. ▲

Example 5

Solution

Solve $\sqrt{5s - 2} = \sqrt{2s + 19}$.

$$\sqrt{5s - 2} = \sqrt{2s + 19}$$
$$(\sqrt{5s - 2})^2 = (\sqrt{2s + 19})^2 \qquad \text{Square both sides.}$$
$$5s - 2 = 2s + 19$$
$$3s = 21$$
$$s = 7$$

✔ Check

$$\sqrt{5s - 2} = \sqrt{2s + 19}$$
$$\sqrt{5(7) - 2} \stackrel{?}{=} \sqrt{2(7) + 19}$$
$$\sqrt{33} = \sqrt{33}$$

The solution set is $\{7\}$. ▲

Example 6

Solution

Solve $\sqrt{2y - 4} = y - 2$.

$$\sqrt{2y - 4} = y - 2$$
$$(\sqrt{2y - 4})^2 = (y - 2)^2 \qquad\qquad \text{Square both sides.}$$
$$2y - 4 = y^2 - 4y + 4$$
$$0 = y^2 - 6y + 8$$
$$0 = (y - 4)(y - 2) \qquad\qquad \text{Factor right side.}$$
$$y - 4 = 0 \quad \text{or} \quad y - 2 = 0 \qquad \text{Remember the property:}$$
$$\qquad\qquad\qquad\qquad\qquad\qquad ab = 0 \text{ if and only if } a = 0 \text{ or } b = 0.$$
$$y = 4 \quad \text{or} \quad y = 2$$

✔ *Check*

$$\sqrt{2y-4} = y-2$$
$$\sqrt{2(4)-4} \overset{?}{=} 4-2$$
$$\sqrt{4} \overset{?}{=} 2$$
$$2 = 2$$

or

$$\sqrt{2y-4} = y-2$$
$$\sqrt{2(2)-4} \overset{?}{=} 2-2$$
$$\sqrt{0} \overset{?}{=} 0$$
$$0 = 0$$

The solution set is $\{2, 4\}$.   ▲

Example 7

Solution

Solve $\sqrt{x} + 6 = x$.

$$\sqrt{x} + 6 = x$$

$$\sqrt{x} = x - 6 \qquad \text{We added } -6 \text{ to both sides so that the term with the radical sign is alone on one side of the equation}$$

$$(\sqrt{x})^2 = (x-6)^2$$

$$x = x^2 - 12x + 36$$

$$0 = x^2 - 13x + 36$$

$$0 = (x-4)(x-9)$$

$$x - 4 = 0 \quad \text{or} \quad x - 9 = 0 \qquad \text{Apply: } ab = 0 \text{ if and only if } a = 0 \text{ or } b = 0.$$

$$x = 4 \quad \text{or} \quad x = 9$$

✔ *Check*

$$\sqrt{x} + 6 = x$$
$$\sqrt{4} + 6 \overset{?}{=} 4$$
$$2 + 6 \overset{?}{=} 4$$
$$8 \neq 4$$

or

$$\sqrt{x} + 6 = x$$
$$\sqrt{9} + 6 \overset{?}{=} 9$$
$$3 + 6 \overset{?}{=} 9$$
$$9 = 9$$

The solution set is $\{9\}$.   ▲

Note in Example 7 that we changed the form of the original equation, $\sqrt{x} + 6 = x$, to $\sqrt{x} = x - 6$ before we squared both sides. Squaring both sides of $\sqrt{x} + 6 = x$ produces $x + 12\sqrt{x} + 36 = x^2$, a more complex equation that still contains a radical. So again, it pays to think ahead a few steps before carrying out the details of the problem.

Another Look at Applications

In Section 9.1 we used the formula $S = \sqrt{30Df}$ to approximate how fast a car was traveling based on the length of skid marks. (Remember that S represents the speed of the car in miles per hour, D the length of the skid marks measured in feet, and f represents a coefficient of friction.) This same formula can be used to estimate the length of skid marks that are produced by cars traveling at different rates on various types of road surfaces. To use the formula for this purpose, let's change the form of the equation by solving for D.

$$\sqrt{30Df} = S$$

$$30Df = S^2 \qquad \text{The result of squaring both sides of the original equation.}$$

$$D = \frac{S^2}{30f} \qquad \begin{array}{l} D, S, \text{ and } f \text{ are positive numbers, so this final} \\ \text{equation and the original one are equivalent.} \end{array}$$

Example 8

Suppose that for a particular road surface the coefficient of friction is .35. How far will a car skid when applying the brakes at 60 miles per hour?

Solution

We can substitute .35 for f and 60 for S in the formula $D = \dfrac{S^2}{30f}$.

$$D = \frac{60^2}{30(.35)} = 343 \text{ to the nearest whole number}$$

The car will skid approximately 343 feet. ▲

REMARK Pause for a moment and think about the result in Example 8. The coefficient of friction of .35 refers to a wet concrete road surface. Note that a car traveling at 60 miles per hour will skid more than the length of a football field. △

Problem Set 9.5

Solve each of the following equations. *Be sure* to check all potential solutions in the original equation.

1. $\sqrt{x} = 7$

2. $\sqrt{x} = 12$

3. $\sqrt{2x} = 6$

4. $\sqrt{3x} = 9$

5. $\sqrt{3x} = -6$

6. $\sqrt{2x} = -8$

7. $\sqrt{4x} = 3$

8. $\sqrt{5x} = 4$

9. $3\sqrt{x} = 2$

10. $4\sqrt{x} = 3$

11. $\sqrt{2n - 3} = 5$

12. $\sqrt{3n + 1} = 7$

13. $\sqrt{5y + 2} = -1$

14. $\sqrt{4n - 3} - 4 = 0$

15. $\sqrt{6x - 5} - 3 = 0$

16. $\sqrt{5x + 3} = -4$

17. $5\sqrt{x} = 30$

18. $6\sqrt{x} = 42$

19. $\sqrt{3a - 2} = \sqrt{2a + 4}$

20. $\sqrt{4a + 3} = \sqrt{5a - 4}$

21. $\sqrt{7x - 3} = \sqrt{4x + 3}$

22. $\sqrt{8x - 6} = \sqrt{4x + 11}$

23. $2\sqrt{y + 1} = 5$

24. $3\sqrt{y - 2} = 4$

25. $\sqrt{x + 3} = x + 3$

26. $\sqrt{x + 7} = x + 7$

27. $\sqrt{-2x + 28} = x - 2$

28. $\sqrt{-2x} = x + 4$

29. $\sqrt{3n - 4} = \sqrt{n}$

30. $\sqrt{5n - 1} = \sqrt{2n}$

31. $\sqrt{3x} = x - 6$

32. $2\sqrt{x} = x - 3$

33. $4\sqrt{x} + 5 = x$

34. $\sqrt{-x} - 6 = x$

35. $\sqrt{x^2 + 27} = x + 3$

36. $\sqrt{x^2 - 35} = x - 5$

37. $\sqrt{x^2 + 2x + 3} = x + 2$

38. $\sqrt{x^2 + x + 4} = x + 3$

39. $\sqrt{8x - 2} = x$

40. $\sqrt{2x - 4} = x - 6$

41. Use the formula given in Example 8 with a coefficient of friction of .95. How far will a car skid at 40 miles per hour? At 55 miles per hour? At 65 miles per hour? Express the answers to the nearest foot.

42. Solve the formula $T = 2\pi\sqrt{\dfrac{L}{32}}$ for L. (Remember that in this formula, which was used in Section 9.1, T represents the period of a pendulum expressed in seconds, and L represents the length of the pendulum in feet.)

43. In Problem 42, you should have obtained the equation $L = \dfrac{8T^2}{\pi^2}$. What is the length of a pendulum that has a period of 2 seconds? Of 2.5 seconds? Of 3 seconds? Use 3.14 as an approximation for π and express the answers to the nearest tenth of a foot.

Further Investigations

For Problems 46–49, solve each of the equations.

46. $\sqrt{x + 1} = 5 - \sqrt{x - 4}$

47. $\sqrt{x - 2} = \sqrt{x + 7} - 1$

48. $\sqrt{2n - 1} + \sqrt{n - 5} = 3$

49. $\sqrt{2n + 1} - \sqrt{n - 3} = 2$

50. Suppose that we solve the equation $x^2 = a^2$ for x as follows:

$$x^2 - a^2 = 0$$
$$(x + a)(x - a) = 0$$
$$x + a = 0 \quad \text{or} \quad x - a = 0$$
$$x = -a \quad \text{or} \quad x = a$$

What is the significance of this result?

THOUGHTS INTO WORDS

44. Explain in your own words why possible solutions for radical equations *must* be checked.

45. Your friend attempts to solve the equation $3 + 2\sqrt{x} = x$ as follows:

$$(3 + 2\sqrt{x})^2 = x^2$$
$$9 + 12\sqrt{x} + 4x = x^2$$

At this step she stops and doesn't know how to proceed. What help would you give her?

SUMMARY

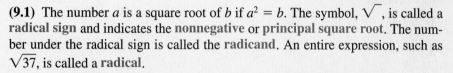

(9.1) The number a is a square root of b if $a^2 = b$. The symbol, $\sqrt{}$, is called a **radical sign** and indicates the **nonnegative** or **principal square root**. The number under the radical sign is called the **radicand**. An entire expression, such as $\sqrt{37}$, is called a **radical**.

The distributive property provides the basis for simplifying a radical expression such as $5\sqrt{2} + 7\sqrt{2}$ to $12\sqrt{2}$.

(9.2) The property, $\sqrt{ab} = \sqrt{a}\sqrt{b}$, provides the basis for changing radicals to **simplest radical form**.

Use the property, $\sqrt{a^2} = a$, when changing radicals involving variables in the radicand to simplest radical form.

(9.3) Use the property, $\sqrt{\dfrac{a}{b}} = \dfrac{\sqrt{a}}{\sqrt{b}}$, when **rationalizing the denominator**.

An algebraic expression that contains square root radicals is said to be in **simplest radical form** if the following conditions are satisfied.

1. No fraction appears within a radical sign.

$$\left(\sqrt{\frac{3}{5}} \text{ violates this condition.}\right)$$

2. No radical appears in the denominator.

$$\left(\frac{3}{\sqrt{10}} \text{ violates this condition.}\right)$$

3. No radicand when expressed in prime factored form contains a factor raised to a power greater than 1.

$$(\sqrt{8} = \sqrt{2^3} \text{ violates this condition.})$$

(9.4) Use the property, $\sqrt{ab} = \sqrt{a}\sqrt{b}$, to multiply square roots and to simplify the resulting radical when appropriate.

(9.5) Use the property of equality, if $a = b$, then $a^2 = b^2$, to solve equations containing radicals.

In general, squaring both sides of an equation will produce an equation that has all of the solutions of the original equation, but it may also have some extra solutions that will not satisfy the original equation. Therefore, when using the "squaring" property, you must check each potential solution in the original equation.

Chapter 9 Review Problem Set

Evaluate each of the following without using a table or a calculator.

1. $\sqrt{64}$

2. $-\sqrt{49}$

3. $\sqrt{1600}$

4. $\sqrt{\dfrac{81}{25}}$

5. $-\sqrt{\dfrac{4}{9}}$

6. $\sqrt{\dfrac{49}{36}}$

Change each of the following to simplest radical form.

7. $\sqrt{20}$

8. $\sqrt{32}$

9. $5\sqrt{8}$

10. $\sqrt{80}$

11. $-3\sqrt{24}$

12. $\dfrac{\sqrt{12}}{\sqrt{49}}$

13. $\dfrac{\sqrt{36}}{\sqrt{7}}$

14. $\sqrt{\dfrac{7}{8}}$

15. $\sqrt{\dfrac{8}{24}}$

16. $\dfrac{3\sqrt{2}}{\sqrt{5}}$

17. $\dfrac{4\sqrt{3}}{\sqrt{12}}$

18. $\dfrac{5\sqrt{2}}{2\sqrt{3}}$

19. $\dfrac{-3\sqrt{2}}{\sqrt{27}}$

20. $\dfrac{4\sqrt{6}}{3\sqrt{12}}$

Use the fact that $\sqrt{3} = 1.73$, to the nearest hundredth, to help evaluate each of the following. Express your final answers to the nearest tenth.

21. $\sqrt{27}$

22. $\dfrac{2}{\sqrt{3}}$

23. $3\sqrt{12} + \sqrt{48}$

24. $2\sqrt{27} - 2\sqrt{75}$

Change each of the following to simplest radical form. All variables represent positive real numbers.

25. $\sqrt{12a^2b^3}$

26. $\sqrt{50xy^4}$

27. $3x\sqrt{2x^5}$

28. $-4t\sqrt{8t^3}$

29. $\dfrac{4}{3}\sqrt{27xy^2}$

30. $\dfrac{3}{4}\sqrt{24x^3y^5}$

31. $\dfrac{\sqrt{2x}}{\sqrt{5y}}$

32. $\dfrac{\sqrt{72x}}{\sqrt{16y}}$

33. $\sqrt{\dfrac{4}{x}}$

34. $\sqrt{\dfrac{2x^3}{9}}$

35. $\dfrac{3\sqrt{x}}{4\sqrt{y^3}}$

36. $\dfrac{-2\sqrt{x^2y}}{5\sqrt{xy}}$

Find the following products and express your answers in simplest radical form.

37. $(\sqrt{6})(\sqrt{12})$

38. $(2\sqrt{3})(3\sqrt{6})$

39. $(-5\sqrt{8})(2\sqrt{2})$

40. $(-3\sqrt{12})(-4\sqrt{7})$

41. $\sqrt{2}(\sqrt{8} + \sqrt{20})$

42. $3\sqrt{5}(\sqrt{8} - 2\sqrt{12})$

43. $(\sqrt{3} + \sqrt{5})(\sqrt{3} + \sqrt{7})$

44. $(2\sqrt{3} + 3\sqrt{2})(\sqrt{3} - 5\sqrt{2})$

45. $(\sqrt{6} + 2\sqrt{7})(3\sqrt{6} - \sqrt{7})$

46. $(3 + 2\sqrt{5})(4 - 3\sqrt{5})$

Rationalize the denominators and simplify.

47. $\dfrac{5}{\sqrt{7} - \sqrt{5}}$

48. $\dfrac{\sqrt{6}}{\sqrt{3} - \sqrt{2}}$

49. $\dfrac{2}{3\sqrt{2} - \sqrt{6}}$

50. $\dfrac{\sqrt{6}}{3\sqrt{7} + 2\sqrt{10}}$

Simplify each of the following radical expressions.

51. $2\sqrt{50} + 3\sqrt{72} - 5\sqrt{8}$

52. $\sqrt{8x} - 3\sqrt{18x}$

53. $5\sqrt{12} - 6\sqrt{27} - \sqrt{48}$

54. $3\sqrt{10} + \sqrt{\dfrac{2}{5}}$

55. $4\sqrt{20} - \dfrac{3}{\sqrt{5}} + \sqrt{45}$

56. $\sqrt{\dfrac{2}{3}} - 2\sqrt{54}$

Solve each of the following equations.

57. $\sqrt{5x + 6} = 6$

58. $\sqrt{6x + 1} = \sqrt{3x + 13}$

59. $3\sqrt{n} = n$

60. $\sqrt{y + 5} = y + 5$

61. $\sqrt{-3a + 10} = a - 2$

62. $3 - \sqrt{2x - 1} = 2$

Use your calculator or the table in the appendix to evaluate each of the following.

63. $\sqrt{2116}$ **64.** $\sqrt{4356}$

65. $\sqrt{5184}$

Use your calculator or the table in the appendix to find a whole number approximation for each of the following.

66. $\sqrt{690}$ **67.** $\sqrt{2185}$

68. $\sqrt{5500}$

CHAPTER 9 TEST

1. Evaluate $-\sqrt{\dfrac{64}{49}}$.

2. Evaluate $\sqrt{.0025}$.

For Problems 3–5, use the fact that $\sqrt{2} = 1.41$, to the nearest hundredth, and evaluate each of the following to the nearest tenth.

3. $\sqrt{8}$

4. $-\sqrt{32}$

5. $\dfrac{3}{\sqrt{2}}$

For Problems 6–14, express each radical expression in simplest radical form. All variables represent positive real numbers.

6. $\sqrt{45}$

7. $-3\sqrt{72}$

8. $\dfrac{2\sqrt{3}}{3\sqrt{6}}$

9. $\sqrt{\dfrac{25}{2}}$

10. $\dfrac{\sqrt{24}}{\sqrt{36}}$

11. $\sqrt{\dfrac{5}{8}}$

12. $\sqrt{80x^2y}$

13. $\dfrac{\sqrt{3x}}{\sqrt{5y}}$

14. $\dfrac{3}{4}\sqrt{48x^3y^2}$

For Problems 15–18, find the indicated products and express answers in simplest radical form.

15. $(\sqrt{8})(\sqrt{12})$

16. $(3\sqrt{6})(5\sqrt{3})$

17. $\sqrt{6}(2\sqrt{12} - 3\sqrt{8})$

18. $(2\sqrt{5} + \sqrt{3})(\sqrt{5} - 3\sqrt{3})$

19. Rationalize the denominator and simplify: $\dfrac{\sqrt{6}}{\sqrt{12} + \sqrt{2}}$.

20. Simplify: $2\sqrt{24} - 4\sqrt{54} + 3\sqrt{96}$.

21. Find a whole number approximation for $\sqrt{500}$.

For Problems 22–25, solve each equation.

22. $\sqrt{3x + 1} = 4$

23. $\sqrt{2x - 5} = -4$

24. $\sqrt{n - 3} = 3 - n$

25. $\sqrt{3x + 6} = x + 2$

Quadratic Equations

The area of a tennis court for singles play is 2106 square feet. The length of the court is $\frac{26}{9}$ times the width. Find the length and width of the court. **The quadratic equation $x\left(\frac{26}{9}x\right) = 2106$ can be used to determine that the court is 27 feet wide and 78 feet long.**

Solving equations has been a central theme throughout this text. Let's pause for a moment and reflect back on the different types of equations that we have solved.

Type of equation	Examples
First-degree equations of one variable	$4x + 3 = 7x + 1$; $3(x - 6) = 9$
Second-degree equations of one variable *that are factorable*	$x^2 + 3x = 0$; $x^2 + 5x + 6 = 0$; $x^2 - 4 = 0$; $x^2 + 10x + 25 = 0$
Fractional equations	$\dfrac{3}{x} + \dfrac{2}{x} = 4$; $\dfrac{5}{a - 2} = \dfrac{6}{a + 3}$; $\dfrac{2}{x^2 - 4} + \dfrac{5}{x + 2} = \dfrac{6}{x - 2}$
Radical equations	$\sqrt{x} = 4$; $\sqrt{y + 2} = 3$; $\sqrt{a + 1} = \sqrt{2a - 7}$
Systems of equations	$\begin{pmatrix} 2x + 3y = 4 \\ 5x - y = 7 \end{pmatrix}$; $\begin{pmatrix} 3a + 5b = 9 \\ 7a - 9b = 12 \end{pmatrix}$

As indicated in the chart, we have learned how to solve some second-degree equations, but only those for which the quadratic polynomial is factorable. In this chapter we will extend our work to include more general types of second-degree equations in one variable and thus enhance our problem solving capabilities.

10.1   Quadratic Equations

A second-degree equation in one variable contains the variable with an exponent of two, but no higher power. Such equations are also called **quadratic equations**. The following are examples of quadratic equations.

$$x^2 = 25, \qquad\qquad y^2 + 6y = 0, \qquad x^2 + 7x - 4 = 0,$$
$$4y^2 + 2y - 1 = 0, \qquad 5x^2 + 2x - 1 = 2x^2 + 6x - 5$$

We can also define a quadratic equation in the variable x as any equation that can be written in the form $ax^2 + bx + c = 0$, where a, b, and c are real numbers and $a \neq 0$. We refer to the form $ax^2 + bx + c = 0$ as the **standard form** of a quadratic equation.

In Chapter 6 you solved quadratic equations (we didn't use the term "quadratic" at that time) by factoring and applying the property *ab = 0 if and only if a = 0 or b = 0*. Let's review a few examples of that type.

Example 1

Solve $x^2 - 13x = 0$.

Solution

$$x^2 - 13x = 0$$
$$x(x - 13) = 0 \qquad\qquad \text{Factor left side of equation}$$
$$x = 0 \quad\text{ or }\quad x - 13 = 0 \qquad \text{Apply } ab = 0 \text{ if and only if } a = 0 \text{ or } b = 0.$$
$$x = 0 \quad\text{ or }\quad x = 13$$

The solution set is $\{0, 13\}$. (Don't forget to check these solutions!)   ▲

Example 2

Solve $n^2 + 2n - 24 = 0$.

Solution

$$n^2 + 2n - 24 = 0$$
$$(n + 6)(n - 4) = 0 \qquad\qquad \text{Factor left side}$$
$$n + 6 = 0 \quad\text{ or }\quad n - 4 = 0 \qquad \text{Apply } ab = 0 \text{ if and only if } a = 0 \text{ or } b = 0.$$
$$n = -6 \quad\text{ or }\quad n = 4$$

The solution set is $\{-6, 4\}$.   ▲

Example 3

Solve $x^2 + 6x + 9 = 0$.

Solution

$$x^2 + 6x + 9 = 0$$
$$(x + 3)(x + 3) = 0 \qquad\qquad \text{Factor left side}$$
$$x + 3 = 0 \quad\text{ or }\quad x + 3 = 0 \qquad \text{Apply } ab = 0 \text{ if and only if } a = 0 \text{ or } b = 0.$$
$$x = -3 \quad\text{ or }\quad x = -3$$

The solution set is $\{-3\}$.   ▲

Example 4

Solution

Solve $y^2 = 49$.

$$y^2 = 49$$
$$y^2 - 49 = 0$$
$$(y + 7)(y - 7) = 0 \qquad \text{Factor left side}$$
$$y + 7 = 0 \quad \text{or} \quad y - 7 = 0 \qquad \text{Apply } ab = 0 \text{ if and only if } a = 0 \text{ or } b = 0.$$
$$y = -7 \quad \text{or} \quad y = 7$$

The solution set is $\{-7, 7\}$. ▲

Note the type of equation that we solved in Example 4. Let's generalize from that example and consider the equation $x^2 = a$, where a is any nonnegative real number. We can solve this equation as follows.

$$x^2 = a$$
$$x^2 = (\sqrt{a})^2 \qquad (\sqrt{a})^2 = a$$
$$x^2 - (\sqrt{a})^2 = 0$$
$$(x - \sqrt{a})(x + \sqrt{a}) = 0 \qquad \text{Factor left side}$$
$$x - \sqrt{a} = 0 \quad \text{or} \quad x + \sqrt{a} = 0 \qquad \text{Apply } ab = 0 \text{ if and only if } a = 0 \text{ or}$$
$$x = \sqrt{a} \quad \text{or} \quad x = -\sqrt{a} \qquad b = 0.$$

The solutions are $\sqrt{a}$ and $-\sqrt{a}$. We shall consider this result as a general property and use it to solve certain types of quadratic equations.

PROPERTY 10.1

For any nonnegative real number a,
$$x^2 = a \quad \text{if and only if } x = \sqrt{a} \text{ or } x = -\sqrt{a}.$$
(The statement "$x = \sqrt{a}$ or $x = -\sqrt{a}$" can be written as $x = \pm\sqrt{a}$.)

Property 10.1 along with our knowledge of square roots makes it very easy to solve quadratic equations of the form $x^2 = a$.

Example 5

Solution

Solve $x^2 = 81$.

$$x^2 = 81$$
$$x = \pm\sqrt{81} \qquad \text{Apply Property 10.1.}$$
$$x = \pm 9$$

The solution set is $\{-9, 9\}$. ▲

Example 6

Solution

Solve $x^2 = 8$.

$$x^2 = 8$$
$$x = \pm\sqrt{8}$$
$$x = \pm 2\sqrt{2} \qquad \sqrt{8} = \sqrt{4}\sqrt{2} = 2\sqrt{2}$$

The solution set is $\{-2\sqrt{2}, 2\sqrt{2}\}$.   ▲

Example 7

Solution

Solve $5n^2 = 12$.

$$5n^2 = 12$$
$$n^2 = \frac{12}{5}$$
$$n = \pm\sqrt{\frac{12}{5}}$$
$$n = \pm\frac{2\sqrt{15}}{5} \qquad \sqrt{\frac{12}{5}} = \frac{\sqrt{12}}{\sqrt{5}} = \frac{\sqrt{12}}{\sqrt{5}} \cdot \frac{\sqrt{5}}{\sqrt{5}} = \frac{\sqrt{60}}{5} = \frac{2\sqrt{15}}{5}$$

The solution set is $\left\{-\dfrac{2\sqrt{15}}{5}, \dfrac{2\sqrt{15}}{5}\right\}$.   ▲

Example 8

Solution

Solve $(x - 2)^2 = 16$.

$$(x - 2)^2 = 16$$
$$x - 2 = \pm 4$$
$$x - 2 = 4 \qquad \text{or} \qquad x - 2 = -4$$
$$x = 6 \qquad \text{or} \qquad x = -2$$

The solution set is $\{-2, 6\}$.   ▲

Example 9

Solution

Solve $(x + 5)^2 = 27$.

$$(x + 5)^2 = 27$$
$$x + 5 = \pm\sqrt{27}$$
$$x + 5 = \pm 3\sqrt{3} \qquad \sqrt{27} = \sqrt{9}\sqrt{3} = 3\sqrt{3}$$
$$x + 5 = 3\sqrt{3} \qquad \text{or} \qquad x + 5 = -3\sqrt{3}$$
$$x = -5 + 3\sqrt{3} \qquad \text{or} \qquad x = -5 - 3\sqrt{3}.$$

The solution set is $\{-5 - 3\sqrt{3}, -5 + 3\sqrt{3}\}$.   ▲

It may be necessary to change form before we can apply Property 10.1. The next example illustrates this point.

Example 10

Solution

Solve $3(2x - 3)^2 + 8 = 44$.

$$3(2x - 3)^2 + 8 = 44$$

$$3(2x - 3)^2 = 36$$

$$(2x - 3)^2 = 12$$

$$2x - 3 = \pm\sqrt{12} \qquad \text{Apply Property 10.1.}$$

$$2x - 3 = \sqrt{12} \quad \text{or} \quad 2x - 3 = -\sqrt{12}$$

$$2x = 3 + \sqrt{12} \quad \text{or} \quad 2x = 3 - \sqrt{12}$$

$$x = \frac{3 + \sqrt{12}}{2} \quad \text{or} \quad x = \frac{3 - \sqrt{12}}{2}$$

$$x = \frac{3 + 2\sqrt{3}}{2} \quad \text{or} \quad x = \frac{3 - 2\sqrt{3}}{2} \qquad \sqrt{12} = \sqrt{4}\sqrt{3} = 2\sqrt{3}$$

The solution set is $\left\{\dfrac{3 - 2\sqrt{3}}{2}, \dfrac{3 + 2\sqrt{3}}{2}\right\}$. ▲

Note that quadratic equations of the form $x^2 = a$, where a is a *negative* number, have no real number solutions. For example, $x^2 = -4$ has no real number solutions since any real number squared is nonnegative. In a like manner an equation such as $(x + 3)^2 = -14$ has no real number solutions.

Using the Pythagorean Theorem

Our work with radicals, Property 10.1, and the Pythagorean theorem merge to form a basis for solving a variety of problems that pertain to right triangles. First, let's restate the Pythagorean theorem.

Pythagorean Theorem

If for a right triangle, a and b are the measures of the legs, and c is the measure of the hypotenuse, then

$$a^2 + b^2 = c^2.$$

(The *hypotenuse* is the side opposite the right angle and the *legs* are the other two sides as shown in Figure 10.1.)

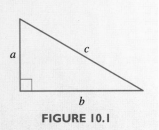

FIGURE 10.1

Example 11

Find c in Figure 10.2.

FIGURE 10.2

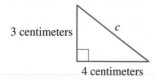

3 centimeters

c

4 centimeters

Solution

Applying the Pythagorean theorem we obtain

$$c^2 = a^2 + b^2$$
$$c^2 = 3^2 + 4^2$$
$$c^2 = 9 + 16$$
$$c^2 = 25$$
$$c = 5.$$

The length of c is 5 centimeters.   ▲

REMARK Don't forget that the equation $c^2 = 25$ does have two solutions, 5 and -5. However, since we are finding the lengths of line segments, we can disregard the negative solutions.   △

Example 12

A 50-foot rope hangs from the top of a flagpole. When pulled taut to its full length, the rope reaches a point on the ground 18 feet from the base of the pole. Find the height of the pole to the nearest tenth of a foot.

Solution

Let's sketch a figure (Figure 10.3) and record the given information. Using the Pythagorean theorem, we can solve for p as follows.

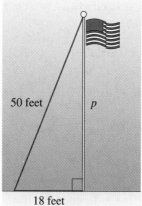

50 feet

p

18 feet

p represents the height
of the flagpole.

FIGURE 10.3

$$p^2 + 18^2 = 50^2$$
$$p^2 + 324 = 2500$$
$$p^2 = 2176$$
$$p = \sqrt{2176} = 46.6 \quad \text{to the nearest tenth}$$

The height of the flagpole is approximately 46.6 feet.

▲

An isosceles triangle has two sides of the same length. Thus, an **isosceles right triangle** is a right triangle that has both legs of the same length. Let's consider a problem involving an isosceles right triangle.

Example 13

Find the length of each leg of an isosceles right triangle if the hypotenuse is 8 meters long.

Solution

Let's sketch an isosceles right triangle (Figure 10.4) and let x represent the length of each leg. Then we can determine x by applying the Pythagorean theorem.

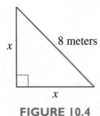

$$x^2 + x^2 = 64$$
$$2x^2 = 64$$
$$x^2 = 32$$
$$x = \sqrt{32} = \sqrt{16}\sqrt{2} = 4\sqrt{2}.$$

Each leg is $4\sqrt{2}$ meters long.

FIGURE 10.4

REMARK In Example 12 we made no attempt to express $\sqrt{2176}$ in simplest radical form because the answer was to be given as a rational approximation to the nearest tenth. However, in Example 13 we left the final answer in radical form and therefore expressed it in simplest radical form.

Another special kind of right triangle is one that contains acute angles of 30° and 60°. In such a right triangle, often referred to as a 30°–60° right triangle, *the side opposite the 30° angle is equal in length to one-half of the length of the hypotenuse.* This relationship, along with the Pythagorean theorem, provides us with another problem solving technique.

Example 14

Suppose that a 20-foot ladder is leaning against a building and makes an angle of 60° with the ground. How far up on the building does the top of the ladder reach? Express your answer to the nearest tenth of a foot.

Solution

Figure 10.5 depicts this situation. The side opposite the 30° angle equals one-half of the hypotenuse, so that side is of length $\frac{1}{2}(20) = 10$ feet. Now we can apply the Pythagorean theorem.

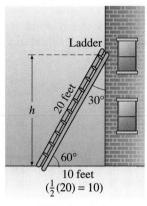

$$h^2 + 10^2 = 20^2$$
$$h^2 + 100 = 400$$
$$h^2 = 300$$
$$h = \sqrt{300} = 17.3 \quad \text{to the nearest}$$
tenth

The top of the ladder touches the building at approximately 17.3 feet from the ground.

FIGURE 10.5   ▲

Problem Set 10.1

For Problems 1–18, solve each quadratic equation by factoring and applying the property $ab = 0$ if and only if $a = 0$ or $b = 0$.

1. $x^2 + 15x = 0$

2. $x^2 - 11x = 0$

3. $n^2 = 12n$

4. $n^2 = -21n$

5. $3y^2 = 15y$

6. $8y^2 = -56y$

7. $x^2 - 9x + 8 = 0$

8. $x^2 + 16x + 48 = 0$

9. $x^2 - 5x - 14 = 0$

10. $x^2 - 5x - 36 = 0$

11. $n^2 + 5n - 6 = 0$

12. $n^2 + 3n - 28 = 0$

13. $6y^2 + 7y - 5 = 0$

14. $4y^2 - 21y - 18 = 0$

15. $30x^2 - 37x + 10 = 0$

16. $42x^2 + 67x + 21 = 0$

17. $4x^2 - 4x + 1 = 0$

18. $9x^2 + 12x + 4 = 0$

For Problems 19–56, use Property 10.1 to help solve each quadratic equation. Express irrational solutions in simplest radical form.

19. $x^2 = 64$

20. $x^2 = 169$

21. $x^2 = \dfrac{25}{9}$

22. $x^2 = \dfrac{4}{81}$

23. $4x^2 = 64$

24. $5x^2 = 500$

25. $n^2 = 14$

26. $n^2 = 22$

27. $n^2 + 16 = 0$

28. $n^2 = 24$

29. $y^2 = 32$

30. $y^2 + 25 = 0$

31. $3x^2 - 54 = 0$

32. $4x^2 - 108 = 0$

33. $2x^2 = 9$

34. $3x^2 = 16$

35. $8n^2 = 25$

36. $12n^2 = 49$

37. $(x - 1)^2 = 4$

38. $(x - 2)^2 = 9$

39. $(x + 3)^2 = 25$

40. $(x + 5)^2 = 36$

41. $(3x - 2)^2 = 49$

42. $(4x + 3)^2 = 1$

43. $(x + 6)^2 = 5$

44. $(x - 7)^2 = 6$

45. $(n - 1)^2 = 8$

46. $(n + 1)^2 = 12$

47. $(2n + 3)^2 = 20$

48. $(3n - 2)^2 = 28$

49. $(4x - 1)^2 = -2$

50. $(5x + 3)^2 - 32 = 0$

51. $(3x - 5)^2 - 40 = 0$

52. $(2x + 9)^2 + 6 = 0$

53. $2(7x - 1)^2 + 5 = 37$

54. $3(4x - 5)^2 - 50 = 25$

55. $2(x + 8)^2 - 9 = 91$

56. $2(x - 7)^2 - 7 = 101$

For Problems 57–62, a and b represent the lengths of the legs of a right triangle and c represents the length of the hypotenuse. Express your answers in simplest radical form.

57. Find c if $a = 1$ inch and $b = 7$ inches.

58. Find c if $a = 2$ inches and $b = 6$ inches.

59. Find a if $c = 8$ meters and $b = 6$ meters.

60. Find a if $c = 11$ centimeters and $b = 7$ centimeters.

61. Find b if $c = 12$ feet and $a = 10$ feet.

62. Find b if $c = 10$ yards and $a = 4$ yards.

For Problems 63–68, use Figure 10.6. Express your answers in simplest radical form.

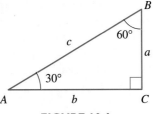

FIGURE 10.6

63. If $c = 8$ inches, find a and b.

64. If $c = 6$ inches, find a and b.

65. If $a = 6$ feet, find b and c.

66. If $a = 5$ feet, find b and c.

67. If $b = 12$ meters, find a and c.

68. If $b = 5$ centimeters, find a and c.

For Problems 69–72, use the isosceles right triangle in Figure 10.7. Express answers in simplest radical form.

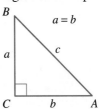

FIGURE 10.7

69. If $b = 10$ inches, find a and c.

70. If $a = 7$ inches, find b and c.

71. If $c = 9$ meters, find a and b.

72. If $c = 5$ meters, find a and b.

73. An 18-foot ladder resting against a house reaches a windowsill 16 feet above the ground. How far is the base of the ladder from the foundation of the house? Express your answer to the nearest tenth of a foot.

74. A 42-foot guy-wire makes an angle of 60° with the ground and is attached to a telephone pole, as in Figure 10.8. Find the distance from the base of the pole to the point on the pole where the wire is attached. Express your answer to the nearest tenth of a foot.

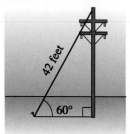

FIGURE 10.8

75. A rectangular plot measures 18 meters by 24 meters. Find the distance, to the nearest meter, from one corner of the plot to the diagonally opposite corner.

76. Consecutive bases of a square-shaped baseball diamond (Figure 10.9) are 90 feet apart. Find the distance, to the nearest tenth of a foot, from first base diagonally across the diamond to third base.

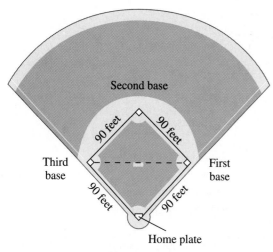

FIGURE 10.9

77. A diagonal of a square parking lot is 50 meters. Find, to the nearest meter, the length of a side of the lot.

78. Explain why the equation $(x - 4)^2 + 14 = 2$ has no real number solutions.

79. Suppose that your friend solved the equation $(x + 3)^2 = 25$ as follows:

$$(x + 3)^2 = 25$$
$$x^2 + 6x + 9 = 25$$
$$x^2 + 6x - 16 = 0$$
$$(x + 8)(x - 2) = 0$$
$$x + 8 = 0 \quad \text{or} \quad x - 2 = 0$$
$$x = -8 \quad \text{or} \quad x = 2$$

Is this a correct approach to the problem? Would you suggest an easier approach to the problem?

Further Investigations

80. Sometimes we simply need to determine whether a particular radical expression is positive or negative. For example, is $-6 + \sqrt{39}$ a positive or negative number? We can determine this by approximating a value for $\sqrt{39}$. Since $6^2 = 36$, we know that $\sqrt{39}$ is a little larger than 6. Therefore, $-6 + \sqrt{39}$ has to be positive.

Determine whether each of the following is positive or negative.

(a) $-8 + \sqrt{56}$
(b) $-7 + \sqrt{47}$
(c) $9 - \sqrt{77}$
(d) $12 - \sqrt{130}$
(e) $-6 + 5\sqrt{2}$
(f) $-10 + 6\sqrt{3}$
(g) $-13 + \sqrt{150}$
(h) $-14 + \sqrt{200}$

81. Find the length of an altitude of an equilateral triangle if each side of the triangle is 6 centimeters long. Express your answer to the nearest tenth of a centimeter.

82. Suppose that we are given a cube with edges of length 12 centimeters. Find the length of a diagonal from a lower corner to the diagonally opposite upper corner. Express your answer to the nearest tenth of a centimeter.

83. Suppose that we are given a rectangular box with a length of 8 centimeters, a width of 6 centimeters, and a height of 4 centimeters. Find the length of a diagonal from a lower corner to the diagonally opposite upper corner. Express your answer to the nearest tenth of a centimeter.

84. The converse of the Pythagorean theorem is also true. It states that *if the measures a, b, and c of the* *sides of a triangle are such that* $a^2 + b^2 = c^2$, *then* *the triangle is a right triangle with a and b the measures of the legs and c the measure of the hypotenuse.* Use the converse of the Pythagorean theorem to determine which of the triangles having sides with the following measures are right triangles.

(a) 9, 40, 41
(b) 20, 48, 52
(c) 19, 21, 26
(d) 32, 37, 49
(e) 65, 156, 169
(f) 21, 72, 75

10.2   Completing the Square

Thus far we have solved quadratic equations by factoring or by applying Property 10.1 (if $x^2 = a$, then $x = \sqrt{a}$ or $x = -\sqrt{a}$). In this section we will consider another method called **completing the square**, which will give us the power to solve *any* quadratic equation.

We studied a factoring technique in Chapter 6 that was based on recognizing **perfect square trinomials**. In each of the following, the trinomial on the right side, which is the result of squaring a binomial on the left side, is a perfect square trinomial.

$$(x + 5)^2 = x^2 + 10x + 25,$$
$$(x + 7)^2 = x^2 + 14x + 49,$$
$$(x - 3)^2 = x^2 - 6x + 9,$$
$$(x - 6)^2 = x^2 - 12x + 36$$

We need to pay attention to the following special relationship. In each of the perfect square trinomials previously listed, **the constant term is equal to the square of one-half of the coefficient of the x-term.** For example,

$$x^2 + 10x + 25 \quad \frac{1}{2}(10) = 5 \quad \text{and} \quad 5^2 = 25,$$

$$x^2 - 12x + 36 \quad \frac{1}{2}(12) = 6 \quad \text{and} \quad 6^2 = 36.$$

This relationship allows us to form a perfect square trinomial by adding the proper constant term. For example, suppose that we want to form a perfect square trinomial from $x^2 + 8x$. Since $\frac{1}{2}(8) = 4$ and $4^2 = 16$, we can form the perfect square trinomial $x^2 + 8x + 16$.

Let's use the previous ideas to help solve some quadratic equations.

Example 1

Solve $x^2 + 8x - 1 = 0$ by the method of completing the square.

Solution

$$x^2 + 8x - 1 = 0$$
$$x^2 + 8x = 1$$
$$x^2 + 8x + 16 = 1 + 16 \qquad \text{16 is added to the left side to form a perfect}$$

16 is added to the left side to form a perfect square trinomial; therefore 16 also has to be added to the right side.

$$(x + 4)^2 = 17$$

Now we can proceed as we did with such equations in the last section.

$$x + 4 = \pm\sqrt{17}$$
$$x + 4 = \sqrt{17} \qquad \text{or} \qquad x + 4 = -\sqrt{17}$$
$$x = -4 + \sqrt{17} \qquad \text{or} \qquad x = -4 - \sqrt{17}$$

The solution set is $\{-4 - \sqrt{17}, -4 + \sqrt{17}\}$.
▲

Observe that the method of completing the square to solve a quadratic equation is just what the name implies. A perfect square trinomial is formed; then the equation can be changed to the necessary form for using the property: if $x^2 = a$, then $x = \sqrt{a}$ or $x = -\sqrt{a}$. Let's consider another example.

Example 2

Solve $x^2 - 2x - 11 = 0$ by the method of completing the square.

Solution

$$x^2 - 2x - 11 = 0$$
$$x^2 - 2x = 11$$
$$x^2 - 2x + 1 = 11 + 1$$
$$(x - 1)^2 = 12$$
$$x - 1 = \pm\sqrt{12}$$
$$x - 1 = \pm 2\sqrt{3}$$
$$x - 1 = 2\sqrt{3} \qquad \text{or} \qquad x - 1 = -2\sqrt{3}$$
$$x = 1 + 2\sqrt{3} \qquad \text{or} \qquad x = 1 - 2\sqrt{3}$$

The solution set is $\{1 - 2\sqrt{3}, 1 + 2\sqrt{3}\}$.
▲

In the next example the coefficient of the x-term is odd, which means that taking one-half of it puts us in the realm of fractions. The use of common fractions rather than decimals makes our previous work with radicals applicable.

Example 3

Solve $x^2 - 3x + 1 = 0$ by the method of completing the square.

Solution

$$x^2 - 3x + 1 = 0$$

$$x^2 - 3x = -1$$

$$x^2 - 3x + \frac{9}{4} = -1 + \frac{9}{4}$$

$$\left(x - \frac{3}{2}\right)^2 = \frac{5}{4}$$

$$x - \frac{3}{2} = \pm\sqrt{\frac{5}{4}}$$

$$x - \frac{3}{2} = \pm\frac{\sqrt{5}}{2}$$

$$x - \frac{3}{2} = \frac{\sqrt{5}}{2} \quad \text{or} \quad x - \frac{3}{2} = -\frac{\sqrt{5}}{2}$$

$$x = \frac{3}{2} + \frac{\sqrt{5}}{2} \quad \text{or} \quad x = \frac{3}{2} - \frac{\sqrt{5}}{2}$$

$$x = \frac{3 + \sqrt{5}}{2} \quad \text{or} \quad x = \frac{3 - \sqrt{5}}{2}$$

The solution set is $\left\{\dfrac{3 - \sqrt{5}}{2}, \dfrac{3 + \sqrt{5}}{2}\right\}$. ▲

The relationship for a perfect square trinomial that states **the constant term is equal to the square of one-half of the coefficient of the x-term** holds only if the coefficient of x^2 is 1. Thus, we need to make a slight adjustment when solving quadratic equations that have a coefficient of x^2 other than 1. The next example shows how to make this adjustment.

Example 4

Solve $2x^2 + 12x - 3 = 0$ by the method of completing the square.

Solution

$$2x^2 + 12x - 3 = 0$$

$$2x^2 + 12x = 3$$

$$x^2 + 6x = \frac{3}{2} \qquad \text{Multiply both sides by } \frac{1}{2}.$$

$$x^2 + 6x + 9 = \frac{3}{2} + 9$$

$$(x + 3)^2 = \frac{21}{2}$$

$$x + 3 = \pm\sqrt{\frac{21}{2}}$$

$$x + 3 = \pm\frac{\sqrt{42}}{2} \qquad \left(\sqrt{\frac{21}{2}} = \frac{\sqrt{21}}{\sqrt{2}} = \frac{\sqrt{21}}{\sqrt{2}} \cdot \frac{\sqrt{2}}{\sqrt{2}} = \frac{\sqrt{42}}{\sqrt{2}}\right)$$

$$x + 3 = \frac{\sqrt{42}}{2} \quad \text{or} \quad x + 3 = -\frac{\sqrt{42}}{2}$$

$$x = -3 + \frac{\sqrt{42}}{2} \quad \text{or} \quad x = -3 - \frac{\sqrt{42}}{2}$$

$$x = \frac{-6 + \sqrt{42}}{2} \quad \text{or} \quad x = \frac{-6 - \sqrt{42}}{2}$$

The solution set is $\left\{ \dfrac{-6 - \sqrt{42}}{2}, \dfrac{-6 + \sqrt{42}}{2} \right\}$.

As we mentioned earlier, we can use the method of completing the square to solve *any* quadratic equation. To illustrate this point, let's use this method to solve an equation that we could also solve by factoring.

Example 5

Solve $x^2 + 2x - 8 = 0$ by the method of completing the square and by factoring.

Solution

By completing the square
$$x^2 + 2x - 8 = 0$$
$$x^2 + 2x = 8$$
$$x^2 + 2x + 1 = 8 + 1$$
$$(x + 1)^2 = 9$$
$$(x + 1) = \pm 3$$
$$x + 1 = 3 \quad \text{or} \quad x + 1 = -3$$
$$x = 2 \quad \text{or} \quad x = -4$$

The solution set is $\{-4, 2\}$.

Solution

By factoring
$$x^2 + 2x - 8 = 0$$
$$(x + 4)(x - 2) = 0$$
$$x + 4 = 0 \quad \text{or} \quad x - 2 = 0$$
$$x = -4 \quad \text{or} \quad x = 2$$

The solution set is $\{-4, 2\}$.

We don't claim that using the method of completing the square with an equation such as the one in Example 5 is easier than the factoring technique. However, it is important for you to recognize that the method of completing the square will work with any quadratic equation.

Our final example of this section demonstrates that the method of completing the square will identify those quadratic equations that have no real number solutions.

Example 6

Solution

Solve $x^2 + 10x + 30 = 0$ by the method of completing the square.

$$x^2 + 10x + 30 = 0$$
$$x^2 + 10x = -30$$
$$x^2 + 10x + 25 = -30 + 25$$
$$(x + 5)^2 = -5$$

We can stop here and reason as follows: Any value of x will yield a nonnegative value for $(x + 5)^2$; thus, it cannot equal -5. The original equation, $x^2 + 10x + 30 = 0$, has *no solutions in the set of real numbers*. ▲

Problem Set 10.2

For Problems 1–32, use the method of completing the square to help solve each quadratic equation.

1. $x^2 + 8x - 1 = 0$

2. $x^2 - 4x - 1 = 0$

3. $x^2 + 10x + 2 = 0$

4. $x^2 + 8x + 3 = 0$

5. $x^2 - 4x - 4 = 0$

6. $x^2 + 6x - 11 = 0$

7. $x^2 + 6x + 12 = 0$

8. $n^2 - 10n = 7$

9. $n^2 + 2n = 17$

10. $n^2 + 12n + 40 = 0$

11. $x^2 + x - 3 = 0$

12. $x^2 + 3x - 5 = 0$

13. $a^2 - 5a = 2$

14. $a^2 - 7a = 4$

15. $2x^2 + 8x - 3 = 0$

16. $2x^2 - 12x + 1 = 0$

17. $3x^2 + 12x - 2 = 0$

18. $3x^2 - 6x + 2 = 0$

19. $2t^2 - 4t + 1 = 0$

20. $4t^2 + 8t + 5 = 0$

21. $5n^2 + 10n + 6 = 0$

22. $2n^2 + 5n - 1 = 0$

23. $-n^2 + 9n = 4$

24. $-n^2 - 7n = 2$

25. $2x^2 + 3x - 1 = 0$

26. $3x^2 - x - 3 = 0$

27. $3x^2 + 2x - 2 = 0$

28. $9x = 3x^2 - 1$

29. $n(n + 2) = 168$

30. $n(n + 4) = 140$

31. $n(n - 4) = 165$

32. $n(n - 2) = 288$

For Problems 33–42, solve each quadratic equation by using (a) the factoring method, and (b) the method of completing the square.

33. $x^2 + 4x - 12 = 0$

34. $x^2 - 6x - 40 = 0$

35. $x^2 + 12x + 27 = 0$

36. $x^2 + 18x + 77 = 0$

37. $n^2 - 3n - 40 = 0$

38. $n^2 + 9n - 36 = 0$

39. $2n^2 - 9n + 4 = 0$

40. $6n^2 - 11n - 10 = 0$

41. $4n^2 + 4n - 15 = 0$

42. $4n^2 + 12n - 7 = 0$

THOUGHTS INTO WORDS

43. Give a step-by-step description of how to solve the equation $3x^2 + 10x - 8 = 0$ by completing the square.

44. An error has been made in the following solution. Find it and explain how to correct it.

$$4x^2 - 4x + 1 = 0$$
$$4x^2 - 4x = -1$$
$$4x^2 - 4x + 4 = -1 + 4$$
$$(2x - 2)^2 = 3$$
$$2x - 2 = \pm\sqrt{3}$$
$$2x - 2 = \sqrt{3} \quad \text{or} \quad 2x - 2 = -\sqrt{3}$$
$$2x = 2 + \sqrt{3} \quad \text{or} \quad 2x = 2 - \sqrt{3}$$
$$x = \frac{2 + \sqrt{3}}{2} \quad \text{or} \quad x = \frac{2 - \sqrt{3}}{2}$$

The solution set is $\left\{\dfrac{2 + \sqrt{3}}{2}, \dfrac{2 - \sqrt{3}}{2}\right\}$.

Further Investigations

45. Use the method of completing the square to solve $ax^2 + bx + c = 0$ for x, where a, b, and c are real numbers and $a \neq 0$.

46. Suppose that in Example 4 we wanted to express the solutions to the nearest tenth. Then we would probably proceed from the step $x + 3 = \pm\sqrt{\dfrac{21}{2}}$ as follows.

$$x + 3 = \pm\sqrt{\frac{21}{2}}$$
$$x = -3 \pm\sqrt{\frac{21}{2}}$$
$$x = -3 + \sqrt{\frac{21}{2}} \quad \text{or} \quad x = -3 - \sqrt{\frac{21}{2}}$$

Now use your calculator to evaluate each of these expressions to the nearest tenth. The solution set is $\{-6.2, .2\}$.

Solve each of the following equations and express the solutions to the nearest tenth.

(a) $x^2 - 6x - 4 = 0$

(b) $x^2 - 8x + 4 = 0$

(c) $x^2 + 4x - 4 = 0$

(d) $x^2 + 2x - 5 = 0$

(e) $x^2 - 14x - 2 = 0$

(f) $x^2 + 12x - 1 = 0$

10.3 Quadratic Formula

We can use the method of completing the square to solve *any* quadratic equation. The equation $ax^2 + bx + c = 0$, where a, b, and c are real numbers with $a \neq 0$, can represent *any* quadratic equation. These two ideas merge to produce the *quadratic formula*, a formula that can be used to solve *any* quadratic equation. The merger is accomplished by using the method of completing the square to solve the equation $ax^2 + bx + c = 0$ as follows.

$$ax^2 + bx + c = 0$$
$$ax^2 + bx = -c$$
$$x^2 + \frac{b}{a}x = -\frac{c}{a} \qquad \text{Multiply both sides by } \frac{1}{a}.$$

$$x^2 + \frac{b}{a}x + \frac{b^2}{4a^2} = -\frac{c}{a} + \frac{b^2}{4a^2}$$

Complete the square by adding $\frac{b^2}{4a^2}$ to both sides.

$$\left(x + \frac{b}{2a}\right)^2 = \frac{b^2 - 4ac}{4a^2}$$

The right side is combined into a single term.

$$x + \frac{b}{2a} = \pm\sqrt{\frac{b^2 - 4ac}{4a^2}}$$

$$x + \frac{b}{2a} = \pm\frac{\sqrt{b^2 - 4ac}}{\sqrt{4a^2}}$$

$$x + \frac{b}{2a} = \pm\frac{\sqrt{b^2 - 4ac}}{2a}$$

$\sqrt{4a^2} = |2a|$ but $2a$ can be used because of the use of $\pm$.

$$x = -\frac{b}{2a} \pm \frac{\sqrt{b^2 - 4ac}}{2a}$$

$$x = \frac{-b \pm \sqrt{b^2 - 4ac}}{2a}$$

The solutions are $\dfrac{-b + \sqrt{b^2 - 4ac}}{2a}$ and $\dfrac{-b - \sqrt{b^2 - 4ac}}{2a}$.

We usually state the **quadratic formula** as follows.

$$\boxed{x = \frac{-b \pm \sqrt{b^2 - 4ac}}{2a}}$$

We can use it to solve any quadratic equation by expressing the equation in standard form and substituting the values for a, b, and c into the formula. Consider the following examples.

Example 1

Solve $x^2 + 7x + 10 = 0$ by using the quadratic formula.

Solution

The equation $x^2 + 7x + 10 = 0$ is in standard form, so $a = 1$, $b = 7$, and $c = 10$. Substituting these values into the formula we obtain

$$x = \frac{-b \pm \sqrt{b^2 - 4ac}}{2a}$$

$$x = \frac{-7 \pm \sqrt{7^2 - 4(1)(10)}}{2(1)}$$

$$x = \frac{-7 \pm \sqrt{9}}{2}$$

$$x = \frac{-7 \pm 3}{2}$$

$$x = \frac{-7 + 3}{2} \quad \text{or} \quad x = \frac{-7 - 3}{2}$$

$$x = -2 \quad \text{or} \quad x = -5.$$

The solution set is $\{-5, -2\}$.

Example 2

Solve $x^2 - 3x = 1$ by using the quadratic formula.

Solution

$$x^2 - 3x = 1$$

$$x^2 - 3x - 1 = 0 \qquad \text{Change to standard form}$$

$$x = \frac{-(-3) \pm \sqrt{(-3)^2 - 4(1)(-1)}}{2(1)}$$

$$x = \frac{3 \pm \sqrt{9 + 4}}{2}$$

$$x = \frac{3 \pm \sqrt{13}}{2}$$

The solution set is $\left\{ \dfrac{3 - \sqrt{13}}{2}, \dfrac{3 + \sqrt{13}}{2} \right\}$.

Example 3

Solve $15n^2 - n - 2 = 0$ by using the quadratic formula.

Solution

Remember that although we commonly use the variable x in the statement of the quadratic formula, any variable could be used.

$$15n^2 - n - 2 = 0$$

$$n = \frac{-(-1) \pm \sqrt{(-1)^2 - 4(15)(-2)}}{2(15)}$$

$$n = \frac{1 \pm \sqrt{1 + 120}}{30}$$

$$n = \frac{1 \pm \sqrt{121}}{30}$$

$$n = \frac{1 \pm 11}{30}$$

$$n = \frac{1 + 11}{30} \quad \text{or} \quad n = \frac{1 - 11}{30}$$

$$n = \frac{12}{30} \quad \text{or} \quad n = \frac{-10}{30}$$

$$n = \frac{2}{5} \quad \text{or} \quad n = -\frac{1}{3}$$

The solution set is $\left\{ -\dfrac{1}{3}, \dfrac{2}{5} \right\}$.

Example 4

Solution

Solve $t^2 - 5t - 84 = 0$ by using the quadratic formula.

$$t^2 - 5t - 84 = 0$$

$$t = \frac{-(-5) \pm \sqrt{(-5)^2 - 4(1)(-84)}}{2(1)}$$

$$t = \frac{5 \pm \sqrt{25 + 336}}{2}$$

$$t = \frac{5 \pm \sqrt{361}}{2}$$

$$t = \frac{5 \pm 19}{2}$$

$$t = \frac{5 + 19}{2} \quad \text{or} \quad t = \frac{5 - 19}{2}$$

$$t = \frac{24}{2} \quad \text{or} \quad t = \frac{-14}{2}$$

$$t = 12 \quad \text{or} \quad t = -7$$

The solution set is $\{-7, 12\}$. ▲

We can easily identify quadratic equations that have no real number solutions when we use the quadratic formula. The final example of this section illustrates this point.

Example 5

Solution

Solve $x^2 - 2x + 8 = 0$ by using the quadratic formula.

$$x^2 - 2x + 8 = 0$$

$$x = \frac{-(-2) \pm \sqrt{(-2)^2 - 4(1)(8)}}{2(1)}$$

$$x = \frac{2 \pm \sqrt{4 - 32}}{2}$$

$$x = \frac{2 \pm \sqrt{-28}}{2}$$

Since $\sqrt{-28}$ is not a real number, we conclude that the given equation has no real number solutions. (More work will be done with this type of equation in Section 11.7.) ▲

Problem Set 10.3

Use the quadratic formula to solve each of the following quadratic equations.

1. $x^2 - 5x - 6 = 0$
2. $x^2 + 3x - 4 = 0$
3. $x^2 + 5x = 36$
4. $x^2 - 8x = -12$
5. $n^2 - 2n - 5 = 0$
6. $n^2 - 4n - 1 = 0$
7. $a^2 - 5a - 2 = 0$
8. $a^2 + 3a + 1 = 0$
9. $x^2 - 2x + 6 = 0$
10. $x^2 - 8x + 16 = 0$
11. $y^2 + 4y + 2 = 0$
12. $n^2 + 6n + 11 = 0$
13. $x^2 - 6x = 0$
14. $x^2 + 8x = 0$
15. $2x^2 = 7x$
16. $3x^2 = -10x$
17. $n^2 - 34n + 288 = 0$
18. $n^2 + 27n + 182 = 0$
19. $x^2 + 2x - 80 = 0$
20. $x^2 - 15x + 54 = 0$
21. $t^2 + 4t + 4 = 0$
22. $t^2 + 6t - 5 = 0$
23. $6x^2 + x - 2 = 0$
24. $4x^2 - x - 3 = 0$
25. $5x^2 + 3x - 2 = 0$
26. $6x^2 - x - 2 = 0$
27. $12x^2 + 19x = -5$
28. $2x^2 + 7x - 6 = 0$
29. $2x^2 + 5x - 6 = 0$
30. $2x^2 + 3x - 3 = 0$
31. $3x^2 + 4x - 1 = 0$
32. $3x^2 + 2x - 4 = 0$
33. $16x^2 + 24x + 9 = 0$

34. $9x^2 - 30x + 25 = 0$
35. $4n^2 + 8n - 1 = 0$
36. $4n^2 + 6n - 1 = 0$
37. $6n^2 + 9n + 1 = 0$
38. $5n^2 + 8n + 1 = 0$
39. $2y^2 - y - 4 = 0$
40. $3t^2 + 6t + 5 = 0$
41. $4t^2 + 5t + 3 = 0$
42. $5x^2 + x - 1 = 0$
43. $7x^2 + 5x - 4 = 0$
44. $6x^2 + 2x - 3 = 0$
45. $7 = 3x^2 - x$
46. $-2x^2 + 3x = -4$
47. $n^2 + 23n = -126$
48. $n^2 + 2n = 195$

THOUGHTS INTO WORDS

49. Explain how to use the quadratic formula to solve the equation $x^2 = 2x + 6$.

50. Your friend states that the equation $-x^2 - 6x + 16 = 0$ must be changed to $x^2 + 6x - 16 = 0$ (by multiplying both sides by -1) before the quadratic formula can be applied. Is he right about this and if not how would you convince him?

51. Another of your friends claims that the quadratic formula can be used to solve the equation $x^2 - 4 = 0$. How would you react to this claim?

Further Investigations

Use the quadratic formula to solve each of the following equations. Express the solutions to the nearest hundredth.

52. $x^2 - 7x - 13 = 0$
53. $x^2 - 5x - 19 = 0$

54. $x^2 + 9x - 15 = 0$

55. $x^2 + 6x - 17 = 0$

56. $2x^2 + 3x - 7 = 0$

57. $3x^2 + 7x - 13 = 0$

58. $5x^2 - 11x - 14 = 0$

59. $4x^2 - 9x - 19 = 0$

60. $-3x^2 + 2x + 11 = 0$

61. $-5x^2 + x + 21 = 0$

62. Let x_1 and x_2 be the two solutions of $ax^2 + bx + c = 0$ obtained by the quadratic formula. Thus, we have

$$x_1 = \frac{-b + \sqrt{b^2 - 4ac}}{2a} \quad \text{and}$$

$$x_2 = \frac{-b - \sqrt{b^2 - 4ac}}{2a}.$$

Find the sum $x_1 + x_2$ and the product $(x_1)(x_2)$. Your answers should be

$$x_1 + x_2 = -\frac{b}{a} \quad \text{and} \quad (x_1)(x_2) = \frac{c}{a}.$$

These relationships provide another way of checking potential solutions when solving quadratic equations. For example, back in Example 3 we solved the equation $15n^2 - n - 2 = 0$ and obtained solutions of $-\frac{1}{3}$ and $\frac{2}{5}$. Let's check these solutions using the sum and product relationships.

Sum of solutions: $\qquad -\frac{1}{3} + \frac{2}{5} = -\frac{5}{15}$

$\qquad +\frac{6}{15} = \frac{1}{15} \quad \text{and} \quad -\frac{b}{a} = -\frac{-1}{15} = \frac{1}{15}$

Product of Solutions: $\quad \left(-\frac{1}{3}\right)\left(\frac{2}{5}\right) = -\frac{2}{15}$

$\quad \text{and} \quad \frac{c}{a} = \frac{-2}{15} = -\frac{2}{15}$

Use the sum and product relationships to check at least ten of the problems that you worked in this problem set.

10.4 Solving Quadratic Equations — Which Method?

Let's summarize the three basic methods of solving quadratic equations we presented in this chapter by solving a specific quadratic equation using each technique. Consider the equation $x^2 + 4x - 12 = 0$.

Factoring Method

$$x^2 + 4x - 12 = 0$$
$$(x + 6)(x - 2) = 0$$
$$x + 6 = 0 \quad \text{or} \quad x - 2 = 0$$
$$x = -6 \quad \text{or} \quad x = 2$$

The solution set is $\{-6, 2\}$.

Completing the Square Method

$$x^2 + 4x - 12 = 0$$
$$x^2 + 4x = 12$$
$$x^2 + 4x + 4 = 12 + 4$$

$$(x + 2)^2 = 16$$
$$x + 2 = \pm\sqrt{16}$$

$$x + 2 = 4 \quad \text{or} \quad x + 2 = -4$$
$$x = 2 \quad \text{or} \quad x = -6$$

The solution set is $\{-6, 2\}$.

Quadratic Formula Method

$$x^2 + 4x - 12 = 0$$

$$x = \frac{-4 \pm \sqrt{4^2 - 4(1)(-12)}}{2(1)}$$

$$x = \frac{-4 \pm \sqrt{64}}{2}$$

$$x = \frac{-4 \pm 8}{2}$$

$$x = \frac{-4 + 8}{2} \quad \text{or} \quad x = \frac{-4 - 8}{2}$$

$$x = 2 \quad \text{or} \quad x = -6$$

The solution set is $\{-6, 2\}$.

We also discussed the use of the property, $x^2 = a$ if and only if $x = \pm\sqrt{a}$, for certain types of quadratic equations. For example, $x^2 = 4$ can be solved easily by applying the property and obtaining $x = \sqrt{4}$ or $x = -\sqrt{4}$; thus, the solutions are 2 and -2.

Which method should you use to solve a particular quadratic equation? Let's consider some examples in which the different techniques are used. However, keep in mind that usually this is a decision you must make as the need arises. So become as familiar as you can with the strengths and weaknesses of each method.

Example 1

Solve $2x^2 + 12x - 54 = 0$.

Solution

First, it is very helpful to recognize a factor of 2 in each of the terms on the left side.

$$2x^2 + 12x - 54 = 0$$
$$x^2 + 6x - 27 = 0 \qquad \text{Multiply both sides by } \tfrac{1}{2}.$$

Now we should recognize that the left side can be factored. Thus, we can proceed as follows.

$$(x + 9)(x - 3) = 0$$
$$x + 9 = 0 \quad \text{or} \quad x - 3 = 0$$
$$x = -9 \quad \text{or} \quad x = 3$$

The solution set is $\{-9, 3\}$.

▲

Example 2

Solve $(4x + 3)^2 = 16$.

Solution

The form of this equation lends itself to the use of the property, $x^2 = a$ if and only if $x = \pm\sqrt{a}$.

$$(4x + 3)^2 = 16$$
$$4x + 3 = \pm\sqrt{16}$$
$$4x + 3 = 4 \quad \text{or} \quad 4x + 3 = -4$$
$$4x = 1 \quad \text{or} \quad 4x = -7$$
$$x = \frac{1}{4} \quad \text{or} \quad x = -\frac{7}{4}$$

The solution set is $\left\{-\frac{7}{4}, \frac{1}{4}\right\}$.

▲

Example 3

Solve $n + \dfrac{1}{n} = 5$.

Solution

First, we need to *clear the equation of fractions* by multiplying both sides by n.

$$n + \frac{1}{n} = 5, \quad n \neq 0$$
$$n\left(n + \frac{1}{n}\right) = 5(n)$$
$$n^2 + 1 = 5n$$

Now, let's change it to standard form.

$$n^2 - 5n + 1 = 0$$

Since the left side cannot be factored using integers, we must solve the equation by using either the method of completing the square or the quadratic formula. Using the formula we obtain

$$n = \frac{-(-5) \pm \sqrt{(-5)^2 - 4(1)(1)}}{2(1)}$$
$$n = \frac{5 \pm \sqrt{21}}{2}.$$

The solution set is $\left\{\dfrac{5 - \sqrt{21}}{2}, \dfrac{5 + \sqrt{21}}{2}\right\}$

▲

Example 4

Solution

Solve $t^2 = \sqrt{2}t$.

A quadratic equation without a constant term can be easily solved by the factoring method.

$$t^2 = \sqrt{2}t$$

$$t^2 - \sqrt{2}t = 0$$

$$t(t - \sqrt{2}) = 0$$

$$t = 0 \quad \text{or} \quad t - \sqrt{2} = 0$$

$$t = 0 \quad \text{or} \quad t = \sqrt{2}$$

The solution set is $\{0, \sqrt{2}\}$. (Check each of these solutions in the given equation.)

▲

Example 5

Solution

Solve $x^2 - 28x + 192 = 0$.

Determining whether or not the left side is factorable presents a bit of a problem because of the size of the constant term. Therefore, let's not concern ourselves with trying to factor, instead we will use the quadratic formula.

$$x^2 - 28x + 192 = 0$$

$$x = \frac{-(-28) \pm \sqrt{(-28)^2 - 4(1)(192)}}{2(1)}$$

$$x = \frac{28 \pm \sqrt{784 - 768}}{2}$$

$$x = \frac{28 \pm \sqrt{16}}{2}$$

$$x = \frac{28 + 4}{2} \quad \text{or} \quad x = \frac{28 - 4}{2}$$

$$x = 16 \quad \text{or} \quad x = 12$$

The solution set is $\{12, 16\}$.

▲

Example 6

Solution

Solve $x^2 + 12x = 17$.

The form of this equation and the fact that the coefficient of x is even makes the method of completing the square a reasonable approach.

$$x^2 + 12x = 17$$

$$x^2 + 12x + 36 = 17 + 36$$

$$(x + 6)^2 = 53$$

$$x + 6 = \pm\sqrt{53}$$

$$x = -6 \pm \sqrt{53}$$

The solution set is $\{-6 - \sqrt{53}, -6 + \sqrt{53}\}$.

▲

Problem Set 10.4

Solve each of the following quadratic equations using the method that seems most appropriate to you.

1. $x^2 + 4x = 45$

2. $x^2 + 4x = 60$

3. $(5n + 6)^2 = 49$

4. $(3n - 1)^2 = 25$

5. $t^2 - t - 2 = 0$

6. $t^2 + 2t - 3 = 0$

7. $8x = 3x^2$

8. $5x^2 = 7x$

9. $9x^2 - 6x + 1 = 0$

10. $4x^2 + 36x + 81 = 0$

11. $5n^2 = \sqrt{8}n$

12. $\sqrt{3}n = 2n^2$

13. $n^2 - 14n = 19$

14. $n^2 - 10n = 14$

15. $5x^2 - 2x - 7 = 0$

16. $3x^2 - 4x - 2 = 0$

17. $15x^2 + 28x + 5 = 0$

18. $20y^2 - 7y - 6 = 0$

19. $x^2 - \sqrt{8}x - 7 = 0$

20. $x^2 + \sqrt{5}x - 5 = 0$

21. $y^2 + 5y = 84$

22. $y^2 + 7y = 60$

23. $2n = 3 + \dfrac{3}{n}$

24. $n + \dfrac{1}{n} = 7$

25. $3x^2 - 9x - 12 = 0$

26. $2x^2 + 10x - 28 = 0$

27. $2x^2 - 3x + 7 = 0$

28. $3x^2 - 2x + 5 = 0$

29. $n(n - 46) = -480$

30. $n(n + 42) = -432$

31. $n - \dfrac{3}{n} = -1$

32. $n - \dfrac{2}{n} = \dfrac{3}{4}$

33. $x + \dfrac{1}{x} = \dfrac{25}{12}$

34. $x + \dfrac{1}{x} = \dfrac{65}{8}$

35. $t^2 + 12t + 36 = 49$

36. $t^2 - 10t + 25 = 16$

37. $x^2 - 28x + 187 = 0$

38. $x^2 - 33x + 266 = 0$

39. $\dfrac{x^2}{3} - x = -\dfrac{1}{2}$

40. $\dfrac{2x - 1}{3} = \dfrac{5}{x + 2}$

41. $\dfrac{2}{x + 2} - \dfrac{1}{x} = 3$

42. $\dfrac{3}{x - 1} + \dfrac{2}{x} = \dfrac{3}{2}$

43. $\dfrac{2}{3n - 1} = \dfrac{n + 2}{6}$

44. $\dfrac{x^2}{2} = x + \dfrac{1}{4}$

45. $(n - 2)(n + 4) = 7$

46. $(n + 3)(n - 8) = -30$

THOUGHTS INTO WORDS

47. Which method would you use to solve the equation $x^2 + 30x = -216$? Explain your reasons for making this choice.

48. Explain how you would solve the equation $0 = -x^2 - x + 6$.

49. How can you tell by inspection that the equation $x^2 + x + 4 = 0$ has no real number solutions?

10.5 Solving Problems Using Quadratic Equations

The following diagram indicates the common thread of this text.

Develop skills $\longrightarrow$ Use skills to solve equations $\longrightarrow$ Use equations to solve word problems

Now you should be ready to use your skills relative to solving systems of equations (Chapter 8) and quadratic equations to help with additional types of word problems. Before you consider such problems, let's review and update the problem solving suggestions we offered in Chapter 3.

Suggestions for Solving Word Problems

1. Read the problem carefully and make certain that you understand the meanings of all the words. Be especially alert for any technical terms used in the statement of the problem.

2. Read the problem a second time (perhaps even a third time) to get an overview of the situation being described and to determine the known facts, as well as what is to be found.

3. Sketch any figure, diagram, or chart that might be helpful in analyzing the problem.

*4. Choose *meaningful* variables to represent the unknown quantities. Use one or two variables, whichever seems easiest. The term "meaningful" refers to the choice of letters to use as variables. Choose letters that have some significance for the problem under consideration. For example, if the problem deals with the length and width of a rectangle, then l and w would seem like natural choices for the variables.

*5. Look for *guidelines* that can be used to help set up equations. A guideline might be a formula such as *area of a rectangular region equals length times width*, or a statement of a relationship such as *the product of the two numbers is 98*.

*6. (a) Form an equation which contains the variable that translates the conditions of the guideline from English to algebra; or

   (b) form two equations which contain the two variables that translate the guidelines from English to algebra.

*7. Solve the equation (system of equations) and use the solution (solutions) to determine all facts requested in the problem.

8. Check all answers back to the original statement of the problem.

The asterisks indicate those suggestions that have been revised to include using systems of equations to solve problems. Keep these suggestions in mind as you study the examples and work the problems in this section.

Problem 1

The length of a rectangular region is 2 centimeters longer than its width. The area of the region is 35 square centimeters. Find the length and width of the rectangle.

Solution

Let l represent the length. Let w represent the width. (Figure 10.10)

FIGURE 10.10

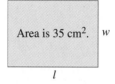

Area is 35 cm². w

l

We can use the area formula for a rectangle, $A = lw$, and the statement "the length of a rectangular region is 2 centimeters longer than its width" as guidelines to form a system of equations.

$$\begin{pmatrix} lw = 35 \\ l = w + 2 \end{pmatrix}$$

The second equation indicates that $w + 2$ can be substituted for l. Making this substitution in the first equation yields

$$(w + 2)(w) = 35.$$

Solving this quadratic equation by factoring, we obtain

$$w^2 + 2w = 35$$
$$w^2 + 2w - 35 = 0$$
$$(w + 7)(w - 5) = 0$$
$$w + 7 = 0 \qquad \text{or} \qquad w - 5 = 0$$
$$w = -7 \qquad \text{or} \qquad w = 5.$$

The width of a rectangle cannot be a negative number, so we discard the solution -7. Thus, the width of the rectangle is 5 centimeters and the length $(w + 2)$ is 7 centimeters. ▲

Problem 2

Find two consecutive whole numbers whose product is 506.

Solution

Let n represent the smaller whole number. Then $n + 1$ represents the next larger whole number. The phrase "whose product is 506" translates into the equation

$$n(n + 1) = 506.$$

Changing this quadratic equation to standard form produces

$$n^2 + n = 506$$
$$n^2 + n - 506 = 0.$$

Because of the size of the constant term, let's not try to factor; instead, use the quadratic formula.

$$n = \frac{-1 \pm \sqrt{1^2 - 4(1)(-506)}}{2(1)}$$

$$n = \frac{-1 \pm \sqrt{2025}}{2}$$

$$n = \frac{-1 \pm 45}{2} \qquad \sqrt{2025} = 45$$

$$n = \frac{-1 + 45}{2} \qquad \text{or} \qquad n = \frac{-1 - 45}{2}$$

$$n = 22 \qquad \text{or} \qquad n = -23$$

Since we are looking for whole numbers, we discard the solution -23. Therefore, the whole numbers are 22 and 23. ▲

Problem 3

The perimeter of a rectangular lot is 100 meters and its area is 616 square meters. Find the length and width of the lot.

Solution

Let l represent the length. Let w represent the width. (Figure 10.11)

FIGURE 10.11

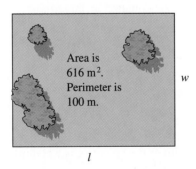

Area is 616 m².
Perimeter is 100 m.

$$\begin{pmatrix} lw = 616 \\ 2l + 2w = 100 \end{pmatrix} \quad \longleftarrow \text{ Area is } 616 \, m^2$$
$$\longleftarrow \text{ Perimeter is } 100 \, m$$

Multiplying the second equation by $\frac{1}{2}$ produces $l + w = 50$, which can be changed to $l = 50 - w$. Substituting $50 - w$ for l in the first equation produces the quadratic equation

$$(50 - w)(w) = 616$$
$$50w - w^2 = 616$$
$$w^2 - 50w = -616.$$

Using the method of completing the square we obtain

$$w^2 - 50w + 625 = -616 + 625$$
$$(w - 25)^2 = 9$$
$$w - 25 = \pm 3$$
$$w - 25 = 3 \quad \text{or} \quad w - 25 = -3$$
$$w = 28 \quad \text{or} \quad w = 22.$$

If $w = 28$, then $l = 50 - w = 22$. If $w = 22$, then $l = 50 - w = 28$.

The rectangle is 28 meters by 22 meters or 22 meters by 28 meters. ▲

Problem 4

Find two numbers such that their sum is 2 and their product is -1.

Solution

Let n represent one of the numbers. Let m represent the other number.

$$\begin{pmatrix} n + m = 2 \\ nm = -1 \end{pmatrix} \begin{array}{l} \leftarrow \text{Their sum is 2.} \\ \leftarrow \text{Their product is } -1. \end{array}$$

We can change the first equation to $m = 2 - n$; then we can substitute $2 - n$ for m in the second equation.

$$n(2 - n) = -1$$
$$2n - n^2 = -1$$
$$-n^2 + 2n + 1 = 0$$
$$n^2 - 2n - 1 = 0$$
$$n = \frac{-(-2) \pm \sqrt{(-2)^2 - 4(1)(-1)}}{2(1)}$$
$$n = \frac{2 \pm \sqrt{8}}{2} = \frac{2 \pm 2\sqrt{2}}{2} = 1 \pm \sqrt{2}$$

If $n = 1 + \sqrt{2}$, then $m = 2 - (1 + \sqrt{2})$
$$= 2 - 1 - \sqrt{2}$$
$$= 1 - \sqrt{2}.$$

If $n = 1 - \sqrt{2}$, then $m = 2 - (1 - \sqrt{2})$
$$= 2 - 1 + \sqrt{2}$$
$$= 1 + \sqrt{2}.$$

The numbers are $1 + \sqrt{2}$ and $1 - \sqrt{2}$.   Perhaps you should check these numbers in the original statement of the problem!   ▲

Finally, let's consider a uniform motion problem similar to those we solved in Chapter 7. Now we have the flexibility of using two equations in two variables.

Problem 5

Larry drive 156 miles in one hour more than it took Mike to drive 108 miles. Mike drove at an average rate of 2 miles per hour faster than Larry. How fast did each one travel?

Solution

We can represent the unknown rates and times in the following ways.

Let r represent Larry's rate.

Let t represent Larry's time.

Then $r + 2$ represents Mike's rate

and $t - 1$ represents Mike's time.

Since *distance equals rate times time*, the following system can be set up.

$$\left(\begin{array}{l} rt = 156 \\ (r + 2)(t - 1) = 108 \end{array} \right)$$

Solving the first equation for r produces $r = \dfrac{156}{t}$. Substituting $\dfrac{156}{t}$ for r in the second equation and simplifying, we obtain

$$\left(\frac{156}{t} + 2 \right)(t - 1) = 108$$

$$156 - \frac{156}{t} + 2t - 2 = 108$$

$$2t - \frac{156}{t} + 154 = 108$$

$$2t - \frac{156}{t} + 46 = 0$$

$$2t^2 - 156 + 46t = 0$$

$$2t^2 + 46t - 156 = 0$$

$$t^2 + 23t - 78 = 0.$$

We can solve this quadratic equation by factoring.

$$(t + 26)(t - 3) = 0$$

$$t + 26 = 0 \quad \text{or} \quad t - 3 = 0$$

$$t = -26 \quad \text{or} \quad t = 3$$

We must disregard the negative solution. So, Mike's time is $3 - 1 = 2$ hours. Larry's rate is $\dfrac{156}{3} = 52$ miles per hour and Mike's rate is $\dfrac{108}{2} = 54$ miles per hour. ▲

Problem Set 10.5

Solve each of the following problems.

1. Find two consecutive whole numbers whose product is 306.

2. Find two consecutive whole numbers whose product is 702.

3. Suppose that the sum of two positive integers is 44 and their product is 475. Find the integers.

4. Two positive integers differ by 6. Their product is 616. Find the integers.

5. Find two numbers such that their sum is 6 and their product is 4.

6. Find two numbers such that their sum is 4 and their product is 1.

7. The sum of a number and its reciprocal is $\dfrac{3\sqrt{2}}{2}$. Find the number.

8. The sum of a number and its reciprocal is $\dfrac{73}{24}$. Find the number.

9. Each of three consecutive even whole numbers is squared. The three results are added and the sum is 596. Find the numbers.

10. Each of three consecutive whole numbers is squared. The three results are added and the sum is 245. Find the three whole numbers.

11. The sum of the square of a number and the square of one-half of the number is 80. Find the number.

12. The difference between the square of a positive number and the square of one-half the number is 243. Find the number.

13. Find the length and width of a rectangle if its length is 4 meters less than twice the width and the area of the rectangle is 96 square meters.

14. Suppose that the length of a rectangular region is 4 centimeters longer than its width. The area of the region is 45 square centimeters. Find the length and width of the rectangle.

15. The perimeter of a rectangle is 80 centimeters and its area is 375 square centimeters. Find the length and width of the rectangle.

16. The perimeter of a rectangle is 132 yards and its area is 1080 square yards. Find the length and width of the rectangle.

17. The area of a tennis court is 2106 square feet. The length of the court is $\dfrac{26}{9}$ times the width. Find the length and width of a tennis court.

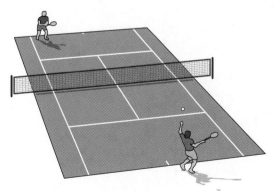

FIGURE 10.12

18. The area of a badminton court is 880 square feet. The length of the court is 2.2 times the width. Find the length and width of the court.

19. An auditorium in a local high school contains 300 seats. There are 5 fewer rows than the number of seats per row. Find the number of rows and the number of seats per row.

20. Three hundred seventy-five trees were planted in rows in an orchard. The number of trees per row was ten more than the number of rows. How many rows of trees are in the orchard?

21. The area of a rectangular region is 63 square feet. If the length and width are each increased by 3 feet, the area is increased by 57 square feet. Find the length and width of the original rectangle.

22. The area of a circle is numerically equal to twice the circumference of the circle. Find the length of a radius of the circle.

23. The sum of the lengths of the two legs of a right triangle is 14 inches. If the length of the hypotenuse is 10 inches, find the length of each leg.

24. A page for a magazine contains 70 square inches of type. The height of a page is twice the width. If the margin around the type is to be 2 inches uniformly, what are the dimensions of the page?

25. A 5-by-7-inch picture is surrounded by a frame of uniform width (Figure 10.13). The area of the picture and frame together is 80 square inches. Find the width of the frame.

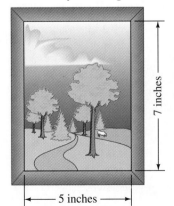

7 inches

5 inches

FIGURE 10.13

26. A rectangular piece of cardboard is 3 inches longer than it is wide. From each corner, a square piece 2 inches on a side is cut out. The flaps are then turned up to form an open box that has a volume of 140 cubic inches. Find the length and width of the original piece of cardboard.

27. A class trip was to cost $3000. If there had been ten more students, it would have cost each student $25 less. How many students took the trip?

28. Simon mowed some lawns and earned $40. It took him 3 hours longer than he anticipated and thus he earned $3 per hour less than he anticipated. How long did he expect it to take him?

29. A piece of wire 56 inches long is cut into two pieces and each piece is bent into the shape of a square. If the sum of the areas of the two squares is 100 square inches, find the length of each piece of wire.

30. Suppose that by increasing the speed of a car by 10 miles per hour, it is possible to make a trip of 200 miles in 1 hour less time. What was the original speed for the trip?

31. On a 50-mile bicycle ride, Irene averaged 4 miles per hour faster for the first 36 miles than she did for the last 14 miles. The entire trip of 50 miles took 3 hours. Find her rate for the first 36 miles.

32. One side of a triangle is 1 foot more than twice the length of the altitude to that side. If the area of the triangle is 18 square feet, find the length of a side and the length of the altitude to that side.

THOUGHTS INTO WORDS

33. Return to Example 1 of this section and explain how the problem could be solved using one variable and one equation.

34. Write a page or two on the topic "Using Algebra to Solve Problems."

SUMMARY

(10.1) A quadratic equation in the variable x is any equation that can be written in the form $ax^2 + bx + c = 0$, where a, b, and c are real numbers and $a \neq 0$. We can solve quadratic equations that are factorable using integers by factoring and applying the property, $ab = 0$ *if and only if* $a = 0$ *or* $b = 0$.

Use the property, $x^2 = a$ *if and only if* $x = \pm\sqrt{a}$, to solve certain types of quadratic equations.

Many times equations of the form $x^2 = a$ are produced when working with the Pythagorean theorem.

Don't forget: (1) In an isosceles right triangle, the lengths of the two legs are equal, and (2) In a 30°–60° right triangle, the length of the leg opposite the 30° angle is one-half of the length of the hypotenuse.

(10.2) You should be able to solve quadratic equations by the method of completing the square. To review this method, look back over the examples in Section 10.2.

(10.3) We usually state the quadratic formula as

$$x = \frac{-b \pm \sqrt{b^2 - 4ac}}{2a}.$$

We can use it to solve any quadratic equation after it has been written in the form $ax^2 + bx + c = 0$.

(10.4) To review the strengths and weaknesses of the three basic methods for solving a quadratic equation (factoring, completing the square, the quadratic formula), go back over the examples in Section 10.4.

(10.5) Our knowledge of systems of equations and quadratic equations provides us with a stronger basis for solving word problems.

Chapter 10 Review Problem Set

Solve each of the following quadratic equations.

1. $(2x + 7)^2 = 25$

2. $x^2 + 8x = -3$

3. $21x^2 - 13x + 2 = 0$

4. $x^2 = 17x$

5. $n - \dfrac{4}{n} = -3$

6. $n^2 - 26n + 165 = 0$

7. $3a^2 + 7a - 1 = 0$

8. $4x^2 - 4x + 1 = 0$

9. $5x^2 + 6x + 7 = 0$

10. $3x^2 + 18x + 15 = 0$

11. $3(x - 2)^2 - 2 = 4$

12. $x^2 + 4x - 14 = 0$

13. $y^2 = 45$

14. $x(x - 6) = 27$

15. $x^2 = x$

16. $n^2 - 4n - 3 = 6$

17. $n^2 - 44n + 480 = 0$

18. $\dfrac{x^2}{4} = x + 1$

19. $\dfrac{5x - 2}{3} = \dfrac{2}{x + 1}$

20. $\dfrac{-1}{3x - 1} = \dfrac{2x + 1}{-2}$

21. $\dfrac{5}{x - 3} + \dfrac{4}{x} = 6$

22. $\dfrac{1}{x + 2} - \dfrac{2}{x} = 3$

Set up an equation or a system of equations to help solve each of the following problems.

23. The perimeter of a rectangle is 42 inches and its area is 108 square inches. Find the length and width of the rectangle.

24. Find two consecutive whole numbers whose product is 342.

25. Each of three consecutive odd whole numbers is squared. The three results are added and the sum is 251. Find the numbers.

26. The combined area of two squares is 50 square meters. Each side of the larger square is three times as long as a side of the smaller square. Find the lengths of the sides of each square.

27. The difference of the lengths of the two legs of a right triangle is 2 yards. If the length of the hypotenuse is $2\sqrt{13}$ yards find the length of each leg.

28. Tony bought a number of shares of stock for a total of $720. A month later the value of the stock increased by $8 per share and he sold all but 20 shares and regained his original investment plus a profit of $80. How many shares did he sell and at what price per share?

29. A company has a rectangular parking lot 40 meters wide and 60 meters long. They plan to increase the area of the lot by 1100 square meters by adding a strip of equal width to one side and one end. Find the width of the strip to be added.

30. Jay traveled 225 miles in two hours less time than it took Jean to travel 336 miles. If Jay's rate was 3 miles per hour less than Jean's rate, find each rate.

31. The length of the hypotenuse of an isosceles right triangle is 12 inches. Find the length of each leg.

32. In a 30°–60° right triangle the side opposite the 60° angle is 8 centimeters long. Find the length of the hypotenuse.

CHAPTER 10 TEST

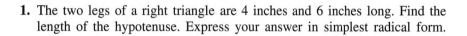

1. The two legs of a right triangle are 4 inches and 6 inches long. Find the length of the hypotenuse. Express your answer in simplest radical form.

2. A diagonal of a rectangular plot of ground measures 14 meters. If the width of the rectangle is 5 meters, find the length to the nearest meter.

3. A diagonal of a square piece of paper measures 10 inches. Find, to the nearest inch, the length of a side of the square.

4. In a 30°–60° right triangle, the side opposite the 30° angle is 4 centimeters long. Find the length of the side opposite the 60° angle. Express your answer in simplest radical form.

For Problems 5–20, solve each equation.

5. $(3x + 2)^2 = 49$ 6. $4x^2 = 64$

7. $8x^2 - 10x + 3 = 0$

8. $x^2 - 3x - 5 = 0$

9. $n^2 + 2n = 9$

10. $(2x - 1)^2 = -16$

11. $y^2 + 10y = 24$

12. $2x^2 - 3x - 4 = 0$

13. $\dfrac{x - 2}{3} = \dfrac{4}{x + 1}$

14. $\dfrac{2}{x - 1} + \dfrac{1}{x} = \dfrac{5}{2}$

15. $n(n - 28) = -195$

16. $n + \dfrac{3}{n} = \dfrac{19}{4}$

17. $(2x + 1)(3x - 2) = -2$

18. $(7x + 2)^2 - 4 = 21$

19. $(4x - 1)^2 = 27$

20. $n^2 - 5n + 7 = 0$

For Problems 21–25, set up an equation or a system of equations to help solve each problem.

21. A room contains 120 seats. The number of seats per row is one less than twice the number of rows. Find the number of seats per row.

(continued on next page)

CHAPTER 10 TEST *(continued)*

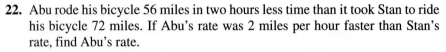

22. Abu rode his bicycle 56 miles in two hours less time than it took Stan to ride his bicycle 72 miles. If Abu's rate was 2 miles per hour faster than Stan's rate, find Abu's rate.

23. Find two consecutive odd whole numbers whose product is 255.

24. The combined area of two squares is 97 square feet. Each side of the larger square is one foot more than twice the length of a side of the smaller square. Find the length of a side of the larger square.

25. Dee bought a number of shares of stock for a total of $160. Two weeks later the value of the stock had increased two dollars per share and she sold all but 4 shares and regained her initial investment of $160. How many shares did she originally buy?

Cumulative Review Problem Set

For Problems 1–6, evaluate each of the numerical expressions.

1. -2^7

2. $\left(\dfrac{1}{4}\right)^{-3}$

3. $\left(\dfrac{1}{3} - \dfrac{1}{4}\right)^{-2}$

4. $-\sqrt{64}$

5. $\sqrt{\dfrac{4}{9}}$

6. $3^0 + 3^{-1} + 3^{-2}$

For Problems 7–10, evaluate each algebraic expression for the given values of the variables.

7. $3(2x - 1) - 4(2x + 3) - (x + 6)$
for $x = -4$

8. $(3x^2 - 4x - 6) - (3x^2 + 3x + 1)$
for $x = 6$

9. $2(a - b) - 3(2a + b) + 2(a - 3b)$
for $a = -2$ and $b = 3$

10. $x^2 - 2xy + y^2$ for $x = 5$ and $y = -2$

For Problems 11–20, perform the indicated operations and express your answers in simplest form.

11. $\dfrac{3}{4x} + \dfrac{5}{2x} - \dfrac{7}{x}$

12. $\dfrac{3}{x - 2} - \dfrac{4}{x + 3}$

13. $\dfrac{3x}{7y} \div \dfrac{6x}{35y^2}$

14. $\dfrac{x - 2}{x^2 + x - 6} \cdot \dfrac{x^2 + 6x + 9}{x^2 - x - 12}$

15. $\dfrac{7}{x^2 + 3x - 18} - \dfrac{8}{x - 3}$

16. $(-3xy)(-4y^2)(5x^3y)$

17. $(9x - 2)(3x + 4)$

18. $(-x - 1)(5x + 7)$

19. $(3x + 1)(2x^2 - x - 4)$

20. $(10x^3 - 8x^2 - 17x - 3) \div (5x + 1)$

For Problems 21–26, factor each expression completely.

21. $12x^3 + 14x^2 - 40x$

22. $12x^2 - 27$

23. $xy + 3x - 2y - 6$

24. $30 + 19x - 5x^2$

25. $4x^4 - 4$

26. $21x^2 + 22x - 8$

For Problems 27–38, change each radical expression to simplest radical form.

27. $4\sqrt{28}$

28. $-3\sqrt{48}$

29. $-\sqrt{45}$

30. $\sqrt{\dfrac{18}{49}}$

31. $\sqrt{\dfrac{36}{5}}$

32. $\dfrac{2\sqrt{3}}{3\sqrt{2}}$

33. $\dfrac{5\sqrt{8}}{6\sqrt{12}}$

34. $\sqrt{27x^3y^4}$

35. $\sqrt{72xy^5}$

36. $\dfrac{3\sqrt{x^3y}}{4\sqrt{xy}}$

37. $\dfrac{-2\sqrt{ab^2}}{5\sqrt{b}}$

38. $\dfrac{2}{3}\sqrt{40x^5}$

For Problems 39–43, find each product and express your answers in simplest radical form.

39. $(3\sqrt{8})(4\sqrt{2})$

40. $(-2\sqrt{6})(5\sqrt{3})$

41. $6\sqrt{2}(9\sqrt{8} - 3\sqrt{12})$

42. $(2\sqrt{3} + \sqrt{5})(\sqrt{3} - 2\sqrt{5})$

43. $(3\sqrt{2} - \sqrt{7})(3\sqrt{2} + \sqrt{7})$

For Problems 44–46, rationalize the denominators and simplify.

44. $\dfrac{4}{\sqrt{3} + \sqrt{2}}$

45. $\dfrac{-6}{3\sqrt{5} - \sqrt{6}}$

46. $\dfrac{\sqrt{2}}{3\sqrt{3} + 2\sqrt{2}}$

For Problems 47–50, simplify each of the radical expressions.

47. $3\sqrt{50} - 7\sqrt{72} + 4\sqrt{98}$

48. $4\sqrt{27} - 2\sqrt{50} - 6\sqrt{48} - 3\sqrt{18}$

49. $\dfrac{2}{3}\sqrt{20} - \dfrac{3}{4}\sqrt{45} + \sqrt{80}$

50. $\dfrac{1}{2}\sqrt{75} - \dfrac{2}{3}\sqrt{27} + \dfrac{1}{4}\sqrt{12}$

For Problems 51–54, graph each of the equations.

51. $3x - 6y = -6$

52. $y = -2x^2 + 1$

53. $y = -2x^3$

54. $y = -x$

For Problems 55–60, solve each of the problems.

55. Find the slope of the line determined by the points $(-3, 6)$ and $(2, -4)$.

56. Find the slope of the line determined by the equation $4x - 7y = 12$.

57. Write the equation of the line that has a slope of $\frac{2}{3}$ and contains the point $(7, 2)$.

58. Write the equation of the line that contains the points $(-4, 1)$ and $(-1, -3)$.

59. Write the equation of the line that has a slope of $-\frac{1}{4}$ and a y-intercept of -3.

60. Write the equation of the line that is perpendicular to the x-axis and contains the point $(4, -5)$.

For Problems 61–64, solve each of the systems by using either the substitution method or the addition method.

61. $\begin{pmatrix} y = 3x - 5 \\ 3x + 4y = -5 \end{pmatrix}$

62. $\begin{pmatrix} 4x - 3y = -20 \\ 3x + 5y = 14 \end{pmatrix}$

63. $\begin{pmatrix} \frac{1}{2}x - \frac{2}{3}y = -11 \\ \frac{1}{3}x + \frac{5}{6}y = 8 \end{pmatrix}$

64. $\begin{pmatrix} 2x + 7y = 22 \\ 4x - 5y = -13 \end{pmatrix}$

For Problems 65–86, solve each of the equations.

65. $-2(n - 1) + 4(2n - 3) = 4(n + 6)$

66. $\dfrac{4}{x - 1} = \dfrac{-1}{x + 6}$

67. $|2x - 6| = 3$

68. $\dfrac{t - 1}{3} - \dfrac{t + 2}{4} = -\dfrac{5}{12}$

69. $(3x - 1)^2 = 9$

70. $x^2 + 4x = 8$

71. $-7 - 2n - 6n = 7n - 5n + 12$

72. $\dfrac{n - 5}{2} = 3 - \dfrac{n + 4}{5}$

73. $.11x + .14(x + 400) = 181$

74. $(2t + 1)(t - 2) = 7$

75. $\dfrac{x}{60 - x} = 7 + \dfrac{4}{60 - x}$

76. $1 + \dfrac{x + 1}{2x} = \dfrac{3}{4}$

77. $\dfrac{4}{x - 2} - \dfrac{2x - 3}{x^2 - 4} = \dfrac{5}{x + 2}$

78. $n - \dfrac{5n}{n - 2} = \dfrac{-10}{n - 2}$

79. $\sqrt{3n - 2} = 7$

80. $\sqrt{x + 4} = x + 4$

81. $4\sqrt{x} + 5 = x$

82. $(x + 1)^2 = 12$

83. $x^2 - 6x - 3 = 0$

84. $3n^2 - n - 3 = 0$

85. $2n^2 + 5n - 6 = 0$

86. $7 = 3t^2 - t$

For Problems 87–92, solve each of the inequalities.

87. $-3n - 4 \le 11$

88. $-5 > 3n - 4 - 7n$

89. $2(x - 2) + 3(x + 4) > 6$

90. $\dfrac{1}{2}n - \dfrac{2}{3}n < -1$

91. $\dfrac{x + 1}{2} + \dfrac{x - 2}{6} < \dfrac{3}{8}$

92. $\dfrac{x - 3}{7} - \dfrac{x - 2}{4} \le \dfrac{9}{14}$

For Problems 93–108, set up an equation, an inequality, or a system of equations to help solve each problem.

93. If two angles are supplementary and the larger angle is $15°$ less than twice the smaller angle, find the measure of each angle.

94. The sum of two numbers is 50. If the larger number is 2 less than 3 times the smaller number, find the numbers.

95. Suppose that Nick has 47 coins consisting of nickels, dimes, and quarters. The number of dimes is 1 more than twice the number of nickels and the number of quarters is 4 more than 3 times the number of nickels. Find the number of coins of each denomination.

96. If a home valued at $70,000 is assessed $1050 in real estate taxes, at the same rate how much are the taxes on a home assessed at $90,000?

97. A retailer has some skirts that cost her $30 each. She wants to sell them at a profit of 60% of the cost. What price should she charge for the skirts?

98. Rosa leaves a town traveling in her car at a rate of 45 miles per hour. One hour later, Polly leaves the same town traveling the same route at a rate of 55 miles per hour. How long will it take Polly to overtake Rosa?

99. How many milliliters of pure acid must be added to 100 milliliters of a 10% acid solution to obtain a 20% solution?

100. Suppose that Andy has scores of 85, 90, and 86 on his first three algebra tests. What score must he make on the fourth algebra test to have an average of 88 or better for the four tests?

101. The Cubs have won 70 games and lost 72 games. They have 20 more games to play. To win more than 50% of all their games, how many of the 20 games remaining must they win?

102. The area of a rectangle is twice the area of a square. If the rectangle is 16 inches long and the width of the rectangle is the same as the length of a square, find the dimensions of both the rectangle and the square.

103. The length of a rectangle is one and one-half times its width. If the area of the rectangle is 54 square feet, find its length and width.

104. An apple orchard contains 112 trees. The number of trees per row is 6 more than the number of rows. Find the number of rows.

105. The sum of two numbers is 66. If the larger number is divided by the smaller, the quotient is 8 and the remainder is 3. Find the numbers.

106. Seth can do a job in 20 minutes. Butch can do the same job in 30 minutes. If they work together, how long will it take them to complete the job?

107. The perimeter of a rectangle is 36 meters and its area is 72 square meters. Find its length and width.

108. Doris traveled 180 miles in 2 hours less time than it took Ellen to travel 250 miles. Doris drove 10 miles per hour faster than Ellen. How fast did each one travel?

Additional Topics

This chapter was included to give you the opportunity to expand your knowledge of topics we presented in earlier chapters. From the list of section titles, the topics may appear disconnected. Actually, each section is a continuation of a topic we presented in a previous chapter. Sections 11.1 and 11.2 continue the development of techniques for solving equations and inequalities that was the focus of Chapter 3.

Sections 11.3 and 11.4 expand our work with coordinate geometry that we started in Chapter 8. Section 11.5 is an extension of the work with exponents we did in Chapter 5 and radicals in Chapter 9. Sections 11.6 and 11.7 enhance the study of quadratic equations from Chapter 10.

11.1 Equations and Inequalities Involving Absolute Value

In Chapter 1 we used the concept of absolute value to describe addition and multiplication of integers. We defined the absolute value of a number to be the distance between the number and zero on a number line. For example, the absolute value of 3 is 3. The absolute value of -3 is 3. The absolute value of 0 is 0. Symbolically, absolute value is denoted with vertical bars.

$$|3| = 3, \qquad |-3| = 3, \qquad |0| = 0$$

In general, we can say that the absolute value of any number, except 0, is positive.

If we interpret absolute value as distance on a number line, we can solve a variety of equations and inequalities that involve absolute value. First, let's consider some equations.

Example I

Solve and graph the solutions for $|x| = 2$.

Solution

When we think in terms of "distance between the number and zero" we can see that x must be 2 or -2. Thus,

$$|x| = 2 \quad \text{implies}$$
$$x = 2 \quad \text{or} \quad x = -2$$

The solution set is $\{-2, 2\}$ and its graph is shown in Figure 11.1.

FIGURE 11.1

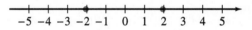

Example 2

Solve and graph the solutions for $|x + 4| = 1$.

Solution

The number, $x + 4$, must be 1 or -1. Thus,

$$|x + 4| = 1 \quad \text{implies}$$
$$x + 4 = 1 \quad \text{or} \quad x + 4 = -1$$
$$x = -3 \quad \text{or} \quad x = -5.$$

The solution set is $\{-5, -3\}$ and its graph is shown in Figure 11.2.

FIGURE 11.2

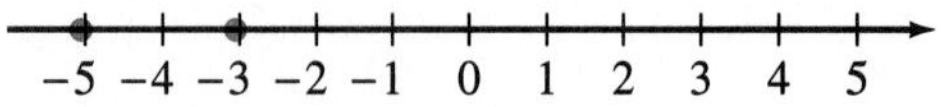

Example 3

Solve and graph the solutions for $|3x - 2| = 4$.

Solution

The number, $3x - 2$, must be 4 or -4. Thus,

$$|3x - 2| = 4 \quad \text{implies}$$

$$3x - 2 = 4 \quad \text{or} \quad 3x - 2 = -4$$
$$3x = 6 \quad \text{or} \quad 3x = -2$$
$$x = 2 \quad \text{or} \quad x = -\frac{2}{3}.$$

The solution set is $\left\{-\dfrac{2}{3}, 2\right\}$ and its graph is shown in Figure 11.3.

FIGURE 11.3

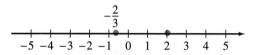

The *distance interpretation* for absolute value also provides a good basis for solving inequalities involving absolute value. Consider the following examples.

Example 4

Solve and graph the solutions for $|x| < 2$.

Solution

The number, x, must be *less than two units away from zero*. Thus,

$$|x| < 2 \quad \text{implies}$$
$$x > -2 \quad \text{and} \quad x < 2.$$

The solution set is $\{x \mid x > -2 \text{ and } x < 2\}$ and its graph is shown in Figure 11.4.

FIGURE 11.4

Example 5

Solve and graph the solutions for $|x - 1| < 2$.

Solution

The number, $x - 1$, must be *less than two units away from zero*. Thus,

$$|x - 1| < 2 \quad \text{implies}$$
$$x - 1 > -2 \quad \text{and} \quad x - 1 < 2$$
$$x > -1 \quad \text{and} \quad x < 3.$$

The solution set is $\{x \mid x > -1 \text{ and } x < 3\}$ and its graph is shown in Figure 11.5.

FIGURE 11.5

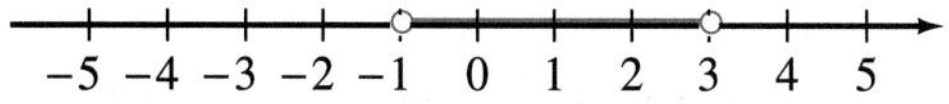

Example 6

Solve and graph the solutions for $|2x + 5| \leq 1$.

Solution

The number, $2x + 5$, must be *equal to or less than one unit away from zero*. Therefore,

$$|2x + 5| \leq 1 \quad \text{implies}$$

$$2x + 5 \geq -1 \quad \text{and} \quad 2x + 5 \leq 1$$
$$2x \geq -6 \quad \text{and} \quad 2x \leq -4$$
$$x \geq -3 \quad \text{and} \quad x \leq -2.$$

The solution set is $\{x \mid x \geq -3 \text{ and } x \leq -2\}$ and its graph is shown in Figure 11.6.

FIGURE 11.6

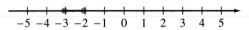

Example 7

Solve and graph the solutions for $|x| > 2$.

Solution

The number, x, must be *greater than two units away from zero*. Thus,

$$|x| > 2 \quad \text{implies}$$
$$x < -2 \quad \text{or} \quad x > 2.$$

The solution set is $\{x \mid x < -2 \text{ or } x > 2\}$ and its graph is shown in Figure 11.7.

FIGURE 11.7

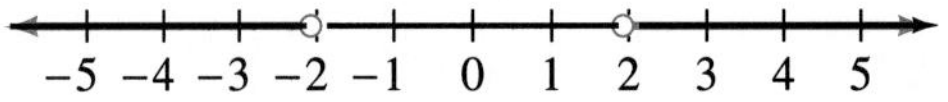

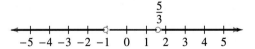

Example 8

Solve and graph the solutions for $|3x - 1| > 4$.

Solution

The number, $3x - 1$, must be *greater than four units away from zero*. Thus

$$|3x - 1| > 4 \quad \text{implies}$$
$$3x - 1 < -4 \quad \text{or} \quad 3x - 1 > 4$$
$$3x < -3 \quad \text{or} \quad 3x > 5$$
$$x < -1 \quad \text{or} \quad x > \frac{5}{3}.$$

The solution set is $\left\{x \mid x < -1 \text{ or } x > \dfrac{5}{3}\right\}$ and its graph is shown in Figure 11.8.

FIGURE 11.8

$$\frac{5}{3}$$

The solutions for equations and inequalities such as $|3x - 7| = -4$, $|x + 5| < -3$, and $|2x - 3| > -7$ can be found by *inspection*. Notice that in each of these examples the right side is a negative number. Therefore, using the fact that the **absolute value of any number is nonnegative**, we can reason as follows.

$|3x - 7| = -4$ has *no solutions* since the absolute value of a number cannot be negative.

$|x + 5| < -3$ has *no solutions* since we cannot obtain an absolute value less than -3.

$|2x - 3| > -7$ is *satisfied by all real numbers* since the absolute value of $2x - 3$, regardless of what number is substituted for x, will always be greater than -7.

Problem Set 11.1

For Problems 1–26, solve and graph the solutions for each of the following.

1. $|x| = 4$
2. $|x| = 3$
3. $|x| < 1$
4. $|x| < 4$
5. $|x| \geq 2$
6. $|x| \geq 1$
7. $|x + 2| = 1$
8. $|x + 3| = 2$
9. $|x - 1| = 2$
10. $|x - 2| = 1$
11. $|x - 2| \leq 2$
12. $|x + 1| \leq 3$
13. $|x + 1| > 3$
14. $|x - 3| > 1$
15. $|2x + 1| = 3$
16. $|3x - 1| = 5$
17. $|5x - 2| = 4$
18. $|4x + 3| = 8$
19. $|2x - 3| \geq 1$
20. $|2x + 1| \geq 3$
21. $|4x + 3| < 2$
22. $|5x - 2| < 8$
23. $|3x + 6| = 0$
24. $|4x - 3| = 0$
25. $|3x - 2| > 0$
26. $|2x + 7| < 0$

For Problems 27–42, solve each of the following.

27. $|3x - 1| = 17$
28. $|4x + 3| = 27$
29. $|2x + 1| > 9$
30. $|3x - 4| > 20$
31. $|3x - 5| < 19$
32. $|5x + 3| < 14$
33. $|-3x - 1| = 17$
34. $|-4x + 7| = 26$
35. $|4x - 7| \leq 31$
36. $|5x - 2| \leq 21$
37. $|5x + 3| \geq 18$
38. $|2x - 11| \geq 4$
39. $|-x - 2| < 4$
40. $|-x - 5| < 7$
41. $|-2x + 1| > 6$
42. $|-3x + 2| > 8$

For Problems 43–50, solve each of the following *by inspection*.

43. $|7x| = 0$
44. $|3x - 1| = -4$
45. $|x - 6| > -4$
46. $|3x + 1| > -3$
47. $|x + 4| < -7$
48. $|5x - 2| < -2$
49. $|x + 6| \leq 0$
50. $|x + 7| > 0$

51. Explain why the equation $|3x + 2| = -6$ has no real number solutions.

52. Explain why the inequality $|x + 6| < -4$ has no real number solutions.

Further Investigations

A conjunction such as "$x > -2$ and $x < 4$" can be written in a more compact form $-2 < x < 4$, which is read as "-2 is less than x and x is less than 4." In other words, x is clamped between -2 and 4. The compact form is very convenient for solving conjunctions as follows.

$$-3 < 2x - 1 < 5$$
$$-2 < 2x < 6 \quad \text{Add 1 to the left side, middle, and right side.}$$
$$-1 < x < 3 \quad \text{Divide through by 2.}$$

Thus, the solution set can be expressed as $\{x | -1 < x < 3\}$.

For Problems 53–62, solve the compound inequalities using the compact form.

53. $-2 < x - 6 < 8$

54. $-1 < x + 3 < 9$

55. $1 \le 2x + 3 \le 11$

56. $-2 \le 3x - 1 \le 14$

57. $-4 < \dfrac{x - 1}{3} < 2$

58. $2 < \dfrac{x + 1}{4} < 5$

59. $|x + 4| < 3$ [*Hint*: $|x + 4| < 3$ implies $-3 < x + 4 < 3$.]

60. $|x - 6| < 5$

61. $|2x - 5| < 7$

62. $|3x + 2| < 14$

11.2 Graphing Linear Inequalities

Linear inequalities in two variables are of the form $Ax + By > C$ or $Ax + By < C$, where A, B, and C are real numbers. (Combined linear equality and inequality statements are of the form $Ax + By \ge C$ or $Ax + By \le C$.) Graphing linear inequalities is almost as easy as graphing linear equations. The following discussion leads to a simple step-by-step process. Let's consider the following equation and related inequalities.

$$x - y = 2$$
$$x - y > 2$$
$$x - y < 2$$

The graph of $x - y = 2$ is shown in Figure 11.9. The line divides the plane into two half-planes, one above the line and one below the line. In Figure 11.10(a) we have indicated coordinates for several points above the line. Note that for each point, the ordered pair of real numbers satisfies the inequality $x - y < 2$. This is true for *all points* in the half-plane above the line. Therefore, the graph of

$x - y < 2$ is the half-plane above the line, indicated by the shaded region in Figure 11.10(b). We use a dashed line to indicate that points on the line do not satisfy $x - y < 2$.

FIGURE 11.9

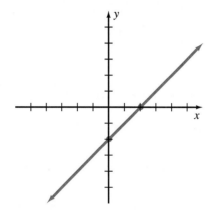

FIGURE 11.10

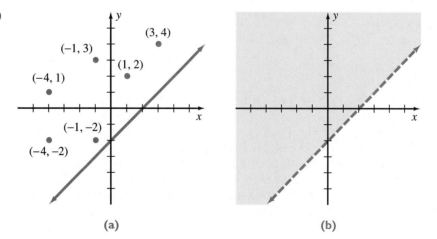

(a)   (b)

In Figure 11.11(a) the coordinates of several points below the line $x - y = 2$ have been indicated. Note that for each point, the ordered pair of real numbers satisfies the inequality $x - y > 2$. This is true for *all points* in the half-plane below the line. Therefore, the graph of $x - y > 2$ is the half-plane below the line, indicated by the shaded region in Figure 11.11(b).

FIGURE 11.11

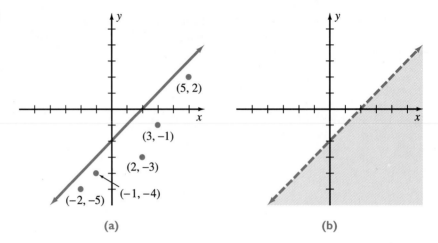

(a)   (b)

Based on the previous discussion, we suggest the following steps for graphing linear inequalities.

1. Graph the corresponding equality. Use a solid line if equality is included in the given statement and a dashed line if equality is not included.

2. Choose a "test point" not on the line and substitute its coordinates into the inequality statement. (The origin is a convenient point to use if it is not on the line.)

3. The graph of the given inequality is

   (a) the half-plane that contains the test point if the inequality is satisfied by the coordinates of the point, or,

   (b) the half-plane that does not contain the test point if the inequality is not satisfied by the coordinates of the point.

Let's apply these steps to some examples.

Example 1

Graph $2x + y > 4$.

Solution

STEP 1   Graph $2x + y = 4$ as a dashed line since equality is not included in the given statement $2x + y > 4$ (Figure 11.12(a)).

STEP 2   Choose the origin as a test point and substitute its coordinates into the inequality.

$$2x + y > 4 \quad \text{becomes } 2(0) + 0 > 4,$$

which is a false statement.

FIGURE 11.12

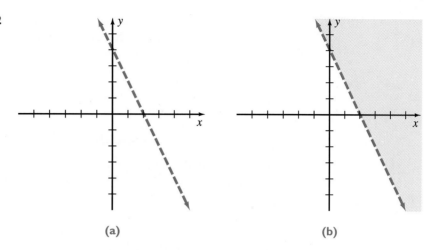

(a)　　　　　　　　　　(b)

STEP 3　Since the test point did not satisfy the given inequality, the graph is the half-plane that does not contain the test point. Thus, the graph of $2x + y > 4$ is the half-plane above the line, indicated in Figure 11.12(b).　▲

Example 2　　　　Graph $y \leq 2x$.

Solution

STEP 1　Graph $y = 2x$ as a solid line since equality is included in the given statement (Figure 11.13(a)).

STEP 2　Since the origin is on the line, we need to choose another point as a test point. Let's use (3, 2).

$$y \leq 2x \quad \text{becomes} \quad 2 \leq 2(3),$$

which is a true statement.

FIGURE 11.13

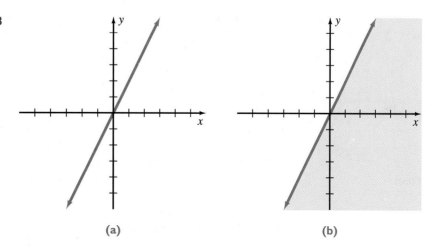

(a)　　　　　　　　　　(b)

STEP 3   Since the test point satisfies the given inequality, the graph is the half-plane that contains the test point. Thus, the graph of $y \leq 2x$ is the line, along with the half-plane below the line, indicated in Figure 11.13(b).   ▲

Systems of Linear Inequalities

It is now easy to use a graphing approach to solve a system of linear inequalities. For example, the solution set of a system of linear inequalities, such as

$$\begin{pmatrix} x + y < 1 \\ x - y > 1 \end{pmatrix},$$

is the intersection of the solution sets of the individual inequalities. In Figure 11.14(a) we indicated the solution set for $x + y < 1$ and in Figure 11.14(b) we indicated the solution set for $x - y > 1$. Then, in Figure 11.14(c) we shaded the region that represents the intersection of the two shaded regions in (a) and (b); thus, it is the solution of the given system. The shaded region in Figure 11.14(c) consists of all points that are below the line $x + y = 1$ *and also* below the line $x - y = 1$.

FIGURE 11.14

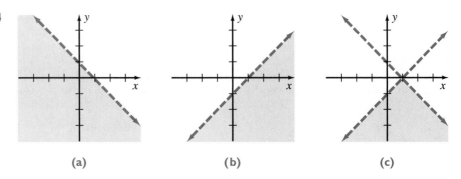

(a)   (b)   (c)

Let's solve another system of linear inequalities.

Example 3

Solve the system

$$\begin{pmatrix} 2x + 3y \leq 6 \\ x - 4y < 4 \end{pmatrix}.$$

Solution

Let's first graph the individual inequalities. The solution set for $2x + 3y \leq 6$ is shown in Figure 11.15(a) and the solution set for $x - 4y < 4$ is shown in Figure 11.15(b). (Note that a solid line is used in (a) and a dashed line in (b).) Then, in Figure 11.15(c) we shaded the intersection of the graphs in (a) and (b). Thus, we represented the solution set for the given system by the shaded region in Figure 11.15(c). This region consists of all points that are on or below the line $2x + 3y = 6$ *and also* are above the line $x - 4y = 4$.

FIGURE 11.15

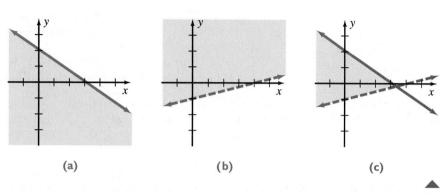

(a)   (b)   (c)

REMARK   Remember that the shaded region in Figure 11.15(c) represents the solution set of the given system. Parts (a) and (b) were drawn only to help determine the final shaded region. With some practice, you may be able to go directly to part (c) without actually sketching the graphs of the individual inequalities.   △

Problem Set 11.2

For Problems 1–20, graph each of the inequalities.

1. $x + y > 1$
2. $2x + y > 4$
3. $3x + 2y < 6$
4. $x + 3y < 3$
5. $2x - y \geq 4$
6. $x - 2y \geq 2$
7. $4x - 3y \leq 12$
8. $3x - 4y \leq 12$
9. $y > -x$
10. $y < x$
11. $2x - y \geq 0$
12. $3x - y \leq 0$
13. $-x + 2y < -2$

14. $-2x + y > -2$
15. $y \leq \dfrac{1}{2}x - 2$
16. $y \geq -\dfrac{1}{2}x + 1$
17. $y \geq -x + 4$
18. $y \leq -x - 3$
19. $3x + 4y > -12$
20. $4x + 3y > -12$

For Problems 21–30, indicate the solution set for each system of linear inequalities by shading the appropriate region.

21. $\begin{pmatrix} 2x + 3y > 6 \\ x - y < 2 \end{pmatrix}$

22. $\begin{pmatrix} x - 2y < 4 \\ 3x + \ y > 3 \end{pmatrix}$

23. $\begin{pmatrix} x - 3y \geq 3 \\ 3x + \ y \leq 3 \end{pmatrix}$

24. $\begin{pmatrix} 4x + 3y \leq 12 \\ 4x - \ y \geq \ 4 \end{pmatrix}$

25. $\begin{pmatrix} y \geq 2x \\ y < \ x \end{pmatrix}$

26. $\begin{pmatrix} y \leq -x \\ y > -3x \end{pmatrix}$

27. $\begin{pmatrix} y < -x + 1 \\ y > -x - 1 \end{pmatrix}$

28. $\begin{pmatrix} y > x - 2 \\ y < x + 3 \end{pmatrix}$

29. $\begin{pmatrix} y < \dfrac{1}{2}x + 2 \\ y < \dfrac{1}{2}x - 1 \end{pmatrix}$

30. $\begin{pmatrix} y > -\dfrac{1}{2}x - 2 \\ y > -\dfrac{1}{2}x + 1 \end{pmatrix}$

32. Why is the point $(3, -2)$ not a good test point to use when graphing the inequality $3x - 2y \leq 13$?

Further Investigations

For Problems 33–36, indicate the solution set for each system of linear inequalities by shading the appropriate region.

33. $\begin{pmatrix} y > x + 1 \\ y < x - 1 \end{pmatrix}$

34. $\begin{pmatrix} x \geq \ 0 \\ y \geq \ 0 \\ 3x + 4y \leq 12 \\ 2x + \ y \leq \ 4 \end{pmatrix}$

35. $\begin{pmatrix} x \geq 0 \\ y \geq 0 \\ 2x + \ y \leq 4 \\ 2x - 3y \leq 6 \end{pmatrix}$

36. $\begin{pmatrix} x \geq 0 \\ y \geq 0 \\ 3x + 5y \geq 15 \\ 5x + 3y \geq 15 \end{pmatrix}$

THOUGHTS INTO WORDS

31. Explain how you would graph the inequality $-x - 2y > 4$.

11.3 Relations and Functions

In the next two sections of this chapter we will work with a concept that plays an important role throughout much of mathematics, namely, the concept of a *function*. Functions are used to unify mathematics and also to serve as a meaningful way of applying mathematics to many real world problems. They provide us with a means of studying quantities that vary with one another, that is, when a change in one quantity causes a corresponding change in another. We will consider the general concept of a function in this section and then deal with two special kinds of functions in the next section.

Mathematically, a function is a special kind of relation. Thus, we will begin with a simple definition for a relation.

DEFINITION 11.1 A relation is a set of ordered pairs of real numbers.

We can generate the ordered pairs of a relation by various means, such as a graph or chart. However, the most common way of generating ordered pairs is by the use of equations. Thus, we say that we use equations to describe relations. Each of the following equations describes a relation between the variables x and y, and we listed some of the ordered pairs of that relation.

1. $x^2 + y^2 = 4$: $(1, \sqrt{3}), (1, -\sqrt{3}), (0, 2), (0, -2), \ldots$

2. $x = y^2$: $(0, 0), (4, 2), (4, -2), (9, 3), (9, -3), \ldots$

3. $y - 2x = 3$: $(0, 3), (1, 5), (-1, 1), (2, 7), (3, 9), \ldots$

4. $y = \dfrac{1}{x - 2}$: $\left(0, -\dfrac{1}{2}\right), (1, -1), (3, 1), \left(-2, -\dfrac{1}{4}\right), \left(5, \dfrac{1}{3}\right), \ldots$

5. $y = x^2$: $(0, 0), (1, 1), (-1, 1), (2, 4), (-2, 4), \ldots$

Now direct your attention to the last three equations above and the ordered pairs associated with each equation. Note that for each equation no two ordered pairs have the same first coordinate. We call such a set of ordered pairs a function.

DEFINITION 11.2 A function is a relation where no two ordered pairs have the same first coordinate.

The domain of a function is the set of all first coordinates and the range is the set of all second coordinates. Usually we can determine the domain and range of a function from the equation used to describe the function. For example, the domain of the function described by $y = 2x + 3$ is the set of all real numbers since we can substitute any real number for x. The range is also all real numbers since y will take on all real number values. The domain of the function described by $y = x^2$ is the set of all real numbers, but the range is only the nonnegative real numbers because squaring a number always produces a nonnegative result.

The domain of a function is frequently of more concern than the range. We should be aware of any necessary restrictions on x. Consider the following examples.

Example 1

Specify the domain for each of the following.

(a) $y = \dfrac{1}{x - 2}$ (b) $y = \dfrac{1}{x^2 - 9}$ (c) $y = \sqrt{x - 1}$

Solution

(a) We can replace x with any real number except 2, since 2 makes the denominator 0. Thus, the domain is all real numbers except 2.

(b) We want to eliminate any value of x that will make the denominator 0. Thus, we solve the equation $x^2 - 9 = 0$.

$$x^2 - 9 = 0$$
$$(x + 3)(x - 3) = 0$$
$$x + 3 = 0 \quad \text{or} \quad x - 3 = 0$$
$$x = -3 \quad \text{or} \quad x = 3$$

The domain is the set of all real numbers except -3 and 3.

(c) Because of the radical sign the radicand, $x - 1$, must be nonnegative.

$$x - 1 \geq 0$$
$$x \geq 1$$

The domain is the set of all real numbers greater than or equal to 1.

▲

Functional Notation

Thus far we have been using the regular notation for writing equations to describe functions; that is, we have used equations such as $y = x^2$ and $y = \dfrac{1}{x - 2}$, where y is expressed in terms of x, to specify certain functions. There is a special **functional notation** that is very convenient to use when working with the function concept.

The notation $f(x)$ is read "f of x" and is defined to be the value of the function f at x. (Do not interpret $f(x)$ to mean f times x!) Instead of writing $y = x^2$ we can write $f(x) = x^2$. Therefore, $f(2)$ means the value of the function f at 2, which is $2^2 = 4$. So we write $f(2) = 4$. It makes a convenient way of expressing various values of the function. Let's illustrate that idea with another example.

Example 2

Solution

If $f(x) = x^2 - 6$, find $f(0), f(1), f(2), f(3), f(-1),$ and $f(h)$.

$$f(x) = x^2 - 6.$$
$$f(0) = 0^2 - 6 = -6.$$
$$f(1) = 1^2 - 6 = -5.$$
$$f(2) = 2^2 - 6 = -2.$$
$$f(3) = 3^2 - 6 = 3.$$
$$f(-1) = (-1)^2 - 6 = -5.$$
$$f(h) = h^2 - 6.$$

▲

When we are working with more than one function in the same problem, we use different letters to designate the different functions, as the next example demonstrates.

Example 3

If $f(x) = 2x + 5$ and $g(x) = x^2 - 2x + 1$, find $f(2), f(-3), g(-1)$, and $g(4)$.

Solution

$f(x) = 2x + 5$. $\qquad\qquad$ $g(x) = x^2 - 2x + 1$.

$f(2) = 2(2) + 5 = 9$. $\qquad$ $g(-1) = (-1)^2 - 2(-1) + 1 = 4$.

$f(-3) = 2(-3) + 5 = -1$. $\qquad$ $g(4) = 4^2 - 2(4) + 1 = 9$. ▲

▼ Problem Set 11.3

Determine the domain and range for each of the following equations.

1. $y = x + 2$

2. $y = 3x - 7$

3. $y = x^2 + 2$

4. $y = x^2 - 1$

5. $y = x^3$

6. $y = x^4$

7. $y = \sqrt{x}$

8. $y = \sqrt{x - 3}$

Determine the domain for each of the following functions.

9. $y = x^2 + 2x - 1$

10. $y = x^2 - 3x - 1$

11. $y = \sqrt{x + 4}$

12. $y = \sqrt{2x - 3}$

13. $f(x) = 2x + 7$

14. $f(x) = x^3 + 2$

15. $f(x) = \dfrac{2}{x - 3}$

16. $f(x) = \dfrac{3x}{2x - 1}$

17. $y = \dfrac{3x}{x^2 - 4}$

18. $y = \dfrac{5}{x^2 - 25}$

19. $f(x) = \dfrac{-2}{x^2 + 4}$

20. $f(x) = \dfrac{3x}{x^2 + 9}$

21. $f(x) = \dfrac{4}{(x + 2)(x - 3)}$

22. $f(x) = \dfrac{5}{(2x - 3)(x - 1)}$

23. $f(x) = \dfrac{1}{x^2 - 5x + 6}$

24. $f(x) = \dfrac{2}{x^2 + 7x + 12}$

25. $y = \dfrac{-2}{x^2 + 4x}$

26. $y = \dfrac{-5}{x^2 - 7x}$

27. If $f(x) = 3x + 4$, find $f(0), f(1), f(-1)$, and $f(6)$.

28. If $f(x) = -2x + 5$, find $f(2), f(-2), f(-3)$, and $f(5)$.

29. If $f(x) = -5x - 1$, find $f(3), f(-4), f(-5)$, and $f(t)$.

30. If $f(x) = 7x - 3$, find $f(-1), f(0), f(4)$, and $f(a)$.

31. If $g(x) = \dfrac{2}{3}x + \dfrac{3}{4}$, find $g(3)$, $g\left(\dfrac{1}{2}\right)$, $g\left(-\dfrac{1}{3}\right)$, and $g(-2)$.

32. If $g(x) = -\dfrac{1}{2}x + \dfrac{5}{6}$, find $g(1)$, $g(-1)$, $g\left(\dfrac{2}{3}\right)$, and $g\left(-\dfrac{1}{3}\right)$.

33. If $f(x) = x^2 - 4$, find $f(2), f(-2), f(7)$, and $f(0)$.

34. If $f(x) = 2x^2 + x - 1$, find $f(2)$, $f(-3)$, $f(4)$, and $f(-1)$.

35. If $f(x) = -x^2 + 1$, find $f(-1)$, $f(2)$, $f(-2)$, and $f(-3)$.

36. If $f(x) = -2x^2 - 3x - 1$, find $f(1), f(0), f(-1)$, and $f(-2)$.

37. If $f(x) = \sqrt{x}$, find $f(4), f(25), f(12)$, and $f(18)$.

38. If $f(x) = \sqrt{x - 1}$, find $f(10)$, $f(17)$, $f(25)$, and $f(45)$.

39. If $f(x) = 4x + 3$ and $g(x) = x^2 - 2x$, find $f(5)$, $f(-6)$, $g(-1)$, and $g(4)$.

40. If $f(x) = -2x - 7$ and $g(x) = 2x^2 + 1$, find $f(-2), f(4), g(-2)$, and $g(4)$.

41. If $f(x) = 3x^2 - x + 4$ and $g(x) = -3x + 5$, find $f(-1)$, $f(4)$, $g(-1)$, and $g(4)$.

42. If $f(x) = -2x^2 + 3x + 1$ and $g(x) = -6x$, find $f(3)$, $f(-5)$, $g(2)$, and $g(-4)$.

43. A car rental agency charges \$35 per day plus \$.30 a mile. Therefore, the daily charge for renting a car is a function of the number of miles traveled (m) and can be expressed as $C(m) = 35 + .3m$. Compute $C(75)$, $C(185)$, $C(275)$, and $C(450)$.

44. The equation $I(r) = 750r$ expresses the amount of simple interest earned by an investment of \$750 for one year as a function of the rate of interest (r). Compute $I(.075)$, $I(.0825)$, $I(.0875)$, and $I(.095)$.

THOUGHTS INTO WORDS

45. Explain how to determine the domain of the function $f(x) = \sqrt{x - 4}$.

46. What meanings do we give to the word "function" in our everyday activities? Are any of these meanings closely related to the use of "function" in mathematics?

11.4 Special Functions

As you continue to study mathematics you will frequently encounter the concept of a function. Functions are a part of almost all branches of mathematics. Basically, the study of different functions centers around (1) the equations that describe them, (2) the graphs they produce, and (3) the real world problems they help to solve. In this section we shall consider three special types of functions: constant functions, linear functions, and quadratic functions.

Any function that can be written in the form

$$f(x) = c,$$

where c is a real number, is called a **constant function**. The following are examples of constant functions.

$$f(x) = 2, \qquad f(x) = -\frac{1}{3}, \qquad f(x) = 0$$

Graphing constant functions is quite easy because the graph of every constant function is a horizontal line.

Example 1

Graph $f(x) = 2$.

Solution

The function $f(x) = 2$ means that the value of 2 is assigned to every value of x. Thus,

$$f(-3) = 2, \quad f(0) = 2, \quad f(2) = 2, \quad \text{and so on.}$$

Therefore, the points $(-3, 2)$, $(0, 2)$, $(2, 2)$, and so on, are all points of the graph, which is the horizontal line in Figure 11.16.

FIGURE 11.16

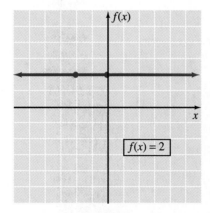

Note that in Figure 11.16 we labeled the vertical axis $f(x)$. We could also label it y since $f(x) = 2$ and $y = 2$ mean the same thing. We will continue to use $f(x)$ merely to help you adjust to the functional notation.

Linear Function

Any function that can be written in the form

$$f(x) = ax + b,$$

where a and b are real numbers, $a \neq 0$, is called a **linear function**. The following are examples of linear functions.

$$f(x) = 2x + 5,$$
$$f(x) = -3x + 6,$$
$$f(x) = -\frac{1}{2}x + \frac{3}{4},$$
$$f(x) = 3x$$

Graphing linear functions is also quite easy since the graph of every linear function is a straight line.

Example 2

Graph $f(x) = 2x + 1$.

Solution

$$f(0) = 2(0) + 1 = 1$$
$$f(2) = 2(2) + 1 = 5$$
$$f(-1) = 2(-1) + 1 = -1$$

The points $(0, 1)$, $(2, 5)$, and $(-1, -1)$ are on the graph and the line can be drawn as in Figure 11.17.

FIGURE 11.17

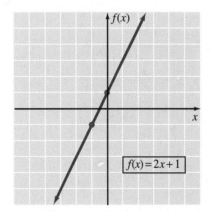

$$f(x) = 2x + 1$$

In Section 8.4, we referred to $y = mx + b$ as the slope-intercept form of the equation of a straight line. In this section we have used the equation $f(x) = ax + b$ to describe linear functions. In both equations, the coefficient of x is the slope of the line and the constant term is the y-intercept. We can use this fact to help graph linear functions.

Example 3

Graph $f(x) = -3x - 2$.

Solution

The y-intercept is -2; therefore, the point $(0, -2)$ is on the graph. Since the slope is $-3\left(-3 = \dfrac{-3}{1}\right)$, we can locate another point by moving 3 units down and 1 unit to the right of $(0, -2)$. This point is $(1, -5)$. We can draw the line as in Figure 11.18.

FIGURE 11.18

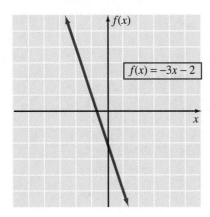

$f(x) = -3x - 2$

Quadratic Functions

Any function that we can write in the form

$$f(x) = ax^2 + bx + c,$$

where a, b, and c are real numbers, $a \neq 0$, is called a **quadratic function**. The following are examples of quadratic functions.

$$f(x) = 2x^2, \qquad f(x) = x^2 + 5x - 7, \qquad f(x) = -3x^2 - 2$$

Graphing all of the variations of quadratic functions requires considerably more effort than did graphing either the constant or linear functions. We shall not consider all such variations in this text but will use a few examples to illustrate some of the basic characteristics of quadratic functions. Let's begin with the following example.

Example 4

Graph $f(x) = x^2$.

Solution

We can generate some points of the graph by finding some functional values.

$$f(0) = 0^2 = 0 \quad \longrightarrow \quad (0, 0),$$
$$f(1) = 1^2 = 1 \quad \longrightarrow \quad (1, 1),$$
$$f(2) = 2^2 = 4 \quad \longrightarrow \quad (2, 4),$$
$$f(-1) = (-1)^2 = 1 \quad \longrightarrow \quad (-1, 1),$$
$$f(-2) = (-2)^2 = 4 \quad \longrightarrow \quad (-2, 4)$$

Plot these points and join them with a smooth curve to produce the graph, called a **parabola**, in Figure 11.19.

FIGURE 11.19

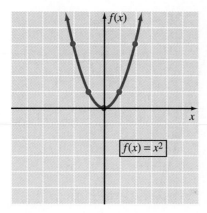

The graph of every quadratic function is a parabola. As we work with parabolas, the vocabulary we indicated in Figure 11.20 will be helpful.

FIGURE 11.20

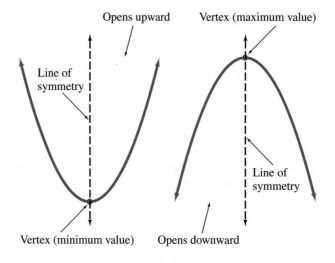

We can conveniently explain the graphing of parabolas by studying various forms of equations used to describe quadratic functions. We shall only begin this process in this text by studying equations of the form $f(x) = ax^2$, where $a \neq 0$. Consider the following examples.

Example 5

Graph $f(x) = 2x^2$.

Solution

Let's compare the graph of $f(x) = 2x^2$ to the graph $f(x) = x^2$ by setting up a table of values. The functional values of $f(x) = 2x^2$ are twice the corresponding functional values of $f(x) = x^2$. Thus, the parabola associated with $f(x) = 2x^2$ has the *same vertex*, the origin, as the graph of $f(x) = x^2$, but it is *narrower*, as indicated in Figure 11.21.

x	$f(x) = x^2$	$f(x) = 2x^2$
0	0	0
1	1	2
2	4	8
−1	1	2
−2	4	8

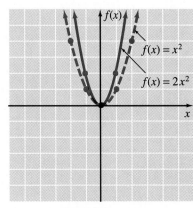

FIGURE 11.21 ▲

Example 6

Graph $f(x) = \frac{1}{2}x^2$.

Solution

Again a table of values is very revealing. The functional values of $f(x) = \frac{1}{2}x^2$ are one-half of the corresponding functional values of $f(x) = x^2$. Thus, the parabola associated with $f(x) = \frac{1}{2}x^2$ has the same vertex as the graph of $f(x) = x^2$, but it is *wider*, as indicated in Figure 11.22.

x	$f(x) = x^2$	$f(x) = \frac{1}{2}x^2$
0	0	0
1	1	$\frac{1}{2}$
2	4	2
−1	1	$\frac{1}{2}$
−2	4	2

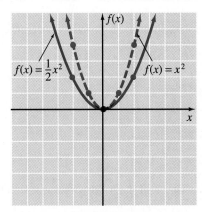

FIGURE 11.22 ▲

Example 7

Graph $f(x) = -x^2$.

Solution

Consider the following table of values. Functional values of $f(x) = -x^2$ are opposites of corresponding functional values of $f(x) = x^2$. Therefore, the graph of $f(x) = -x^2$ is a reflection across the x-axis of the graph of $f(x) = x^2$, as shown in Figure 11.23.

x	$f(x) = x^2$	$f(x) = -x^2$
0	0	0
1	1	-1
2	4	-4
-1	1	-1
-2	4	-4

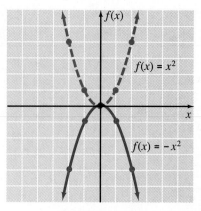

FIGURE 11.23

In general, the graph of a quadratic function of the form $f(x) = ax^2$ has its vertex at the origin and opens upward if a is positive and downward if a is negative. The parabola is *narrower* than the parabola associated with $f(x) = x^2$ if $|a| > 1$ and *wider* if $|a| < 1$.

Problem Set 11.4

Graph each of the following constant functions.

1. $f(x) = 4$

2. $f(x) = 1$

3. $f(x) = -2$

4. $f(x) = -3$

5. $f(x) = \dfrac{1}{2}$

6. $f(x) = 0$

Graph each of the following linear functions.

7. $f(x) = 3x + 2$

8. $f(x) = 4x + 5$

9. $f(x) = 2x - 5$

10. $f(x) = 3x - 7$

11. $f(x) = -4x + 1$

12. $f(x) = -2x + 3$

13. $f(x) = -5x - 1$

14. $f(x) = -x - 2$

15. $f(x) = 3x$

16. $f(x) = x$

17. $f(x) = -x$

18. $f(x) = -4x$

Graph each of the following quadratic functions.

19. $f(x) = 3x^2$

20. $f(x) = \dfrac{1}{2}x^2$

21. $f(x) = \dfrac{1}{4}x^2$

22. $f(x) = -2x^2$

23. $f(x) = -3x^2$

24. $f(x) = -\dfrac{1}{2}x^2$

25. $f(x) = -\dfrac{1}{4}x^2$

26. Your friend sets up the following table of values when graphing the function $f(x) = -x^2$. Therefore, his graph for $f(x) = -x^2$ is the same as the graph for $f(x) = x^2$. Where has he gone wrong and how would you help him?

27. Suppose that someone shows you the graph of $f(x) = x^3$. How could you then sketch the graph of $f(x) = -x^3$?

x	f(x)
0	0
1	1
2	4
3	9
−1	1
−2	4
−3	9

11.5 Fractional Exponents

At the beginning of Chapter 9 we defined and discussed the concept of square root. For your convenience and for the sake of continuity, let's repeat a small portion of that material at this time.

To **square a number** means to raise it to the second power—that is, to use the number as a factor twice.

$$3^2 = 3 \cdot 3 = 9,$$
$$7^2 = 7 \cdot 7 = 49,$$
$$(-3)^2 = (-3)(-3) = 9,$$
$$(-7)^2 = (-7)(-7) = 49$$

A **square root of a number** is one of its two equal factors. Thus, 3 is a square root of 9 because $3 \cdot 3 = 9$. Likewise, -3 is also a square root of 9 because $(-3)(-3) = 9$. The concept of square root is defined as follows.

DEFINITION 11.3 a is a square root of b if $a^2 = b$.

The following generalizations are a direct consequence of Definition 11.3.

1. Every positive real number has two square roots; one is positive and the other is negative. They are opposites of each other.

2. Negative real numbers have no real number square roots. (This follows from Definition 11.3 since any nonzero real number is positive when squared.)

3. The square root of 0 is 0.

We use the symbol $\sqrt{}$, called **a radical sign**, to indicate the nonnegative square root. Consider the following examples.

$$\sqrt{49} = 7 \qquad \sqrt{49} \text{ indicates the nonnegative or principal square root of 49.}$$
$$-\sqrt{49} = -7 \qquad -\sqrt{49} \text{ indicates the negative square root of 49.}$$
$$\sqrt{0} = 0 \qquad \text{Zero has only one square root.}$$
$$\sqrt{-4} \qquad \sqrt{-4} \text{ is not a real number.}$$
$$-\sqrt{-4} \qquad -\sqrt{-4} \text{ is not a real number.}$$

To **cube a number** means to raise it to the third power, that is, to use the number as a factor three times.

$$2^3 = 2 \cdot 2 \cdot 2 = 8,$$
$$4^3 = 4 \cdot 4 \cdot 4 = 64,$$
$$\left(\frac{2}{3}\right)^3 = \frac{2}{3} \cdot \frac{2}{3} \cdot \frac{2}{3} = \frac{8}{27},$$
$$(-2)^3 = (-2)(-2)(-2) = -8$$

A **cube root of a number** is one of its three equal factors. Thus, -2 is a cube root of -8 because $(-2)(-2)(-2) = -8$. In general, the concept of cube root can be defined as follows.

DEFINITION 11.4 a is a cube root of b if $a^3 = b$.

The following generalizations are a direct consequence of Definition 11.4.

1. Every positive real number has one positive real number cube root.

2. Every negative real number has one negative real number cube root.

3. The cube root of 0 is 0.

(Technically, every nonzero real number has three cube roots, but only one of them is a real number. The other two cube roots are complex numbers.)

The symbol, $\sqrt[3]{}$, is used to designate the real number cube root. Thus, we can write

$$\sqrt[3]{8} = 2, \qquad \sqrt[3]{\frac{1}{27}} = \frac{1}{3}, \qquad \sqrt[3]{-8} = -2, \qquad \sqrt[3]{-\frac{1}{27}} = -\frac{1}{3}.$$

We can extend the concept of *root* to fourth roots, fifth roots, sixth roots, and in general, nth roots. We can make the following generalizations.

If n is an even positive integer, then the following statements are true.

1. Every positive real number has exactly two real nth roots, one positive and one negative. For example, the real fourth roots of 16 are 2 and -2. We use the symbol $\sqrt[n]{}$ to designate the positive root. Thus, we write $\sqrt[4]{16} = 2$.

2. Negative real numbers do not have real nth roots. For example, there are no real fourth roots of -16.

If n is an odd positive integer greater than one, then the following statements are true.

1. Every real number has exactly one real nth root and we designate this root by the symbol $\sqrt[n]{}$.

2. The real nth root of a positive number is positive. For example, the fifth root of 32 is 2 and we write $\sqrt[5]{32} = 2$.

3. The nth root of a negative number is negative. For example, the fifth root of -32 is -2 and we write $\sqrt[5]{-32} = -2$.

To complete our terminology, we call the n in the radical $\sqrt[n]{b}$ the **index** of the radical. If $n = 2$, we commonly write $\sqrt{b}$ instead of $\sqrt[2]{b}$.

Merging of Exponents and Roots

In Section 5.6 we used the basic properties of positive integral exponents to motivate a definition for the use of zero and negative integers as exponents. Now let's use the properties of integral exponents to motivate definitions for the use of all rational numbers as exponents. This material is commonly referred to as "fractional exponents."

Consider the following comparisons.

From the meaning of root we know that	If $(b^n)^m = b^{mn}$ is to hold when n equals a rational number of the form $\frac{1}{p}$, where p is a positive integer greater than one, then
$(\sqrt{5})^2 = 5,$	$\left(5^{\frac{1}{2}}\right)^2 = 5^{2\left(\frac{1}{2}\right)} = 5^1 = 5,$
$(\sqrt[3]{8})^3 = 8,$	$\left(8^{\frac{1}{3}}\right)^3 = 8^{3\left(\frac{1}{3}\right)} = 8^1 = 8,$
$(\sqrt[4]{21})^4 = 21.$	$\left(21^{\frac{1}{4}}\right)^4 = 21^{4\left(\frac{1}{4}\right)} = 21^1 = 21.$

The following definition is motivated by such examples.

DEFINITION 11.5   If b is a real number, n a positive integer greater than one, and $\sqrt[n]{b}$ exists, then

$$b^{\frac{1}{n}} = \sqrt[n]{b}.$$

Definition 11.5 states that $b^{\frac{1}{n}}$ means the nth root of b. The following examples illustrate this definition.

$$25^{\frac{1}{2}} = \sqrt{25} = 5, \qquad\qquad 16^{\frac{1}{4}} = \sqrt[4]{16} = 2,$$

$$8^{\frac{1}{3}} = \sqrt[3]{8} = 2, \qquad\qquad \left(\frac{36}{49}\right)^{\frac{1}{2}} = \sqrt{\frac{36}{49}} = \frac{6}{7},$$

$$(-27)^{\frac{1}{3}} = \sqrt[3]{-27} = -3, \qquad (-32)^{\frac{1}{5}} = \sqrt[5]{-32} = -2$$

Now the following definition provides the basis for the use of *all* rational numbers as exponents.

DEFINITION 11.6   If $\dfrac{m}{n}$ is a rational number, where n is a positive integer greater than one, and b is a real number such that $\sqrt[n]{b}$ exists, then

$$b^{\frac{m}{n}} = \sqrt[n]{b^m} = (\sqrt[n]{b})^m. \qquad \frac{m}{n} \text{ is in reduced form}$$

Whether we use the form $\sqrt[n]{b^m}$ or $(\sqrt[n]{b})^m$ for computational purposes depends somewhat on the magnitude of the problem. Let's use both forms on two problems to illustrate this point.

$$\begin{aligned}
8^{\frac{2}{3}} &= \sqrt[3]{8^2} \qquad \text{or} \qquad 8^{\frac{2}{3}} = (\sqrt[3]{8})^2 \\
&= \sqrt[3]{64} \qquad\qquad\qquad\quad = (2)^2 \\
&= 4 \qquad\qquad\qquad\qquad = 4
\end{aligned}$$

$$\begin{aligned}
27^{\frac{2}{3}} &= \sqrt[3]{27^2} \qquad \text{or} \qquad 27^{\frac{2}{3}} = (\sqrt[3]{27})^2 \\
&= \sqrt[3]{729} \qquad\qquad\qquad\quad = 3^2 \\
&= 9 \qquad\qquad\qquad\qquad = 9
\end{aligned}$$

To compute $8^{\frac{2}{3}}$, either form seems to work about as well as the other one. However, to compute $27^{\frac{2}{3}}$, it should be obvious that $(\sqrt[3]{27})^2$ is much easier to handle than $\sqrt[3]{27^2}$.

Remember that in Section 5.6 we used the definition $b^{-n} = \dfrac{1}{b^n}$ as a basis for our work with negative integral exponents. Definition 11.6 and this definition, extended to all rational numbers, are used in the following examples.

$$25^{\frac{3}{2}} = (\sqrt{25})^3 = 5^3 = 125,$$

$$16^{\frac{3}{4}} = (\sqrt[4]{16})^3 = 2^3 = 8,$$

$$36^{-\frac{1}{2}} = \frac{1}{36^{\frac{1}{2}}} = \frac{1}{\sqrt{36}} = \frac{1}{6},$$

$$(-8)^{-\frac{1}{3}} = \frac{1}{(-8)^{\frac{1}{3}}} = \frac{1}{\sqrt[3]{-8}} = \frac{1}{-2} = -\frac{1}{2},$$

$$-8^{\frac{2}{3}} = -(\sqrt[3]{8})^2 = -(2)^2 = -4$$

The basic properties of exponents we discussed in Chapter 5 are true for all rational numbers. They provide the basis for simplifying algebraic expressions that contain rational exponents, as the following examples illustrate. Our objective is to simplify and express the final result using only positive exponents.

1. $x^{\frac{1}{2}} \cdot x^{\frac{1}{3}} = x^{\frac{1}{2}+\frac{1}{3}} = x^{\frac{3}{6}+\frac{2}{6}} = x^{\frac{5}{6}}$

2. $\left(2a^{\frac{3}{4}}\right)\left(3a^{\frac{1}{3}}\right) = 2 \cdot 3 \cdot a^{\frac{3}{4}} \cdot a^{\frac{1}{3}} = 6a^{\frac{3}{4}+\frac{1}{3}} = 6a^{\frac{9}{12}+\frac{4}{12}} = 6a^{\frac{13}{12}}$

3. $\left(3n^{\frac{1}{2}}\right)\left(5n^{-\frac{1}{6}}\right) = 3 \cdot 5 \cdot n^{\frac{1}{2}} \cdot n^{-\frac{1}{6}}$

$$= 15n^{\frac{1}{2}+\left(-\frac{1}{6}\right)}$$

$$= 15n^{\frac{3}{6}-\frac{1}{6}}$$

$$= 15n^{\frac{2}{6}}$$

$$= 15n^{\frac{1}{3}}$$

4. $\left(2x^{\frac{1}{2}}y^{\frac{1}{3}}\right)^2 = (2)^2\left(x^{\frac{1}{2}}\right)^2\left(y^{\frac{1}{3}}\right)^2$ $(b^n)^m = b^{mn}$

$$= 4xy^{\frac{2}{3}}$$

5. $\dfrac{x^{\frac{1}{3}}}{x^{\frac{1}{2}}} = x^{\frac{1}{3}-\frac{1}{2}} = x^{\frac{2}{6}-\frac{3}{6}} = x^{-\frac{1}{6}} = \dfrac{1}{x^{\frac{1}{6}}}$ $\dfrac{b^n}{b^m} = b^{n-m}$

6. $\dfrac{12x^{\frac{3}{4}}}{2x^{-\frac{1}{4}}} = \dfrac{12}{2}x^{\frac{3}{4}-\left(-\frac{1}{4}\right)} = 6x^{\frac{4}{4}} = 6x$

7. $\left(\dfrac{2x^{\frac{1}{2}}}{3y^{\frac{2}{3}}}\right)^2 = \dfrac{(2)^2\left(x^{\frac{1}{2}}\right)^2}{(3)^2\left(y^{\frac{2}{3}}\right)^2} = \dfrac{4x}{9y^{\frac{4}{3}}}$

8. $3^{\frac{1}{2}} \cdot 3^{\frac{1}{3}} = 3^{\frac{1}{2}+\frac{1}{3}} = 3^{\frac{3}{6}+\frac{2}{6}} = 3^{\frac{5}{6}}$

9. $\dfrac{2^{\frac{3}{4}}}{2^{\frac{1}{4}}} = 2^{\frac{3}{4}-\frac{1}{4}} = 2^{\frac{2}{4}} = 2^{\frac{1}{2}}$

Problem Set 11.5

Evaluate each of the following numerical expressions.

1. $\sqrt{81}$

2. $\sqrt{\dfrac{49}{4}}$

3. $-\sqrt{100}$

4. $-\sqrt{121}$

5. $\sqrt[3]{125}$

6. $\sqrt[3]{\dfrac{27}{8}}$

7. $\sqrt[3]{-64}$

8. $\sqrt[3]{-\dfrac{8}{27}}$

9. $\dfrac{\sqrt[3]{64}}{\sqrt{49}}$

10. $\dfrac{\sqrt{144}}{\sqrt[3]{8}}$

11. $\sqrt[4]{81}$

12. $\sqrt[4]{\dfrac{16}{81}}$

13. $\sqrt[5]{-243}$

14. $\sqrt[5]{1}$

15. $64^{\frac{1}{2}}$

16. $64^{\frac{1}{3}}$

17. $64^{\frac{2}{3}}$

18. $(-27)^{\frac{1}{3}}$

19. $(-64)^{\frac{2}{3}}$

20. $16^{\frac{3}{2}}$

21. $4^{\frac{5}{2}}$

22. $16^{\frac{1}{4}}$

23. $32^{-\frac{1}{5}}$

24. $-16^{\frac{1}{2}}$

25. $-27^{\frac{1}{3}}$

26. $8^{-\frac{2}{3}}$

27. $16^{-\frac{3}{4}}$

28. $\left(\dfrac{1}{2}\right)^{-2}$

29. $\left(\dfrac{2}{3}\right)^{-3}$

30. $\left(-\dfrac{1}{8}\right)^{-\frac{1}{3}}$

31. $\left(\dfrac{16}{64}\right)^{-\frac{1}{2}}$

32. $81^{\frac{3}{4}}$

33. $125^{\frac{4}{3}}$

34. $-16^{\frac{5}{2}}$

35. $-16^{\frac{5}{4}}$

36. $\left(\dfrac{1}{27}\right)^{-\frac{2}{3}}$

37. $\left(\dfrac{1}{32}\right)^{\frac{3}{5}}$

38. $(-8)^{\frac{4}{3}}$

39. $2^{\frac{1}{3}} \cdot 2^{\frac{2}{3}}$

40. $2^{\frac{3}{4}} \cdot 2^{\frac{5}{4}}$

41. $3^{\frac{4}{3}} \cdot 3^{\frac{5}{3}}$

42. $2^{\frac{1}{2}} \cdot 2^{-\frac{1}{2}}$

43. $\dfrac{2^{\frac{1}{2}}}{2^{\frac{1}{2}}}$

44. $\dfrac{3^{\frac{1}{3}}}{3^{-\frac{2}{3}}}$

45. $\dfrac{3^{-\frac{2}{3}}}{3^{\frac{1}{3}}}$

46. $\dfrac{2^{\frac{3}{4}}}{2^{-\frac{1}{4}}}$

47. $\dfrac{2^{\frac{9}{4}}}{2^{\frac{1}{4}}}$

48. $\dfrac{3^{-\frac{1}{2}}}{3^{-\frac{1}{2}}}$

49. $\dfrac{7^{\frac{4}{3}}}{7^{-\frac{2}{3}}}$

50. $\dfrac{5^{\frac{3}{5}}}{5^{-\frac{7}{5}}}$

Simplify each of the following and express final results using positive exponents only. For example,

$$\left(2x^{\frac{1}{3}}\right)\left(3x^{\frac{1}{2}}\right) = 6x^{\frac{5}{6}}.$$

51. $x^{\frac{1}{2}} \cdot x^{\frac{1}{4}}$

52. $x^{\frac{1}{3}} \cdot x^{\frac{1}{4}}$

53. $a^{\frac{2}{3}} \cdot a^{\frac{3}{4}}$

54. $a^{\frac{1}{3}} \cdot a^{-\frac{1}{4}}$

55. $\left(3x^{\frac{1}{4}}\right)\left(5x^{\frac{1}{3}}\right)$

56. $\left(2x^{\frac{1}{2}}\right)\left(5x^{\frac{1}{3}}\right)$

57. $\left(4x^{\frac{2}{3}}\right)\left(6x^{\frac{1}{4}}\right)$

58. $\left(3x^{\frac{2}{5}}\right)\left(6x^{\frac{1}{4}}\right)$

59. $\left(2y^{\frac{2}{3}}\right)\left(y^{-\frac{1}{4}}\right)$

60. $\left(y^{-\frac{1}{3}}\right)\left(4y^{\frac{2}{5}}\right)$

61. $\left(5n^{\frac{3}{4}}\right)\left(2n^{-\frac{1}{2}}\right)$

62. $\left(7n^{-\frac{1}{3}}\right)\left(8n^{\frac{5}{6}}\right)$

63. $\left(2x^{\frac{1}{3}}\right)\left(x^{-\frac{1}{2}}\right)$

64. $\left(x^{\frac{2}{5}}\right)\left(3x^{-\frac{1}{2}}\right)$

65. $\left(5x^{\frac{1}{2}}y\right)^{2}$

66. $\left(2x^{\frac{1}{3}}y^{2}\right)^{3}$

67. $\left(4x^{\frac{1}{4}}y^{\frac{1}{2}}\right)^{3}$

68. $(9x^{2}y^{4})^{\frac{1}{2}}$

69. $(8x^{6}y^{3})^{\frac{1}{3}}$

70. $\dfrac{18x^{\frac{1}{2}}}{9x^{\frac{1}{3}}}$

71. $\dfrac{24x^{\frac{3}{5}}}{6x^{\frac{1}{3}}}$

72. $\dfrac{56a^{\frac{1}{6}}}{7a^{\frac{1}{4}}}$

73. $\dfrac{48b^{\frac{1}{3}}}{12b^{\frac{3}{4}}}$

74. $\dfrac{16n^{\frac{1}{3}}}{8n^{-\frac{2}{3}}}$

75. $\dfrac{27n^{-\frac{1}{3}}}{9n^{-\frac{1}{3}}}$

76. $\dfrac{5x^{\frac{2}{5}}}{3x^{-\frac{1}{3}}}$

77. $\left(\dfrac{3x^{\frac{1}{3}}}{2x^{\frac{1}{2}}}\right)^{2}$

78. $\left(\dfrac{2a^{\frac{1}{2}}}{3a^{\frac{1}{4}}}\right)^{3}$

79. $\left(\dfrac{5x^{\frac{1}{2}}}{6y^{\frac{1}{3}}}\right)^{3}$

80. $\left(\dfrac{4y^{\frac{2}{5}}}{3x^{\frac{1}{3}}}\right)^{2}$

THOUGHTS INTO WORDS

81. Why is $\sqrt[4]{-16}$ not a real number?
82. Explain how you would evaluate $-4^{\frac{7}{2}}$.

Further Investigations

83. Use your calculator to evaluate each of the following.

(a) $\sqrt[3]{21,952}$
(b) $\sqrt[3]{42,875}$
(c) $\sqrt[4]{83,521}$
(d) $\sqrt[4]{3,111,696}$

84. Use your calculator to evaluate each of the following.

(a) $16^{\frac{5}{2}}$
(b) $36^{\frac{3}{2}}$
(c) $16^{\frac{7}{4}}$
(d) $27^{\frac{5}{3}}$
(e) $343^{\frac{4}{3}}$
(f) $81^{\frac{3}{4}}$

85. Use your calculator to estimate each of the following to the nearest hundredth.

(a) $5^{\frac{3}{2}}$
(b) $8^{\frac{4}{5}}$
(c) $17^{\frac{2}{5}}$
(d) $19^{\frac{5}{2}}$
(e) $12^{\frac{3}{4}}$
(f) $14^{\frac{2}{3}}$

11.6 Complex Numbers

In Chapter 10 we found that some quadratic equations have no real number solutions. For example, the equation $x^2 = -4$ has no real number solutions because $2^2 = 4$ and $(-2)^2 = 4$. In this section we will consider a set of numbers that contains some numbers whose squares are negative real numbers. Then in the next section we will see that this set of numbers, called the set of **complex numbers**, provides solutions for not only equations such as $x^2 = -4$, but also for any quadratic equation with real number coefficients in one variable.

Our work with complex numbers is based on the following definition.

DEFINITION 11.7

The number i is such that

$$i = \sqrt{-1} \quad \text{and} \quad i^2 = -1.$$

The number i is not a real number and is often called the **imaginary unit**, but the number i^2 is the real number -1.

In Chapter 9 we used the property $\sqrt{a}\sqrt{b} = \sqrt{ab}$ to multiply radicals and to express radicals in simplest radical form. This property also holds if *only one of a or b* is negative. Thus, we can simplify square root radicals that contain negative numbers as radicands as follows.

$$\sqrt{-4} = \sqrt{-1}\sqrt{4} = i(2), \qquad \text{Usually written as } 2i$$
$$\sqrt{-13} = \sqrt{-1}\sqrt{13} = i\sqrt{13},$$
$$\sqrt{-12} = \sqrt{-1}\sqrt{12} = i\sqrt{12} = i\sqrt{4}\sqrt{3} = 2i\sqrt{3},$$
$$\sqrt{-18} = \sqrt{-1}\sqrt{18} = i\sqrt{18} = i\sqrt{9}\sqrt{2} = 3i\sqrt{2}$$

The imaginary unit i is used to define a complex number as follows.

DEFINITION 11.8

A **complex number** is any number that can be expressed in the form

$a + bi$

where a and b are real numbers.

The form $a + bi$ is called the **standard from** of a complex number. We call the real number a the **real part** of the complex number and we call b the **imaginary part**. The following examples demonstrate this terminology.

1. The number $3 + 4i$ is a complex number in standard form that has a real part of 3 and an imaginary part of 4.

2. We can write the number $-5 - 2i$ in the standard form $-5 + (-2i)$, thus it is a complex number that has a real part of -5 and an imaginary part of -2. (We often use the form $-5 - 2i$, knowing that it means $-5 + (-2i)$.)

3. We can write the number $-7i$ in the standard form $0 + (-7i)$; thus it is a complex number that has a real part of 0 and an imaginary part of -7.

4. We can write the number 9 in the standard form $9 + 0i$ and it is thus a complex number that has a real part of 9 and an imaginary part of 0.

Number 4 shows us that all real numbers can be considered complex numbers.

The commutative, associative, and distributive properties hold for all complex numbers and form the basis for manipulating with complexes. The following two statements describe addition and subtraction of complex numbers.

$$(a + bi) + (c + di) = (a + c) + (b + d)i, \qquad \textit{Addition}$$

$$(a + bi) - (c + di) = (a - c) + (b - d)i \qquad \textit{Subtraction}$$

To add complex numbers, add their real parts and add their imaginary parts. To subtract complex numbers, subtract their real parts and subtract their imaginary parts. Consider the following examples.

1. $(3 + 5i) + (4 + 7i) = (3 + 4) + (5 + 7)i = 7 + 12i$

2. $(-5 + 2i) + (7 - 8i) = (-5 + 7) + (2 + (-8))i$
$$= 2 + (-6i) \qquad \text{Usually written as } 2 - 6i$$

3. $\left(\dfrac{1}{2} + \dfrac{3}{4}i\right) + \left(\dfrac{2}{3} + \dfrac{1}{5}i\right) = \left(\dfrac{1}{2} + \dfrac{2}{3}\right) + \left(\dfrac{3}{4} + \dfrac{1}{5}\right)i$
$$= \left(\dfrac{3}{6} + \dfrac{4}{6}\right) + \left(\dfrac{15}{20} + \dfrac{4}{20}\right)i$$
$$= \dfrac{7}{6} + \dfrac{19}{20}i$$

4. $(8 + 9i) - (5 + 4i) = (8 - 5) + (9 - 4)i = 3 + 5i$

5. $(2 - 3i) - (5 + 9i) = (2 - 5) + (-3 - 9)i = -3 + (-12i)$
$$= -3 - 12i$$

6. $(-1 + 6i) - (-2 - i) = (-1 - (-2)) + (6 - (-1))i = 1 + 7i$

Multiplying Complex Numbers

Since complex numbers have a *binomial form*, we find the product of two complex numbers in the same way that we find the product of two binomials. Consider the following examples.

Example 1

Multiply $(3 + 2i)(4 + 5i)$.

Solution

When we multiply each term of the first number times each term of the second number and simplify we get

$$(3 + 2i)(4 + 5i) = 3(4) + 3(5i) + 2i(4) + 2i(5i)$$
$$= 12 + 15i + 8i + 10i^2$$
$$= 12 + (15 + 8)i + 10(-1) \qquad i^2 = -1$$
$$= 2 + 23i.$$

▲

Example 2

Multiply $(6 - 3i)(-5 + 2i)$.

Solution

$$(6 - 3i)(-5 + 2i) = 6(-5) + 6(2i) - (3i)(-5) - (3i)(2i)$$
$$= -30 + 12i + 15i - 6i^2$$
$$= -30 + 27i - 6(-1)$$
$$= -24 + 27i$$

▲

Example 3

Find the indicated product $(5 - i)^2$.

Solution

Remember that $(a - b)^2$ means $(a - b)(a - b)$.

$$(5 - i)^2 = (5 - i)(5 - i) = 5(5) - 5(i) - i(5) - (i)(-i)$$
$$= 25 - 5i - 5i + i^2$$
$$= 25 - 10i + (-1)$$
$$= 24 - 10i$$

▲

To find products such as $(4i)(3i)$ or $2i(4 + 6i)$, we could change each number to standard *binomial form* and then proceed as in the previous examples, but it is easier to handle them as follows.

$$(4i)(3i) = 4 \cdot 3 \cdot i \cdot i = 12i^2 = 12(-1) = -12,$$
$$2i(4 + 6i) = 2i(4) + 2i(6i)$$
$$= 8i + 12i^2$$
$$= 8i + 12(-1)$$
$$= -12 + 8i$$

Problem Set 11.6

For Problems 1–12, write each radical in terms of i and simplify. For example, $\sqrt{-20} = i\sqrt{20} = i\sqrt{4} \cdot \sqrt{5} = 2i\sqrt{5}$.

1. $\sqrt{-64}$ **2.** $\sqrt{-81}$

3. $\sqrt{-\dfrac{25}{9}}$ **4.** $\sqrt{-\dfrac{49}{16}}$

5. $\sqrt{-11}$ **6.** $\sqrt{-17}$

7. $\sqrt{-50}$ **8.** $\sqrt{-32}$

9. $\sqrt{-48}$ **10.** $\sqrt{-45}$

11. $\sqrt{-54}$ **12.** $\sqrt{-28}$

For Problems 13–34, add or subtract the complex numbers as indicated.

13. $(3 + 8i) + (5 + 9i)$

14. $(7 + 10i) + (2 + 3i)$

15. $(7 - 6i) + (3 - 4i)$

16. $(8 - 2i) + (-7 + 3i)$

17. $(10 + 4i) - (6 + 2i)$

18. $(12 + 7i) - (5 + i)$

19. $(5 + 2i) - (7 + 8i)$

20. $(3 + i) - (7 + 4i)$

21. $(-2 - i) - (3 - 4i)$

22. $(-7 - 3i) - (8 - 9i)$

23. $(-4 - 7i) + (-8 - 9i)$

24. $(-1 - 2i) + (-6 - 6i)$

25. $(0 - 6i) + (-10 + 2i)$

26. $(4 - i) + (4 + i)$

27. $(-9 + 7i) - (-8 - 5i)$

28. $(-12 + 6i) - (-7 + 2i)$

29. $(-10 - 4i) - (10 + 4i)$

30. $(-8 - 2i) - (8 + 2i)$

31. $\left(\dfrac{1}{2} + \dfrac{2}{3}i\right) + \left(\dfrac{1}{3} - \dfrac{1}{4}i\right)$

32. $\left(\dfrac{3}{4} - \dfrac{1}{2}i\right) + \left(\dfrac{1}{2} + i\right)$

33. $\left(\dfrac{3}{5} - \dfrac{1}{4}i\right) - \left(\dfrac{2}{3} - \dfrac{5}{6}i\right)$

34. $\left(\dfrac{1}{6} + \dfrac{5}{4}i\right) - \left(-\dfrac{1}{3} - \dfrac{3}{5}i\right)$

For Problems 35–54, find each product and express it in the standard form of a complex number ($a + bi$).

35. $(7i)(8i)$ **36.** $(-6i)(7i)$

37. $2i(6 + 3i)$

38. $3i(-4 + 9i)$

39. $-4i(-5 - 6i)$

40. $-5i(7 - 8i)$

41. $(2 + 3i)(5 + 4i)$

42. $(4 + 2i)(6 + 5i)$

43. $(7 - 3i)(8 + i)$

44. $(9 - 3i)(2 - 5i)$

45. $(-2 - 3i)(6 - 3i)$

46. $(-3 - 8i)(1 - i)$

47. $(-1 - 4i)(-2 - 7i)$

48. $(-3 - 7i)(-6 - 10i)$

49. $(4 + 5i)^2$

50. $(7 - 2i)^2$

51. $(5 - 6i)(5 + 6i)$

52. $(7 - 3i)(7 + 3i)$

53. $(-2 + i)(-2 - i)$

54. $(-5 - 8i)(-5 + 8i)$

THOUGHTS INTO WORDS

55. Why is the set of real numbers a subset of the set of complex numbers?

56. Is it possible for the product of two nonreal complex numbers to be a real number? Defend your answer.

Quadratic Equations: Complex Solutions

As we stated in the previous section, the set of complex numbers provides solutions for all quadratic equations that have real number coefficients. In other words, every quadratic equation of the form $ax^2 + bx + c = 0$, where a, b, and c are real numbers and $a \neq 0$, has a solution (or solutions) from the set of complex numbers.

To find solutions for quadratic equations we continue to use the techniques of factoring, completing the square, the quadratic formula, and the property, if $x^2 = a$, then $x = \pm\sqrt{a}$. Let's consider some examples.

Example 1

Solve $x^2 = -4$.

Solution

Use the property, *if $x^2 = a$, then $x = \pm\sqrt{a}$*, and proceed as follows.

$$x^2 = -4$$
$$x = \pm\sqrt{-4}$$
$$x = \pm 2i \qquad \sqrt{-4} = \sqrt{-1}\sqrt{4} = 2i$$

 Check

$x^2 = -4$	$x^2 = -4$
$(2i)^2 \stackrel{?}{=} -4$	$(-2i)^2 \stackrel{?}{=} -4$
$4i^2 \stackrel{?}{=} -4$	$4i^2 \stackrel{?}{=} -4$
$4(-1) \stackrel{?}{=} -4$	$4(-1) \stackrel{?}{=} -4$
$-4 = -4$	$-4 = -4$

The solution set is $\{-2i, 2i\}$. ▲

Example 2

Solve $(x - 2)^2 = -7$.

Solution

$$(x - 2)^2 = -7$$
$$x - 2 = \pm\sqrt{-7}$$
$$x - 2 = \pm i\sqrt{7} \qquad \sqrt{-7} = \sqrt{-1}\sqrt{7} = i\sqrt{7}$$
$$x = 2 \pm i\sqrt{7}$$

 Check

$(x - 2)^2 = -7$	$(x - 2)^2 = -7$
$(2 + i\sqrt{7} - 2)^2 \stackrel{?}{=} -7$	$(2 - i\sqrt{7} - 2)^2 \stackrel{?}{=} -7$
$(i\sqrt{7})^2 \stackrel{?}{=} -7$	$(-i\sqrt{7})^2 \stackrel{?}{=} -7$
$7i^2 \stackrel{?}{=} -7$	$7i^2 \stackrel{?}{=} -7$

$$7(-1) \overset{?}{=} -7 \qquad\qquad 7(-1) \overset{?}{=} -7$$
$$-7 = -7 \qquad\qquad -7 = -7$$

The solution set is $\{2 - i\sqrt{7}, 2 + i\sqrt{7}\}$. ▲

Example 3

Solve $x^2 + 2x = -10$.

Solution

The form of the equation lends itself to **completing the square** so let's proceed as follows.

$$x^2 + 2x = -10$$
$$x^2 + 2x + 1 = -10 + 1$$
$$(x + 1)^2 = -9$$
$$x + 1 = \pm 3i$$
$$x = -1 \pm 3i$$

✓ **Check**

$$x^2 + 2x = -10 \qquad\qquad\qquad x^2 + 2x = -10$$
$$(-1 + 3i)^2 + 2(-1 + 3i) \overset{?}{=} -10 \qquad (-1 - 3i)^2 + 2(-1 - 3i) \overset{?}{=} -10$$
$$1 - 6i + 9i^2 - 2 + 6i \overset{?}{=} -10 \qquad 1 + 6i + 9i^2 - 2 - 6i \overset{?}{=} -10$$
$$-1 + 9i^2 \overset{?}{=} -10 \qquad\qquad\qquad -1 + 9i^2 \overset{?}{=} -10$$
$$-1 + 9(-1) \overset{?}{=} -10 \qquad\qquad\qquad -1 + 9(-1) \overset{?}{=} -10$$
$$-10 = -10 \qquad\qquad\qquad\qquad -10 = -10$$

The solution set is $\{-1 - 3i, -1 + 3i\}$. ▲

Example 4

Solve $x^2 - 2x + 2 = 0$.

Solution

Use the quadratic formula to obtain the following solutions.

$$x^2 - 2x + 2 = 0$$

$$x = \frac{2 \pm \sqrt{(-2)^2 - 4(1)(2)}}{2} \qquad x = \frac{-b \pm \sqrt{b^2 - 4ac}}{2a}$$

$$x = \frac{2 \pm \sqrt{4 - 8}}{2}$$

$$x = \frac{2 \pm \sqrt{-4}}{2}$$

$$x = \frac{2 \pm 2i}{2}$$

$$x = \frac{2(1 \pm i)}{2}$$

$$x = 1 \pm i$$

✔ **Check**

$$x^2 - 2x + 2 = 0 \qquad\qquad x^2 - 2x + 2 = 0$$
$$(1 + i)^2 - 2(1 + i) + 2 \overset{?}{=} 0 \qquad (1 - i)^2 - 2(1 - i) + 2 \overset{?}{=} 0$$
$$1 + 2i + i^2 - 2 - 2i + 2 = 0 \qquad 1 - 2i + i^2 - 2 + 2i + 2 \overset{?}{=} 0$$
$$1 + i^2 \overset{?}{=} 0 \qquad\qquad 1 + i^2 \overset{?}{=} 0$$
$$1 - 1 \overset{?}{=} 0 \qquad\qquad 1 - 1 \overset{?}{=} 0$$
$$0 = 0 \qquad\qquad 0 = 0$$

The solution set is $\{1 - i, 1 + i\}$.   ▲

REMARK   If you worked on Problem 62 of Problem Set 10.3, then you are familiar with the sum and product relationships of solutions of quadratic equations. These relationships can be used to check complex solutions. The check for Example 4 is as follows:

Sum of solutions: $(1 + i) + (1 - i) = 2$ and $-\dfrac{b}{a} = -\dfrac{-2}{1} = 2$

Product of solutions: $(1 + i)(1 - i) = 1 - i^2 = 1 - (-1) = 2$ and
$$\dfrac{c}{a} = \dfrac{2}{1} = 2$$
△

Example 5   Solve $x^2 + 3x - 10 = 0$.

Solution   We can factor $x^2 + 3x - 10$ and proceed as follows.

$$x^2 + 3x - 10 = 0$$
$$(x + 5)(x - 2) = 0$$
$$x + 5 = 0 \quad \text{or} \quad x - 2 = 0$$
$$x = -5 \quad \text{or} \quad x = 2$$

The solution set is $\{-5, 2\}$. (Don't forget that all real numbers are complex numbers. That is, -5 and 2 can be written as $-5 + 0i$ and $2 + 0i$.)   ▲

To summarize our work with quadratic equations in Chapter 10 and this section, use the following approach to solve a quadratic equation.

1. If the equation is in a form where the property *if $x^2 = a$, then $x = \pm\sqrt{a}$* applies, use it. (See Examples 1 and 2.)
2. If the quadratic expression can be factored using integers, factor it and apply the property *if $ab = 0$, then $a = 0$ or $b = 0$.* (See Example 5.)
3. If numbers 1 and 2 don't apply, use either the quadratic formula or the process of completing the square. (See Examples 3 and 4.)

Problem Set 11.7

Solve each of the following quadratic equations and check your solutions.

1. $x^2 = -64$

2. $x^2 = -49$

3. $(x - 2)^2 = -1$

4. $(x + 3)^2 = -16$

5. $(x + 5)^2 = -13$

6. $(x - 7)^2 = -21$

7. $(x - 3)^2 = -18$

8. $(x + 4)^2 = -28$

9. $(5x - 1)^2 = 9$

10. $(7x + 3)^2 = 1$

11. $a^2 - 3a - 4 = 0$

12. $a^2 + 2a - 35 = 0$

13. $t^2 + 6t = -12$

14. $t^2 - 4t = -9$

15. $n^2 - 6n + 13 = 0$

16. $n^2 - 4n + 5 = 0$

17. $x^2 - 4x + 20 = 0$

18. $x^2 + 2x + 5 = 0$

19. $3x^2 - 2x + 1 = 0$

20. $2x^2 + x + 1 = 0$

21. $2x^2 - 3x - 5 = 0$

22. $3x^2 - 5x - 2 = 0$

23. $y^2 - 2y = -19$

24. $y^2 + 8y = -24$

25. $x^2 - 4x + 7 = 0$

26. $x^2 - 2x + 3 = 0$

27. $4x^2 - x + 2 = 0$

28. $5x^2 + 2x + 1 = 0$

29. $6x^2 + 2x + 1 = 0$

30. $7x^2 + 3x + 3 = 0$

THOUGHTS INTO WORDS

31. Which method would you use to solve the equation $x^2 + 4x = -5$? Explain your reasons for making that choice.

32. Explain why the expression $b^2 - 4ac$ from the quadratic formula will determine if the solutions of a particular quadratic equation are imaginary.

SUMMARY

(11.1) Interpreting **absolute value** to mean *the distance between a number and zero on a number line* allows us to solve a variety of equations and inequalities involving absolute value.

(11.2) **Linear inequalities** in two variables are of the form $Ax + By > C$ or $Ax + By < C$. To **graph a linear inequality**, we suggest the following steps.

1. First, graph the corresponding equality. Use a solid line if equality is included in the original statement and a dashed line if equality is not included.

2. Choose a test point not on the line and substitute its coordinates into the inequality.

3. The graph of the original inequality is
   (a) the half-plane that contains the test point, if the inequality is satisfied by that point, or
   (b) the half-plane that does not contain the test point, if the inequality is not satisfied by the point.

The solution set of a system of linear inequalities is the intersection of the solution sets of the individual inequalities.

(11.3) A **relation** is a set of ordered pairs. A **function** is a relation where no two ordered pairs have the same first coordinate.

The notation $f(x)$ is read "f of x" and we define it to be the value of the function f at x. Therefore, if $f(x) = 2x + 3$, then $f(1) = 2(1) + 3 = 5$.

(11.4) Any function that can be written in the form $f(x) = c$, where c is a real number, is a **constant function**. The graph of every constant function is a **horizontal line**.

Any function that can be written in the form $f(x) = ax + b$, where a and b are real numbers and $a \neq 0$, is a **linear function**. The graph of every linear function is a **straight line**.

Any function that can be written in the form $f(x) = ax^2 + bx + c$, where a, b, and c are real numbers and $a \neq 0$, is a **quadratic function**. The graph of every quadratic function is a **parabola**. The graph of every quadratic function of the form $f(x) = ax^2$ is a parabola that has its vertex at the origin and is symmetrical about the y-axis.

(11.5) The following definitions merge the concepts of *root* and *fractional exponents*.

1. a is a square root of b if $a^2 = b$.
2. a is a cube root of b if $a^3 = b$.
3. a is an nth root of b if $a^n = b$.
4. $b^{\frac{1}{n}}$ means $\sqrt[n]{b}$.
5. $b^{\frac{m}{n}}$ means $\sqrt[n]{b^m}$, which equals $(\sqrt[n]{b})^m$.

(11.6) A **complex number** is any number that we can express in the form $a + bi$, where a and b are real numbers and $i = \sqrt{-1}$.

The property, $\sqrt{ab} = \sqrt{a}\sqrt{b}$, holds if only one of a or b is negative and therefore we use it as a basis for simplifying radicals containing negative radicands.

We describe **addition** and **subtraction** of complex numbers as follows.

$$(a + bi) + (c + di) = (a + c) + (b + d)i,$$

$$(a + bi) - (c + di) = (a - c) + (b - d)i$$

We find the **product** of two complex numbers in the same way that we find the product of two binomials.

(11.7) Every quadratic equation of the form $ax^2 + bx + c = 0$, where a, b, and c are real numbers and $a \neq 0$, has a solution (solutions) in the set of complex numbers.

Chapter 11 Review Problem Set

Write each of the following in terms of i and simplify as much as possible.

1. $\sqrt{-64}$
2. $\sqrt{-29}$
3. $\sqrt{-54}$
4. $\sqrt{-\dfrac{9}{4}}$
5. $\sqrt{-108}$
6. $\sqrt{-96}$

Perform the following indicated operations on complex numbers.

7. $(5 - 7i) + (-4 + 9i)$
8. $(-3 + 2i) + (-4 - 7i)$
9. $(6 - 9i) - (4 - 5i)$
10. $(-5 + 3i) - (-8 + 7i)$
11. $(7 - 2i) - (6 - 4i) + (-2 + i)$
12. $(-4 + i) - (-4 - i) - (6 - 8i)$
13. $(2 + 5i)(3 + 8i)$
14. $(4 - 3i)(1 - 2i)$
15. $(-1 + i)(-2 + 6i)$
16. $(-3 - 3i)(7 + 8i)$

17. $(2 + 9i)(2 - 9i)$
18. $(-3 + 7i)(-3 - 7i)$
19. $(-3 - 8i)(3 + 8i)$
20. $(6 + 9i)(-1 - i)$

Solve each of the following quadratic equations.

21. $(x - 6)^2 = -25$
22. $n^2 + 2n = -7$
23. $x^2 - 2x + 17 = 0$
24. $x^2 - x + 7 = 0$
25. $2x^2 - x + 3 = 0$
26. $6x^2 - 11x + 3 = 0$
27. $-x^2 + 5x - 7 = 0$
28. $-2x^2 - 3x - 6 = 0$

29. $3x^2 + x + 5 = 0$

30. $x(4x + 1) = -3$

Evaluate each of the following numerical expressions.

31. $\sqrt{\dfrac{64}{36}}$

32. $-\sqrt{1}$

33. $\sqrt[3]{\dfrac{27}{64}}$

34. $\sqrt[3]{-125}$

35. $\sqrt[4]{\dfrac{81}{16}}$

36. $25^{\frac{3}{2}}$

37. $8^{\frac{5}{3}}$

38. $(-8)^{\frac{5}{3}}$

39. 4^{-2}

40. $4^{-\frac{1}{2}}$

41. $(32)^{-\frac{2}{5}}$

42. $\left(\dfrac{2}{3}\right)^{-1}$

43. $2^{\frac{7}{4}} \cdot 2^{\frac{5}{4}}$

44. $3^{\frac{1}{3}} \cdot 3^{\frac{5}{3}}$

45. $\dfrac{3^{\frac{1}{3}}}{3^{\frac{4}{3}}}$

For Problems 46−53, simplify and express final results with positive exponents only.

46. $x^{\frac{5}{6}} \cdot x^{\frac{5}{6}}$

47. $\left(3x^{\frac{1}{4}}\right)\left(2x^{\frac{3}{5}}\right)$

48. $\left(9a^{\frac{1}{2}}\right)\left(4a^{-\frac{1}{3}}\right)$

49. $\left(3x^{\frac{1}{3}}y^{\frac{2}{3}}\right)^{3}$

50. $\left(25x^{4}y^{6}\right)^{\frac{1}{2}}$

51. $\dfrac{39n^{\frac{3}{5}}}{3n^{\frac{1}{4}}}$

52. $\dfrac{64n^{\frac{5}{8}}}{16n^{\frac{7}{8}}}$

53. $\left(\dfrac{6x^{\frac{2}{7}}}{3x^{-\frac{5}{7}}}\right)^{3}$

54. If $f(x) = x^2 - 3x - 1$, find $f(2), f(-1)$, and $f(-4)$.

55. If $f(x) = -2x^2 - 7x + 2$, find $f(3), f(-3)$, and $f(-4)$.

For Problems 56−60, graph each of the functions.

56. $f(x) = 3$

57. $f(x) = 2x - 3$

58. $f(x) = -3x - 1$

59. $f(x) = 4x^2$

60. $f(x) = -4x^2$

For Problems 61−66, solve and graph the solutions for each problem.

61. $|3x - 5| = 7$

62. $|x - 4| < 1$

63. $|2x - 1| \geq 3$

64. $|3x - 2| \leq 4$

65. $|2x - 1| = 9$

66. $|5x - 2| \geq 6$

67. Graph the inequality $-2x + y < 4$.

68. Graph the inequality $3x + 2y \geq -6$.

69. Solve, by graphing, the system of inequalities $\begin{pmatrix} -x + 2y > 2 \\ 3x - \ y > 3 \end{pmatrix}$.

CHAPTER 11 TEST

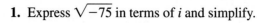

1. Express $\sqrt{-75}$ in terms of i and simplify.

2. Evaluate $36^{\frac{3}{2}}$.

3. Evaluate $9^{-\frac{1}{2}}$.

4. Evaluate $\left(\frac{2}{3}\right)^{-3}$.

5. Evaluate $16^{\frac{5}{4}}$.

6. Simplify $\left(2x^{\frac{1}{4}}\right)\left(5x^{\frac{2}{3}}\right)$.

7. Simplify $\dfrac{30n^{\frac{1}{2}}}{6n^{\frac{2}{5}}}$.

8. If $f(x) = 2x^2 - x + 3$, find $f(-3)$.

9. If $f(x) = -x^2 - 6x - 1$, find $f(2)$.

For Problems 10–18, solve each of the equations.

10. $(x - 2)^2 = -16$

11. $x^2 - 2x + 3 = 0$

12. $x^2 + 6x = -21$

13. $x^2 - 3x + 5 = 0$

14. $|x - 2| = 6$

15. $|4x + 5| = 2$

16. $|3x - 1| = -4$

17. $2x^2 - x + 1 = 0$

18. $3x^2 + 5x - 28 = 0$

19. Solve the inequality $|x + 3| \geq 2$.

20. Solve the inequality $|2x - 1| < 7$.

21. Solve the inequality $|5x - 2| > -3$.

22. Graph the inequality $3x - 2y > -6$.

23. Solve, by graphing, the system of inequalities $\begin{pmatrix} x - 2y \leq 4 \\ 2x + \ y \leq 4 \end{pmatrix}$.

24. Graph the function $f(x) = -2x - 2$.

25. Graph the function $f(x) = -x^2 + 3$.

Appendix

Table of Squares and Approximate Square Roots

n	%n²	√n	n	n²	√n
1	1	1.000	26	676	5.099
2	4	1.414	27	729	5.196
3	9	1.732	28	784	5.292
4	16	2.000	29	841	5.385
5	25	2.236	30	900	5.477
6	36	2.449	31	961	5.568
7	49	2.646	32	1024	5.657
8	64	2.828	33	1089	5.745
9	81	3.000	34	1156	5.831
10	100	3.162	35	1225	5.916
11	121	3.317	36	1296	6.000
12	144	3.464	37	1369	6.083
13	169	3.606	38	1444	6.164
14	196	3.742	39	1521	6.245
15	225	3.873	40	1600	6.325
16	256	4.000	41	1681	6.403
17	289	4.123	42	1764	6.481
18	324	4.243	43	1849	6.557
19	361	4.359	44	1936	6.633
20	400	4.472	45	2025	6.708
21	441	4.583	46	2116	6.782
22	484	4.690	47	2209	6.856
23	529	4.796	48	2304	6.928
24	576	4.899	49	2401	7.000
25	625	5.000	50	2500	7.071

(continued)

Squares and
Approximate
Square Roots
(continued)

n	n^2	$\sqrt{n}$	n	n^2	$\sqrt{n}$
51	2601	7.141	76	5776	8.718
52	2704	7.211	77	5929	8.775
53	2809	7.280	78	6084	8.832
54	2916	7.348	79	6241	8.888
55	3025	7.416	80	6400	8.944
56	3136	7.483	81	6561	9.000
57	3249	7.550	82	6724	9.055
58	3364	7.616	83	6889	9.110
59	3481	7.681	84	7056	9.165
60	3600	7.746	85	7225	9.220
61	3721	7.810	86	7396	9.274
62	3844	7.874	87	7569	9.327
63	3969	7.937	88	7744	9.381
64	4096	8.000	89	7921	9.434
65	4225	8.062	90	8100	9.487
66	4356	8.124	91	8281	9.539
67	4489	8.185	92	8464	9.592
68	4624	8.246	93	8649	9.644
69	4761	8.307	94	8836	9.695
70	4900	8.367	95	9025	9.747
71	5041	8.426	96	9216	9.798
72	5184	8.485	97	9409	9.849
73	5329	8.544	98	9604	9.899
74	5476	8.602	99	9801	9.950
75	5625	8.660	100	10000	10.000

From the table we can find rational approximations as follows.

$$\sqrt{31} = 5.568 \qquad \text{Rounded to three decimal places}$$

Locate 31
in the column
labeled n.

Locate this value for
$\sqrt{31}$ in the column
labeled $\sqrt{n}$.

Be sure that you agree with the following values taken from the table. Each of these is rounded to three decimal places.

$$\sqrt{14} = 3.742, \qquad \sqrt{46} = 6.782, \qquad \sqrt{65} = 8.062$$

The column labeled n^2 contains the squares of the whole numbers from 1 through 100, inclusive. For example, from the table we obtain $25^2 = 625$, $62^2 = 3844$, and $89^2 = 7921$. Since $25^2 = 625$, we can state that $\sqrt{625} = 25$. Thus, the column labeled n^2 also provides us with additional square root facts.

$$\sqrt{1936} = 44$$

Locate 1936 in
the column
labeled n^2.

Locate this value for
$\sqrt{1936}$ in the
column labeled n.

Now suppose that you want to find a value for $\sqrt{500}$. Scanning the column labeled n^2, we see that 500 does not appear; however, it is between two values given in the table. We can reason as follows.

$$22^2 = 484$$
$$23^2 = 529 \quad \longleftarrow \quad 500$$

Since 500 is closer to 484 than to 529, we approximate $\sqrt{500}$ to be closer to 22 than to 23. Thus, we write $\sqrt{500} = 22$, to the nearest whole number. This is merely a whole number approximation, but for some purposes it may be sufficiently precise.

Answers to Odd-Numbered Problems and All Chapter Review, Chapter Test, and Cumulative Review Problems

CHAPTER 1

Problem Set 1.1 (page 9)

1. 16   **3.** 35   **5.** 51   **7.** 72   **9.** 82
11. 55   **13.** 60   **15.** 66   **17.** 26
19. 2   **21.** 47   **23.** 21   **25.** 11
27. 15   **29.** 14   **31.** 79   **33.** 6
35. 74   **37.** 12   **39.** 187   **41.** 884
43. 9   **45.** 18   **47.** 55   **49.** 99
51. 72   **53.** 11   **55.** 48   **57.** 21
59. 40   **61.** 81   **63.** 100   **65.** 84
67. 170   **69.** 164   **71.** 153

Problem Set 1.2 (page 14)

1. True   **3.** False   **5.** True   **7.** True
9. True   **11.** False   **13.** True   **15.** False
17. True   **19.** False   **21.** Prime
23. Prime   **25.** Composite   **27.** Prime
29. Composite   **31.** $2 \cdot 13$ **33.** $2 \cdot 2 \cdot 3 \cdot 3$
35. $7 \cdot 7$ **37.** $2 \cdot 2 \cdot 2 \cdot 7$ **39.** $2 \cdot 2 \cdot 2 \cdot 3 \cdot 5$
41. $3 \cdot 3 \cdot 3 \cdot 5$ **43.** 4   **45.** 8   **47.** 9
49. 12   **51.** 18   **53.** 12   **55.** 24
57. 48   **59.** 140   **61.** 392   **63.** 168
65. 90
69. All other even numbers are divisible by 2.
71. 61   **73.** x **75.** xy **77.** $2 \cdot 2 \cdot 19$
79. $3 \cdot 41$ **81.** $5 \cdot 23$ **83.** $3 \cdot 3 \cdot 7 \cdot 7$
85. $3 \cdot 3 \cdot 17$

Problem Set 1.3 (page 22)

1. 2   **3.** -4 **5.** -7 **7.** 6   **9.** -6
11. 8   **13.** -11 **15.** -15 **17.** -7
19. -31 **21.** -19 **23.** 9   **25.** -61

27. -18 **29.** -92 **31.** -5 **33.** -13
35. 12   **37.** 6   **39.** -1 **41.** -45
43. -29 **45.** 27   **47.** -65 **49.** -29
51. -11 **53.** -1 **55.** -8 **57.** -13
59. -35 **61.** -15 **63.** -32 **65.** 2
67. -4 **69.** -31 **71.** -9 **73.** 18
75. 8   **77.** -29 **79.** -7 **81.** 15
83. 1   **85.** 36   **87.** -39 **89.** -24
91. 7   **93.** -1 **95.** 10   **97.** 9
99. -17 **101.** -3 **103.** -10 **105.** -3
107. 11   **109.** 5   **111.** -65 **113.** -100
115. -25 **117.** 130   **119.** 80
121. $-17 + 14 = -3 \ (-3°F)$
123. $3 + (-2) + (-3) + (-5) = -7$ (7 under par)

Problem Set 1.4 (page 29)

1. -30 **3.** -9 **5.** 7   **7.** -56
9. 60   **11.** -12 **13.** -126 **15.** 154
17. -9 **19.** 11   **21.** 225   **23.** -14
25. 0   **27.** 23   **29.** -19 **31.** 90
33. 14   **35.** Undefined   **37.** -4
39. -972 **41.** -47 **43.** 18   **45.** 69
47. 4   **49.** 4   **51.** -6 **53.** 31   **55.** 4
57. 28   **59.** -7 **61.** 10   **63.** -59
65. 66   **67.** 7   **69.** 69   **71.** -7
73. 126   **75.** -70 **77.** 15   **79.** -10
81. -25 **83.** 77   **85.** 104   **87.** 14
89. $800(19) + 800(2) + 800(4)(-1) = 13,600$
($13,600)

Problem Set 1.5 (page 37)

1. Distributive property
3. Associative property for addition
5. Commutative property for multiplication

7. Additive inverse property
9. Identity property for addition
11. Associative property for multiplication
13. 56 **15.** 7 **17.** 1800 **19.** $-14,400$
21. -3700 **23.** 5900 **25.** -338
27. -38 **29.** 7 **31.** $-5x$ **33.** $-3m$
35. $-11y$ **37.** $-3x - 2y$ **39.** $-16a - 4b$
41. $-7xy + 3x$ **43.** $10x + 5$ **45.** $6xy - 4$
47. $-6a - 5b$ **49.** $5ab - 11a$ **51.** $8x + 36$
53. $11x + 28$ **55.** $8x + 44$ **57.** $5a + 29$
59. $3m + 29$ **61.** $-8y + 6$ **63.** -5
65. -40 **67.** 72 **69.** -18 **71.** 37
73. -74 **75.** 180 **77.** 34 **79.** -65

Chapter 1 Review Problem Set (page 40)

1. -3 **2.** -25 **3.** -5 **4.** -15
5. -1 **6.** 2 **7.** -156 **8.** 252
9. 6 **10.** -13 **11.** Prime
12. Composite **13.** Composite
14. Composite **15.** Composite
16. $2 \cdot 2 \cdot 2 \cdot 3$ **17.** $3 \cdot 3 \cdot 7$ **18.** $3 \cdot 19$
19. $2 \cdot 2 \cdot 2 \cdot 2 \cdot 2 \cdot 2$ **20.** $2 \cdot 2 \cdot 3 \cdot 7$
21. 18 **22.** 12 **23.** 180 **24.** 945
25. 66 **26.** -7 **27.** -2 **28.** 4
29. -18 **30.** 12 **31.** -34 **32.** -27
33. -38 **34.** -93 **35.** 2 **36.** 3
37. 35 **38.** 27 **39.** $8x$ **40.** $-5y - 9$
41. $-5x + 4y$ **42.** $13a - 6b$ **43.** $-ab - 2a$
44. $-3xy - y$ **45.** $10x + 74$ **46.** $2x + 7$
47. $-7x - 18$ **48.** $-3x + 12$ **49.** $-2a + 4$
50. $-2a - 4$ **51.** -59 **52.** -57 **53.** 2
54. 1 **55.** 12 **56.** 13 **57.** 22
58. 32 **59.** -9 **60.** 37 **61.** -39
62. -32 **63.** 9 **64.** -44

Chapter 1 Test (page 41)

1. 7 **2.** 45 **3.** 38 **4.** -11
5. -58 **6.** -58 **7.** 4 **8.** -1
9. -20 **10.** -7 **11.** 26 **12.** -36
13. 9 **14.** -57 **15.** -47 **16.** -4
17. 5 **18.** Prime **19.** $2 \cdot 2 \cdot 2 \cdot 3 \cdot 3 \cdot 5$
20. 12 **21.** 72

22. Associative property of addition
23. Distributive property **24.** $-13x + 6y$
25. $-13x - 21$

CHAPTER 2

Problem Set 2.1 (page 49)

1. $\frac{2}{3}$ **3.** $\frac{2}{3}$ **5.** $\frac{5}{3}$ **7.** $-\frac{1}{6}$ **9.** $-\frac{3}{4}$
11. $\frac{27}{28}$ **13.** $\frac{6}{11}$ **15.** $\frac{3x}{7y}$ **17.** $\frac{2x}{5}$
19. $-\frac{5a}{13c}$ **21.** $\frac{8z}{7x}$ **23.** $\frac{5b}{7}$ **25.** $\frac{15}{28}$
27. $\frac{10}{21}$ **29.** $\frac{3}{10}$ **31.** $-\frac{4}{3}$ **33.** $\frac{7}{5}$
35. $-\frac{3}{10}$ **37.** $\frac{1}{4}$ **39.** -27 **41.** $\frac{35}{27}$
43. $\frac{8}{21}$ **45.** $-\frac{5}{6y}$ **47.** $2a$ **49.** $\frac{2}{5}$
51. $\frac{y}{2x}$ **53.** $\frac{20}{13}$ **55.** $-\frac{7}{9}$ **57.** $\frac{2}{9}$
59. $\frac{2}{5}$ **61.** $\frac{13}{28}$ **63.** $\frac{8}{5}$ **65.** -4
67. $\frac{36}{49}$ **69.** 1 **71.** $\frac{2}{3}$ **73.** $\frac{20}{9}$
75. $\frac{1}{4}$ **77.** $\frac{3}{8}$ of a cup **83.** (a) 8 (c) 40
(e) 5 **85.** (a) $\frac{11}{13}$ (c) $-\frac{37}{41}$ (e) $\frac{6}{11}$ (g) $\frac{7}{11}$

Problem Set 2.2 (page 57)

1. $\frac{5}{7}$ **3.** $\frac{5}{9}$ **5.** 3 **7.** $\frac{2}{3}$ **9.** $-\frac{1}{2}$
11. $\frac{2}{3}$ **13.** $\frac{15}{x}$ **15.** $\frac{2}{y}$ **17.** $\frac{8}{15}$
19. $\frac{9}{16}$ **21.** $\frac{37}{30}$ **23.** $\frac{59}{96}$ **25.** $-\frac{19}{72}$
27. $-\frac{1}{24}$ **29.** $-\frac{1}{3}$ **31.** $-\frac{1}{6}$ **33.** $-\frac{31}{7}$
35. $-\frac{21}{4}$ **37.** $\frac{3y + 4x}{xy}$ **39.** $\frac{7b - 2a}{ab}$
41. $\frac{11}{2x}$ **43.** $\frac{4}{3x}$ **45.** $-\frac{2}{5x}$ **47.** $\frac{19}{6y}$

49. $\dfrac{1}{24y}$ **51.** $-\dfrac{17}{24n}$ **53.** $\dfrac{5y + 7x}{3xy}$

55. $\dfrac{32y + 15x}{20xy}$ **57.** $\dfrac{63y - 20x}{36xy}$

59. $\dfrac{-6y - 5x}{4xy}$ **61.** $\dfrac{3x + 2}{x}$ **63.** $\dfrac{4x - 3}{2x}$

65. $\dfrac{1}{4}$ **67.** $\dfrac{37}{30}$ **69.** $\dfrac{1}{3}$ **71.** $-\dfrac{12}{5}$

73. $-\dfrac{1}{30}$ **75.** 14   **77.** 68   **79.** $\dfrac{7}{26}$

81. $\dfrac{11}{15}x$ **83.** $\dfrac{5}{24}a$ **85.** $\dfrac{4}{3}x$ **87.** $\dfrac{13}{20}n$

89. $\dfrac{20}{9}n$ **91.** $-\dfrac{79}{36}n$ **93.** $\dfrac{13}{14}x + \dfrac{9}{8}y$

95. $-\dfrac{11}{45}x - \dfrac{9}{20}y$ **97.** \$12.50   **99.** $1\dfrac{3}{4}$ miles

Problem Set 2.3 (page 67)

1. .62   **3.** 1.45   **5.** 3.8   **7.** -3.3
9. 7.5   **11.** 7.8   **13.** $-.9$ **15.** -7.8
17. 1.16   **19.** $-.272$ **21.** -24.3
23. 44.8   **25.** .0156   **27.** 1.2   **29.** -7.4
31. .38   **33.** 7.2   **35.** $-.42$ **37.** .76
39. 4.7   **41.** 4.3   **43.** -14.8 **45.** 1.3
47. $-1.2x$ **49.** $3n$ **51.** $.5t$
53. $-5.8x + 2.8y$ **55.** $.1x + 1.2$

57. $-3x - 2.3$ **59.** $4.6x - 8$ **61.** $\dfrac{11}{12}$

63. $\dfrac{4}{3}$ **65.** 17.3   **67.** -97.8 **69.** 2.2

71. 13.75   **73.** .6   **75.** \$9931.25
77. 19.1 centimeters
85. **(a)** The denominator has only factors of 2.
(c) $\dfrac{7}{8}, \dfrac{11}{16}, \dfrac{13}{32}, \dfrac{17}{40}, \dfrac{9}{20},$ and $\dfrac{3}{64}$

Problem Set 2.4 (page 73)

1. 64   **3.** 81   **5.** -8 **7.** -9 **9.** 16
11. $\dfrac{16}{81}$ **13.** $-\dfrac{1}{8}$ **15.** $\dfrac{9}{4}$ **17.** .027
19. -1.44 **21.** -47 **23.** -33 **25.** 11
27. -75 **29.** -60 **31.** 31   **33.** -13
35. $9x^2$ **37.** $12xy^2$ **39.** $-18x^4y$

41. $15xy$ **43.** $12x^4$ **45.** $8a^5$
47. $-8x^2$ **49.** $4y^3$ **51.** $-2x^2 + 6y^2$
53. $-\dfrac{11}{60}n^2$ **55.** $-2x^2 - 6x$

57. $7x^2 - 3x + 8$ **59.** $\dfrac{3y}{5}$ **61.** $\dfrac{11}{3y}$

63. $\dfrac{7b^2}{17a}$ **65.** $-\dfrac{3ac}{4}$ **67.** $\dfrac{x^2y^2}{4}$ **69.** $\dfrac{4x}{9}$

71. $\dfrac{5}{12ab}$ **73.** $\dfrac{6y^2 + 5x}{xy^2}$ **75.** $\dfrac{5 - 7x^2}{x^4}$

77. $\dfrac{3 + 12x^2}{2x^3}$ **79.** $\dfrac{13}{12x^2}$ **81.** $\dfrac{11b^2 - 14a^2}{a^2b^2}$

83. $\dfrac{3 - 8x}{6x^3}$ **85.** $\dfrac{3y - 4x - 5}{xy}$ **87.** 79

89. $\dfrac{23}{36}$ **91.** $\dfrac{25}{4}$ **93.** -64 **95.** -25

97. -33 **99.** .45

Problem Set 2.5 (page 80)

Answers may vary somewhat for Problems 1–12.
1. The difference of a and b
3. One-third of the product of B and h
5. Two times the quantity, l plus w
7. The quotient of A divided by w
9. The quantity, a plus b, divided by 2
11. Two more than three times y **13.** $l + w$
15. ab **17.** $\dfrac{d}{t}$ **19.** lwh **21.** $y - x$

23. $xy + 2$ **25.** $7 - y^2$ **27.** $\dfrac{x - y}{4}$

29. $10 - x$ **31.** $10(n + 2)$ **33.** $xy - 7$
35. $xy - 12$ **37.** $35 - n$ **39.** $n + 45$
41. $y + 10$ **43.** $2x - 3$ **45.** $10d + 25q$

47. $\dfrac{d}{t}$ **49.** $\dfrac{d}{p}$ **51.** $\dfrac{d}{12}$ **53.** $n + 1$

55. $n + 2$ **57.** $3y - 2$ **59.** $36y + 12f$

61. $\dfrac{f}{3}$ **63.** $8w$ **65.** $3l - 4$

67. $48f + 72$

Chapter 2 Review Problem Set (page 83)

1. 64   **2.** -27 **3.** -16 **4.** $\dfrac{9}{16}$

5. $\dfrac{49}{36}$ **6.** .216 **7.** .0144 **8.** .0036

9. $-\dfrac{8}{27}$ **10.** $\dfrac{1}{16}$ **11.** $\dfrac{19}{24}$ **12.** $\dfrac{39}{70}$

13. $\dfrac{1}{15}$ **14.** $\dfrac{14y + 9x}{2xy}$ **15.** $\dfrac{5x - 8y}{x^2y}$

16. $\dfrac{7y}{20}$ **17.** $\dfrac{4x^3}{5y^2}$ **18.** $\dfrac{2}{7}$ **19.** 1

20. $\dfrac{27n^2}{28}$ **21.** $\dfrac{1}{24}$ **22.** $-\dfrac{13}{8}$ **23.** $\dfrac{7}{9}$

24. $\dfrac{29}{12}$ **25.** $\dfrac{1}{2}$ **26.** .67 **27.** .49

28. 2.4 **29.** $-.11$ **30.** 1.76

31. $\dfrac{5}{56}x^2 + \dfrac{7}{20}y^2$ **32.** $-.58ab + .36bc$

33. $\dfrac{11x}{24}$ **34.** $2.2a + 1.7b$ **35.** $-\dfrac{1}{10}n$

36. $\dfrac{41}{20}n$ **37.** $\dfrac{19}{42}$ **38.** $-\dfrac{1}{72}$ **39.** $-.75$

40. $-.35$ **41.** $\dfrac{1}{17}$ **42.** -8

43. $72 - n$ **44.** $p + 10d$ **45.** $\dfrac{x}{60}$

46. $2y - 3$ **47.** $5n + 3$ **48.** $36y + 12f$
49. $100m$ **50.** $5n + 10d + 25q$ **51.** $n - 5$
52. $5 - n$ **53.** $10(x - 2)$ **54.** $10x - 2$

55. $x - 3$ **56.** $\dfrac{d}{r}$ **57.** $x^2 + 9$

58. $(x + 9)^2$ **59.** $x^3 + y^3$ **60.** $xy - 4$

Chapter 2 Test (page 85)

1. 81 and -64 **2.** $\dfrac{7}{9}$ **3.** $\dfrac{9xy}{16}$ **4.** -2.6

5. 3.04 **6.** $-.56$ **7.** .09 **8.** $\dfrac{2}{9}$

9. $-\dfrac{5}{24}$ **10.** $\dfrac{187}{60}$ or $3\dfrac{7}{60}$ **11.** $-\dfrac{13}{48}$

12. $\dfrac{4y}{5}$ **13.** $2x^2$ **14.** $\dfrac{4y^2 - 5x}{xy^2}$ **15.** $\dfrac{8}{3x}$

16. $\dfrac{35y + 27}{21y^2}$ **17.** $\dfrac{10a^2b}{9}$

18. $\dfrac{3 - 4xy + 6x^2}{x^2y}$ **19.** $\dfrac{26 - 25x}{20x}$ **20.** $\dfrac{37}{36}$

21. $-.48$ **22.** $-\dfrac{31}{40}$ **23.** 2.85

24. $5n + 10d + 25q$ **25.** $4n - 3$

Cumulative Review Problem Set (page 86)

1. 10 **2.** -30 **3.** 1 **4.** -26

5. -29 **6.** 17 **7.** $\dfrac{1}{2}$ **8.** $-\dfrac{7}{6}$

9. $\dfrac{1}{36}$ **10.** -64 **11.** $-\dfrac{11}{16}$ **12.** .173

13. -142 **14.** 136 **15.** $\dfrac{19}{9}$ **16.** $-.01$

17. -2.4 **18.** $\dfrac{79}{40}$ **19.** $\dfrac{7}{50}$ **20.** $\dfrac{3}{5}$

21. $2 \cdot 3 \cdot 3 \cdot 3$ **22.** $2 \cdot 3 \cdot 13$ **23.** $7 \cdot 13$
24. $3 \cdot 3 \cdot 17$ **25.** 14 **26.** 9 **27.** 4
28. 6 **29.** 140 **30.** 200 **31.** 108

32. 80 **33.** $-\dfrac{1}{12}x - \dfrac{11}{12}y$ **34.** $-\dfrac{1}{15}n$

35. $-3a + 1.9b$ **36.** $-2n + 6$

37. $-x - 15$ **38.** $-9a - 13$ **39.** $\dfrac{11}{48}$

40. $-\dfrac{31}{36}$ **41.** $\dfrac{5 - 2y + 3x}{xy}$ **42.** $\dfrac{-7y + 9x}{x^2y}$

43. $\dfrac{2x}{3}$ **44.** $\dfrac{8a^2}{21b}$ **45.** $\dfrac{4x^2}{3y}$ **46.** $-\dfrac{27}{16}$

47. $p + 5n + 10d$ **48.** $4n - 5$
49. $36y + 12f + i$
50. $200x + 200y$ or $200(x + y)$

CHAPTER 3

Problem Set 3.1 (page 95)

1. $\{8\}$ **3.** $\{-6\}$ **5.** $\{-9\}$ **7.** $\{-6\}$
9. $\{13\}$ **11.** $\{48\}$ **13.** $\{23\}$ **15.** $\{-7\}$

17. $\left\{\dfrac{17}{12}\right\}$ **19.** $\left\{-\dfrac{4}{15}\right\}$ **21.** $\{.27\}$

23. $\{-3.5\}$ **25.** $\{-17\}$ **27.** $\{-35\}$

29. $\{-8\}$ **31.** $\{-17\}$ **33.** $\left\{\dfrac{37}{5}\right\}$ **35.** $\{-3\}$

37. $\left\{\dfrac{13}{2}\right\}$ **39.** $\{144\}$ **41.** $\{24\}$ **43.** $\{-15\}$

45. $\{24\}$ **47.** $\{-35\}$ **49.** $\left\{\dfrac{3}{10}\right\}$

51. $\left\{-\dfrac{9}{10}\right\}$ **53.** $\left\{\dfrac{1}{2}\right\}$ **55.** $\left\{-\dfrac{1}{3}\right\}$

57. $\left\{\dfrac{27}{32}\right\}$ **59.** $\left\{-\dfrac{5}{14}\right\}$ **61.** $\left\{-\dfrac{7}{5}\right\}$

63. $\left\{-\dfrac{1}{12}\right\}$ **65.** $\left\{-\dfrac{3}{20}\right\}$ **67.** $\{.3\}$

69. $\{9\}$ **71.** $\{-5\}$

Problem Set 3.2 (page 100)

1. $\{4\}$ **3.** $\{6\}$ **5.** $\{8\}$ **7.** $\{11\}$

9. $\left\{\dfrac{17}{6}\right\}$ **11.** $\left\{\dfrac{19}{2}\right\}$ **13.** $\{6\}$ **15.** $\{-1\}$

17. $\{-5\}$ **19.** $\{-6\}$ **21.** $\left\{\dfrac{11}{2}\right\}$ **23.** $\{-2\}$

25. $\left\{\dfrac{10}{7}\right\}$ **27.** $\{18\}$ **29.** $\left\{-\dfrac{25}{4}\right\}$

31. $\{-7\}$ **33.** $\left\{-\dfrac{24}{7}\right\}$ **35.** $\left\{\dfrac{5}{2}\right\}$

37. $\left\{\dfrac{4}{17}\right\}$ **39.** $\left\{-\dfrac{12}{5}\right\}$ **41.** 9

43. 22   **45.** $18   **47.** 35 years old
49. $6.50   **51.** 6   **53.** 5   **55.** 11
57. 8   **59.** 3   **61.** $300   **63.** 4 meters
65. 14 cars   **67.** 5 hours

Problem Set 3.3 (page 107)

1. $\{5\}$ **3.** $\{-8\}$ **5.** $\left\{\dfrac{8}{5}\right\}$ **7.** $\{-11\}$

9. $\left\{-\dfrac{5}{2}\right\}$ **11.** $\{-9\}$ **13.** $\{2\}$ **15.** $\{-3\}$

17. $\left\{\dfrac{13}{2}\right\}$ **19.** $\left\{\dfrac{5}{3}\right\}$ **21.** $\{17\}$ **23.** $\left\{-\dfrac{13}{2}\right\}$

25. $\left\{\dfrac{16}{3}\right\}$ **27.** $\{2\}$ **29.** $\left\{-\dfrac{1}{3}\right\}$

31. $\left\{-\dfrac{19}{10}\right\}$ **33.** 17   **35.** 35 and 37

37. 36, 38, and 40   **39.** $\dfrac{3}{2}$ **41.** -6

43. 32° and 58°   **45.** 50° and 130°
47. 65° and 75°   **49.** $26 per share
51. $9 per hour   **53.** 150 males and 450 females
55. 690 votes   **57.** 6 feet

Problem Set 3.4 (page 115)

1. $\{1\}$ **3.** $\{10\}$ **5.** $\{-9\}$ **7.** $\left\{\dfrac{29}{4}\right\}$

9. $\left\{-\dfrac{17}{3}\right\}$ **11.** $\{10\}$ **13.** $\{44\}$ **15.** $\{26\}$

17. $\{-38\}$ **19.** $\left\{\dfrac{11}{6}\right\}$ **21.** $\{3\}$ **23.** $\{-1\}$

25. $\{-2\}$ **27.** $\{16\}$ **29.** $\left\{\dfrac{22}{3}\right\}$ **31.** $\{-2\}$

33. $\left\{-\dfrac{1}{6}\right\}$ **35.** $\{-57\}$ **37.** $\left\{-\dfrac{7}{5}\right\}$ **39.** $\{2\}$

41. $\{-3\}$ **43.** $\left\{\dfrac{27}{10}\right\}$ **45.** $\left\{\dfrac{3}{28}\right\}$ **47.** $\left\{\dfrac{18}{5}\right\}$

49. $\left\{\dfrac{24}{7}\right\}$ **51.** $\{5\}$ **53.** $\{0\}$ **55.** $\left\{-\dfrac{51}{10}\right\}$

57. $\{-12\}$ **59.** $\{15\}$ **61.** 7 and 8
63. 14, 15, and 16   **65.** 6 and 11   **67.** 48
69. 12 and 18   **71.** 8 feet and 12 feet
73. 12 nickels, 17 dimes, and 40 quarters
75. 40 nickels, 80 dimes, and 90 quarters
77. 8 dimes and 10 quarters
79. 75 pennies, 150 nickels, and 225 dimes
81. 30°   **83.** 20°, 50°, and 110°
85. 40°   **91.** Any three consecutive integers

Problem Set 3.5 (page 124)

1. True   **3.** False   **5.** False   **7.** True
9. True

11. $\{x \mid x > -2\}$

13. $\{x \mid x \le 3\}$

15. $\{x \mid x > 2\}$

17. $\{x \mid x \le -2\}$

19. $\{x \mid x < -1\}$

21. $\{x \mid x < 2\}$

23. $\{x \mid x < -20\}$ 25. $\{x \mid x \geq -9\}$

27. $\{x \mid x > 9\}$ 29. $\{x \mid x < \frac{10}{3}\}$

31. $\{x \mid x < -8\}$ 33. $\{n \mid n \geq 8\}$

35. $\{n \mid n > -\frac{24}{7}\}$ 37. $\{n \mid n > 7\}$

39. $\{x \mid x > 5\}$ 41. $\{x \mid x \leq 6\}$

43. $\{x \mid x \leq -21\}$ 45. $\{x \mid x < \frac{8}{3}\}$

47. $\{x \mid x < \frac{5}{4}\}$ 49. $\{x \mid x < 1\}$

51. $\{t \mid t \geq 4\}$ 53. $\{x \mid x > 14\}$ 55. $\{x \mid x > \frac{3}{2}\}$

57. $\{t \mid t \geq \frac{1}{4}\}$ 59. $\{x \mid x < -\frac{9}{4}\}$

65. All real numbers   67. $\varnothing$

69. All real numbers   71. $\varnothing$

Problem Set 3.6 (page 129)

1. $\{x \mid x > 2\}$ 3. $\{x \mid x < -1\}$

5. $\{x \mid x > -\frac{10}{3}\}$ 7. $\{n \mid n \geq -11\}$

9. $\{t \mid t \leq 11\}$ 11. $\{x \mid x > -\frac{11}{5}\}$

13. $\{x \mid x < \frac{5}{2}\}$ 15. $\{x \mid x \leq 8\}$ 17. $\{n \mid n > \frac{3}{2}\}$

19. $\{y \mid y > -3\}$ 21. $\{x \mid x < \frac{5}{2}\}$

23. $\{x \mid x < 8\}$ 25. $\{x \mid x < 21\}$ 27. $\{x \mid x < 6\}$

29. $\{n \mid n > -\frac{17}{2}\}$ 31. $\{n \mid n \leq 42\}$

33. $\{n \mid n > -\frac{9}{2}\}$ 35. $\{x \mid x > \frac{4}{3}\}$

37. $\{n \mid n \geq 4\}$ 39. $\{t \mid t > 300\}$

41. $\{x \mid x \leq 50\}$ 43. $\{x \mid x > 0\}$

45. $\{x \mid x > 64\}$ 47. $\{n \mid n > \frac{33}{5}\}$

49. $\{x \mid x \geq -\frac{16}{3}\}$

51. $-1 \quad 2$

53. $-2 \quad 1$

55. $-2 \quad 2$

57. 2

59. -4

61. $\varnothing$

63. $0 \quad 2$

65. all reals

67. All numbers greater than 7   69. 15 inches

71. 158 or better   73. Better than 90

75. More than 12%   77. 77 or less

Chapter 3 Review Problem Set (page 133)

1. $\{-3\}$ 2. $\{1\}$ 3. $\{-\frac{3}{4}\}$ 4. $\{9\}$

5. $\{-4\}$ 6. $\{\frac{40}{3}\}$ 7. $\{\frac{9}{4}\}$ 8. $\{-\frac{15}{8}\}$

9. $\{-7\}$ 10. $\{\frac{2}{41}\}$ 11. $\{\frac{19}{7}\}$ 12. $\{\frac{1}{2}\}$

13. $\{-32\}$ 14. $\{-12\}$ 15. $\{21\}$

16. $\{-60\}$ 17. $\{10\}$ 18. $\{-\frac{11}{4}\}$

19. $\{-\frac{8}{5}\}$ 20. $\{\frac{5}{21}\}$ 21. $\{x \mid x > 4\}$

22. $\{x \mid x > -4\}$ 23. $\{x \mid x \geq 13\}$

24. $\{x \mid x \geq \frac{11}{2}\}$ 25. $\{x \mid x > 35\}$

26. $\{x \mid x < \frac{26}{5}\}$ 27. $\{n \mid n < 2\}$

28. $\{n \mid n > \frac{5}{11}\}$ 29. $\{y \mid y < 24\}$

30. $\{x \mid x > 10\}$ 31. $\{n \mid n < \frac{2}{11}\}$

32. $\{n \mid n > 33\}$ 33. $\{n \mid n \leq 120\}$

34. $\{n \mid n \leq -\frac{180}{13}\}$ 35. $\{x \mid x > \frac{9}{2}\}$

36. $\left\{x \mid x < -\dfrac{43}{3}\right\}$

37.

38.

39. all reals

40.

41. 24　　**42.** 7　　**43.** 33　　**44.** 8
45. 89 or better　　**46.** 16 and 24　　**47.** 18
48. 88 or better　　**49.** 8 nickels and 22 dimes
50. 8 nickels, 25 dimes, and 50 quarters　　**51.** 52°
52. 700 miles

Chapter 3 Test (page 134)

1. $\{2\}$　　**2.** $\{3\}$　　**3.** $\{-9\}$　　**4.** $\{-5\}$
5. $\{-53\}$　　**6.** $\{-18\}$　　**7.** $\left\{-\dfrac{5}{2}\right\}$　　**8.** $\left\{\dfrac{35}{18}\right\}$
9. $\{12\}$　　**10.** $\left\{\dfrac{11}{5}\right\}$　　**11.** $\{22\}$　　**12.** $\left\{\dfrac{31}{2}\right\}$
13. $\{x \mid x < 5\}$　　**14.** $\{x \mid x \le 1\}$　　**15.** $\{x \mid x \ge -9\}$
16. $\{x \mid x < 0\}$　　**17.** $\left\{x \mid > -\dfrac{23}{2}\right\}$
18. $\{n \mid n \ge 12\}$

19.

20.

21. $18.50 per hour
22. 15 meters, 25 meters, and 30 meters
23. 96 or better
24. 17 nickels, 33 dimes, and 53 quarters
25. 20°

CHAPTER 4

Problem Set 4.1 (page 144)

1. $\{9\}$　　**3.** $\{10\}$　　**5.** $\left\{\dfrac{15}{2}\right\}$　　**7.** $\{-22\}$
9. $\{-4\}$　　**11.** $\{6\}$　　**13.** $\{-28\}$　　**15.** $\{34\}$
17. $\{6\}$　　**19.** $\left\{-\dfrac{8}{5}\right\}$　　**21.** $\{7\}$　　**23.** $\left\{\dfrac{9}{2}\right\}$
25. $\left\{-\dfrac{53}{2}\right\}$　　**27.** $\{50\}$　　**29.** $\{120\}$
31. $\left\{\dfrac{9}{7}\right\}$　　**33.** 55%　　**35.** 60%　　**37.** $16\dfrac{2}{3}$%
39. $37\dfrac{1}{2}$%　　**41.** 150%　　**43.** 240%
45. 2.66　　**47.** 42　　**49.** 80%　　**51.** 60
53. 115%　　**55.** 90　　**57.** 15 feet by $19\dfrac{1}{2}$ feet
59. 330 miles　　**61.** 60 centimeters
63. 7.5 pounds　　**65.** $33\dfrac{1}{3}$ pounds　　**67.** 90,000
69. 15 inches by 10 inches　　**71.** $300
73. $150,000　　**77.** All real numbers except 2
79. $\{0\}$　　**81.** All real numbers

Problem Set 4.2 (page 151)

1. $\{1.11\}$　　**3.** $\{6.6\}$　　**5.** $\{.48\}$　　**7.** $\{80\}$
9. $\{3\}$　　**11.** $\{50\}$　　**13.** $\{70\}$　　**15.** $\{200\}$
17. $\{450\}$　　**19.** $\{150\}$　　**21.** $\{2200\}$
23. $50　　**25.** $36　　**27.** $20.80
29. 30%　　**31.** $8.50　　**33.** $4.65
35. $1000　　**37.** 40%
39. $400 at 9% and $650 at 10%
41. $500 at 9% and $700 at 12%　　**43.** $4000
45. $800 at 10% and $1500 at 12%
47. $3000 at 8% and $2400 at 10%
51. Yes, if the profit is figured as a percent of the selling price.
53. Yes　　**55.** $\{1.625\}$　　**57.** $\{350\}$
59. $\{.06\}$　　**61.** $\{15.4\}$

Problem Set 4.3 (page 160)

1. 7   **3.** 500   **5.** 20   **7.** 48   **9.** 9

11. 46 centimeters   **13.** 15 inches

15. 504 square feet   **17.** $2   **19.** 7 inches

21. 150π square centimeters

23. $\frac{1}{4}\pi$ square yards

25. $S = 324\pi$ square inches and $V = 972\pi$ cubic inches

27. $V = 1152\pi$ cubic feet and $S = 416\pi$ square feet

29. 12 inches   **31.** 8 feet   **33.** $h = \dfrac{V}{B}$

35. $B = \dfrac{3V}{h}$ **37.** $w = \dfrac{P - 21}{2}$ **39.** $h = \dfrac{3V}{\pi r^2}$

41. $C = \dfrac{5}{9}(F - 32)$ **43.** $h = \dfrac{A - 2\pi r^2}{2\pi r}$

45. $x = \dfrac{9 - 7y}{3}$ **47.** $y = \dfrac{9x - 13}{6}$

49. $x = \dfrac{11y - 14}{2}$ **51.** $x = \dfrac{-y - 4}{3}$

53. $y = \dfrac{3}{2}x$ **55.** $y = \dfrac{ax - c}{b}$

57. $x = \dfrac{2y - 22}{5}$ **59.** $y = mx + b$

65. 125.6 square centimeters

67. 245 square centimeters   **69.** 65 cubic inches

Problem Set 4.4 (page 168)

1. $\left\{8\frac{1}{3}\right\}$ **3.** $\{16\}$ **5.** $\{25\}$ **7.** $\{7\}$

9. $\{24\}$ **11.** $\{4\}$ **13.** $12\frac{1}{2}$ years

15. 20 years

17. The width is 14 inches and the length is 42 inches.

19. The width is 12 centimeters and the length is 34 centimeters.

21. 80 square inches

23. 24 feet, 31 feet, and 45 feet

25. 6 centimeters, 19 centimeters, and 21 centimeters

27. 12 centimeters   **29.** 7 centimeters

31. 9 hours   **33.** $2\frac{1}{2}$ hours

35. 55 miles per hour

37. 64 and 72 miles per hour   **39.** 60 miles

Problem Set 4.5 (page 173)

1. $\{15\}$ **3.** $\left\{\dfrac{20}{7}\right\}$ **5.** $\left\{\dfrac{15}{4}\right\}$ **7.** $\left\{\dfrac{5}{3}\right\}$

9. $\{2\}$ **11.** $\left\{\dfrac{33}{10}\right\}$ **13.** 12.5 milliliters

15. 15 centiliters

17. $7\frac{1}{2}$ quarts of the 30% solution and $2\frac{1}{2}$ quarts of the 50% solution

19. 5 gallons   **21.** 3 quarts   **23.** 12 gallons

25. 16.25%

27. The square is 6 inches by 6 inches and the rectangle is 9 inches long and 3 inches wide.

29. 40 minutes   **31.** Pam is 9 and Bill is 18.

33. Abby is 14 and her mother is 35.

35. 1 year old and 10 years old   **37.** 56 miles

Chapter 4 Review Problem Set (page 176)

1. $\left\{\dfrac{17}{12}\right\}$ **2.** $\{5\}$ **3.** $\{800\}$ **4.** $\{16\}$

5. $\{73\}$ **6.** 6   **7.** 25   **8.** $t = \dfrac{A - P}{Pr}$

9. $x = \dfrac{13 + 3y}{2}$ **10.** 77 square inches

11. 6 centimeters   **12.** 15 feet   **13.** 60%

14. 40 and 56   **15.** 40

16. 6 meters by 17 meters   **17.** $1\frac{1}{2}$ hours

18. 20 liters

19. 15 centimeters by 40 centimeters

20. 29 yards by 10 yards   **21.** 20°

22. 30 gallons   **23.** $900 at 9% and $1200 at 11%

24. $40   **25.** 35%   **26.** 34° and 99°

27. 5 hours   **28.** 18 gallons   **29.** 26%

30. 10 years ago

Chapter 4 Test (page 177)

1. $\{-22\}$ **2.** $\left\{-\dfrac{17}{18}\right\}$ **3.** $\{-77\}$ **4.** $\left\{\dfrac{4}{3}\right\}$

5. $\{14\}$ **6.** $\left\{\dfrac{12}{5}\right\}$ **7.** $\{100\}$ **8.** $\{70\}$

25. Not factorable　　**27.** $(7x + 3)(2x + 7)$

29. $(4x - 3)(5x - 4)$　　**31.** $(4n + 3)(4n - 5)$

33. $(6x - 5)(4x - 5)$　　**35.** $(2x + 9)(x + 8)$

37. $(3a + 1)(7a - 2)$　　**39.** $(12a + 5)(a - 3)$

41. $(2x + 3)(2x + 3)$　　**43.** $(2x - y)(3x - y)$

45. $(4x + 3y)(5x - 2y)$　　**47.** $(5x - 2)(x - 6)$

49. $(8x + 1)(x - 7)$　　**51.** $\left\{-6, -\dfrac{1}{2}\right\}$

53. $\left\{-\dfrac{2}{3}, -\dfrac{1}{4}\right\}$　　**55.** $\left\{\dfrac{1}{3}, 8\right\}$　　**57.** $\left\{\dfrac{2}{5}, \dfrac{7}{3}\right\}$

59. $\left\{-7, \dfrac{5}{6}\right\}$　　**61.** $\left\{-\dfrac{3}{8}, \dfrac{3}{2}\right\}$　　**63.** $\left\{-\dfrac{2}{3}, \dfrac{4}{3}\right\}$

65. $\left\{\dfrac{2}{5}, \dfrac{5}{2}\right\}$　　**67.** $\left\{-\dfrac{5}{2}, -\dfrac{2}{3}\right\}$　　**69.** $\left\{-\dfrac{5}{4}, \dfrac{1}{4}\right\}$

71. $\left\{-\dfrac{3}{7}, \dfrac{7}{5}\right\}$　　**73.** $\left\{\dfrac{5}{4}, 10\right\}$　　**75.** $\left\{-7, \dfrac{3}{7}\right\}$

77. $\left\{-\dfrac{5}{12}, 4\right\}$　　**79.** $\left\{-\dfrac{7}{2}, \dfrac{4}{9}\right\}$

Problem Set 6.5 (page 260)

1. $(x + 2)^2$　　**3.** $(x - 5)^2$　　**5.** $(3n + 2)^2$

7. $(4a - 1)^2$　　**9.** $(2 + 9x)^2$　　**11.** $(4x - 3y)^2$

13. $(2x + 1)(x + 8)$　　**15.** $2x(x - 6)(x + 6)$

17. $(n - 12)(n + 5)$　　**19.** Not factorable

21. $8(x^2 + 9)$　　**23.** $(3x + 5)^2$

25. $5(x + 2)(3x + 7)$　　**27.** $(4x - 3)(6x + 5)$

29. $(x + 5)(y - 8)$　　**31.** $(5x - y)(4x + 7y)$

33. $3(2x - 3)(4x + 9)$　　**35.** $6(2x^2 + x + 5)$

37. $5(x - 2)(x + 2)(x^2 + 4)$　　**39.** $(x + 6y)^2$

41. $\{0, 5\}$　　**43.** $\{-3, 12\}$　　**45.** $\{-2, 0, 2\}$

47. $\left\{-\dfrac{2}{3}, \dfrac{11}{2}\right\}$　　**49.** $\left\{\dfrac{1}{3}, \dfrac{3}{4}\right\}$　　**51.** $\{-3, -1\}$

53. $\{0, 6\}$　　**55.** $\{-4, 0, 6\}$　　**57.** $\left\{\dfrac{2}{5}, \dfrac{6}{5}\right\}$

59. $\{12, 16\}$　　**61.** $\left\{-\dfrac{10}{3}, 1\right\}$　　**63.** $\{0, 6\}$

65. $\left\{\dfrac{4}{3}\right\}$　　**67.** $\{-5, 0\}$　　**69.** $\left\{-\dfrac{4}{3}, \dfrac{5}{8}\right\}$

71. $\dfrac{5}{4}$ and 12 or -5 and -3

73. -1 and 1 or $-\dfrac{1}{2}$ and 2

75. 4 and 9 or $-\dfrac{24}{5}$ and $-\dfrac{43}{5}$

77. 6 rows and 9 chairs per row

79. One square is 6 feet by 6 feet and the other one is 18 feet by 18 feet.

81. 11 centimeters long and 5 centimeters wide

83. The side is 17 inches long and the altitude to that side is 6 inches long.

85. $1\dfrac{1}{2}$ inches　　**87.** 6 inches and 12 inches

Chapter 6 Review Problem Set (page 264)

1. $(x - 2)(x - 7)$　　**2.** $3x(x + 7)$

3. $(3x + 2)(3x - 2)$　　**4.** $(2x - 1)(2x + 5)$

5. $(5x - 6)^2$　　**6.** $n(n + 5)(n + 8)$

7. $(y + 12)(y - 1)$　　**8.** $3xy(y + 2x)$

9. $(x + 1)(x - 1)(x^2 + 1)$　　**10.** $(6n + 5)(3n - 1)$

11. Not factorable　　**12.** $(4x - 7)(x + 1)$

13. $3(n + 6)(n - 5)$　　**14.** $x(x + y)(x - y)$

15. $(2x - y)(x + 2y)$　　**16.** $2(n - 4)(2n + 5)$

17. $(x + y)(5 + a)$　　**18.** $(7t - 4)(3t + 1)$

19. $2x(x + 1)(x - 1)$　　**20.** $3x(x + 6)(x - 6)$

21. $(4x + 5)^2$　　**22.** $(y - 3)(x - 2)$

23. $(5x + y)(3x - 2y)$　　**24.** $n^2(2n - 1)(3n - 1)$

25. $\{-6, 2\}$　　**26.** $\{0, 11\}$　　**27.** $\left\{-4, \dfrac{5}{2}\right\}$

28. $\left\{-\dfrac{8}{3}, \dfrac{1}{3}\right\}$　　**29.** $\{-2, 2\}$　　**30.** $\left\{-\dfrac{5}{4}\right\}$

31. $\{-1, 0, 1\}$　　**32.** $\left\{-\dfrac{2}{7}, -\dfrac{9}{4}\right\}$　　**33.** $\{-7, 4\}$

34. $\{-5, 5\}$　　**35.** $\left\{-6, \dfrac{3}{5}\right\}$　　**36.** $\left\{-\dfrac{7}{2}, 1\right\}$

37. $\{-2, 0, 2\}$　　**38.** $\{8, 12\}$　　**39.** $\left\{-5, \dfrac{3}{4}\right\}$

40. $\{-2, 3\}$　　**41.** $\left\{-\dfrac{7}{3}, \dfrac{5}{2}\right\}$　　**42.** $\{-9, 6\}$

43. $\left\{-5, \dfrac{3}{2}\right\}$　　**44.** $\left\{\dfrac{4}{3}, \dfrac{5}{2}\right\}$

45. $-\dfrac{8}{3}$ and $-\dfrac{19}{3}$ or 4 and 7

46. The length is 8 centimeters and the width is 2 centimeters.

47. A 2-by-2-inch square and a 10-by-10-inch square

48. 8 by 15 by 17

49. $-\dfrac{13}{6}$ and -12 or 2 and 13   **50.** 7, 9, and 11

51. 4 shelves

52. A 5-by-5-yard square and a 5-by-40-yard rectangle

53. -18 and -17 or 17 and 18   **54.** 6 units

55. 2 meters and 7 meters   **56.** 9 and 11

57. 2 centimeters   **58.** 15 feet

Chapter 6 Test (page 265)

1. $(x + 5)(x - 2)$ **2.** $(x + 3)(x - 8)$

3. $2x(x + 1)(x - 1)$ **4.** $(x + 9)(x + 12)$

5. $3(2n + 1)(3n + 2)$ **6.** $(x + y)(a + 2b)$

7. $(4x - 3)(x + 5)$ **8.** $6(x^2 + 4)$

9. $2x(5x - 6)(3x - 4)$ **10.** $(7 - 2x)(4 + 3x)$

11. $\{-3, 3\}$ **12.** $\{-6, 1\}$ **13.** $\{0, 8\}$

14. $\left\{-\dfrac{5}{2}, \dfrac{2}{3}\right\}$ **15.** $\{-6, 2\}$ **16.** $\{-12, -4, 0\}$

17. $\{5, 9\}$ **18.** $\left\{-12, \dfrac{1}{3}\right\}$ **19.** $\left\{-4, \dfrac{2}{3}\right\}$

20. $\{-5, 0, 5\}$ **21.** $\left\{\dfrac{7}{5}\right\}$ **22.** 14 inches

23. 12 centimeters   **24.** 16 chairs per row

25. 12 units

Cumulative Review Problem Set (page 266)

1. 81   **2.** -32 **3.** $\dfrac{3}{2}$ **4.** 16

5. 36   **6.** $1\dfrac{3}{4}$ **7.** 0   **8.** $\dfrac{13}{40}$ **9.** $-\dfrac{2}{13}$

10. 5   **11.** -1 **12.** -33 **13.** $-15x^3y^7$

14. $12ab^7$ **15.** $-8x^6y^{15}$

16. $-6x^2y + 15xy^2$ **17.** $15x^2 - 11x + 2$

18. $21x^2 + 25x - 4$ **19.** $-2x^2 - 7x - 6$

20. $49 - 4y^2$ **21.** $3x^3 - 7x^2 - 2x + 8$

22. $2x^3 - 3x^2 - 13x + 20$

23. $8n^3 + 36n^2 + 54n + 27$

24. $1 - 6n + 12n^2 - 8n^3$

25. $2x^4 + x^3 - 4x^2 + 42x - 36$ **26.** $-4x^2y^2$

27. $14ab^2$ **28.** $7y - 8x^2 - 9x^3y^3$

29. $2x^2 - 4x - 7$ **30.** $x^2 + 6x + 4$

31. $-\dfrac{6}{x}$ **32.** $\dfrac{2}{x}$ **33.** $\dfrac{xy^2}{3}$ **34.** $\dfrac{z^2}{x^2y^4}$

35. .12   **36.** .0000000018   **37.** 200

38. $7(x + 2)(x - 2)$ **39.** $(2c - d)(a + b)$

40. $(2x + 5)(3x - 7)$ **41.** $3(x + 6)(x - 2)$

42. Not factorable

43. $(2x + 1)(2x - 1)(4x^2 + 1)$ **44.** $\{13\}$

45. $\left\{\dfrac{9}{14}\right\}$ **46.** $\{500\}$ **47.** $\left\{0, \dfrac{3}{2}\right\}$

48. $\{-5, 5\}$ **49.** $\{-8, 6\}$ **50.** $\left\{-\dfrac{2}{7}, \dfrac{1}{2}\right\}$

51. $\left\{-\dfrac{3}{4}, 0, 1\right\}$ **52.** $\{-6, 4\}$ **53.** $\left\{-\dfrac{10}{3}, \dfrac{3}{5}\right\}$

54. $\{-2, 4\}$ **55.** $\{x \mid x < -2\}$ **56.** $\left\{x \mid \geq \dfrac{12}{7}\right\}$

57. $\{x \mid x \geq 300\}$ **58.** 3   **59.** 40

60. 8 dimes and 10 quarters

61. \$700 at 8% and \$800 at 9%   **62.** 3 gallons

63. $3\dfrac{1}{2}$ hours

64. The length is 15 meters and the width is 7 meters.

65. The length is 16 feet and the width is 9 feet.

66. 4 inches and 7 inches

67. 12 centimeters, 16 centimeters, and 20 centimeters

CHAPTER 7

Problem Set 7.1 (page 274)

1. $\dfrac{3x}{7y}$ **3.** $\dfrac{3y}{8}$ **5.** $-\dfrac{3xy}{5}$ **7.** $\dfrac{3y}{4x^2}$

9. $-\dfrac{2b^2}{9}$ **11.** $\dfrac{4xy}{9z}$ **13.** $\dfrac{y}{x - 2}$

15. $\dfrac{2x + 3y}{3}$ **17.** $\dfrac{x + 2}{x - 7}$ **19.** -1

21. -3 **23.** $-4x$ **25.** $\dfrac{x + 1}{3x}$

27. $\dfrac{x + y}{x}$ **29.** $\dfrac{x(2x - 5y)}{2(x + 4y)}$ **31.** $\dfrac{n}{n + 1}$

33. $\dfrac{2n - 1}{n - 3}$ **35.** $\dfrac{2x + 7}{3x + 4}$ **37.** $\dfrac{3}{4(x - 1)}$

39. $\dfrac{x + 9}{x + 3}$ **41.** $\dfrac{2a - 1}{3a - 1}$ **43.** $\dfrac{x + 3y}{2x + y}$

45. $-\dfrac{x-3}{x}$ **47.** $\dfrac{n+7}{8}$ **49.** $\dfrac{2n-3}{n+1}$

51. $\dfrac{y-12}{y-14}$ **53.** $\dfrac{1+x}{x}$ **55.** $\dfrac{2+x}{4+5x}$

57. $-\dfrac{x+9}{x+6}$ **59.** $-\dfrac{1}{2}$ **63.** $\dfrac{y-3}{y+5}$

65. $\dfrac{x+1}{x+5}$ **67.** $\dfrac{1}{x^6}$ **69.** $\dfrac{1}{x^3y^2}$ **71.** $-\dfrac{4}{a^3}$

Problem Set 7.2 (page 279)

1. $\dfrac{1}{6}$ **3.** $-\dfrac{9}{14}$ **5.** $-\dfrac{17}{19}$ **7.** $\dfrac{2x^2}{7y}$

9. $-\dfrac{3n^2}{10}$ **11.** $2a$ **13.** $\dfrac{10b^3}{27}$ **15.** $\dfrac{3x^2y}{2}$

17. $-\dfrac{4}{5b^3}$ **19.** $\dfrac{s}{17}$ **21.** $\dfrac{x-y}{x}$ **23.** $\dfrac{1}{5}$

25. $\dfrac{3a}{14}$ **27.** $\dfrac{5(x+6)}{x+9}$ **29.** $\dfrac{3y(2x-y)}{2(x+y)}$

31. $\dfrac{a}{(5a+2)(3a+1)}$ **33.** $\dfrac{2x(x+4)}{5y(x+8)}$

35. $\dfrac{5}{x+y}$ **37.** $\dfrac{4t+1}{3(4t+3)}$ **39.** 4

41. $\dfrac{y^2}{3}$ **43.** $\dfrac{x-1}{y^2(1-y)(x-y)}$ **45.** $\dfrac{x+6}{x}$

Problem Set 7.3 (page 284)

1. $\dfrac{17}{x}$ **3.** $\dfrac{2}{3x}$ **5.** $\dfrac{4}{n}$ **7.** $-\dfrac{1}{x^2}$

9. $\dfrac{x+4}{x}$ **11.** $-\dfrac{3}{x-1}$ **13.** 1

15. $\dfrac{5t+2}{4}$ **17.** $\dfrac{3a+8}{3}$ **19.** $\dfrac{5n+4}{4}$

21. $-n-1$ **23.** $\dfrac{-3x-5}{7x}$ **25.** $\dfrac{9}{4}$

27. 3 **29.** $-\dfrac{1}{7x}$ **31.** $a-2$ **33.** $\dfrac{3}{x-6}$

35. $\dfrac{13x}{8}$ **37.** $-\dfrac{3n}{4}$ **39.** $\dfrac{11y}{12}$ **41.** $\dfrac{47x}{21}$

43. $\dfrac{14x}{15}$ **45.** $\dfrac{13n}{24}$ **47.** $\dfrac{7x-14}{10}$

49. $\dfrac{4x}{9}$ **51.** $\dfrac{18n+11}{12}$ **53.** $\dfrac{9n-14}{18}$

55. $\dfrac{7x}{24}$ **57.** $\dfrac{-23x-18}{60}$ **59.** $\dfrac{19}{24x}$

61. $\dfrac{1}{18y}$ **63.** $\dfrac{20x-33}{48x^2}$ **65.** $\dfrac{25}{12x}$

67. $\dfrac{10x-35}{x(x-5)}$ **69.** $\dfrac{-n+3}{n(n-1)}$ **71.** $\dfrac{-2n+16}{n(n+4)}$

73. $\dfrac{6}{x(2x+1)}$ **75.** $\dfrac{10x+12}{(x+4)(x-3)}$

77. $\dfrac{-6x+21}{(x-2)(x+1)}$ **79.** $\dfrac{x+7}{(2x-1)(3x+1)}$

85. $\dfrac{4}{x-3}$ **87.** $\dfrac{-6}{a-1}$ **89.** $\dfrac{n+3}{2n-1}$

Problem Set 7.4 (page 292)

1. $\dfrac{3x-8}{x(x-4)}$ **3.** $\dfrac{-5x-3}{x(x+2)}$ **5.** $\dfrac{8n-50}{n(n-6)}$

7. $-\dfrac{4}{n+1}$ **9.** $\dfrac{5x-7}{2x(x-1)}$

11. $\dfrac{5x-17}{(x+4)(x-4)}$ **13.** $\dfrac{4}{x+1}$

15. $\dfrac{11a-6}{a(a-2)(a+2)}$ **17.** $\dfrac{12}{x(x-6)(x+6)}$

19. $\dfrac{n+8}{3(n+4)(n-4)}$ **21.** $\dfrac{19x}{6(3x+2)}$

23. $\dfrac{-2x+17}{15(x+1)}$ **25.** $\dfrac{5x+6}{(x+3)(x+4)(x-3)}$

27. $\dfrac{x^2-10x-20}{(x+2)(x+4)(x-5)}$ **29.** $\dfrac{a-b}{ab}$

31. $\dfrac{8x-14}{(x-5)(x+5)}$ **33.** $\dfrac{15x+4}{x(x-2)(x+2)}$

35. $\dfrac{4x-4}{(x+5)(x+2)}$ **37.** $\dfrac{-2x+6}{(3x-5)(x+4)}$

39. $\dfrac{3}{x+4}$ **41.** $-\dfrac{1}{2}$ **43.** $\dfrac{10}{3}$ **45.** $\dfrac{28}{27}$

47. $\dfrac{y}{3x}$ **49.** $\dfrac{2y+3x}{5y-x}$ **51.** $\dfrac{x^2-4y}{7xy-3x^2}$

53. $\dfrac{6+2x}{3+4x}$ **55.** $\dfrac{9-24x}{10x+42}$ **57.** $\dfrac{x^2+2x}{6x+4}$

59. $\dfrac{-2x+3}{4x-1}$ **61.** $\dfrac{m}{40}$ **63.** $\dfrac{k}{r}$ **65.** $\dfrac{d}{l}$

67. $\dfrac{34}{n}$ **69.** $\dfrac{47}{l}$ **71.** $\dfrac{96}{b}$

73. $\dfrac{-n^2 + n - 1}{n - 1}$ **75.** $\dfrac{3x^2 - 4x + 2}{4x - 2}$

49. 40 words per minute for Paul and 60 words per minute for Amelia

53. $\{0\}$

Problem Set 7.5 (page 299)

1. $\{12\}$ **3.** $\left\{-\dfrac{2}{21}\right\}$ **5.** $\{4\}$ **7.** $\{-1\}$

9. $\{3\}$ **11.** $\{-38\}$ **13.** $\left\{-\dfrac{13}{8}\right\}$ **15.** $\{2\}$

17. $\{6\}$ **19.** $\left\{\dfrac{5}{18}\right\}$ **21.** $\left\{-\dfrac{7}{10}\right\}$

23. $\left\{-\dfrac{1}{3}\right\}$ **25.** $\{8\}$ **27.** $\{37\}$ **29.** $\varnothing$

31. $\{39\}$ **33.** $\left\{-\dfrac{5}{4}\right\}$ **35.** $\{-6\}$

37. $\left\{-\dfrac{2}{3}\right\}$ **39.** $\left\{\dfrac{12}{7}\right\}$ **41.** $\dfrac{40}{48}$ **43.** 10

45. 7 and 58   **47.** $\dfrac{14}{10}$ **49.** 15 miles per hour

51. 50 miles per hour for Dave and 54 miles per hour for Kent

57. All real numbers except 0

59. All real numbers except -2 and 3

Problem Set 7.6 (page 308)

1. $\{-2\}$ **3.** $\{5\}$ **5.** $\left\{\dfrac{9}{2}\right\}$ **7.** $\left\{-\dfrac{49}{10}\right\}$

9. $\left\{\dfrac{2}{3}\right\}$ **11.** $\left\{\dfrac{4}{3}\right\}$ **13.** $\left\{\dfrac{3}{2}\right\}$ **15.** $\left\{\dfrac{11}{3}\right\}$

17. $\left\{\dfrac{13}{4}\right\}$ **19.** $\left\{-1, -\dfrac{5}{8}\right\}$ **21.** $\left\{\dfrac{1}{4}, 4\right\}$

23. $\left\{-\dfrac{5}{2}, 6\right\}$ **25.** $\{5\}$ **27.** $\{-21\}$ **29.** $\{2\}$

31. $\{-8, 1\}$ **33.** $\dfrac{1}{2}$ or 4   **35.** $-\dfrac{2}{5}$ or $\dfrac{5}{2}$

37. 17 miles per hour for Tom and 20 miles per hour for Celia

39. 16 miles per hour for the trip out and 12 miles per hour for the return trip

41. 30 minutes

43. 60 minutes for Mike and 120 minutes for Barry

45. 15 hours   **47.** 4 minutes

Chapter 7 Review Problem Set (page 311)

1. $\dfrac{7x^2}{9y^2}$ **2.** $\dfrac{x}{x + 3}$ **3.** $\dfrac{3n + 5}{n + 2}$

4. $\dfrac{4a + 3}{5a - 2}$ **5.** $\dfrac{3x}{8}$ **6.** $x(x - 3)$

7. $\dfrac{n - 7}{n^2}$ **8.** $\dfrac{2a + 1}{a + 6}$ **9.** $\dfrac{22x - 19}{20}$

10. $\dfrac{43x - 3}{12x^2}$ **11.** $\dfrac{10n - 7}{n(n - 1)}$

12. $\dfrac{-a + 8}{(a - 4)(a - 2)}$ **13.** $\dfrac{5x + 9}{4x(x - 3)}$

14. $\dfrac{5x - 4}{(x + 5)(x - 5)(x + 2)}$ **15.** $\dfrac{6x - 37}{(x - 7)(x + 3)}$

16. $\dfrac{3y^2 - 4x}{4xy + 5y^2}$ **17.** $\dfrac{2y - xy}{3xy + 5x}$ **18.** $\left\{\dfrac{20}{17}\right\}$

19. $\left\{-\dfrac{61}{60}\right\}$ **20.** $\{9\}$ **21.** $\left\{\dfrac{28}{3}\right\}$ **22.** $\left\{\dfrac{3}{4}\right\}$

23. $\{1\}$ **24.** $\{-7\}$ **25.** $\left\{\dfrac{1}{7}\right\}$ **26.** $\left\{\dfrac{1}{2}, 2\right\}$

27. $\{-5, 10\}$ **28.** $\left\{-\dfrac{1}{5}\right\}$ **29.** $\{-1\}$

30. 7 and 68

31. Becky $2\dfrac{2}{3}$ hours, Nancy 8 hours   **32.** 1 or 2

33. $\dfrac{36}{72}$

34. Todd's rate is 15 miles per hour and Lanette's rate is 22 miles per hour.

35. 8 miles per hour   **36.** 60 minutes

37. Corinne's rate is 56 words per minute and Sue's rate is 50 words per minute.

Chapter 7 Test (page 313)

1. $\dfrac{8x^2y}{9}$ **2.** $\dfrac{x}{x - 6}$ **3.** $\dfrac{2n + 1}{3n + 4}$

4. $\dfrac{2x - 3}{x - 5}$ **5.** $2x^2y^2$ **6.** $\dfrac{x - 2}{x + 3}$

7. $\dfrac{(x + 4)^2}{x(x + 7)}$ **8.** $\dfrac{6x + 5}{24}$ **9.** $\dfrac{9n - 4}{30}$

10. $\dfrac{41 - 15x}{18x}$ **11.** $\dfrac{2n - 6}{n(n - 1)}$ **12.** $\dfrac{5x - 18}{4x(x + 6)}$

13. $\dfrac{5x - 11}{(x - 4)(x + 8)}$ **14.** $\dfrac{-13x + 43}{(2x - 5)(3x + 4)(x - 6)}$

15. $\{-5\}$ **16.** $\left\{-\dfrac{19}{16}\right\}$ **17.** $\left\{\dfrac{4}{3}, 3\right\}$

18. $\{-6, 8\}$ **19.** $\left\{-\dfrac{1}{5}, 2\right\}$ **20.** $\{-23\}$

21. $\{2\}$ **22.** $\left\{-\dfrac{3}{2}\right\}$ **23.** $\dfrac{2}{3}$ or 3

24. 14 miles per hour   **25.** 12 minutes

Cumulative Review Problem Set (page 314)

1. $\dfrac{5}{2}$ **2.** 6   **3.** $\dfrac{17}{12}$ **4.** .6   **5.** 20

6. 0   **7.** 2   **8.** -3 **9.** $\dfrac{1}{27}$ **10.** $\dfrac{3}{2}$

11. 1   **12.** $\dfrac{12}{7}$ **13.** $-\dfrac{1}{16}$ **14.** $\dfrac{16}{9}$

15. $\dfrac{4}{25}$ **16.** $-\dfrac{1}{27}$ **17.** $\dfrac{19}{10x}$ **18.** $\dfrac{2y}{3x}$

19. $\dfrac{7x - 2}{(x - 6)(x + 4)}$ **20.** $\dfrac{-x + 12}{x^2(x - 4)}$

21. $\dfrac{x - 7}{3y}$ **22.** $\dfrac{-3x - 4}{(x - 4)(x + 3)}$ **23.** $-35x^5y^5$

24. $81a^2b^6$ **25.** $-15n^4 - 18n^3 + 6n^2$

26. $15x^2 + 17x - 4$ **27.** $4x^2 + 20x + 25$

28. $2x^3 + x^2 - 7x - 2$

29. $x^4 + x^3 - 6x^2 + x + 3$ **30.** $-6x^2 + 11x + 7$

31. $3xy - 6x^3y^3$ **32.** $7x + 4$

33. $3x(x^2 + 5x + 9)$ **34.** $(x + 10)(x - 10)$

35. $(5x - 2)(x - 4)$ **36.** $(4x + 7)(2x - 9)$

37. $(n + 16)(n + 9)$ **38.** $(x + y)(n - 2)$

39. $3x(x + 1)(x - 1)$ **40.** $2x(x - 9)(x + 6)$

41. $(6x - 5)^2$ **42.** $(3x + y)(x - 2y)$

43. $\left\{\dfrac{16}{3}\right\}$ **44.** $\{-11, 0\}$ **45.** $\left\{\dfrac{1}{14}\right\}$

46. $\{15\}$ **47.** $\{-1, 1\}$ **48.** $\{-6, 1\}$

49. $\{2\}$ **50.** $\left\{\dfrac{11}{12}\right\}$ **51.** $\{1, 2\}$ **52.** $\left\{-\dfrac{1}{18}\right\}$

53. $\left\{-\dfrac{7}{2}, \dfrac{1}{3}\right\}$ **54.** $\left\{\dfrac{1}{2}, 8\right\}$ **55.** $\{-9, 2\}$

56. $\left\{\dfrac{16}{5}\right\}$ **57.** $\left\{\dfrac{1}{2}\right\}$

CHAPTER 8

Problem Set 8.1 (page 323)

1. $y = \dfrac{13 - 3x}{7}$ **3.** $x = 3y + 9$

5. $y = \dfrac{x + 14}{5}$ **7.** $x = \dfrac{y - 7}{3}$ **9.** $y = \dfrac{2x - 5}{3}$

11.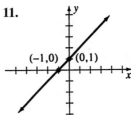

13.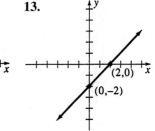

15.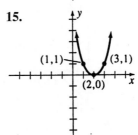

17.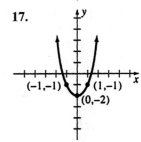

19.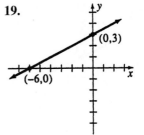

21.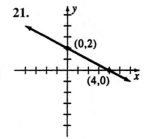

23.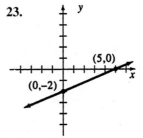

25.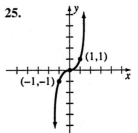

27.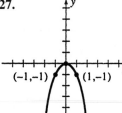

29.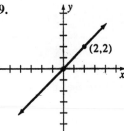
(2,2)

13.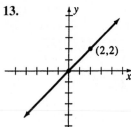
(2,2)

15.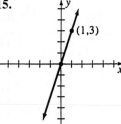
(1,3)

31.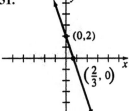
(0,2)
$\left(\frac{2}{3}, 0\right)$

33.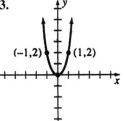
(−1,2) (1,2)

17.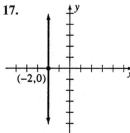
(−2,0)

19.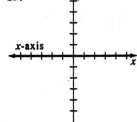
x-axis

Problem Set 8.2 (page 329)

1.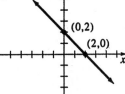
(0,2)
(2,0)

3.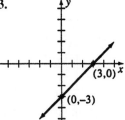
(3,0)
(0,−3)

21.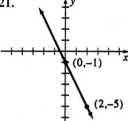
(0,−1)
(2,−5)

23.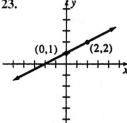
(0,1) (2,2)

5.
(0,4)
(−4,0)

7.
(0,1)
(2,0)

25.
(0,−2) (3,−3)

27.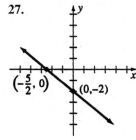
$\left(-\frac{5}{2}, 0\right)$
(0,−2)

9.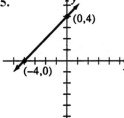
(2,0)
(0,−6)

11.
(2,0)
(0,−3)

29.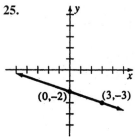
(2,0)
(0,−4)

31.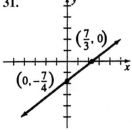
$\left(\frac{7}{3}, 0\right)$
$\left(0, -\frac{7}{4}\right)$

33.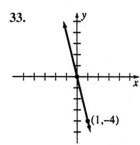

35.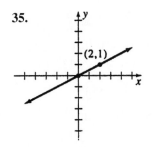

41.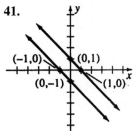

43.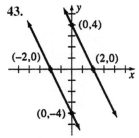

45. They appear to be parallel lines.

Problem Set 8.3 (page 335)

1. $\dfrac{3}{4}$ **3.** $\dfrac{7}{5}$ **5.** $-\dfrac{6}{5}$ **7.** $-\dfrac{10}{3}$ **9.** $\dfrac{3}{4}$

11. 0   **13.** $-\dfrac{3}{2}$ **15.** Undefined   **17.** 1

19. $\dfrac{b-d}{a-c}$ or $\dfrac{d-b}{c-a}$ **21.** 4   **23.** -6

25–31. Answers will vary   **33.** Negative
35. Positive   **37.** Zero   **39.** Negative

41. $-\dfrac{3}{2}$ **43.** $\dfrac{5}{4}$ **45.** $-\dfrac{1}{5}$ **47.** 2

49. 0   **51.** $\dfrac{2}{5}$ **53.** $\dfrac{6}{5}$ **55.** -3 **57.** 4

59. $\dfrac{2}{3}$ **61.** 5.1%   **63.** 32 cm

65. 1.0 feet

Problem Set 8.4 (page 343)

1. $2x - 3y = -5$ **3.** $x - 2y = 7$
5. $x + 3y = 20$ **7.** $y = -7$ **9.** $4x + 9y - 0$
11. $3x - y = -16$ **13.** $7x - 5y = -1$
15. $3x + 2y = 5$ **17.** $x - y = 1$
19. $5x - 3y = 0$ **21.** $4x + 7y = 28$
23. $y = \dfrac{3}{5}x + 2$ **25.** $y = 2x - 1$

27. $y = -\dfrac{1}{6}x - 4$ **29.** $y = -x + \dfrac{5}{2}$

31. $y = -\dfrac{5}{9}x - \dfrac{1}{2}$ **33.** $m = -2, b = -5$

35. $m = \dfrac{3}{5}, b = -3$ **37.** $m = \dfrac{4}{9}, b = 2$

39. $m = \dfrac{3}{4}, b = -4$ **41.** $m = -\dfrac{2}{11}, b = -1$

43. $m = -\dfrac{9}{7}, b = 0$ **49.** Perpendicular

51. Intersecting lines that are not perpendicular
53. Parallel   **55.** $2x - 3y = -1$

Problem Set 8.5 (page 349)

1. No   **3.** Yes   **5.** Yes   **7.** Yes
9. No   **11.** $\{(2, -1)\}$ **13.** $\{(2, 1)\}$ **15.** $\varnothing$
17. $\{(0, 0)\}$ **19.** $\{(1, -1)\}$
21. Infinitely many   **23.** $\{(1, 3)\}$
25. $\{(3, -2)\}$ **27.** $\{(2, 4)\}$ **29.** $\{(-2, -3)\}$

Problem Set 8.6 (page 356)

1. $\{(6, 8)\}$ **3.** $\{(-5, -4)\}$ **5.** $\{(-6, 12)\}$

7. $\{(5, -2)\}$ **9.** $\left\{\left(\dfrac{11}{4}, \dfrac{9}{8}\right)\right\}$ **11.** $\left\{\left(-\dfrac{2}{3}, \dfrac{2}{3}\right)\right\}$

13. $\{(-4, 5)\}$ **15.** $\{(4, 1)\}$ **17.** $\left\{\left(\dfrac{3}{2}, -3\right)\right\}$

19. $\left\{\left(-\dfrac{18}{71}, \dfrac{5}{71}\right)\right\}$ **21.** $\{(250, 500)\}$

23. $\{(100, 200)\}$ **25.** 9 and 21   **27.** 8 and 15
29. 6 and 12
31. \$.15 per lemon and \$.30 per apple
33. 7 dimes and 3 quarters
35. 14 books at \$12 each and 21 books at \$14 each
37. 4 gallons of 10% and 6 gallons of 15%
39. \$500 at 10% and \$800 at 12%   **43.** $\{(12, 24)\}$
45. $\varnothing$

Problem Set 8.7 (page 366)

1. $\{(5, 9)\}$ **3.** $\{(-4, 10)\}$ **5.** $\left\{\left(\dfrac{3}{2}, 4\right)\right\}$

7. $\left\{\left(-\dfrac{1}{3}, \dfrac{4}{3}\right)\right\}$ **9.** $\{(18, 24)\}$ **11.** $\{(-10, -15)\}$

13. $\varnothing$ **15.** $\{(-2, 6)\}$ **17.** $\{(-6, -13)\}$

19. $\left\{\left(\dfrac{9}{8}, \dfrac{7}{12}\right)\right\}$ **21.** $\left\{\left(\dfrac{6}{31}, \dfrac{9}{31}\right)\right\}$

23. $\{(100, 400)\}$ **25.** $\{(3, 10)\}$ **27.** $\{(-2, -6)\}$

29. $\left\{\left(-\dfrac{6}{7}, \dfrac{6}{7}\right)\right\}$ **31.** $\{(10, 12)\}$

33. $\left\{\left(\dfrac{3}{5}, \dfrac{12}{5}\right)\right\}$ **35.** $\left\{\left(-\dfrac{5}{2}, 6\right)\right\}$ **37.** $\left\{\left(\dfrac{1}{2}, 4\right)\right\}$

39. Infinitely many solutions   **41.** $\left\{\left(5, \dfrac{5}{2}\right)\right\}$

43. $\{(12, 4)\}$ **45.** $\left\{\left(-\dfrac{1}{11}, -\dfrac{10}{11}\right)\right\}$ **47.** 12 and 34

49. 35 double rooms and 15 single rooms   **51.** 45

53. 18 dimes and 41 quarters   **55.** 93

57. $150 at 8% and $400 at 9%

59. 7.5 liters of 30% and 2.5 liters of 70%

65. $\{0, 0\}$ **67.** $\varnothing$

7. $-\dfrac{9}{5}$ **8.** $\dfrac{5}{6}$ **9.** $5x + 7y = -11$

10. $8x - 3y = 1$ **11.** $2x - 9y = 9$

12. $x = 2$ **13.** $\{(3, -2)\}$ **14.** $\{(7, 13)\}$

15. $\{(16, -5)\}$ **16.** $\left\{\left(\dfrac{41}{23}, \dfrac{19}{23}\right)\right\}$ **17.** $\{(10, 25)\}$

18. $\{(-6, -8)\}$ **19.** $\{(400, 600)\}$ **20.** $\varnothing$

21. $\left\{\left(\dfrac{5}{16}, -\dfrac{17}{16}\right)\right\}$ **22.** $t = 4$ and $u = 8$

23. $t = 8$ and $u = 4$ **24.** $t = 3$ and $u = 7$

25. $\{(-9, 6)\}$ **26.** 38 and 75

27. $250 at 9% and $300 at 11%

28. 18 nickels and 25 dimes

29. Length of 19 inches and width of 6 inches

30. 49   **31.** 36   **32.** 32° and 58°

33. 50° and 130°

34. $1.15 for a cheeseburger and $.75 for a milkshake

35. $.89 for prune juice and $.59 for tomato juice

Chapter 8 Review Problem Set (page 370)

1.   **2.**

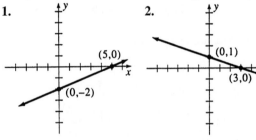

3.   **4.**

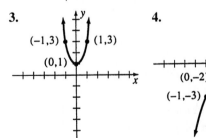

5.   **6.**

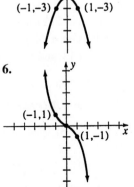

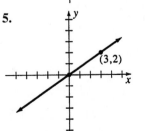

Chapter 8 Test (page 372)

1.   **2.**

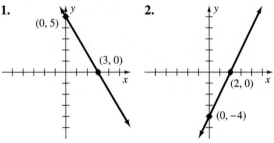

3.   **4.**

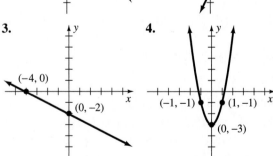

5. Yes   **6.** Yes   **7.** -3 **8.** -4 and 4

9. $-\dfrac{3}{7}$ **10.** $\dfrac{4}{5}$ **11.** $3x + 5y = 20$

12. $4x - 9y = 34$ **13.** $3x - 2y = 0$

14.

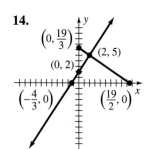

15. $\{(3, 4)\}$ **16.** $\{(-4, 6)\}$ **17.** $\{(-1, -4)\}$
18. $\{(3, -6)\}$ **19.** 29   **20.** 13 inches

CHAPTER 9

Problem Set 9.1 (page 379)

1. 7   **3.** -8 **5.** 11   **7.** 60   **9.** -40
11. 80   **13.** 18   **15.** $\frac{5}{3}$ **17.** .4
19. 24   **21.** 48   **23.** 28   **25.** 65
27. 58   **29.** 4.36   **31.** 7.07   **33.** 8.66
35. 9.75   **37.** 66   **39.** 34   **41.** 97
43. 81   **45.** 58   **47.** $21\sqrt{2}$ **49.** $8\sqrt{7}$
51. $-9\sqrt{3}$ **53.** $6\sqrt{5}$ **55.** $-\sqrt{2} + 2\sqrt{3}$
57. $-9\sqrt{7} - 2\sqrt{10}$ **59.** 17.3   **61.** 13.4
63. -1.4 **65.** 26.5   **67.** 4.1   **69.** 2.1
71. -29.8
73. 1.6 seconds; 2.1 seconds; 2.2 seconds
75. 2.2 seconds; 2.8 seconds; 18.2 seconds

Problem Set 9.2 (page 385)

1. $2\sqrt{6}$ **3.** $3\sqrt{2}$ **5.** $3\sqrt{3}$ **7.** $2\sqrt{10}$
9. $2\sqrt{7}$ **11.** $4\sqrt{5}$ **13.** $3\sqrt{13}$
15. $24\sqrt{2}$ **17.** $15\sqrt{3}$ **19.** $-10\sqrt{5}$
21. $-32\sqrt{6}$ **23.** $3\sqrt{2}$ **25.** $\frac{3}{2}\sqrt{3}$
27. $-2\sqrt{5}$ **29.** $-5\sqrt{6}$ **31.** $xy\sqrt{y}$
33. $x\sqrt{2y}$ **35.** $2x\sqrt{2}$ **37.** $3a\sqrt{3ab}$
39. $12b^2\sqrt{ab}$ **41.** $3x^2y\sqrt{7}$ **43.** $12x\sqrt{3}$

45. $-36x^3\sqrt{2x}$ **47.** $\frac{2}{3}\sqrt{6xy}$
49. $\frac{3}{2}x^2y^3\sqrt{13x}$ **51.** $-\frac{26}{3}a^4$ **53.** $33\sqrt{2}$
55. $3\sqrt{5}$ **57.** $27\sqrt{2}$ **59.** $-2\sqrt{7}$
61. $9\sqrt{3}$ **63.** $2\sqrt{5}$ **65.** $-15\sqrt{2} - 55\sqrt{5}$
69. (a) $9\sqrt{2}$ (c) $5\sqrt{11}$

Problem Set 9.3 (page 390)

1. $\frac{4}{5}$ **3.** -3 **5.** $\frac{1}{8}$ **7.** $\frac{13}{12}$ **9.** $-\frac{5}{16}$
11. $\frac{\sqrt{19}}{5}$ **13.** $\frac{2\sqrt{2}}{7}$ **15.** $\frac{2\sqrt{7}}{3}$
17. $\frac{\sqrt{3}}{3}$ **19.** $\frac{\sqrt{6}}{2}$ **21.** $\frac{\sqrt{10}}{4}$ **23.** $\sqrt{7}$
25. 3   **27.** $\frac{\sqrt{10}}{6}$ **29.** $\frac{2\sqrt{3}}{9}$ **31.** $\frac{\sqrt{6}}{12}$
33. $\frac{2\sqrt{15}}{5}$ **35.** $\frac{4\sqrt{6}}{9}$ **37.** $\frac{\sqrt{21}}{8}$
39. $\frac{\sqrt{37}}{3}$ **41.** $\frac{3\sqrt{x}}{x}$ **43.** $\frac{5\sqrt{2x}}{2x}$
45. $\frac{\sqrt{3x}}{x}$ **47.** $\frac{2\sqrt{3}}{x}$ **49.** $\frac{\sqrt{10xy}}{5y}$
51. $\frac{\sqrt{15xy}}{9y}$ **53.** $\frac{x\sqrt{xy}}{2y}$ **55.** $\frac{3\sqrt{x}}{x^2}$
57. $\frac{4\sqrt{x}}{x^4}$ **59.** $\frac{3\sqrt{xy}}{2y^2}$ **61.** $\frac{22\sqrt{3}}{3}$
63. $\frac{19\sqrt{10}}{5}$ **65.** $-3\sqrt{5}$ **67.** $-\frac{14\sqrt{6}}{3}$
69. $-6\sqrt{3}$ **71.** $-7\sqrt{15}$

Problem Set 9.4 (page 396)

1. $\sqrt{35}$ **3.** $4\sqrt{3}$ **5.** $5\sqrt{2}$ **7.** 9
9. $4\sqrt{6}$ **11.** $15\sqrt{21}$ **13.** $-6\sqrt{14}$
15. 72   **17.** $40\sqrt{6}$ **19.** $24\sqrt{5}$
21. $\sqrt{6} + \sqrt{10}$ **23.** $2\sqrt{3} - 5\sqrt{6}$
25. $3\sqrt{2} + \sqrt{21}$ **27.** $6\sqrt{2} - 4\sqrt{6}$
29. $4\sqrt{6} - 8\sqrt{15}$ **31.** $56 + 15\sqrt{2}$
33. $-9 - 2\sqrt{6}$ **35.** $3\sqrt{2} + 2\sqrt{6} + 4\sqrt{3} + 6$

37. 15 **39.** 15 **41.** 51 **43.** $x\sqrt{y}$

45. $3\sqrt{2xy}$ **47.** $12a\sqrt{b}$

49. $x\sqrt{6} - 2\sqrt{3xy}$ **51.** $x + 2\sqrt{x} - 15$

53. $x - 49$ **55.** $\dfrac{-3\sqrt{2} + 12}{14}$ **57.** $4\sqrt{6} + 8$

59. $\sqrt{5} - \sqrt{3}$ **61.** $\dfrac{-20 - 30\sqrt{3}}{23}$

63. $\dfrac{4\sqrt{x} + 8}{x - 4}$ **65.** $\dfrac{x - 3\sqrt{x}}{x - 9}$

67. $\dfrac{a + 7\sqrt{a} + 10}{a - 25}$

69. $\dfrac{6 + 2\sqrt{2} + 3\sqrt{3} + \sqrt{6}}{7}$

Problem Set 9.5 (page 401)

1. $\{49\}$ **3.** $\{18\}$ **5.** $\varnothing$ **7.** $\left\{\dfrac{9}{4}\right\}$

9. $\left\{\dfrac{4}{9}\right\}$ **11.** $\{14\}$ **13.** $\varnothing$ **15.** $\left\{\dfrac{7}{3}\right\}$

17. $\{36\}$ **19.** $\{6\}$ **21.** $\{2\}$ **23.** $\left\{\dfrac{21}{4}\right\}$

25. $\{-3, -2\}$ **27.** $\{6\}$ **29.** $\{2\}$ **31.** $\{12\}$

33. $\{25\}$ **35.** $\{3\}$ **37.** $\left\{-\dfrac{1}{2}\right\}$ **39.** $\{2\}$

41. 56 feet; 106 feet; 148 feet

43. 3.2 feet; 5.1 feet; 7.3 feet **47.** $\{18\}$

49. $\{4, 12\}$

Chapter 9 Review Problem Set (page 404)

1. 8 **2.** -7 **3.** 40 **4.** $\dfrac{9}{5}$ **5.** $-\dfrac{2}{3}$

6. $\dfrac{7}{6}$ **7.** $2\sqrt{5}$ **8.** $4\sqrt{2}$ **9.** $10\sqrt{2}$

10. $4\sqrt{5}$ **11.** $-6\sqrt{6}$ **12.** $\dfrac{2\sqrt{3}}{7}$

13. $\dfrac{6\sqrt{7}}{7}$ **14.** $\dfrac{\sqrt{14}}{4}$ **15.** $\dfrac{\sqrt{3}}{3}$

16. $\dfrac{3\sqrt{10}}{5}$ **17.** 2 **18.** $\dfrac{5\sqrt{6}}{6}$ **19.** $\dfrac{-\sqrt{6}}{3}$

20. $\dfrac{2\sqrt{2}}{3}$ **21.** 5.2 **22.** 1.2 **23.** 17.3

24. -6.9 **25.** $2ab\sqrt{3b}$ **26.** $5y^2\sqrt{2x}$

27. $3x^3\sqrt{2x}$ **28.** $-8t^2\sqrt{2t}$ **29.** $4y\sqrt{3x}$

30. $\dfrac{3xy^2\sqrt{6xy}}{2}$ **31.** $\dfrac{\sqrt{10xy}}{5y}$ **32.** $\dfrac{3\sqrt{2xy}}{2y}$

33. $\dfrac{2\sqrt{x}}{x}$ **34.** $\dfrac{x\sqrt{2x}}{3}$ **35.** $\dfrac{3\sqrt{xy}}{4y^2}$

36. $\dfrac{-2\sqrt{x}}{5}$ **37.** $6\sqrt{2}$ **38.** $18\sqrt{2}$

39. -40 **40.** $24\sqrt{21}$ **41.** $4 + 2\sqrt{10}$

42. $6\sqrt{10} - 12\sqrt{15}$

43. $3 + \sqrt{21} + \sqrt{15} + \sqrt{35}$ **44.** $-24 - 7\sqrt{6}$

45. $4 + 5\sqrt{42}$ **46.** $-18 - \sqrt{5}$

47. $\dfrac{5(\sqrt{7} + \sqrt{5})}{2}$ **48.** $3\sqrt{2} + 2\sqrt{3}$

49. $\dfrac{3\sqrt{2} + \sqrt{6}}{6}$ **50.** $\dfrac{3\sqrt{42} - 4\sqrt{15}}{23}$

51. $18\sqrt{2}$ **52.** $-7\sqrt{2x}$ **53.** $-12\sqrt{3}$

54. $\dfrac{16\sqrt{10}}{5}$ **55.** $\dfrac{52\sqrt{5}}{5}$ **56.** $\dfrac{-17\sqrt{6}}{3}$

57. $\{6\}$ **58.** $\{4\}$ **59.** $\{0, 9\}$ **60.** $\{-5, -4\}$

61. $\{3\}$ **62.** $\{1\}$ **63.** 46 **64.** 66

65. 72 **66.** 26 **67.** 47 **68.** 74

Chapter 9 Test (page 406)

1. $-\dfrac{8}{7}$ **2.** .05 **3.** 2.8 **4.** -5.6

5. 2.1 **6.** $3\sqrt{5}$ **7.** $-18\sqrt{2}$ **8.** $\dfrac{\sqrt{2}}{3}$

9. $\dfrac{5\sqrt{2}}{2}$ **10.** $\dfrac{\sqrt{6}}{3}$ **11.** $\dfrac{\sqrt{10}}{4}$

12. $4x\sqrt{5y}$ **13.** $\dfrac{\sqrt{15xy}}{5y}$ **14.** $3xy\sqrt{3x}$

15. $4\sqrt{6}$ **16.** $45\sqrt{2}$ **17.** $12\sqrt{2} - 12\sqrt{3}$

18. $1 - 5\sqrt{15}$ **19.** $\dfrac{3\sqrt{2} - \sqrt{3}}{5}$ **20.** $4\sqrt{6}$

21. 22 **22.** $\{5\}$ **23.** $\varnothing$ **24.** $\{3\}$

25. $\{-2, 1\}$

CHAPTER 10

Problem Set 10.1 (page 416)

1. $\{-15, 0\}$ **3.** $\{0, 12\}$ **5.** $\{0, 5\}$ **7.** $\{1, 8\}$

9. $\{-2, 7\}$ **11.** $\{-6, 1\}$ **13.** $\left\{-\dfrac{5}{3}, \dfrac{1}{2}\right\}$

15. $\left\{\dfrac{2}{5}, \dfrac{5}{6}\right\}$ **17.** $\left\{\dfrac{1}{2}\right\}$ **19.** $\{-8, 8\}$

21. $\left\{-\dfrac{5}{3}, \dfrac{5}{3}\right\}$ **23.** $\{-4, 4\}$ **25.** $\{-\sqrt{14}, \sqrt{14}\}$

27. No real number solutions **29.** $\{-4\sqrt{2}, 4\sqrt{2}\}$

31. $\{-3\sqrt{2}, 3\sqrt{2}\}$ **33.** $\left\{-\dfrac{3\sqrt{2}}{2}, \dfrac{3\sqrt{2}}{2}\right\}$

35. $\left\{-\dfrac{5\sqrt{2}}{4}, \dfrac{5\sqrt{2}}{4}\right\}$ **37.** $\{-1, 3\}$ **39.** $\{-8, 2\}$

41. $\left\{-\dfrac{5}{3}, 3\right\}$ **43.** $\{-6 - \sqrt{5}, -6 + \sqrt{5}\}$

45. $\{1 - 2\sqrt{2}, 1 + 2\sqrt{2}\}$

47. $\left\{\dfrac{-3 - 2\sqrt{5}}{2}, \dfrac{-3 + 2\sqrt{5}}{2}\right\}$

49. No real number solutions

51. $\left\{\dfrac{5 - 2\sqrt{10}}{3}, \dfrac{5 + 2\sqrt{10}}{3}\right\}$ **53.** $\left\{-\dfrac{3}{7}, \dfrac{5}{7}\right\}$

55. $\{-8 - 5\sqrt{2}, -8 + 5\sqrt{2}\}$ **57.** $5\sqrt{2}$ inches

59. $2\sqrt{7}$ meters **61.** $2\sqrt{11}$ feet

63. $a = 4$ inches and $b = 4\sqrt{3}$ inches

65. $c = 12$ feet and $b = 6\sqrt{3}$ feet

67. $a = 4\sqrt{3}$ meters and $c = 8\sqrt{3}$ meters

69. $a = 10$ inches and $c = 10\sqrt{2}$ inches

71. $a = b = \dfrac{9\sqrt{2}}{2}$ meters **73.** 8.2 feet

75. 30 meters **77.** 35 meters

81. 5.2 centimeters **83.** 10.8 centimeters

Problem Set 10.2 (page 423)

1. $\{-4 - \sqrt{17}, -4 + \sqrt{17}\}$

3. $\{-5 - \sqrt{23}, -5 + \sqrt{23}\}$

5. $\{2 - 2\sqrt{2}, 2 + 2\sqrt{2}\}$

7. No real number solutions

9. $\{-1 - 3\sqrt{2}, -1 + 3\sqrt{2}\}$

11. $\left\{\dfrac{-1 - \sqrt{13}}{2}, \dfrac{-1 + \sqrt{13}}{2}\right\}$

13. $\left\{\dfrac{5 - \sqrt{33}}{2}, \dfrac{5 + \sqrt{33}}{2}\right\}$

15. $\left\{\dfrac{-4 - \sqrt{22}}{2}, \dfrac{-4 + \sqrt{22}}{2}\right\}$

17. $\left\{\dfrac{-6 - \sqrt{42}}{3}, \dfrac{-6 + \sqrt{42}}{3}\right\}$

19. $\left\{\dfrac{2 - \sqrt{2}}{2}, \dfrac{2 + \sqrt{2}}{2}\right\}$

21. No real number solutions

23. $\left\{\dfrac{9 - \sqrt{65}}{2}, \dfrac{9 + \sqrt{65}}{2}\right\}$

25. $\left\{\dfrac{-3 - \sqrt{17}}{4}, \dfrac{-3 + \sqrt{17}}{4}\right\}$

27. $\left\{\dfrac{-1 - \sqrt{7}}{3}, \dfrac{-1 + \sqrt{7}}{3}\right\}$ **29.** $\{-14, 12\}$

31. $\{-11, 15\}$ **33.** $\{-6, 2\}$ **35.** $\{-9, -3\}$

37. $\{-5, 8\}$ **39.** $\left\{\dfrac{1}{2}, 4\right\}$ **41.** $\left\{-\dfrac{5}{2}, \dfrac{3}{2}\right\}$

45. $\left\{\dfrac{-b + \sqrt{b^2 - 4ac}}{2a}, \dfrac{-b - \sqrt{b^2 - 4ac}}{2a}\right\}$

Problem Set 10.3 (page 428)

1. $\{-1, 6\}$ **3.** $\{-9, 4\}$ **5.** $\{1 - \sqrt{6}, 1 + \sqrt{6}\}$

7. $\left\{\dfrac{5 - \sqrt{33}}{2}, \dfrac{5 + \sqrt{33}}{2}\right\}$

9. No real number solutions

11. $\{-2 - \sqrt{2}, -2 + \sqrt{2}\}$ **13.** $\{0, 6\}$

15. $\left\{0, \dfrac{7}{2}\right\}$ **17.** $\{16, 18\}$ **19.** $\{-10, 8\}$

21. $\{-2\}$ **23.** $\left\{-\dfrac{2}{3}, \dfrac{1}{2}\right\}$ **25.** $\left\{-1, \dfrac{2}{5}\right\}$

27. $\left\{-\dfrac{5}{4}, -\dfrac{1}{3}\right\}$ **29.** $\left\{\dfrac{-5 - \sqrt{73}}{4}, \dfrac{-5 + \sqrt{73}}{4}\right\}$

31. $\left\{\dfrac{-2 - \sqrt{7}}{3}, \dfrac{-2 + \sqrt{7}}{3}\right\}$ **33.** $\left\{-\dfrac{3}{4}\right\}$

35. $\left\{\dfrac{-2 - \sqrt{5}}{2}, \dfrac{-2 + \sqrt{5}}{2}\right\}$

37. $\left\{\dfrac{-9 - \sqrt{57}}{12}, \dfrac{-9 + \sqrt{57}}{12}\right\}$

39. $\left\{\dfrac{1 - \sqrt{33}}{4}, \dfrac{1 + \sqrt{33}}{4}\right\}$

41. No real number solutions

43. $\left\{\dfrac{-5 - \sqrt{137}}{14}, \dfrac{-5 + \sqrt{137}}{14}\right\}$

45. $\left\{\dfrac{1 - \sqrt{85}}{6}, \dfrac{1 + \sqrt{85}}{6}\right\}$ **47.** $\{-14, -9\}$

53. $\{-2.52, 7.52\}$ **55.** $\{-8.10, 2.10\}$

57. $\{-3.55, 1.22\}$ **59.** $\{-1.33, 3.58\}$

61. $\{-1.95, 2.15\}$

Problem Set 10.4 (page 433)

1. $\{-9, 5\}$ **3.** $\left\{-\dfrac{13}{5}, \dfrac{1}{5}\right\}$ **5.** $\{-1, 2\}$

7. $\left\{0, \dfrac{8}{3}\right\}$ **9.** $\left\{\dfrac{1}{3}\right\}$ **11.** $\left\{0, \dfrac{2\sqrt{2}}{5}\right\}$

13. $\{7 - 2\sqrt{17}, 7 + 2\sqrt{17}\}$ **15.** $\left\{-1, \dfrac{7}{5}\right\}$

17. $\left\{-\dfrac{5}{3}, -\dfrac{1}{5}\right\}$ **19.** $\{\sqrt{2} - 3, \sqrt{2} + 3\}$

21. $\{-12, 7\}$ **23.** $\left\{\dfrac{3 - \sqrt{33}}{4}, \dfrac{3 + \sqrt{33}}{4}\right\}$

25. $\{-1, 4\}$ **27.** No real number solutions

29. $\{16, 30\}$ **31.** $\left\{\dfrac{-1 - \sqrt{13}}{2}, \dfrac{-1 + \sqrt{13}}{2}\right\}$

33. $\left\{\dfrac{3}{4}, \dfrac{4}{3}\right\}$ **35.** $\{-13, 1\}$ **37.** $\{11, 17\}$

39. $\left\{\dfrac{3 - \sqrt{3}}{2}, \dfrac{3 + \sqrt{3}}{2}\right\}$ **41.** $\left\{-1, -\dfrac{2}{3}\right\}$

43. $\left\{\dfrac{-5 - \sqrt{193}}{6}, \dfrac{-5 + \sqrt{193}}{6}\right\}$ **45.** $\{-5, 3\}$

Problem Set 10.5 (page 439)

1. 17 and 18   **3.** 19 and 25

5. $3 + \sqrt{5}$ and $3 - \sqrt{5}$ **7.** $\sqrt{2}$ or $\dfrac{\sqrt{2}}{2}$

9. 12, 14, and 16   **11.** -8 or 8

13. 8 meters by 12 meters

15. 15 centimeters by 25 centimeters

17. 27 feet by 78 feet

19. 15 rows and 20 seats per row

21. 7 feet by 9 feet   **23.** 6 inches and 8 inches

25. $1\dfrac{1}{2}$ inches   **27.** 30 students

29. 24 inches and 32 inches

31. 18 miles per hour

Chapter 10 Review Problem Set (page 441)

1. $\{-6, -1\}$ **2.** $\{-4 - \sqrt{13}, -4 + \sqrt{13}\}$

3. $\left\{\dfrac{2}{7}, \dfrac{1}{3}\right\}$ **4.** $\{0, 17\}$ **5.** $\{-4, 1\}$

6. $\{11, 15\}$ **7.** $\left\{\dfrac{-7 - \sqrt{61}}{6}, \dfrac{-7 + \sqrt{61}}{6}\right\}$

8. $\left\{\dfrac{1}{2}\right\}$ **9.** No real number solutions

10. $\{-5, -1\}$ **11.** $\{2 - \sqrt{2}, 2 + \sqrt{2}\}$

12. $\{-2 - 3\sqrt{2}, -2 + 3\sqrt{2}\}$ **13.** $\{-3\sqrt{5}, 3\sqrt{5}\}$

14. $\{-3, 9\}$ **15.** $\{0, 1\}$

16. $\{2 - \sqrt{13}, 2 + \sqrt{13}\}$ **17.** $\{20, 24\}$

18. $\{2 - 2\sqrt{2}, 2 + 2\sqrt{2}\}$ **19.** $\left\{-\dfrac{8}{5}, 1\right\}$

20. $\left\{\dfrac{-1 - \sqrt{73}}{12}, \dfrac{-1 + \sqrt{73}}{12}\right\}$ **21.** $\left\{\dfrac{1}{2}, 4\right\}$

22. $\left\{-\dfrac{4}{3}, -1\right\}$ **23.** 9 inches by 12 inches

24. 18 and 19   **25.** 7, 9, and 11

26. $\sqrt{5}$ meters and $3\sqrt{5}$ meters

27. 4 yard and 6 yards

28. 40 shares at \$20 per share   **29.** 10 meters

30. Jay's rate was 45 miles per hour and Jean's rate was 48 miles per hour; or Jay's rate was $7\dfrac{1}{2}$ miles per hour and Jean's rate was $10\dfrac{1}{2}$ miles per hour.

31. $6\sqrt{2}$ inches   **32.** $\dfrac{16\sqrt{3}}{3}$ centimeters

Chapter 10 Test (page 443)

1. $2\sqrt{13}$ inches   **2.** 13 meters   **3.** 7 inches

4. $4\sqrt{3}$ centimeters   **5.** $\left\{-3, \dfrac{5}{3}\right\}$ **6.** $\{-4, 4\}$

7. $\left\{\dfrac{1}{2}, \dfrac{3}{4}\right\}$ **8.** $\left\{\dfrac{3 - \sqrt{29}}{2}, \dfrac{3 + \sqrt{29}}{2}\right\}$

9. $\left\{-1 - \sqrt{10}, -1 + \sqrt{10}\right\}$ **10.** $\varnothing$

11. $\left\{-12, 2\right\}$ **12.** $\left\{\dfrac{3 - \sqrt{41}}{4}, \dfrac{3 + \sqrt{41}}{4}\right\}$

13. $\left\{\dfrac{1 - \sqrt{57}}{2}, \dfrac{1 + \sqrt{57}}{2}\right\}$ **14.** $\left\{\dfrac{1}{5}, 2\right\}$

15. $\left\{13, 15\right\}$ **16.** $\left\{\dfrac{3}{4}, 4\right\}$ **17.** $\left\{0, \dfrac{1}{6}\right\}$

18. $\left\{-1, \dfrac{3}{7}\right\}$ **19.** $\left\{\dfrac{1 - 3\sqrt{3}}{4}, \dfrac{1 + 3\sqrt{3}}{4}\right\}$

20. $\varnothing$ **21.** 15 seats per row

22. 14 miles per hour **23.** 15 and 17

24. 9 feet **25.** 20 shares

Cumulative Review Problem Set (page 444)

1. -128 **2.** 64 **3.** 144 **4.** -8

5. $\dfrac{2}{3}$ **6.** $\dfrac{13}{9}$ **7.** -9 **8.** -49

9. -29 **10.** 49 **11.** $-\dfrac{15}{4x}$

12. $\dfrac{-x + 17}{(x - 2)(x + 3)}$ **13.** $\dfrac{5y}{2}$ **14.** $\dfrac{1}{x - 4}$

15. $\dfrac{-8x - 41}{(x + 6)(x - 3)}$ **16.** $60x^4y^4$

17. $27x^2 + 30x - 8$ **18.** $-5x^2 - 12x - 7$

19. $6x^3 - x^2 - 13x - 4$ **20.** $2x^2 - 2x - 3$

21. $2x(2x + 5)(3x - 4)$ **22.** $3(2x + 3)(2x - 3)$

23. $(y + 3)(x - 2)$ **24.** $(5 - x)(6 + 5x)$

25. $4(x + 1)(x - 1)(x^2 + 1)$

26. $(7x - 2)(3x + 4)$ **27.** $8\sqrt{7}$

28. $-12\sqrt{3}$ **29.** $-3\sqrt{5}$ **30.** $\dfrac{3\sqrt{2}}{7}$

31. $\dfrac{6\sqrt{5}}{5}$ **32.** $\dfrac{\sqrt{6}}{3}$ **33.** $\dfrac{5\sqrt{6}}{18}$

34. $3xy^2\sqrt{3x}$ **35.** $6y^2\sqrt{2xy}$ **36.** $\dfrac{3x}{4}$

37. $-\dfrac{2\sqrt{ab}}{5}$ **38.** $\dfrac{4x^2\sqrt{10x}}{3}$ **39.** 48

40. $-30\sqrt{2}$ **41.** $216 - 36\sqrt{6}$

42. $-4 - 3\sqrt{15}$ **43.** 11 **44.** $4\sqrt{3} - 4\sqrt{2}$

45. $-\dfrac{2(3\sqrt{5} + \sqrt{6})}{13}$ **46.** $\dfrac{3\sqrt{6} - 4}{19}$

47. $\sqrt{2}$ **48.** $-12\sqrt{3} - 19\sqrt{2}$ **49.** $\dfrac{37\sqrt{5}}{12}$

50. $\sqrt{3}$

51. 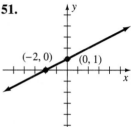 **52.**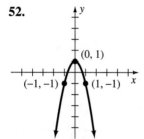

53. 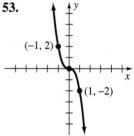 **54.**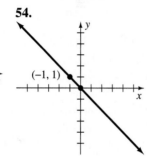

55. -2 **56.** $\dfrac{4}{7}$ **57.** $2x - 3y = 8$

58. $4x + 3y = -13$ **59.** $x + 4y = -12$

60. $x = 4$ **61.** $\left\{(1, -2)\right\}$ **62.** $\left\{(-2, 4)\right\}$

63. $\left\{(-6, 12)\right\}$ **64.** $\left\{\left(\dfrac{1}{2}, 3\right)\right\}$ **65.** $\left\{17\right\}$

66. $\left\{-\dfrac{23}{5}\right\}$ **67.** $\left\{\dfrac{3}{2}, \dfrac{9}{2}\right\}$ **68.** $\left\{5\right\}$

69. $\left\{-\dfrac{2}{3}, \dfrac{4}{3}\right\}$ **70.** $\left\{-2 - 2\sqrt{3}, -2 + 2\sqrt{3}\right\}$

71. $\left\{-\dfrac{19}{10}\right\}$ **72.** $\left\{\dfrac{47}{7}\right\}$ **73.** $\left\{500\right\}$

74. $\left\{-\dfrac{3}{2}, 3\right\}$ **75.** $\left\{53\right\}$ **76.** $\left\{-\dfrac{2}{3}\right\}$

77. $\left\{7\right\}$ **78.** $\left\{5\right\}$ **79.** $\left\{27\right\}$ **80.** $\left\{-4, -3\right\}$

81. $\left\{25\right\}$ **82.** $\left\{-1 - 2\sqrt{3}, -1 + 2\sqrt{3}\right\}$

83. $\left\{3 - 2\sqrt{3}, 3 + 2\sqrt{3}\right\}$

84. $\left\{\dfrac{1 - \sqrt{37}}{6}, \dfrac{1 + \sqrt{37}}{6}\right\}$

85. $\left\{\dfrac{-5 - \sqrt{73}}{4}, \dfrac{-5 + \sqrt{73}}{4}\right\}$

86. $\left\{\dfrac{1 - \sqrt{85}}{6}, \dfrac{1 + \sqrt{85}}{6}\right\}$ **87.** $\{n \mid n \geq -5\}$

88. $\left\{n \mid n > \dfrac{1}{4}\right\}$ **89.** $\left\{x \mid x > -\dfrac{2}{5}\right\}$

90. $\{n \mid n > 6\}$ **91.** $\left\{x \mid x < \dfrac{5}{16}\right\}$

92. $\left\{x \mid x \geq -\dfrac{16}{3}\right\}$ **93.** $65°$ and $115°$

94. 13 and 37

95. 7 nickels, 15 dimes, and 25 quarters

96. \$1350   **97.** \$48   **98.** $4\dfrac{1}{2}$ hours

99. 12.5 milliliters   **100.** 91 or better

101. More than 11

102. The rectangle is 8 inches by 16 inches and the square is 8 inches by 8 inches.

103. 6 feet by 9 feet

104. 8 rows and 14 trees per row   **105.** 7 and 59

106. 12 minutes   **107.** 6 meters by 12 meters

108. 60 miles per hour for Doris and 50 miles per hour for Ellen

CHAPTER 11

Problem Set 11.1 (page 453)

1. $\{-4, 4\}$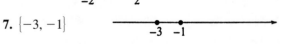

3. $\{x \mid x > -1 \text{ and } x < 1\}$

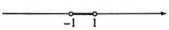

5. $\{x \mid x \leq -2 \text{ or } x \geq 2\}$

7. $\{-3, -1\}$

9. $\{-1, 3\}$

11. $\{x \mid x \geq 0 \text{ and } x \leq 4\}$

13. $\{x \mid x < -4 \text{ or } x > 2\}$

15. $\{-2, 1\}$

17. $\left\{-\dfrac{2}{5}, \dfrac{6}{5}\right\}$

19. $\{x \mid x \leq 1 \text{ or } x \geq 2\}$

21. $\left\{x \mid x > -\dfrac{5}{4} \text{ and } x < -\dfrac{1}{4}\right\}$

23. $\{-2\}$

25. $\left\{x \mid x \neq \dfrac{2}{3}\right\}$

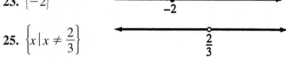

27. $\left\{-\dfrac{16}{3}, 6\right\}$ **29.** $\{x \mid x < -5 \text{ or } x > 4\}$

31. $\left\{x \mid x > -\dfrac{14}{3} \text{ and } x < 8\right\}$ **33.** $\left\{-6, \dfrac{16}{3}\right\}$

35. $\left\{x \mid x \geq -6 \text{ and } x \leq \dfrac{19}{2}\right\}$

37. $\left\{x \mid x \leq -\dfrac{21}{5} \text{ or } x \geq 3\right\}$

39. $\{x \mid x > -6 \text{ and } x < 2\}$

41. $\left\{x \mid x < -\dfrac{5}{2} \text{ or } x > \dfrac{7}{2}\right\}$ **43.** $\{0\}$

45. $\{x \mid x \text{ is any real number}\}$ **47.** $\varnothing$

49. $\{-6\}$ **53.** $\{x \mid 4 < x < 14\}$

55. $\{x \mid -1 \leq x \leq 4\}$ **57.** $\{x \mid -11 < x < 7\}$

59. $\{x \mid -7 < x < -1\}$ **61.** $\{x \mid -1 < x < 6\}$

Problem Set 11.2 (page 459)

1. 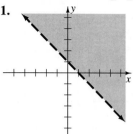 **3.**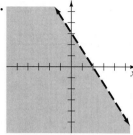

5. 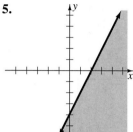 **7.**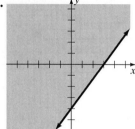

9. 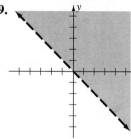 **11.**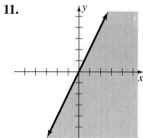

13. 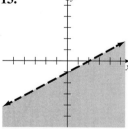 **15.**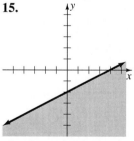

17. 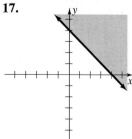 **19.**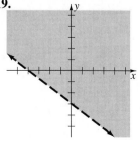

21. 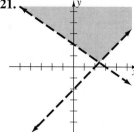 **23.**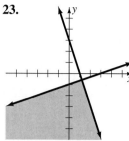

25. 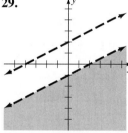 **27.**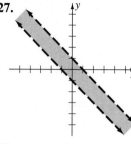

29. 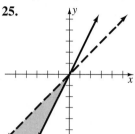 **33.** ϕ

35.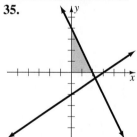

Problem Set 11.3 (page 463)

1. Domain: all reals
Range: all reals

3. Domain: all reals
Range: all reals greater than or equal to 2

5. Domain: all reals
Range: all reals

7. Domain: nonnegative reals
Range: nonnegative reals

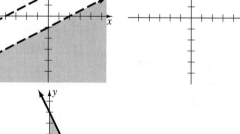

9. All reals

11. All reals greater than or equal to -4

13. All reals   **15.** All reals except 3

17. All reals except 2 and -2 **19.** All reals

21. All reals except -2 and 3

23. All reals except 2 and 3

25. All reals except -4 and 0   **27.** 4; 7; 1; 22

29. -16; 19; 24; $-5t-1$ **31.** $\dfrac{11}{4}$; $\dfrac{13}{12}$; $\dfrac{19}{36}$; $-\dfrac{7}{12}$

33. 0; 0; 45; -4 **35.** 0; -3; -3; -8

37. 2; 5; $2\sqrt{3}$; $3\sqrt{2}$ **39.** 23; -21; 3; 8

41. 8; 48; 8; -7

43. \$57.50; \$90.50; \$117.50; \$170

13.
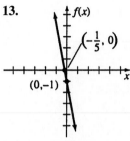

15.
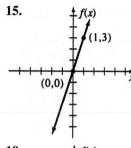

17.
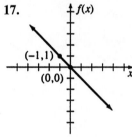

19.
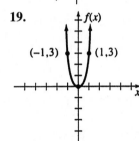

21.
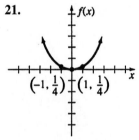

23.
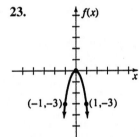

25.
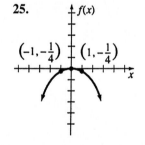

Problem Set 11.4 (page 470)

1.   **3.**
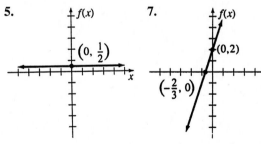

5.   **7.**
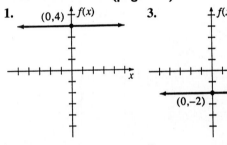

9.   **11.**
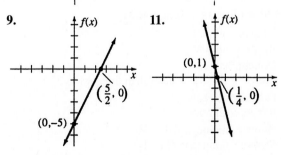

Problem Set 11.5 (page 476)

1. 9   **3.** -10 **5.** 5   **7.** -4 **9.** $\dfrac{4}{7}$

11. 3   **13.** -3 **15.** 8   **17.** 16

19. 16   **21.** 32   **23.** $\dfrac{1}{2}$ **25.** -3

27. $\dfrac{1}{8}$ **29.** $\dfrac{27}{8}$ **31.** 2   **33.** 625

35. -32 **37.** $\dfrac{1}{8}$ **39.** 2   **41.** 27

43. 1   **45.** $\dfrac{1}{3}$ **47.** 4   **49.** 49

51. $x^{\frac{3}{4}}$ **53.** $a^{\frac{17}{12}}$ **55.** $15x^{\frac{7}{12}}$ **57.** $24x^{\frac{11}{12}}$

59. $2y^{\frac{5}{12}}$ **61.** $10n^{\frac{1}{4}}$ **63.** $\dfrac{2}{x^{\frac{1}{6}}}$ **65.** $25xy^2$

67. $64x^{\frac{3}{4}}y^{\frac{3}{2}}$ **69.** $2x^2y$ **71.** $4x^{\frac{4}{15}}$

73. $\dfrac{4}{b^{\frac{5}{12}}}$ **75.** 3   **77.** $\dfrac{9}{4x^{\frac{1}{3}}}$ **79.** $\dfrac{125x^{\frac{3}{2}}}{216y}$

83. (a) 28 **(c)** 17
85. (a) 11.18 **(c)** 3.11   **(e)** 6.45

Problem Set 11.6 (page 481)

1. $8i$ **3.** $\dfrac{5}{3}i$ **5.** $i\sqrt{11}$ **7.** $5i\sqrt{2}$

9. $4i\sqrt{3}$ **11.** $3i\sqrt{6}$ **13.** $8 + 17i$
15. $10 - 10i$ **17.** $4 + 2i$ **19.** $-2 - 6i$
21. $-5 + 3i$ **23.** $-12 - 16i$ **25.** $-10 - 4i$

27. $-1 + 12i$ **29.** $-20 - 8i$ **31.** $\dfrac{5}{6} + \dfrac{5}{12}i$

33. $-\dfrac{1}{15} + \dfrac{7}{12}i$ **35.** $-56 + 0i$

37. $-6 + 12i$ **39.** $-24 + 20i$
41. $-2 + 23i$ **43.** $59 - 17i$ **45.** $-21 - 12i$
47. $-26 + 15i$ **49.** $-9 + 40i$ **51.** $61 + 0i$
53. $5 + 0i$

Problem Set 11.7 (page 485)
1. $\{-8i, 8i\}$ **3.** $\{2 - i, 2 + i\}$
5. $\{-5 - i\sqrt{13}, -5 + i\sqrt{13}\}$

7. $\{3 - 3i\sqrt{2}, 3 + 3i\sqrt{2}\}$ **9.** $\left\{-\dfrac{2}{5}, \dfrac{4}{5}\right\}$

11. $\{-1, 4\}$ **13.** $\{-3 - i\sqrt{3}, -3 + i\sqrt{3}\}$
15. $\{3 - 2i, 3 + 2i\}$ **17.** $\{2 - 4i, 2 + 4i\}$

19. $\left\{\dfrac{1 - i\sqrt{2}}{3}, \dfrac{1 + i\sqrt{2}}{3}\right\}$ **21.** $\left\{-1, \dfrac{5}{2}\right\}$

23. $\{1 - 3i\sqrt{2}, 1 + 3i\sqrt{2}\}$
25. $\{2 - i\sqrt{3}, 2 + i\sqrt{3}\}$

27. $\left\{\dfrac{1 - i\sqrt{31}}{8}, \dfrac{1 + i\sqrt{31}}{8}\right\}$

29. $\left\{\dfrac{-1 - i\sqrt{5}}{6}, \dfrac{-1 + i\sqrt{5}}{6}\right\}$

Chapter 11 Review Problem Set (page 487)

1. $8i$ **2.** $i\sqrt{29}$ **3.** $3i\sqrt{6}$ **4.** $\dfrac{3}{2}i$

5. $6i\sqrt{3}$ **6.** $4i\sqrt{6}$ **7.** $1 + 2i$
8. $-7 - 5i$ **9.** $2 - 4i$ **10.** $3 - 4i$
11. $-1 + 3i$ **12.** $-6 + 10i$ **13.** $-34 + 31i$
14. $-2 - 11i$ **15.** $-4 - 8i$ **16.** $3 - 45i$
17. 85   **18.** 58   **19.** $55 - 48i$
20. $3 - 15i$ **21.** $\{6 - 5i, 6 + 5i\}$
22. $\{-1 - i\sqrt{6}, -1 + i\sqrt{6}\}$ **23.** $\{1 - 4i, 1 + 4i\}$

24. $\left\{\dfrac{1 - 3i\sqrt{3}}{2}, \dfrac{1 + 3i\sqrt{3}}{2}\right\}$

25. $\left\{\dfrac{1 - i\sqrt{23}}{4}, \dfrac{1 + i\sqrt{23}}{4}\right\}$ **26.** $\left\{\dfrac{1}{3}, \dfrac{3}{2}\right\}$

27. $\left\{\dfrac{5 - i\sqrt{3}}{2}, \dfrac{5 + i\sqrt{3}}{2}\right\}$

28. $\left\{\dfrac{-3 - i\sqrt{39}}{4}, \dfrac{-3 + i\sqrt{39}}{4}\right\}$

29. $\left\{\dfrac{-1 - i\sqrt{59}}{6}, \dfrac{-1 + i\sqrt{59}}{6}\right\}$

30. $\left\{\dfrac{-1 - i\sqrt{47}}{8}, \dfrac{-1 + i\sqrt{47}}{8}\right\}$ **31.** $\dfrac{4}{3}$

32. -1 **33.** $\dfrac{3}{4}$ **34.** -5 **35.** $\dfrac{3}{2}$

36. 125   **37.** 32   **38.** -32 **39.** $\dfrac{1}{16}$

40. $\dfrac{1}{2}$ **41.** $\dfrac{1}{4}$ **42.** $\dfrac{3}{2}$ **43.** 8   **44.** 9

45. $\dfrac{1}{3}$ **46.** $x^{\frac{5}{3}}$ **47.** $6x^{\frac{17}{20}}$ **48.** $36a^{\frac{1}{6}}$

49. $27xy^2$ **50.** $5x^2y^3$ **51.** $13n^{\frac{7}{20}}$

52. $\dfrac{4}{n^{\frac{1}{4}}}$ **53.** $8x^3$

54. $f(2) = -3, f(-1) = 3, f(-4) = 27$
55. $f(3) = -37, f(-3) = 5, f(-4) = -2$

56.

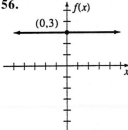

(0,3)

57.

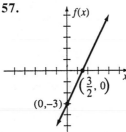

$\left(\frac{3}{2}, 0\right)$

(0,−3)

69.

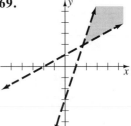

58.

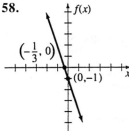

$\left(-\frac{1}{3}, 0\right)$

(0,−1)

59.

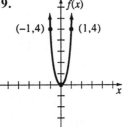

(−1,4)   (1,4)

Chapter 11 Test (page 489)

1. $5i\sqrt{3}$ **2.** 216   **3.** $\frac{1}{3}$ **4.** $\frac{27}{8}$

5. 32   **6.** $10x^{\frac{11}{12}}$ **7.** $5n^{\frac{1}{10}}$ **8.** 24

9. -17 **10.** $\{2 - 4i, 2 + 4i\}$

11. $\{1 - i\sqrt{2}, 1 + i\sqrt{2}\}$

12. $\{-3 - 2i\sqrt{3}, -3 + 2i\sqrt{3}\}$

13. $\left\{\dfrac{3 - i\sqrt{11}}{2}, \dfrac{3 + i\sqrt{11}}{2}\right\}$ **14.** $\{-4, 8\}$

15. $\left\{-\dfrac{7}{4}, -\dfrac{3}{4}\right\}$ **16.** $\varnothing$

17. $\left\{\dfrac{1 - i\sqrt{7}}{4}, \dfrac{1 + i\sqrt{7}}{4}\right\}$ **18.** $\left\{-4, \dfrac{7}{3}\right\}$

19. $\{x \mid x \le -5 \text{ or } x \ge -1\}$

20. $\{x \mid x > -3 \text{ and } x < 4\}$ **21.** All real numbers

60.

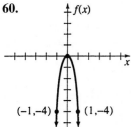

(−1,−4)   (1,−4)

61.

$-\frac{2}{3}$ 4

62.

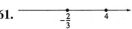

3   5

63.

−1   2

64.

$-\frac{2}{3}$ 2

65.

−4   5

66.

$-\frac{4}{5}$ $\frac{8}{5}$

67.

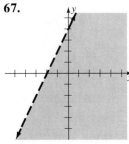

68.

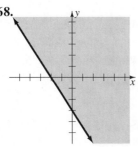

22.

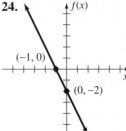

(0, 3)

(−2, 0)

23.

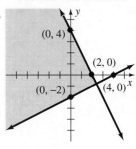

(0, 4)   (2, 0)

(0, −2)   (4, 0)

24.

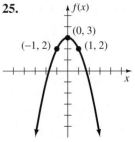

(−1, 0)

(0, −2)

25.

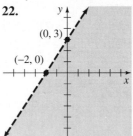

(0, 3)

(−1, 2)   (1, 2)